THE
HANDY
SCIENCE
ANSWER
BOOK

THE HANDY SCIENCE ANSWER BOOK

FOURTH EDITION

Compiled by the Carnegie Library of Pittsburgh

VISIBLE
INK
PRESS

Detroit

500

THE HANDY SCIENCE ANSWER BOOK

Visible Ink Press®
43311 Joy Rd., #414
Canton, MI 48187-2075

Visible Ink Press is a registered trademark of Visible Ink Press LLC.

Most Visible Ink Press books are available at special quantity discounts when purchased in bulk by corporations, organizations, or groups. Customized printings, special imprints, messages, and excerpts can be produced to meet your needs. For more information, contact Special Markets Director, Visible Ink Press, www.visibleinkpress.com, or 734-667-3211.

Managing Editor: Kevin S. Hile
Art Director: Mary Claire Krzewinski
Typesetting: Marco Di Vita
Proofreaders: Sarah Hermsen and Sharon Malinowski
ISBN 978-1-57859-321-7 (pbk.)

Library of Congress Cataloguing-in-Publication Data

The handy science answer book / [edited by] Naomi E. Balaban and James E. Bobick. — 4th ed.
    p. cm. — (The handy answer book series)
Includes bibliographical references and index.
ISBN 978-1-57859-321-7 (pbk.)
1. Science—Miscellanea. I. Balaban, Naomi E. II. Bobick, James E.
Q173.H24 2011
500—dc22   2011000429

Printed in the United States of America

10 9 8 7 6 5 4 3 2 1

Contents

GENERAL SCIENCE, MATHEMATICS, AND TECHNOLOGY ... 1

Introduction ... Societies, Publications, and Awards ... Numbers ... Mathematics ... Computers

PHYSICS AND CHEMISTRY ... 55

Energy, Motion, and Force ... Light, Sound, and Other Waves ... Electricity and Magnetism ... Matter ... Chemical Elements ... Temperature, Measurement, and Methodology

ASTRONOMY AND SPACE ... 93

Universe ... Stars ... Planets and Moons ... Comets and Meteorites ... Observation and Measurement ... Exploration

EARTH ... 145

Air ... Physical Characteristics ... Water ... Land ... Volcanoes and Earthquakes ... Observation and Measurement

CLIMATE AND WEATHER... 181

Temperature ... Air Phenomena ... Wind ... Precipitation ... Weather Prediction

MINERALS, METALS, AND OTHER MATERIALS ... 221

Rocks and Minerals ... Metals ... Natural Substances ... Man-made Products

Introduction

In the years since the first edition of *The Handy Science Answer Book* was published in 1994, innumerable discoveries and advancements have been made in all fields of science and technology. These accomplishments range from the microscopic to the global—from an understanding of how genes interact and ultimately produce proteins to the recent definition of a planet that excludes Pluto. As a society, we have increased our awareness of the environment and the sustainability of resources with a focus on increasing our use of renewable fuels, reducing greenhouse gases, and building "green."

This newly updated fourth edition of *The Handy Science Answer Book* continues to be a fun and educational resource that is both informative and enjoyable. There are nearly 2,000 questions in all areas of science, technology, mathematics, medicine, and other areas. The questions are interesting, unusual, frequently asked, or difficult to answer. Statistical data have been updated for the twenty-first century. Both of us are pleased and excited about the various changes, additions, and improvements in this new edition, which continues to add to and enhance the original publication presented by the Science and Technology Department of the Carnegie Library of Pittsburgh.

ACKNOWLEDGMENTS

The Carnegie Library of Pittsburgh, established in 1902, fields—and answers—more than 60,000 science and technology questions every single year, which is how a library became an author. The most common questions and their answers were collected and became the library's own ready reference file. *The Handy Science Answer Book* is a selection of the most interesting, frequently asked, and unusual of these queries.

This fourth edition of *The Handy Science Answer Book* was revised and updated thanks to the help of James E. Bobick and Naomi E. Balaban, who have worked on the previous editions. Bobick recently retired after sixteen years as Head of the Science and Technology Department at the Carnegie Library of Pittsburgh. During the same time, he taught the science resources course in the School of Information Sciences at the University of Pittsburgh. He co-authored Science and Technology Resources: A Guide for Information Professionals and Researchers with G. Lynn Berard from Carnegie Mellon University. He has master's degrees in both biology and library science.

Balaban, a reference librarian for twenty years at the Carnegie Library of Pittsburgh, has extensive experience in the areas of science and technology. In addition to working on the two earlier editions of *The Handy Science Answer Book* with Bobick, she coauthored *The Handy Biology Answer Book* and *The Handy Anatomy Answer Book* with him. She has a background in linguistics and a master's degree in library science.

Jim and Naomi dedicate this edition to Sandi and Carey: "We owe you a lot!" In addition, the authors thank their families for the ongoing interest, encouragement, support, and especially their understanding while this edition was being revised.

PHOTO CREDITS

All photos and illustrations are from iStock.com, with the following exceptions:

Electronic Illustrators Group: 28, 64, 79, 95, 99, 109, 114, 147, 297, 305, 351, 533, 538, 544, 546, 551, 555, 557, 559, 565, 585, 595, 636.

Library of Congress: 19, 33, 77, 104, 395, 414.

National Aeronautics and Space Administration: 182.

National Oceanic and Atmospheric Administration: 193, 307.

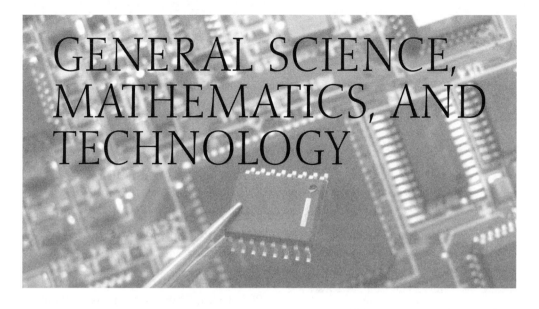

GENERAL SCIENCE, MATHEMATICS, AND TECHNOLOGY

INTRODUCTION

What is the **difference** between **science** and **technology**?

Science and technology are related disciplines, but have different goals. The basic goal of science is to acquire a fundamental knowledge of the natural world. Outcomes of scientific research are the theorems, laws, and equations that explain the natural world. It is often described as a pure science. Technology is the quest to solve problems in the natural world with the ultimate goal of improving humankind's control of their environment. Technology is, therefore, often described as applied science; applying the laws of science to specific problems. The distinction between science and technology blurs since many times researchers investigating a scientific problem will discover a practical application for the knowledge they acquire.

What is the **scientific method**?

The scientific method is the basis of scientific investigation. A scientist will pose a question and formulate a hypothesis as a potential explanation or answer to the question. The hypothesis will be tested through a series of experiments. The results of the experiments will either prove or disprove the hypothesis. Hypotheses that are consistent with available data are conditionally accepted.

What are the **steps** of the **scientific method**?

Research scientists follow these steps:

1. State a hypothesis.
2. Design an experiment to "prove" the hypothesis.

1

Abu Ali al-Hasan ibn al-Haytham (or al Haitham), also known as Alhazen or Alhacen, is considered the "father of optics." His image appears here on a postage stamp from Qatar.

3. Assemble the materials and set up the experiment.
4. Do the experiment and collect data.
5. Analyze the data using quantitative methods.
6. Draw conclusions.
7. Write up and publish the results.

Who is one of the **first individuals associated** with the **scientific method**?

Abu Ali al-Hasan ibn al-Haytham (c. 966–1039), whose name is usually Latinized to Alhazen or Alhacen, is known as the "father" of the science of optics and was also one of the earliest experimental scientists. Between the tenth and fourteenth centuries, Muslim scholars were responsible for the development of the scientific method. These individuals were the first to use experiments and observation as the basis of science, and many historians regard science as starting during this period. Alhazen is considered as the architect of the scientific method. His scientific method involved the following steps:

1. Observation of the natural world
2. Stating a definite problem
3. Formulating a hypothesis
4. Test the hypothesis through experimentation
5. Assess and analyze the results
6. Interpret the data and draw conclusions
7. Publish the findings

What is a **variable**?

A variable is something that is changed or altered in an experiment. For example, to determine the effect of light on plant growth, growing one plant in a sunny window and one in a dark closet will provide evidence as to the effect of light on plant growth. The variable is light.

How does an **independent variable** differ from a **dependent variable**?

An independent variable is manipulated and controlled by the researcher. A dependent variable is the variable that the researcher watches and/or measures. It is called a dependent variable because it depends upon and is affected by the independent variable. For example, a researcher may investigate the effect of sunlight on plant growth by exposing some plants to eight hours of sunlight per day and others to only four hours of sunlight per day. The plant growth rate is dependent upon the amount of sunlight, which is controlled by the researcher.

What is a **control group**?

A control group is the experimental group tested without changing the variable. For example, to determine the effect of temperature on seed germination, one group of seeds may be heated to a certain temperature. The percent of seeds in this group that germinates and the time it takes them to germinate is then compared to another group of seeds (the control group) that has not been heated. All other variables, such as light and water, will remain the same for each group.

What is a **double-blind study**?

In a double-blind study, neither the subjects of the experiment nor the persons administering the experiment know the critical aspects of the experiment. This method is used to guard against both experimenter bias and placebo effects.

How does **deductive reasoning** differ from **inductive reasoning**?

Deductive reasoning, often used in mathematics and philosophy, uses general principles to examine specific cases. Inductive reasoning is the method of discovering general principles by close examination of specific cases. Inductive reasoning first became important to science in the 1600s, when Francis Bacon (1561–1626), Sir Isaac Newton (1642–1727), and their contemporaries began to use the results of specific experiments to infer general scientific principles.

How do **scientific laws** differ from **theories**?

A scientific law is a statement of how something in nature behaves, which has proven to be true every time it is tested. Unlike the general usage of the term "theory," which often means an educated guess, a scientific theory explains a phenomenon that is based on observation, experimentation, and reasoning. Scientific laws do not become theories. A scientific theory may explain a law, but theories do not become laws.

What is **high technology** or high tech?

This buzz term used mainly by the lay media (as opposed to scientific, medical, or technological media) appeared in the late 1970s. It was initially used to identify the newest, "hottest" application of technology to fields such as medical research, genetics, automation, communication systems, and computers. It usually implied a distinction between technology to meet the information needs of society and traditional heavy industry, which met more material needs. By the mid–1980s, the term had become a catch-all applying primarily to the use of electronics (especially computers) to accomplish everyday tasks.

What is **nanotechnology**?

Nanotechnology is a relatively new field of science that aims to understand matter at dimensions between 1 and 100 nanometers. Nanomaterials may be engineered or occur in nature. Some of the different types of nanomaterials, named for their individual shape and dimensions, are nanoparticles, nanotubes, and nanofilms. Nanoparticles are bits of material where all the dimensions are nanosized. Nanotubes are long cylindrical strings of molecules whose diameter is nanosized. Nanofilms have a thickness that is nanosized, but the other dimensions may be larger. Researchers are developing ways to apply nanotechnology to a wide variety of fields, including transportation, sports, electronics, and medicine. Specific applications of nanotechnology include fabrics with added insulation without additional bulk. Other fabrics are treated with coatings to make them stain proof. Nanorobots are being used in medicine to help diagnose and treat health problems. In the field of electronics, nanotechnology could shrink the size of many electronic products. Researchers in the food industry are investigating the use of nanotechnology to enhance the flavor of food. They are also searching for ways to introduce antibacterial nanostructures into food packaging.

How large is a **nanometer**?

A nanometer equals one-billionth of a meter. A sheet of paper is about 100,000 nanometers thick. As a comparison, a single-walled carbon nanotube, measuring one nanometer in diameter, is 100,000 times smaller than a single strand of human hair which measures 100 micrometers in diameter.

What is a **patent**?

A patent grants the property rights of an invention to the inventor. Once a patent is issued, it excludes others from making, using, or selling the invention in the United States. The U.S. Patent and Trademark Office issues three types of patents:

1. Utility patents are granted to anyone who invents or discovers any new and useful process, machine, manufactured article, compositions of matter, or any new and useful improvement in any of the above.
2. Design patents are granted to anyone who invents a new, original, and ornamental design for an article of manufacture.
3. Plant patents are granted to anyone who has invented or discovered and asexually reproduced any distinct and new variety of plant.

When was the **first patent issued** in the **United States**?

The first U.S. patent was granted on July 31, 1790 to Samuel Hopkins (1743–1818) of Philadelphia for making "pot ash and pearl ash"—a cleaning formula called potash. It was a key ingredient for making glass, dyeing fabrics, baking, making saltpeter for gun powder, and most importantly for making soap.

How many **patents** have been issued by the **U.S. Patent Office**?

Over seven million patents have been granted by the U.S. Patent Office since its inception in 1790. In recent years, the number of patents issued on a yearly basis has risen dramatically. The following chart shows the numbers of patents of all types (utility, design, plant, and reissue) issued for selected years:

Year	Total Number of Patents Granted
1970	67,964
1975	76,810
1980	66,170
1985	77,245
1990	99,077
1995	113,834
2000	175,979
2005	157,718
2008	185,224

Who is the **only U.S. president** to receive a **patent**?

On May 22, 1849, 12 years before he became the sixteenth U.S. president, Abraham Lincoln (1809–1865), was granted U.S. patent number 6,469 for a device to help steam-

Abraham Lincoln is the only U.S. president to have received a patent for an invention. He created a device designed to help steamboats pass over shoals and sandbars.

boats pass over shoals and sandbars. The device, never tested or manufactured, had a set of adjustable buoyancy chambers (made from metal and waterproof cloth) attached to the ship's sides below the waterline. Bellows could fill the chambers with air to float the vessel over the shoals and sand bars. It was the only patent ever held by a United States president.

What is a **trademark**?

A trademark protects a word, phrase, name, symbol, sound, or color that identifies and distinguishes the source of the goods or services of one party (individual or company) from those of another party.

What is the **purpose** of a **trade secret**?

A trade secret is information a company chooses to protect from its competitors. Perhaps the most famous trade secret is the formula for Coca-Cola.

SOCIETIES, PUBLICATIONS, AND AWARDS

What was the **first important scientific society** in the United States?

The first significant scientific society in the United States was the American Philosophical Society, organized in 1743 in Philadelphia, Pennsylvania, by Benjamin Franklin (1706–1790). During colonial times, the quest to understand nature and seek information about the natural world was called natural philosophy.

What was the **first national scientific society** organized in the United States?

The first national scientific society organized in the United States was the American Association for the Advancement of Science (AAAS). It was established on September 20, 1848, in Philadelphia, Pennsylvania, for the purpose of "advancing science in every way." The first president of the AAAS was William Charles Redfield (1789–1857).

What was the **first national science institute**?

On March 3, 1863, President Abraham Lincoln signed a congressional charter creating the National Academy of Sciences, which stipulated that "the Academy shall, whenever called upon by any department of the government, investigate, examine, experiment, and report upon any subject of science or art, the actual expense of such investigations, examinations, experiments, and reports to be paid from appropriations which may be made for the purpose, but the Academy shall receive no compensation whatever for any services to the Government of the United States." The Academy's first president was Alexander Dallas Bache (1806–1867). Today, the Academy and its sister organizations—the National Academy of Engineering, established in 1964, and the Institute of Medicine, established in 1970—serve as the country's preeminent sources of advice on science and technology and their bearing on the nation's welfare.

The National Research Council was established in 1916 by the National Academy of Sciences at the request of President Woodrow Wilson (1856–1924) "to bring into cooperation existing governmental, educational, industrial and other research organizations, with the object of encouraging the investigation of natural phenomena, the increased use of scientific research in the development of American industries, the employment of scientific methods in strengthening the national defense, and such other applications of science as will promote the national security and welfare."

The National Academy of Sciences, the National Academy of Engineering, and the Institute of Medicine work through the National Research Council of the United States, one of world's most important advisory bodies. More than 6,000 scientists, engineers, industrialists, and health and other professionals participating in numerous committees comprise the National Research Council.

What was the **first national physics society** organized in the United States?

The first national physics society in the United States was the American Physical Society, organized on May 20, 1899, at Columbia University in New York City. The first president was physicist Henry Augustus Rowland (1848–1901).

What was the **first national chemical society** organized in the United States?

The first national chemical society in the United States was the American Chemical Society, organized in New York City on April 20, 1876. The first president was John William Draper (1811–1882).

What was the **first mathematical society** organized in the United States?

The first mathematical society in the United States was the American Mathematical Society founded in 1888 to further the interests of mathematics research and scholarship. The first president was John Howard Van Amringe (1835–1915).

Sir Isaac Newton.

What was the **first scientific journal**?

The first scientific journal was *Journal des Sçavans*, published and edited by Denys de Sallo (1626–1669). The first issue appeared on January 5, 1665. It contained reviews of books, obituaries of famous men, experimental findings in chemistry and physics, and other general interest information. Publication was suspended following the thirteenth issue in March 1665.

Although the official reason for the suspension of the publication was that de Sallo was not submitting his proofs for official approval prior to publication, there is speculation that the real reason for the suspension was his criticism of the work of important people, papal policy, and the old orthodox views of science. It was reinstated in January 1666 and continued as a weekly publication until 1724.

The journal was then published on a monthly basis until the French Revolution in 1792. It was published briefly in 1797 under the title *Journal des Savants*. It began regular publication again in 1816 under the auspices of the Institut de France evolving as a general interest publication.

What is the **oldest continuously published** scientific journal?

The *Philosophical Transactions* of the Royal Society of London, first published a few months after the first issue of the *Journal des Sçavans*, on March 6, 1665, is the oldest, continuously published scientific journal.

What was the **first technical report** written in **English**?

Geoffrey Chaucer's (1343–1400) *Treatise on the Astrolabe* was written in 1391.

What **scientific article** has the **most authors**?

The article "First Measurement of the Left-Right Cross Section Asymmetry in Z-Boson Production by e^+e^- Collisions," published in *Physical Review Letters,* Volume 70, issue 17 (26 April 1993), pages 2,515–2,520, listed 406 authors on two pages.

What book is considered the most important and most influential scientific work?

Isaac Newton's 1687 book, *Philosophiae Naturalis Principia Mathematica* (known most commonly as the abbreviated *Principia*). Newton wrote *Principia* in 18 months, summarizing his work and covering almost every aspect of modern science. Newton introduced gravity as a universal force, explaining that the motion of a planet responds to gravitational forces in inverse proportion to the planet's mass. Newton was able to explain tides, and the motion of planets, moons, and comets using gravity. He also showed that spinning bodies such as earth are flattened at the poles. The first printing of *Principia* produced only 500 copies. It was published originally at the expense of his friend, Edmond Halley (1656–1742), because the Royal Society had spent its entire budget on a history of fish.

What is the **most frequently cited** scientific **journal article**?

The most frequently cited scientific article is "Protein Measurement with the Folin Phenol Reagent" by Oliver Howe Lowry (1910–1996) and coworkers, published in 1951 in the *Journal of Biological Chemistry,* Volume 193, issue 1, pages 265–275. As of 2010, this article had been cited 292,968 times since it first appeared.

When was the **Nobel Prize first awarded**?

The Nobel Prize was established by Alfred Nobel (1833–1896) to recognize individuals whose achievements during the preceding year had conferred the greatest benefit to mankind. Five prizes were to be conferred each year in the areas of physics, chemistry, physiology or medicine, economic sciences, and peace. Although Nobel passed away in 1896, the first prizes were not awarded until 1901.

Who are the **youngest** and **oldest Nobel Laureates** in the areas of physics, chemistry, and physiology or medicine?

Youngest Nobel Laureates

Category	Nobel Laureate	Age	Year of Award
Chemistry	Frédéric Joliet (1900–1958)	35	1935
Physics	William Lawrence Bragg (1890–1971)	25	1915
Physiology or Medicine	Frederick Banting (1891–1941)	32	1923

Oldest Nobel Laureates

Category	Nobel Laureate	Age	Year of Award
Chemistry	John B. Fenn (1917–2010)	85	2002
Physics	Raymond Davis Jr. (1914–2006)	88	2002
Physiology or Medicine	Peyton Rous (1879–1970)	87	1966

Are there any **multiple Nobel Prize winners**?

Four individuals have received multiple Nobel prizes. They are Marie Curie (1867–1934, Physics in 1903 and Chemistry in 1911; John Bardeen (1908–1991), Physics in 1956 and 1972; Linus Pauling (1901–1994), Chemistry in 1954 and Peace in 1962; and Frederick Sanger (1918–), Chemistry in 1958 and 1980.

Who was the **first woman** to receive the **Nobel Prize**?

Marie Curie was the first woman to receive the Nobel Prize. She received the Nobel Prize in Physics in 1903 for her work on radioactivity in collaboration with her husband, Pierre Curie (1859–1906) and A.H. Becquerel (1852–1908). The 1903 prize in physics was shared by all three individuals. Marie Curie was also the first person to be awarded two Nobel Prizes and is one of only two individuals who have been awarded a Nobel Prize in two different fields.

How **many women** have been awarded the **Nobel Prize** in Chemistry, Physics, or Physiology or Medicine?

Since 1901, the Nobel Prize in Chemistry, Physics, or Physiology or Medicine has been awarded to women 16 times to 15 different women. Marie Curie (1867–1934) was the only woman and one of the few individuals to receive the Nobel Prize twice.

Year of Award	Nobel Laureate	Category
1903	Marie Curie (1867–1934)	Physics
1911	Marie Curie (1867–1934)	Chemistry
1935	Irène Joliot-Curie (1897–1956)	Chemistry
1947	Gerty Theresa Cori (1896–1957)	Physiology or Medicine
1963	Maria Goeppert-Mayer (1906–1972)	Physics
1964	Dorothy Crowfoot Hodgkin (1910–1994)	Chemistry
1977	Rosalyn Yarrow (1921–)	Physiology or Medicine
1983	Barbara McClintock (1902–1992)	Physiology or Medicine
1986	Rita Levi-Montalcini (1909–)	Physiology or Medicine
1988	Gertrude B. Elion (1918–1999)	Physiology or Medicine

Year of Award	Nobel Laureate	Category
1995	Christianne Nüsslein-Volhard (1942–)	Physiology or Medicine
2004	Linda B. Buck (1947–)	Physiology or Medicine
2008	Françoise Barré-Sinoussi (1947–)	Physiology or Medicine
2009	Ada E. Yonath (1939–)	Chemistry
2009	Carol W. Greider (1961–)	Physiology or Medicine
2009	Elizabeth H. Blackburn (1948–)	Physiology or Medicine

When was the first time **two women shared** the **Nobel Prize** in the **same field**?

It was not until 2009 that two women shared the Nobel Prize in the same field. Carol W. Greider (1961–) and Elizabeth H. Blackburn (1948–) shared the prize in Physiology or Medicine, along with Jack W. Szostak (1952–) for their discovery of how chromosomes are protected by telomeres and the enzyme telomerase.

Is there a **Nobel Prize** in **mathematics**?

We do not know for certain why Alfred Nobel did not establish a prize in mathematics. There are several theories revolving around his relationship and dislike for Gosta Mittag-Leffler (1846–1927), the leading Swedish mathematician in Nobel's time. Most likely it never occurred to Nobel or he decided against another prize. The Fields Medal in mathematics is generally considered as prestigious as the Nobel Prize. The Fields Medal was first awarded in 1936. Its full name is now the CRM-Fields-PIMS prize. The 2009 winner was Martin Barlow (1953–) for his work in probability and in the behavior of diffusions on fractals and other disordered media.

What is the **Turing Award**?

The Turing Award, considered the Nobel Prize in computing, is awarded annually by the Association for Computing Machinery to an individual who has made a lasting contribution of major technical importance in the computer field. The award, named for the British mathematician A.M. Turing (1912–1954), was first presented in 1966. The Intel Corporation and Google Inc. provide financial support for the $250,000 prize that accompanies the award. Recent winners of the Turing Award include:

Year	Award Recipient
2005	Peter Naur (1928–)
2006	Frances E. Allen (1932–)
2007	Edmund M. Clarke (1945–), E. Allen Emerson (1954–), Joseph Sifakis (1946–)
2008	Barbara H. Liskov (1939–)
2009	Charles P. Thacker (1943–)

NUMBERS

When and where did the **concept of "numbers"** and **counting first develop**?

The human adult (including some of the higher animals) can discern the numbers one through four without any training. After that people must learn to count. To count requires a system of number manipulation skills, a scheme to name the numbers, and some way to record the numbers. Early people began with fingers and toes, and progressed to shells and pebbles. In the fourth millennium B.C.E. in Elam (near what is today Iran along the Persian Gulf), accountants began using unbaked clay tokens instead of pebbles. Each represented one order in a numbering system: a stick shape for the number one, a pellet for ten, a ball for 100, and so on. During the same period, another clay-based civilization in Sumer in lower Mesopotamia invented the same system.

When was a symbol for the **concept of zero** first used?

Surprisingly, the symbol for zero emerged later than the concept for the other numbers. Although the Babylonians (600 B.C.E. and earlier) had a symbol for zero, it was merely a placeholder and not used for computational purposes. The ancient Greeks conceived of logic and geometry, concepts providing the foundation for all mathematics, yet they never had a symbol for zero. The Maya also had a symbol for zero as a placeholder in the fourth century, but they also did not use zero in computations. Hindu mathematicians are usually given credit for developing a symbol for the concept "zero." They recognized zero as representing the absence of quantity and developed its use in mathematical calculations. It appears in an inscription at Gwalior dated 870 C.E. However, it is found even earlier than that in inscriptions dating from the seventh century in Cambodia, Sumatra, and Bangka Island (off the coast of Sumatra). Although there is no documented evidence in printed material for the zero in China before 1247, some historians maintain that there was a blank space on the Chinese counting board, representing zero, as early as the fourth century B.C.E.

Hindu-Arabic	Greek	Hebrew	Japanese
1	α′	א	—
2	β′	ב	=
3	γ′	ג	三
4	δ′	ד	四
5	ε′	ה	五
6	ς′	ו	六
7	ζ′	ז	七
8	η′	ח	八
9	θ′	ט	九
10	ι′	י	十

The numbers 1 through 10 as written in Greek, Hebrew, Japanese, and the Arabic-Hindu system used in Western cultures.

What are **Roman numerals**?

Roman numerals are symbols that stand for numbers. They are written using seven basic symbols: I (1), V (5), X (10), L (50), C (100), D (500), and M (1,000). Sometimes a bar is place over a numeral to multiply it by 1,000. A smaller numeral appearing before a larger numeral indicates that the smaller numeral is subtracted from the larger one. This notation is generally used for 4s and 9s; for example, 4 is written IV, 9 is IX, 40 is XL, and 90 is XC.

What are **Fibonacci numbers**?

Fibonacci numbers are a series of numbers where each, after the second term, is the sum of the two preceding numbers—for example, 1, 1, 2, 3, 5, 8, 13, 21, and so on). They were first described by Leonardo Fibonacci (c. 1180–c. 1250), also known as Leonard of Pisa, as part of a thesis on series in his most famous book *Liber abaci* (*The Book of the Calculator*), published in 1202 and later revised by him. Fibonacci numbers are used frequently to illustrate natural sequences, such as the spiral organization of a sunflower's seeds, the chambers of a nautilus shell, or the reproductive capabilities of rabbits.

What is the **largest prime number** presently known?

A prime number is one that is evenly divisible only by itself and 1. The integers 1, 2, 3, 5, 7, 11, 13, 17, and 19 are prime numbers. Euclid (c. 335–270 B.C.E.) proved that there is no "largest prime number," because any attempt to define the largest results in a paradox. If there is a largest prime number (P), adding 1 to the product of all primes up to and including P, 1 1 (1 3 2 3 3 3 5 3 … 3 P), yields a number that is itself a prime number, because it cannot be divided evenly by any of the known primes. In 2003, Michael Shafer discovered the largest known (and the fortieth) prime number: $2^{20996011} - 1$. This is over six million digits long and would take more than three weeks to write out by hand. In July 2010, double-checking proved this was the fortieth Mersenne prime (named after Marin Mersenne, 1588–1648, a French monk who did the first work in this area). Mersenne primes occur where 2^{n-1} is prime.

There is no apparent pattern to the sequence of primes. Mathematicians have been trying to find a formula since the days of Euclid, without success. The fortieth prime was discovered on a personal computer as part of the GIMPS effort (the Great

Internet Mersenne Prime Search), which was formed in January 1996 to discover new world-record-size prime numbers. GIMPS relies on the computing efforts of thousands of small, personal computers around the world. Interested participants can become involved in the search for primes by going to: http://www.mersenne.org/default.php.

What is a **perfect number**?

A perfect number is a number equal to the sum of all its proper divisors (divisors smaller than the number) including 1. The number 6 is the smallest perfect number; the sum of its divisors 1, 2, and 3 equals 6. The next three perfect numbers are 28, 496, and 8,126. No odd perfect numbers are known. The largest known perfect number is

$$(2^{3021376})(2^{3021377} - 1)$$

It was discovered in 2001.

What is the **Sieve of Eratosthenes**?

Eratosthenes (c. 285 –c. 205 B.C.E.) was a Greek mathematician and philosopher who devised a method to identify (or "sift" out) prime numbers from a list of natural numbers arranged in order. It is a simple method, although it becomes tedious to identify large prime numbers. The steps of the sieve are:

1. Write all natural numbers in order, omitting 1.
2. Circle the number 2 and then cross out every other number. Every second number will be a multiple of 2 and hence is not a prime number.
3. Circle the number 3 and then cross out every third number which will be a multiple of 3 and, therefore, not a prime number.
4. The numbers that are circled are prime and those that are crossed out are composite numbers.

How are **names** for **large** and **small quantities** constructed in the **metric system**?

Each prefix listed below can be used in the metric system and with some customary units. For example, centi + meter = centimeter, meaning one-hundredth of a meter.

Prefix	Power	Numerals
Exa-	10^{18}	1,000,000,000,000,000,000
Peta-	10^{15}	1,000,000,000,000,000
Tera-	10^{12}	1,000,000,000,000
Giga-	10^{9}	1,000,000,000
Mega-	10^{6}	1,000,000
Myria-	10^{5}	100,000
Kilo-	10^{3}	1,000
Hecto-	10^{2}	100
Deca-	10^{1}	10
Deci-	10^{-1}	0.1
Centi-	10^{-2}	0.01
Milli-	10^{-3}	0.001
Micro-	10^{-6}	0.000001
Nano-	10^{-9}	0.000000001
Pico-	10^{-12}	0.000000000001
Femto-	10^{-15}	0.000000000000001
Atto-	10^{-18}	0.000000000000000001

Why is the **number ten** considered **important**?

One reason is that the metric system is based on the number ten. The metric system emerged in the late eighteenth century out of a need to bring standardization to measurement, which had up to then been fickle, depending upon the preference of the ruler of the day. But ten was important well before the metric system. Nicomachus of Gerasa (c. 60–c. 120), a second-century neo-Pythagorean from Judea, considered ten a "perfect" number, the figure of divinity present in creation with mankind's fingers and toes. Pythagoreans believed ten to be "the first-born of the numbers, the mother of them all, the one that never wavers and gives the key to all things." Shepherds of West Africa counted sheep in their flocks by colored shells based on ten, and ten had evolved as a "base" of most numbering schemes. Some scholars believe the reason ten developed as a base number had more to do with ease: ten is easily counted on fingers and the rules of addition, subtraction, multiplication, and division for the number ten are easily memorized.

What are some **very large numbers**?

Value in Name	Number powers of 10	Number of groups of 0s	Number of three 0s after 1,000
Billion	10^{9}	9	2
Trillion	10^{12}	12	3
Quadrillion	10^{15}	15	4

Value in Name	Number powers of 10	Number of groups of 0s	Number of three 0s after 1,000
Quintillion	10^{18}	18	5
Sextillion	10^{21}	21	6
Septillion	10^{24}	24	7
Octillion	10^{27}	27	8
Nonillion	10^{30}	30	9
Decillion	10^{33}	33	10
Undecillion	10^{36}	36	11
Duodecillion	10^{39}	39	12
Tredecillion	10^{42}	42	13
Quattuor-decillion	10^{45}	45	14
Quindecillion	10^{48}	48	15
Sexdecillion	10^{51}	51	16
Septen-decillion	10^{54}	54	17
Octodecillion	10^{57}	57	18
Novemdecillion	10^{60}	60	19
Vigintillion	10^{63}	63	20
Centillion	10^{303}	303	100

The British, French, and Germans use a different system for naming denominations above one million. The googol and googolplex are rarely used outside the United States.

How large is a **googol**?

A googol is 10^{100} (the number 1 followed by 100 zeros). Unlike most other names for numbers, it does not relate to any other numbering scale. The American mathematician Edward Kasner (1878–1955) first used the term in 1938; when searching for a term for this large number, Kasner asked his nephew, Milton Sirotta (1911–1981), then about nine years old, to suggest a name. The googolplex is 10 followed by a googol of zeros, represented as 10^{googol}. The popular Web search engine Google.com is named after the concept of a googol.

What is an **irrational number**?

Numbers that cannot be expressed as an exact ratio are called irrational numbers; numbers that can be expressed as an exact ratio are called rational numbers. For instance, 1/2 (one half, or 50 percent of something) is rational; however, 1.61803 (ϕ), 3.14159 (π), 1.41421 ($\sqrt{2}$), are irrational. History claims that Pythagoras in the sixth century B.C.E. first used the term when he discovered that the square root of 2 could not be expressed as a fraction.

Why is seven considered a supernatural number?

In magical lore and mysticism, all numbers are ascribed certain properties and energies. Seven is a number of great power, a magical number, a lucky number, a number of psychic and mystical powers, of secrecy and the search for inner truth. The origin of belief in seven's power lies in the lunar cycle. Each of the moon's four phases lasts about seven days. Life cycles on Earth also have phases demarcated by seven, such as there are said to be seven years to each stage of human growth. There are seven colors to the rainbow; and seven notes are in the musical scale. The seventh son of a seventh son is said to be born with formidable magical and psychic powers.

The number seven is widely held to be a lucky number, especially in matters of love and money, and it also carries great prominence in the old and new testaments. Here are a few examples: the Lord rested on the seventh day; there were seven years of plenty and seven years of famine in Egypt in the story of Joseph; God commanded Joshua to have seven priests carry trumpets, and on the seventh day they were to march around Jericho seven times; Solomon built the temple in seven years; and there are seven petitions of the Lord's Prayer.

What are **imaginary numbers**?

Imaginary numbers are the square roots of negative numbers. Since the square is the product of two equal numbers with like signs it is always positive. Therefore, no number multiplied by itself can give a negative real number. The symbol "i" is used to indicate an imaginary number.

What is the **value of pi** out to 30 digits past the decimal point?

Pi (π) represents the ratio of the circumference of a circle to its diameter, used in calculating the area of a circle (πr^2) and the volume of a cylinder ($\pi r^2 h$) or cone. It is a "transcendental number," an irrational number with an exact value that can be measured to any degree of accuracy, but that can't be expressed as the ratio of two integers. In theory, the decimal extends into infinity, though it is generally rounded to 3.1416. The Welsh-born mathematician William Jones (1675–1749) selected the Greek symbol (π) for pi. Rounded to 30 digits past the decimal point, it equals 3.1415926535 89793238462643383279.

In 1989, Gregory (1952–) and David Chudnovsky (1947–) at Columbia University in New York City calculated the value of pi to 1,011,961,691 decimal places. They performed the calculation twice on an IBM 3090 mainframe and on a CRAY-2 supercomputer with matching results. In 1991, they calculated pi to 2,260,321,336 decimal places.

In 1999, Yasumasa Kanada (1948–) and Daisuke Takahashi of the University of Tokyo calculated pi out to 206,158,430,000 digits. Professor Kanada at the University of Tokyo continues to calculate the value of pi to greater and greater digits. His laboratory's newest record, achieved in 2002 and subsequently verified, calculated pi to 1.2411×10^{12} digits (more than one trillion). The calculation required more than 600 hours of computing time using a Hitachi SR8000 computer with access to a memory of about 1 terabyte.

Mathematicians have also calculated pi in binary format (i.e., 0s and 1s). The five trillionth binary digit of pi was computed by Colin Percival and 25 others at Simon Fraser University. The computation took over 13,500 hours of computer time.

What are some **examples** of numbers and **mathematical concepts** in **nature**?

The world can be articulated with numbers and mathematics. Some numbers are especially prominent. The number six is ubiquitous: every normal snowflake has six sides; every honeybee colony's combs are six-sided hexagons. The curved, gradually decreasing chambers of a nautilus shell are propagating spirals of the golden section and the Fibonacci sequence of numbers. Pine cones also rely on the Fibonacci sequence, as do many plants and flowers in their seed and stem arrangements. Fractals are evident in shorelines, blood vessels, and mountains.

MATHEMATICS

How is **arithmetic different** from **mathematics**?

Arithmetic is the study of positive integers (i.e., 1, 2, 3, 4, 5) manipulated with addition, subtraction, multiplication, and division, and the use of the results in daily life. Mathematics is the study of shape, arrangement, and quantity. It is traditionally viewed as consisting of three fields: algebra, analysis, and geometry. But any lines of division have evaporated because the fields are now so interrelated.

What is the most **enduring mathematical work** of all time?

The Elements of Euclid (c. 300 B.C.E.) has been the most enduring and influential mathematical work of all time. In it, the ancient Greek mathematician presented the work of earlier mathematicians and included many of his own innovations. The *Elements* is divided into thirteen books: the first six cover plane geometry; seven to nine address arithmetic and number theory; ten treats irrational numbers; and eleven to thirteen discuss solid geometry. In presenting his theorems, Euclid used the synthetic approach, in which one proceeds from the known to the unknown by logical steps. This method became the standard procedure for scientific investigation for many centuries, and the *Elements* probably had a greater influence on scientific thinking than any other work.

Who **invented calculus**?

The German mathematician Gottfried Wilhelm Leibniz (1646–1716) published the first paper on calculus in 1684. Most historians agree that Isaac Newton invented calculus eight to ten years earlier, but he was typically very late in publishing his works. The invention of calculus marked the beginning of higher mathematics. It provided scientists and mathematicians with a tool to solve problems that had been too complicated to attempt previously.

Is it possible to **count** to **infinity**?

No. Very large finite numbers are not the same as infinite numbers. Infinite numbers are defined as being unbounded, or without limit. Any number that can be reached by counting or by representation of a number followed by billions of zeros is a finite number.

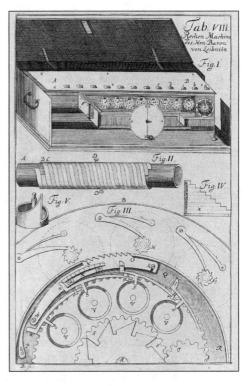

Gottfried Wilhelm Leibniz published the first paper on calculus and also invented this mechanical calculator, which could add, subtract, multiply, and divide.

How long has the **abacus** been used?

The abacus grew out of early counting boards, with hollows in a board holding pebbles or beads used to calculate. It has been documented in Mesopotamia back to around 3500 B.C.E. The current form, with beads sliding on rods, dates back at least to fifteenth-century China. Before the use of decimal number systems, which allowed the familiar paper-and-pencil methods of calculation, the abacus was essential for almost all multiplication and division. Unlike the modern calculator, the abacus does not perform any mathematical computations. The person using the abacus performs calculations in his/her head relying on the abacus as a physical aid to keep track of the sums. It has become a valuable tool for teaching arithmetic to blind students.

What are **Napier's bones**?

In the sixteenth century, the Scottish mathematician John Napier (1550–1617), Baron of Merchiston, developed a method of simplifying the processes of multiplication and division, using exponents of 10, which Napier called logarithms (commonly abbreviated as logs). Using this system, multiplication is reduced to addition and division to subtraction. For example, the log of 100 (10^2) is 2; the log of 1000 (10^3) is 3; the multiplication of 100 by 1000, $100 \times 1000 = 100,000$, can be accomplished by adding their logs: $\log[(100)(1000)] = \log(100) + \log(1000) = 2 + 3 = 5 = \log(100,000)$. Napier published his methodology in *A Description of the Admirable Table of Logarithms* in 1614. In 1617 he published a method of using a device, made up of a series of rods in a frame, marked with the digits 1 through 9, to multiply and divide using the principles of logarithms. This device was commonly called "Napier's bones" or "Napier's rods."

What are **Cuisenaire rods**?

The Cuisenaire method is a teaching system used to help young students independently discover basic mathematical principles. Developed by Emile-Georges Cuisenaire (1891–1976), a Belgian schoolteacher, the method uses rods of ten different colors and lengths that are easy to handle. The rods help students understand mathematical principles rather than merely memorizing them. They are also used to teach elementary arithmetic properties such as associative, commutative, and distributive properties.

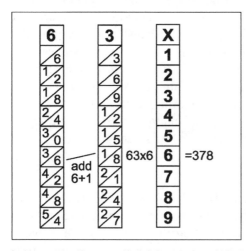

In this example of how to use Napier's Bones, 63 is multiplied by 6 to get the correct result of 378.

What is a **slide rule**, and who invented it?

Up until about 1974, most engineering and design calculations for buildings, bridges, automobiles, airplanes, and roads were done on a slide rule. A slide rule is an apparatus with moveable scales based on logarithms, which were invented by John Napier, Baron of Merchiston, and published in 1614. The slide rule can, among other things, quickly multiply, divide, square root, or find the logarithm of a number. In 1620, Edmund Gunter (1581–1626) of Gresham College, London, England, described an immediate forerunner of the slide rule, his "logarithmic line of numbers." William Oughtred (1574–1660), rector of Aldbury, England, made the first rectilinear slide rule in 1621. This slide rule consisted of two logarithmic scales that could be manipulated together for calculation. His former pupil, Richard Delamain, published a description of a circular slide rule in 1630 (and received a patent about that time for it), three years before Oughtred published a description of his invention (at least one source says that Delamain published in 1620). Oughtred accused Delamain of stealing his idea, but evidence indicates that the inventions were probably arrived at independently.

The earliest existing straight slide rule using the modern design of a slider moving in a fixed stock dates from 1654. A wide variety of specialized slide rules were developed by the end of the seventeeth century for trades such as masonry, carpentry, and excise tax collecting. Peter Mark Roget (1779–1869), best known for his *Thesaurus of English Words and Phrases*, invented a log-log slide rule for calculating the roots and powers of numbers in 1814. In 1967, Hewlett-Packard produced the first pocket calculators. Within a decade, slide rules became the subject of science trivia and collector's books. Interestingly, slide rules were carried on five of the Apollo space missions, including a trip to the moon. They were known to be accurate and efficient in the event of a computer malfunction.

How is **casting out nines** used to check the results of addition or multiplication?

The method of "casting out nines" is based on the excess of nines in digits of whole numbers (the remainder when a sum of digits is divided by 9). Illustrating this process in the multiplication example below, the method begins by adding the digits in both the multiplicand (one of the terms that is being multiplied) and the multiplier (the other term being multiplied). In the example below, this operation leads to the results of "13" and "12," respectively. If these results are greater than 9 (>9), then the operation is repeated until the resulting figures are less than 9 (<9). In the example below, the repeated calculation gives the results as "4" and "3," respectively. Multiply the resulting "excess" from the multiplicand by the excess from the multiplier (4 3 3 below). Add the digits of the result to eventually yield a number equal to or less than 9 (≤9). Repeat the process of casting out nines in the multiplication product (the result of the multiplication process). The result must equal the result of the previous set of transactions, in this case "3." If the two figures disagree, then the original multiplication procedure was done incorrectly. "Casting out nines" can also be applied to check the accuracy of the results of addition.

$$
\begin{array}{r}
328 \\
\underline{624} \\
1312 \\
656 \\
\underline{1968} \\
204672
\end{array}
\quad
\begin{array}{c}
\rightarrow \ 13 \ \rightarrow \ 4 \\
\rightarrow \ 12 \ \rightarrow \ 3 \\
 12 \rightarrow 3
\end{array}
$$

$$204672 \rightarrow 21 \rightarrow 3$$

What is the difference between a **median** and a **mean**?

If a string of numbers is arranged in numerical order, the median is the middle value of the string. If there is an even number of values in the string, the median is found by adding the two middle values and dividing by two. The arithmetic mean, also known as the simple average, is found by taking the sum of the numbers in the string and dividing by the number of items in the string. While easy to calculate for relatively short strings, the arithmetic mean can be misleading, as very large or very small values in the string can distort it. For example, the mean of the salaries of a professional football team would be skewed if one of the players was a high-earning superstar; it could be well above the salaries of any of the other players thus making the mean higher. The mode is the number in a string that appears most often.

For the string 111222234455667, for example, the median is the middle number of the series: 3. The arithmetic mean is the sum of numbers divided by the number of numbers in the series, 51 / 15 = 3.4. The mode is the number that occurs most often, 2.

When did the **concept** of **square root** originate?

A square root of a number is a number that, when multiplied by itself, equals the given number. For instance, the square root of 25 is 5 ($5 \times 5 = 25$). The concept of the square root has been in existence for many thousands of years. Exactly how it was discovered is not known, but several different methods of exacting square roots were used by early mathematicians. Babylonian clay tablets from 1900 to 1600 B.C.E. contain the squares and cubes of integers 1 through 30. The early Egyptians used square roots around 1700 B.C.E., and during the Greek Classical Period (600 to 300 B.C.E.) better arithmetic methods improved square root operations. In the sixteenth century, French mathematician René Descartes (1596–1650) was the first to use the square root symbol, called "the radical sign," $\sqrt{}$.

What are **Venn diagrams**?

Venn diagrams are graphical representations of set theory, which use circles to show the logical relationships of the elements of different sets, using the logical operators (also called in computer parlance "Boolean Operators") and, or, and not. John Venn (1834–1923) first used them in his 1881 *Symbolic Logic*, in which he interpreted and

---

When does 0 × 0 = 1?

Factorials are the product of a given number and all the factors less than that number. The notation n! is used to express this idea. For example, 5! (five factorial) is $5 \times 4 \times 3 \times 2 \times 1 = 120$. For completeness, 0! is assigned the value 1, so $0 \times 0 = 1$.

---

corrected the work of George Boole (1815–1864) and Augustus de Morgan (1806–1871). While his attempts to clarify perceived inconsistencies and ambiguities in Boole's work are not widely accepted, the new method of diagraming is considered to be an improvement. Venn used shading to better illustrate inclusion and exclusion. Charles Dodgson (1832–1898), better known by his pseudonym Lewis Carroll, refined Venn's system, in particular by enclosing the diagram to represent the universal set.

How many **feet** are on each **side** of an **acre** that is square?

An acre that is square in shape has about 208.7 feet (64 meters) on each side.

What are the common **mathematical formulas** for **area**?

Area of a rectangle:

Area = length times width

$A = lw$

Area = altitude times base

$A = ab$

Area of a circle:

Area = pi times the radius squared

$A = \pi r^2$ or $A = 1/4\pi d^2$

Area of a triangle:

Area = one half the altitude times the base

$A = 1/2ab$

Area of the surface of a sphere:

Area = four times pi times the radius squared

$A = 4\pi r^2$ or $A = \pi d^2$

Area of a square:

Area = length times width, or length of one side squared

$A = s^2$

23

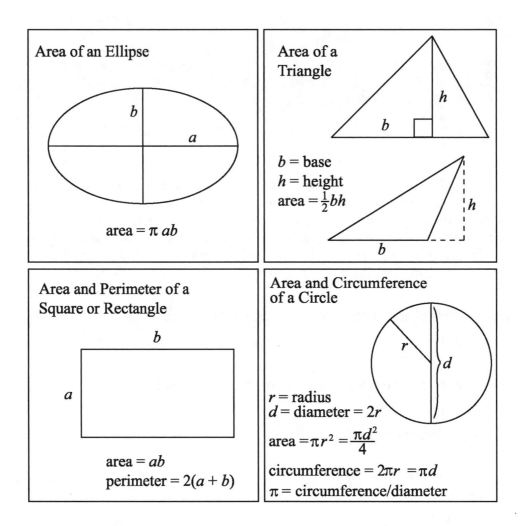

Area of an Ellipse	Area of a Triangle
area = $\pi\, ab$	b = base h = height area = $\frac{1}{2}bh$
Area and Perimeter of a Square or Rectangle	Area and Circumference of a Circle
area = ab perimeter = $2(a + b)$	r = radius d = diameter = $2r$ area = $\pi r^2 = \dfrac{\pi d^2}{4}$ circumference = $2\pi r = \pi d$ π = circumference/diameter

Area of a cube:

 Area = square of the length of one side times 6

 $A = 6s^2$

Area of an ellipse:

 Area = long diameter times short diameter times 0.7854.

What are the common **mathematical formulas** for **volume**?

Volume of a sphere:

 Volume = 4/3 times pi times the cube of the radius

 $V = 4/3 \times \pi r3$

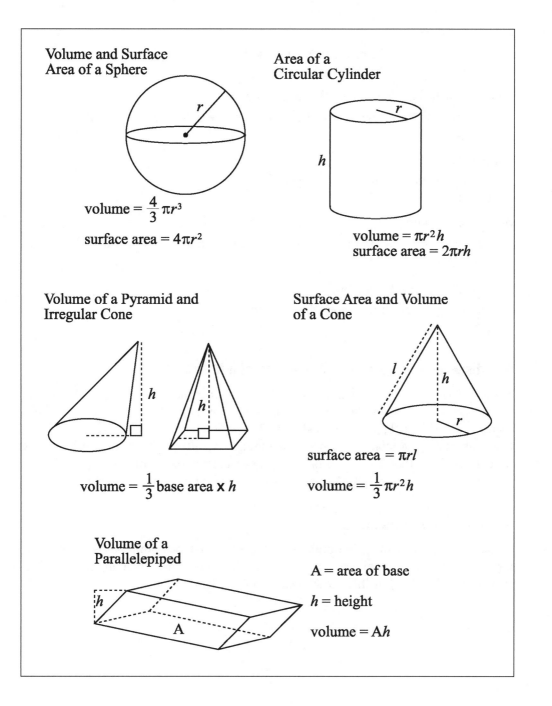

Volume and Surface Area of a Sphere

$$\text{volume} = \frac{4}{3}\pi r^3$$

$$\text{surface area} = 4\pi r^2$$

Area of a Circular Cylinder

$$\text{volume} = \pi r^2 h$$
$$\text{surface area} = 2\pi rh$$

Volume of a Pyramid and Irregular Cone

$$\text{volume} = \frac{1}{3}\text{base area} \times h$$

Surface Area and Volume of a Cone

$$\text{surface area} = \pi rl$$

$$\text{volume} = \frac{1}{3}\pi r^2 h$$

Volume of a Parallelepiped

A = area of base

h = height

$$\text{volume} = Ah$$

Volume of a pyramid:

Volume = 1/3 times the area of the base times the height

V = 1/3bh

25

Volume of a cylinder:
   Volume = area of the base times the height
   V = bh

Volume of a circular cylinder (with circular base):
   Volume = pi times the square of the radius of the base times the height
   $V = \pi r^2 h$

Volume of a cube:
   Volume = the length of one side cubed
   V = S3

Volume of a cone:
   Volume = 1/3 times pi times the square of the radius of the base times the height
   $V = 1/3\ \pi r^2 h$

Volume of a rectangular solid:
   Volume = length times width times height
   V = lwh

Who **discovered** the **formula** for the **area** of a **triangle**?

Heron (or Hero) of Alexandria (first century B.C.E.) is best known in the history of mathematics for the formula that bears his name. This formula calculates the area of a triangle with sides a, b, and c, with s = half the perimeter: $A = \sqrt{}$.
The Arab mathematicians who preserved and transmitted the mathematics of the Greeks reported that this formula was known earlier to Archimedes (c. 287–212 B.C.E.), but the earliest proof now known is that appearing in Heron's *Metrica*.

How is **Pascal's triangle used**?

Pascal's triangle is an array of numbers, arranged so that every number is equal to the sum of the two numbers above it on either side. It can be represented in several slightly different triangles, but this is the most common form:

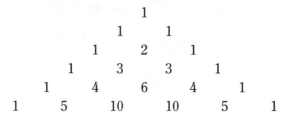

```
              1
           1     1
        1     2     1
     1     3     3     1
   1    4     6     4     1
 1    5    10    10     5    1
```

The triangle is used to determine the numerical coefficients resulting from the computation of higher powers of a binomial (two numbers added together). When a binomial is raised to a higher power, the result is expanded, using the numbers in that

> ## What is the ancient Greek problem of squaring the circle?
>
> This problem was to construct, with a straight-edge and compass, a square having the same area as a given circle. The Greeks were unable to solve the problem because the task is impossible, as was shown by the German mathematician Ferdinand von Lindemann (1852–1939) in 1882.

row of the triangle. For example, $(a + b)1 = a1 + b1$, using the coefficients in the second line of the triangle $(a + b)^2 = a^2 + 2ab + b^2$, using the coefficients in the next line of the triangle. (The first line of the triangle correlates to $[a + b]0$.) While the calculation of coefficients is fairly straightforward, the triangle is useful in calculating them for the higher powers without needing to multiply them out. Binomial coefficients are useful in calculating probabilities; Blaise Pascal was one of the pioneers in developing laws of probability.

As with many other mathematical developments, there is some evidence of a previous appearance of the triangle in China. Around 1100 C.E., the Chinese mathematician Chia Hsien wrote about "the tabulation system for unlocking binomial coefficients"; the first publication of the triangle was probably in a book called *Piling-Up Powers and Unlocking Coefficients* by Liu Ju-Hsieh.

What is the **Pythagorean theorem**?

In a right triangle (one where two of the sides meet in a 90-degree angle), the hypotenuse is the side opposite the right angle. The Pythagorean theorem, also known as the rule of Pythagoras, states that the square of the length of the hypotenuse is equal to the sum of the squares of the other two sides ($h^2 = a^2 + b^2$). If the lengths of the sides are: h = 5 inches, a = 4 inches, and b = 3 inches, then:

$$h = \sqrt{(a^2 + b^2)} = \sqrt{(42 + 32)} = \sqrt{(16 + 9)} = \sqrt{25} = 5$$

The theorem is named for the Greek philosopher and mathematician Pythagoras (c. 580–c. 500 B.C.E.). Pythagoras is credited with the theory of the functional significance of numbers in the objective world and numerical theories of musical pitch. As he left no writings, the Pythagorean theorem may actually have been formulated by one of his disciples.

What are the **Platonic solids**?

The Platonic solids are the five regular polyhedra: the four-sided tetrahedron, the six-sided cube or hexahedron, the eight-sided octahedron, the twelve-sided dodecahedron, and the twenty-sided icosahedron. Although they had been studied as long ago as the

27

time of Pythagorus (around 500 B.C.E.), they are called the Platonic solids because they were first described in detail by Plato (427–347 B.C.E.) around 400 B.C.E. The ancient Greeks gave mystical significance to the Platonic solids: the tetrahedron represented fire, the icosahedron represented water, the stable cube represented the earth, the octahedron represented the air. The twelve faces of the dodecahedron corresponded to the twelve signs of the zodiac, and this figure represented the entire universe.

What does the expression **"tiling the plane"** mean?

It is a mathematical expression describing the process of forming a mosaic pattern (a "tessellation") by fitting together an infinite number of polygons so that they cover an entire plane. Tessellations are the familiar patterns that can be seen in designs for quilts, floor coverings, and bathroom tilework.

What is a **golden section**?

Golden section, also called the divine proportion, is the division of a line segment so that the ratio of the whole segment to the larger part is equal to the ratio of the larger part to the smaller part. The ratio is approximately 1.61803 to 1. The number 1.61803 is called the golden number (also called Phi [with a capital P]). The golden number is the limit of the ratios of consecutive Fibonacci numbers, such as, for instance, 21/13 and 34/21. A golden rectangle is one whose length and width correspond to this ratio. The ancient Greeks thought this shape had the most pleasing proportions. Many famous painters have used the golden rectangle in their paintings, and architects have used it in their design of buildings, the most famous example being the Greek Parthenon.

What is a **Möbius strip**?

A Möbius strip is a surface with only one side, usually made by connecting the two ends of a rectangular strip of paper after putting a half-twist (180 degrees relative to the opposite side) in the strip. Cutting a Möbius strip in half down the center of the length of the strip results in a single band with four half-twists. Devised by the German mathematician August Ferdinand Möbius (1790–1868) to illustrate the properties of one-sided surfaces, it was presented in a paper that was not discovered or published until after his death. Another nineteenth-century German

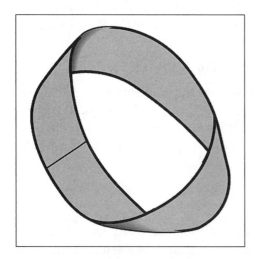

28   A Möbius strip.

> ## How many ways are there to shuffle a deck of cards?
>
> There are a possible 80,658,175,170,943,878,571,660,636,856,403,766,975, 289,505,440,883,277,824,000,000,000,000 ways to shuffle a deck of cards.

mathematician, Johann Benedict Listing (1808–1882), developed the idea independently at the same time.

How is the **rule of 70** used?

This rule is a quick way of estimating the period of time it will take a quantity to double given the percentage of increase. Divide the percentage of increase into 70. For example, if a sum of money is invested at six percent interest, the money will double in value in 70/6 = 11.7 years.

How is **percent of increase calculated**?

To find the percent of increase, divide the amount of increase by the base amount. Multiply the result by 100 percent. For example, a raise in salary from $10,000 to $12,000 would have percent of increase = (2,000/10,000) × 100% = 20%.

How many **different bridge games** are possible?

Roughly 54 octillion different bridge games are possible.

What are **fractals**?

A fractal is a set of points that are too irregular to be described by traditional geometric terms, but that often possess some degree of self-similarity; that is, are made of parts that resemble the whole. They are used in image processing to compress data and to depict apparently chaotic objects in nature such as mountains or coastlines. Scientists also use fractals to better comprehend rainfall trends, patterns formed by clouds and waves, and the distribution of vegetation. Fractals are also used to create computer-generated art.

What is the difference between **simple interest** and **compound interest**?

Simple interest is calculated on the amount of principal only. Compound interest is calculated on the amount of principal plus any previous interest already earned. For example, $100 invested at a rate of five percent for one year will earn $5.00 after one year earning simple interest. The same $100 will earn $5.12 if compounded monthly.

What is the **probability** of a **triple play** occurring in a single baseball game?

The odds against a triple play in a game of baseball are 1,400 to 1.

What is the **law of very large numbers**?

Formulated by Persi Diaconis (1945–) and Frederick Mosteller (1916–2006) of Harvard University, this long-understood law of statistics states that "with a large enough sample, any outrageous thing is apt to happen." Therefore, seemingly amazing coincidences can actually be expected if given sufficient time or a large enough pool of subjects. For example, when a New Jersey woman won the lottery twice in four months, the media publicized it as an incredible long shot of 1 in 17 trillion. However, when statisticians looked beyond this individual's chances and asked what were the odds of the same happening to any person buying a lottery ticket in the United States over a six-month period, the number dropped dramatically to 1 in 30. According to researchers, coincidences arise often in statistical work, but some have hidden causes and therefore are not coincidences at all. Many are simply chance events reflecting the luck of the draw.

What is the **Königsberg Bridge Problem**?

The city of Königsberg was located in Prussia on the Pregel River. Two islands in the river were connected by seven bridges. By the eighteenth century, it had become a tradition for the citizens of Königsberg to go for a walk through the town trying to cross each bridge only once. No one was able to succeed, and the question was asked whether it was possible to do so. In 1736, Leonhard Euler (1707–1783) proved that it was not possible to cross the Königsberg bridges only once. Euler's solution led to the development of two new areas of mathematics: graph theory, which deals with questions about networks of points that are connected by lines; and topology, which is the study of those aspects of the shape of an object that do not depend on length measurements.

How long did it take to **prove** the **four-color map theorem**?

The four-color map problem was first posed by Francis Guthrie (1831–1899) in 1852. While coloring a map of the English counties, Guthrie discovered he could do it with only four colors and no two adjacent counties would be the same color. He extrapolated the question to whether every map, no matter how complicated and how many countries are on the map, could be colored using only four colors with no two adjacent countries being the same color. The theorem was not proved until 1976, 124 years after the question had been raised, by Kenneth Appel (1932–) and Wolfgang Haken (1928–). Their proof is considered correct although it relies on computers for the calculations. There is no known simple way to check the proof by hand.

What is the science of **chaos**?

Chaos or chaotic behavior is the behavior of a system whose final state depends very sensitively on the initial conditions. The behavior is unpredictable and cannot be distinguished from a random process, even though it is strictly determinate in a mathematical sense. Chaos studies the complex and irregular behavior of many systems in nature, such as changing weather patterns, flow of turbulent fluids, and swinging pendulums. Scientists once thought they could make exact predictions about such systems, but found that a tiny difference in starting conditions can lead to greatly different results. Chaotic systems do obey certain rules, which mathematicians have described with equations, but the science of chaos demonstrates the difficulty of predicting the long-range behavior of chaotic systems.

What is **Zeno's paradox**?

Zeno of Elea (c. 490–c. 425 B.C.E.), a Greek philosopher and mathematician, is famous for his paradoxes, which deal with the continuity of motion. One form of the paradox is: If an object moves with constant speed along a straight line from point 0 to point 1, the object must first cover half the distance (1/2), then half the remaining distance (1/4), then half the remaining distance (1/8), and so on without end. The conclusion is that the object never reaches point 1. Because there is always some distance to be covered, motion is impossible. In another approach to this paradox, Zeno used an allegory telling of a race between a tortoise and Achilles (who could run 100 times as fast), where the tortoise started running 10 rods (165 feet) in front of Achilles. Because the tortoise always advanced 1/100 of the distance that Achilles advanced in the same time period, it was theoretically impossible for Achilles to pass him. The English mathematician and writer Charles Dodgson, better known as Lewis Carroll, used the characters of Achilles and the tortoise to illustrate his paradox of infinity.

Are there any **unsolved problems** in mathematics?

The earliest challenges and contests to solve important problems in mathematics date back to the sixteenth and seventeenth centuries. Some of these problems have continued to challenge mathematicians until modern times. For example, Pierre de Fermat (1601–1665) issued a set of mathematical challenges in 1657, many on prime numbers and divisibility. The solution to what is now known as Fermat's Last Theorem was not established until the late 1990s by Andrew Wiles (1953–). David Hilbert (1862–1943), a German mathematician, identified 23 unsolved problems in 1900 with the hope that these problems would be solved in the twenty-first century. Although some of the problems were solved, others remain unsolved to this day. More recently, in 2000 the Clay Mathematics Institute named seven mathematical problems that had not been solved with the hope that they could be solved in the twenty-first century. A $1 million prize will be awarded for solving each of these seven problems.

What are the seven **Millennium Prize Problems**?

The seven Millennium Prize Problems are:

1. Birch and Swinnerton-Dyer Conjecture
2. Hodge Conjecture
3. Poincaré Conjecture
4. Riemann Hypothesis
5. Solution of the Navier-Stokes equations
6. Formulation of the Yang-Mills theory
7. P vs NP

COMPUTERS

What is an **algorithm**?

An algorithm is a set of clearly defined rules and instructions for the solution of a problem. It is not necessarily applied only in computers, but can be a step-by-step procedure for solving any particular kind of problem. A nearly 4,000-year-old Babylonian banking calculation inscribed on a tablet is an algorithm, as is a computer program that consists of step-by-step procedures for solving a problem.

The term is derived from the name of Muhammad ibn Musa al Kharizmi (c. 780–c. 850), a Baghdad mathematician who introduced Hindu numerals (including 0) and decimal calculation to the West. When his treatise was translated into Latin in the twelfth century, the art of computation with Arabic (Hindu) numerals became known as algorism.

Who **invented** the **computer**?

Computers developed from calculating machines. One of the earliest mechanical devices for calculating, still widely used today, is the abacus—a frame carrying parallel rods on which beads or counters are strung. The abacus originated in Egypt in 2000 B.C.E. it reached the Orient about a thousand years later, and arrived in Europe in about the year 300 C.E. In 1617, John Napier invented "Napier's Bones"—marked pieces of ivory for multiples of numbers. In the middle of the same century, Blaise Pascal produced a simple mechanism for adding and subtracting. Multiplication by repeated addition was a feature of a stepped drum or wheel machine of 1694 invented by Gottfried Wilhelm Leibniz.

In 1823, the English visionary Charles Babbage (1792–1871) persuaded the British government to finance an "analytical engine." This would have been a

The Englishman Charles Babbage is credited with inventing the first computer back in the nineteenth century.

machine that could undertake any kind of calculation. It would have been driven by steam, but the most important innovation was that the entire program of operations was stored on a punched tape. Babbage's machine was not completed in his lifetime because the technology available to him was not sufficient to support his design. However, in 1991 a team led by Doron Swade (1946–) at London's Science Museum built the "analytical engine" (sometimes called the "difference engine") based on Babbage's work. Measuring 10 feet (3 meters) wide by 6.5 feet (2 meters) tall, it weighed three tons and could calculate equations down to 31 digits. The feat proved that Babbage was way ahead of his time, even though the device was impractical because one had to turn a crank hundreds of times in order to generate a single calculation. Modern computers use electrons, which travel at nearly the speed of light.

Based on the concepts of British mathematician Alan M. Turing (1912–1954), the earliest programmable electronic computer was the 1,500-valve "Colossus," formulated by Max Newman (1897–1985), built by T.H. Flowers, (1905–1998) and used by the British government in 1943 to crack the German codes generated by the cipher machine "Enigma."

What is an **expert system**?

An expert system is a type of highly specialized software that analyzes complex problems in a particular field and recommends possible solutions based on information previously programmed into it. The person who develops an expert system first analyzes the behavior of a human expert in a given field, then inputs all the explicit rules resulting from their study into the system. Expert systems are used in equipment repair, insurance planning, training, medical diagnosis, and other areas.

What was the **first** major **use** for **punched cards**?

Punched cards were a way of programming, or giving instructions to, a machine. In 1801, Joseph Marie Jacquard (1752–1834) built a device that could do automated pattern weaving. Cards with holes were used to direct threads in the loom, creating predefined patterns in the cloth. The pattern was determined by the arrangement of holes in the cards, with wire hooks passing through the holes to grab and pull through specific threads to be woven into the cloth.

By the 1880s, Herman Hollerith (1860–1929) was using the idea of punched cards to give machines instructions. He built a punched card tabulator that processed the data gathered for the 1890 United States Census in six weeks (three times the speed of previous compilations). Metal pins in the machine's reader passed through holes punched in cards the size of dollar bills, momentarily closing electric circuits.

The resulting pulses advanced counters assigned to details such as income and family size. A sorter could also be programmed to pigeonhole cards according to pattern of holes, an important aid in analyzing census statistics. Later, Hollerith founded Tabulating Machines Co., which in 1924 became IBM. When IBM adopted the 80-column punched card (measuring 7 3/8 × 3 1/4 inches [18.7 × 8.25 centimeters] and 0.007 inches [0.17 millimeters] thick), the de facto industry standard was set, which has endured for decades.

What was **MANIAC**?

MANIAC (mathematical analyzer, numerator, integrator, and computer) was built at the Los Alamos Scientific Laboratory under the direction of Nicholas C. Metropolis (1915–1999) between 1948 and 1952. It was one of several different copies of the high-speed computer built by John von Neumann (1903–1957) for the Institute for Advanced Studies (IAS). It was constructed primarily for use in the development of atomic energy applications, specifically the hydrogen bomb.

It originated with the work on ENIAC (electronic numerical integrator and computer), the first fully operational, large-scale, electronic digital computer. ENIAC was built at the Moore School of Electrical Engineering at the University of Pennsylvania between 1943 and 1946. Its builders, John Presper Eckert Jr. (1919–1995) and John William Mauchly (1907–1980), virtually launched the modern era of the computer with ENIAC.

What was the **name** of the **personal computer** introduced by **Apple** in the early 1980s?

Lisa was the name of the personal computer that Apple introduced. The forerunner of the Macintosh personal computer, Lisa had a graphical user interface and a mouse.

What was the **first computer game**?

Despite the fact that computers were not invented for playing games, the idea that they could be used for games did not take long to emerge. Alan Turing proposed a famous game called "Imitation Game" in 1950. In 1952, Rand Air Defense Lab in Santa Monica created the first military simulation games. In 1953, Arthur Samuel (1901–1990) created a checkers program on the new IBM 701. From these beginnings computer games have today become a multi-billion-dollar industry.

What was the **first** successful **video-arcade game**?

Pong, a simple electronic version of a tennis game, was the first successful video-arcade game. Although it was first marketed in 1972, Pong was actually invented 14 years earlier in 1958 by William Higinbotham (1910–1994), who, at the time, headed instrumentation design at Brookhaven National Laboratory. Invented to amuse visitors touring the laboratory, the game was so popular that visitors would stand in line for hours to play it. Higinbotham dismantled the system two years later, and, considering it a trifle, did not patent it. In 1972, Atari released Pong, an arcade version of Higinbotham's game, and Magnavox released Odyssey, a version that could be played on home televisions.

What was **"The Turk"**?

The Turk was the name for a famous chess-playing automaton. An automaton, such as a robot, is a mechanical figure constructed to act as if it moves by its own power. On a dare in 1770, a civil servant in the Vienna Imperial Court named Wolfgang von Kempelen (1734–1804) created a chess-playing machine. This mustached, man-sized figure carved from wood wore a turban, trousers, and robe, and sat behind a desk. In one hand it held a long Turkish pipe, implying that it had just finished a pre-game smoke, and its innards were filled with gears, pulleys, and cams. The machine seemed a keen chess player and dumbfounded onlookers by defeating all the best human chess players. It was a farce, however: its moves were surreptitiously made by a man hiding inside.

The Turk, so dubbed because of the outfit similar to traditional Turkish garb, is regarded as a forerunner to the industrial revolution because it created a commotion over devices that could complete complex tasks. Historians argue that it inspired people to invent other early devices such as the power loom and the telephone, and it even was a precursor to concepts such as artificial intelligence and computerization. Today, however, computer chess games are so sophisticated that they can defeat even the world's best chess masters. In May 1997, the "Deep Blue" chess computer defeated World Champion Garry Kasparov (1963–). "Deep Blue" was a 32-node IBM RS/6000 SP high-performance computer that used Power Two Super Chip processors (P2SC). Each node had a single microchannel card containing eight dedicated VLSI chess processors for a total of 256 processors working in tandem, allowing "Deep Blue" to calculate 100 to 200 billion chess moves within three minutes.

What are the **components** of a **computer**?

Computers have two major components: the hardware and the software. Hardware consists of all the physical devices needed to actually build and operate a computer. Examples of computer hardware are the central processing unit (CPU), hard drive, memory, modems, and external devices such as the keyboard, monitor, printers, scanners, and other devices that can be physically touched. Software is an integral part of a computer and consists of the various computer programs that allow the user to interact with it and specify the tasks the computer performs. Without software, a computer is merely a collection of circuits and metal in a box unable to perform even the most basic functions.

Why does the **actual** amount of computer storage space differ from the **advertised** amount of storage?

Computer storage space for hard drives and other storage media is calculated in base 2 using binary format with a byte as the basic unit. The common units of computer storage are:

Kilobyte (KB)	1,024 bytes
Megabyte (MB)	1,024 kilobytes or 1,048,576 bytes
Gigabyte (GB)	1,024 megabytes or 1,073,741,824 bytes
Terebyte (TB)	1,024 gigabytes or 1,099,511,627,776 bytes

However, since consumers are more familiar with the decimal, base 10, system of numbers, computer manufacturers describe storage sizes in base 10 where one megabyte is one million bytes and one gigabyte is one billion bytes. Therefore, for each gigabyte they are over reporting storage space by 73,741,824 bytes. The concept is further complicated because some storage media may have the actual amount of advertised storage, but some of the available storage is lost due to formatting.

What is a **silicon chip**?

A silicon chip is an almost pure piece of silicon, usually less than one centimeter square and about half a millimeter thick. It contains hundreds of thousands of miniaturized electronic circuit components, mainly transistors, packed and interconnected in layers beneath the surface. These components can perform control, logic, and memory functions. There is a grid of thin metallic strips on the surface of the chip; these wires are used for electrical connections to other devices. The silicon chip was developed independently by two researchers: Jack Kilby (1923–2005) of Texas Instruments in 1958, and Robert Noyce (1927–1990) of Fairchild Semiconductor in 1959. Jack Kilby was awarded the Nobel Prize in Physics in 2000 for his discovery of the silicon chip.

While silicon chips are essential to almost all computer operations today, a myriad of other devices depend on them as well, including calculators, microwave ovens, automobiles, and VCRs.

What are the **sizes** of **silicon chips**?

Small silicon chips may be no more than 1/16" square by 1/30" thick and hold up to tens of thousands of transistors. Large chips, the size of a postage stamp, can contain hundreds of millions of transistors.

Chip Size	Number of Transistors per Chip
SSI—small-scale integration	Less than 100
MSI—medium-scale integration	Between 100 and 3,000
LSI—large-scale integration	Between 3,000 and 100,000
VLSI—very-large-scale integration	Between 100,000 and 1 million
ULSI—ultra-large-scale integration	More than 1 million

What is the **central processing unit** of a computer?

The central processing unit (CPU) of a computer is where almost all computing takes place in all computers including mainframes, desktops, laptops, and servers. The CPU of almost every computer is contained on a single chip.

How is the **speed** of a **CPU measured**?

Separate from the "real-time clock" which keeps track of the time of day, the CPU clock sets the tempo for the processor and measures the transmission speed of electronic devices. The clock is used to synchronize data pulses between sender

With the invention of the silicon chip, computers went from bulky and expensive machines filled with vacuum tubes to much smaller and affordable pieces of technology that average people could purchase.

and receiver. A one megahertz clock manipulates a set number of bits one million times per second. In general, the higher the clock speed, the quicker data is processed. However, newer versions of software often require quicker computers just to maintain their overall processing speed.

The hertz is named in honor of Heinrich Hertz (1857–1894) who detected electromagnetic waves in 1883. One hertz is equal to the number of electromagnetic waves or cycles in a signal that is one cycle per second.

What is the difference between **RAM** and **ROM**?

Random-access memory (RAM) is where programs and the systems that run the computer are stored until the CPU can access them. RAM may be read and altered by the user. In general, the more RAM, the faster the computer. RAM holds data only when the current is on to the computer. Newer computers have DDR (double data rate) memory chips. Read-only memory (ROM) is memory that can be read, but not altered by the user. ROM stores information, such as operating programs, even when the computer is switched off.

Are any devices being developed to **replace silicon chips**?

When transistors were introduced in 1948, they demanded less power than fragile, high-temperature vacuum tubes; they allowed electronic equipment to become smaller, faster, and more dependable; and they generated less heat. These developments made computers much more economical and accessible; they also made portable

What is Moore's Law?

Gordon Moore (1929–), cofounder of Intel®, a top microchip manufacturer, observed in 1965 that the number of transistors per microchip—and hence a chip's processing power—would double about every year and a half. The press dubbed this Moore's Law. Despite claims that this ever-increasing trend cannot perpetuate, history has shown that microchip advances are, indeed, keeping pace with Moore's prediction.

radios practical. However, the smaller components were harder to wire together, and hand wiring was both expensive and error-prone.

In the early 1960s, circuits on silicon chips allowed manufacturers to build increased power, speed, and memory storage into smaller packages, which required less electricity to operate and generated even less heat. While through most of the 1970s manufacturers could count on doubling the components on a chip every year without increasing the size of the chip, the size limitations of silicon chips are becoming more restrictive. Though components continue to grow smaller, the same rate of shrinking cannot be maintained.

Researchers are investigating different materials to use in making circuit chips. Gallium arsenide is harder to handle in manufacturing, but it has the potential for greatly increased switching speed. Organic polymers are potentially cheaper to manufacture and could be used for liquid-crystal and other flat-screen displays, which need to have their electronic circuits spread over a wide area. Unfortunately, organic polymers do not allow electricity to pass through as well as the silicons do. Several researchers are working on hybrid chips, which could combine the benefits of organic polymers with those of silicon. Researchers are also in the initial stages of developing integrated optical chips, which would use light rather than electric current. Optical chips would generate little or no heat, would allow faster switching, and would be immune to electrical noise.

What is a **hard drive** of a computer?

Hard disks, formerly called hard disk drives and more recently just hard drives, were invented in the 1950s. They are storage devices in desktop computers, laptops, servers, and mainframes. Hard disks use a magnetic recording surface to record, access, and erase data, in much the same way as magnetic tape records, plays, and erases sound or images. A read/write head, suspended over a spinning disk, is directed by the central processing unit (CPU) to the sector where the requested data is stored, or where the data is to be recorded. A hard disk uses rigid aluminum disks coated with iron oxide to store data. Data are stored in files that are named collections of bytes. The bytes could be anything from the ASCII codes for the characters of a text file to instructions for a software application to the records of a database to the pixel colors for an image. Hard drive size ranges from several hundred gigabytes to more than one terabyte.

A hard disk rotates from 5,400 to 7,200 revolutions per minute (rpm) and is constantly spinning (except in laptops, which conserve battery life by spinning the hard disk only when in use). An ultra-fast hard disk has a separate read/write head over each track on the disk, so that no time is lost in positioning the head over the desired track; accessing the desired sector takes only milliseconds, the time it takes for the disk to spin to the sector.

Hard drive performance is measured by data rate and seek time. Data rate is the number of bytes per second that the hard drive can deliver to the CPU. Seek time is the amount of time that elapses from when the CPU requests a file and when the first byte of the file is delivered to the CPU.

What were **floppy disks**?

The first floppy disk drive was invented by Alan Shugart (1930–2006) in 1967 at IBM. Floppy disks, also called diskettes, were made of plastic film covered with a magnetic coating, which were enclosed in a nonremovable plastic protective envelope. Floppy disks varied in storage capacity from one hundred thousand bytes to more than two megabytes. The three common floppy disk (diskette) sizes varied widely in storage capacity.

Envelope size (inches)	Storage capacity
8	100,000–500,000 bytes
5.25	100 kilobytes–1.2 megabytes
3.5	400 kilobytes–more than 2 megabytes

An 8-inch or 5.25-inch diskette was enclosed in a plastic protective envelope, which did not protect the disk from bending or folding; parts of the disk surface were also exposed and could be contaminated by fingerprints or dust. These diskettes became known as "floppy" disks because the packaging of the 5.25-inch disk was a very flexible plastic envelope. The casing on a 3.5-inch floppy disk was rigid plastic, and included a sliding disk guard that protected the disk surface, but allowed it to be exposed when the disk was inserted in the disk drive. This protection, along with the increased data storage capacity, made the 3.5-inch disk the most popular. Zip® disks were very similar to floppy disks but the magnetic coating was of much higher quality. They were able store up to 750 megabytes of data.

By the mid–1990s, floppy disks and Zip® disks had become obsolete as computer files and memory required larger storage and computers were no longer being manufactured with floppy disk drives. These disks can still be accessed by using an external floppy drive reader with a USB connection.

What are some **newer forms** of **portable storage media**?

External hard drives are similar to the internal hard drive of the machine with storage capacity of up to and beyond 2 TB. Often using a USB port to connect, they provide an easy alternative for back-up storage.

Flash memory sticks can hold dozens of gigabytes of memory in an amazingly small space not even dreamed of in the early days of computing.

Compact discs (CDs) and DVDs are optical storage devices. There are three types of CDs and DVDs: read-only, write-once (CD-R and DVD-R), and rewritable (CD-RW and DVD-RW). Although read-only CDs and DVDs are wonderful for prepackaged software, they do not permit a user to save their own material. CD-Rs store 700 MB of data and can be written once. DVD-Rs are similar to CD-Rs except they hold 4 to 28 times more data. Single disc DVD-Rs are available that can store 4.7 GB of data or two hours of video. Read/write CDs use a different chemical compound which allows data to be recorded, erased, and re-written.

Solid state storage technology has no moving parts. One example is flash memory sticks. All cells are set to 0 on the memory chip before data is stored in a flash memory stick. When data are entered, electric charges are applied to certain cells. These charges pierce a thin layer of oxide and become trapped. The trapped charges become 1s. The binary code pattern of 0s and 1s is stored into the memory. Flash memory sticks are available with up to 64 GB of storage providing large amounts of easily transportable storage.

How many **DVDs, CDs**, or 3.5-inch **floppy disks** does it take to **equal** the amount of storage available on one **64GB flash drive**?

A 64GB flash drive can hold the same amount of data as on:

- 4 DVDs
- 90 CDs
- 45,000 3.5 inch floppy disks

Who **invented** the computer **mouse**?

A computer "mouse" is a hand-held input device that, when rolled across a flat surface, causes a cursor to move in a corresponding way on a display screen. A prototype mouse was part of an input console demonstrated by Douglas C. Englehart (1925–) in

1968 at the Fall Joint Computer Conference in San Francisco. Popularized in 1984 by the Macintosh from Apple Computer, the mouse was the result of 15 years devoted to exploring ways to make communicating with computers simpler and more flexible. The physical appearance of the small box with the dangling, tail-like wire suggested the name of "mouse."

In recent years, the mouse has evolved into other shapes and forms. One type is the wireless (or "tailless") mouse which does not have a cord to connect to the computer. Wireless mice use radio signals or infrared to connect to the computer.

What is a **USB port**?

The Universal Serial Bus (USB) connectors first appeared on computers in the late 1990s. It has become the most widely used interface to attach peripherals, such as mice, printers and scanners, external storage drives, digital cameras, and other devices to a computer. Unlike older serial ports and parallel ports, USB ports are easy-to-reach and can easily be plugged in—even while the computer is in use.

What is **Hopper's rule**?

Electricity travels 1 foot (0.3 meter) in a nanosecond (a billionth of a second). This is one of a number of rules compiled for the convenience of computer programmers; it is also considered to be a fundamental limitation on the possible speed of a computer—signals in an electrical circuit cannot move any faster.

Is **assembly language** the same thing as **machine language**?

While the two terms are often used interchangeably, assembly language is a more "user friendly" translation of machine language. Machine language is the collection of patterns of bits recognized by a central processing unit (CPU) as instructions. Each particular CPU design has its own machine language. The machine language of the CPU of a microcomputer generally includes about 75 instructions; the machine language of the CPU of a large mainframe computer may include hundreds of instructions. Each of these instructions is a pattern of 1s and 0s that tells the CPU to perform a specific operation.

Assembly language is a collection of symbolic, mnemonic names for each instruction in the machine language of its CPU. Like the machine language, the assembly language is tied to a particular CPU design. Programming in assembly language requires intimate familiarity with the CPU's architecture, and assembly language programs are difficult to maintain and require extensive documentation.

The computer language C first developed in the late 1980s, is a high-level programming language that can be compiled into machine languages for almost all com-

> ## Who was the first programmer?
>
> According to historical accounts, Lord Byron's (1788–1824) daughter, Augusta Ada Byron (1815–1852), the Countess of Lovelace, was the first person to write a computer program for Charles Babbage's "analytical engine." This machine was to work by means of punched cards that could store partial answers that could later be retrieved for additional operations and would then print the results. Her work with Babbage and the essays she wrote about the possibilities of the "engine" established her as a kind of founding parent of the art and science of programming. The programming language called "Ada" was named in her honor by the U.S. Department of Defense. In modern times, Commodore Grace Murray Hopper (1906–1992) of the U.S. Navy is acknowledged as one of the first programmers of the Mark I computer in 1944.

puters, from microcomputers to mainframes, because of its functional structure. It was the first series of programs that allowed a computer to use higher-level language programs and is the most widely used programming language for personal computer software development. C++ was first released in 1985 and is still widely used today.

Who **invented** the **COBOL** computer language?

COBOL (common business oriented language) is a prominent computer language designed specifically for commercial uses, created in 1960 by a team drawn from several computer makers and the Pentagon. The best-known individual associated with COBOL was then-Lieutenant Grace Murray Hopper who made fundamental contributions to the U.S. Navy's standardization of COBOL. COBOL excels at the most common kinds of data processing for business—simple arithmetic operations performed on huge files of data. The language endures because its syntax is very much like English and because a program written in COBOL for one kind of computer can run on many others without alteration.

Who **invented** the **PASCAL** computer language?

Niklaus Wirth (1934–), a Swiss computer programmer, created the PASCAL computer language in 1970. It was used in the academic setting as a teaching tool for computer scientists and programmers.

Which was the **first widely used** high-level **programming language**?

FORTRAN (FORmula TRANslator) was developed by IBM in the late 1950s. John Backus (1924–2007) was the head of the team that developed FORTRAN. Designed for

43

scientific work containing mathematical formulas, FORTRAN allowed programmers to use algebraic expressions rather than cryptic assembly code. The FORTRAN compiler translated the algebraic expressions into machine-level code. By the late 1960s, FORTRAN was available on almost every computer, especially IBM machines, and utilized by many users.

When was **Java developed**?

Java was released by Sun Microsystems in 1995. A team of developers headed by James Gosling (1955–) began working on a refinement of C++ that ultimately led to Java. Unlike other computer languages, which are either compiled or interpreted, Java compiles the source code into a format called bytecode. The bytecode is then executed by an interpreter. Java was adapted to the emerging World Wide Web and formed the basis of the Netscape Internet browser.

What does **DOS stand for**?

DOS stands for "disk operating system," a program that controls the computer's transfer of data to and from a hard or floppy disk. Frequently, it is combined with the main operating system. The operating system was originally developed at Seattle Computer Products as SCP-DOS. When IBM decided to build a personal computer and needed an operating system, it chose the SCP-DOS after reaching an agreement with the Microsoft Corporation to produce the actual operating system. Under Microsoft, SCP-DOS became MS-DOS, which IBM referred to as PC-DOS (personal computer), which everyone eventually simply called DOS.

What are the **tasks** of an **operating system**?

An operating system is found in all computers and more recently in other electronic devices, such as cell phones. The operating system manages all the hardware and software resources of the computer. Operating systems manage data and devices, such as printers, in the computer. Operating systems today have the ability to multitask, allowing the user to keep several different applications open at the same time. Popular operating systems for computers are Windows (Microsoft), OS × (Macintosh), and Linux.

How did the **Linux operating system** get its name?

The name Linux is a combination of the first name of its principal programmer, Finland's Linus Torvalds (1970–), and the UNIX operating system. Linux (pronounced with a short "i") is an open source computer operating system that is comparable to more powerful, expansive, and usually more costly UNIX systems, of which it resembles in form and function. Linux allows users to run an amalgam of reliable and hearty open-source software tools and interfaces, including powerful web utilities such as the

> ## What is the correct way to face a computer screen?
>
> **C**orrect positioning of the body at a computer, also termed ergonomics, is essential to preventing physical medical conditions such as carpal tunnel syndrome and back pain. You should sit so that your eyes are 18 to 24 inches (45 to 61 centimeters) from the screen, and at a height so that they are six to eight inches (15 to 20 centimeters) above the center of the screen. Your hands should be level with or slightly below the arms.
>
> Correct posture is also necessary. You should sit upright, keeping the spine straight. Sit all the way back in the chair with the knees level with or below the thighs. Both feet should be on the floor. The arms may rest on the desk or chair arms, but make sure you do not slouch. If you need to bend or lean forward, do so from the waist.

popular Apache server, on their home computers. Anyone can download Linux for free or can obtain it on disk for only a marginal fee. Torvalds created the kernel—or heart of the system—"just for fun," and released it freely to the world, where other programmers helped further its development. The world, in turn, has embraced Linux and made Torvalds into a computer folk hero.

What does it mean to **"boot"** a computer?

Booting a computer is starting it, in the sense of turning control over to the operating system. The term comes from bootstrap, because bootstraps allow an individual to pull on boots without help from anyone else. Some people prefer to think of the process in terms of using bootstraps to lift oneself off the ground, impossible in the physical sense, but a reasonable image for representing the process of searching for the operating system, loading it, and passing control to it. The commands to do this are embedded in a read-only memory (ROM) chip that is automatically executed when a microcomputer is turned on or reset. In mainframe or minicomputers, the process usually involves a great deal of operator input. A cold boot powers on the computer and passes control to the operating system; a warm boot resets the operating system without powering off the computer.

What is the difference between **sleep** and **hibernate**?

Sleep and hibernate are power-saving modes that place computers and associated monitors into a low-power setting to conserve energy and potentially save $25 to $75 per PC per year in electricity bills. In sleep mode, all open documents and windows are saved in memory and the system uses a small amount of power. The computer quickly

45

resumes full-power operation when awakened from sleep mode. Sleep is an efficient mode when the computer will not be used for several hours or even overnight.

During hibernation, all open documents and programs are saved to the hard disk and the computer shuts off. The hibernation option uses almost no power but it takes longer to wake up from hibernation than sleep.

When the computer will not be used for several days, it is advisable for it to be shut down.

Where did the **term "bug"** originate?

The slang term "bug" is used to describe problems and errors occurring in computer programs. The term may have originated during the early 1940s at Harvard University, when computer pioneer Grace Murray Hopper discovered that a dead moth had caused the breakdown of a machine on which she was working. When asked what she was doing while removing the corpse with tweezers, she replied, "I'm debugging the machine." The moth's carcass, taped to a page of notes, is preserved with the trouble log notebook at the Virginia Naval Museum.

How did the **term "glitch"** originate?

A glitch is a sudden interruption or fracture in a chain of events, such as in commands to a processor. The stability may or may not be salvageable. The word is thought to have evolved from the German *glitschen*, meaning "to slip," or by the Yiddish *glitshen*, to slide or skid.

One small glitch can lead to a cascade of failure along a network. For instance, in 1997 a small Internet service provider in Virginia unintentionally provided incorrect router (a router is the method by which the network determines the next location for information) information to a backbone operator (a backbone is a major network thoroughfare in which local and regional networks patch into for lengthy interconnections). Because many other Internet service providers rely on the backbone providers, the error echoed around the globe, causing temporary network failures.

What is a **computer virus** and how is it **spread**?

Taken from the obvious analogy with biological viruses, a computer "virus" is a name for a type of computer program that searches out uninfected computers, "infects" them by causing them to execute the virus, and then attempts to spread to other computers. A virus does two things: execute code on a computer, and spread to other computers.

The executed code can accomplish anything that a regular computer program can do; it can delete files, send emails, install programs, and can copy information from one place to another. These actions can happen immediately or after some set delay. It is

46

> ## When did the first computer virus spread via the Internet?
>
> The first known case of large-scale damage caused by a computer virus spread via the Internet was in 1988. Robert Morris Jr. (1965–), a graduate student at Cornell University, crafted a "worm" virus that infected thousands of computers, shutting down many of them and causing millions of dollars of damage.

often not noticed that a virus has infected a computer because it will mimic the actual actions of the infected computer. By the time it is recognized that the computer is infected, much damage may have occurred.

Earliest computer viruses spread via physical media, such as floppy disks. Modern viruses propagate rapidly throughout the Internet. In May 2000, the "ILOVEYOU" virus made international headlines as it spread around the world in a single day, crashing millions of computers and costing approximately $5 billion in economic damages. Since then, criminals have learned to prevent their viruses from crashing computers, making detection of these viruses much more difficult. The highly advanced Conficker worm, first detected in 2008, operates silently, ensuring that the typical computer user won't realize that the virus is present.

What is a **fuzzy search**?

Fuzzy search is an inexact search that allows a user to search for data that are similar to but not exactly the same as what he or she specifies. It can produce results when the exact spelling is unknown, or it can help users obtain information that is loosely related to a topic.

What is **GIGO**?

GIGO is not a computer language despite its similarity to the names of computer languages; instead it is an acronym for the truism that one gets out of something what one puts into it. GIGO stands for the phrase "Garbage In, Garbage Out." The phrase means that a program working with imprecise data produces imprecise results.

What is a **pixel**?

A pixel (from the words pix, for picture, and element) is the smallest element on a video display screen. A screen contains thousands of pixels, each of which can be made up of one or more dots or a cluster of dots. On a simple monochrome screen, a pixel is one dot; the two colors of image and background are created when the pixel is switched either on or off. Some monochrome screen pixels can be energized to create different light intensities to allow a range of shades from light to dark. On color screens, three dot colors are included in each pixel—red, green, and blue. The simplest screens have just one dot of each color, but more elaborate screens have pixels with clusters of each color. These more elaborate displays can show a large number of colors and intensities. On color screens, black is created by leaving all three colors off; white by all three colors on; and a range of grays by equal intensities of all the colors.

The resolution of a computer monitor is expressed as the number of pixels on the horizontal axis and the number of pixels on the vertical axis, For example, a monitor described as 800x600 has 800 pixels on the horizontal axis and 600 pixels on the vertical axis. The higher the numbers, e.g., 1600×1200, the better the resolution.

What is the idea behind "open-source-software"?

Open-source software is computer software where the code (the rules governing its operation) is available for users to modify. This is in contrast to proprietary code, where the software vendor veils the code so users cannot view and, hence, manipulate (or steal) it. The software termed open source is not necessarily free—that is, without charge; authors can charge for its use, usually only nominal fees. According to the Free Software Foundation, "Free software" is a matter of liberty, not price. To understand the concept, you should think of "free" as in "free speech," not as in "free food." Free software is a matter of the users' freedom to run, copy, distribute, study, change, and improve the software. Despite this statement, most of it is available without charge.

Open-source software is usually protected under the notion of "copyleft," instead of "copyright" law. Copyleft does not mean releasing material to the public domain, nor does it mean near absolute prohibition from copying, like the federal copyright law. Instead, according to the Free Software Foundation, copyleft is a form of protection guaranteeing that whoever redistributes software, whether modified or not, "must pass along the freedom to further copy and share it." Open source has evolved into a movement of sharing, cooperation, and mutual innovation, ideas that many believe are necessary in today's cutthroat corporatization of software.

INTERNET

What is the Internet?

The Internet is the world's largest computer network. It links computer terminals together via wires or telephone lines in a web of networks and shared software. With the proper equipment, an individual can access vast amounts of information and search databases on various computers connected to the Internet, or communicate with someone located anywhere in the world as long as he or she has the proper equipment.

Originally created in the late 1960s by the U.S. Department of Defense's Advanced Research Projects Agency (DARPA) to share information with other researchers, the Internet expanded immensely when scientists and academics using the network discovered its great value. Despite its origin, however, the Internet is not owned or funded by the U.S. government or any other organization or institution. A group of volunteers, the Internet Society, addresses such issues as daily operations and technical standards.

Who coined the term "information superhighway"?

A term originally coined by former Vice President Al Gore (1948–), the information superhighway was envisioned as a high-speed electronic communications network to enhance education in America in the twenty-first century. Its goal was to help all citizens regardless of their income level, connecting all users to one another and providing every type of electronic service possible, including shopping, electronic banking, education, medical diagnosis, video conferencing, and game playing. As the World Wide Web grew, it became the "information superhighway."

Who invented the World Wide Web?

Tim Berners-Lee (1955–) is considered the creator of the World Wide Web (WWW). The WWW is a massive collection of interlinked hypertext documents that travel over the Internet and are viewed through a browser. The Internet is a global network of computers developed in the 1960s and 1970s by the U.S. Department of Defense's Advanced Research Project Agency (hence the term "Arpanet"). The idea of the Internet was to provide redundancy of communications in case of a catastrophic event (like a nuclear blast), which might destroy a single connection or computer but not the entire network. The browser is used to translate the hypertext, usually written in Hypertext Markup Language (HTML), so it is human-readable on a computer screen. Along with Gutenberg's invention of the printing press, the inception of the WWW in 1990 and 1991, when Berners-Lee released the tools and protocols onto the Internet, is considered one of humanity's greatest communications achievements.

What is the client/server principle and how does it apply to the Internet?

The client/server principle refers to the two components of a centralized computer network: client and server machines. Clients request information and servers send them the requested infomration. For

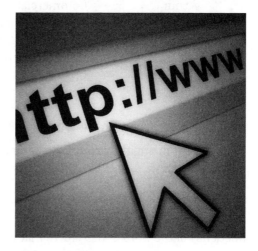

The "www" in URLs stands for "World Wide Web," the invention of Tim Berners-Lee.

example, when an individual uses his computer to look at a Web page, his computer acts as the client, and the computer hosting the Web page is the server. Browsers enable the connection between clients and servers. During the 1990s, Netscape, an outgrowth of the early browser Mosaic, and Internet Explorer were the dominant browsers. Eventually, Microsoft's Internet Explorer became the dominant browser. Other frequently used browsers are Mozilla Firefox and Safari.

What is a **Uniform Resource Locator**, or URL?

A Uniform Resource Locator, URL, can be thought of as the "address" for a given computer on the Internet. A URL consists of two parts: 1) the protocol identifier and 2) the resource name. The protocol identifier indicates which protocol to use, e.g., "http" for Hypertext Transfer Protocol or "ftp" for File Transfer Protocol. Most browsers default to "http" as the protocol identifier so it is not necessary to include that part of the URL when entering an Internet address in the browser's toolbar. The resource identifier specifies the IP (Internet Protocol) address or domain name where the Web page is located. An IP address is a string of numbers separated by periods. More familiar and easier to remember than the IP address is a domain name. Domain names are composed of a name with a top level domain (TLD) suffix. The IP address and domain name both point to the same place; typing "192.0.32.10" in the address bar of a Web browser will bring up the same page as entering the more user-friendly "example.com".

What are some **common top level domain** (TLD) **suffixes**?

Common top level domain suffixes are:

Top Level Domain (TLD)	Meaning of the Suffix
.com	Commercial organization, business, or company
.edu	Educational institution
.org	Nonprofit organization (sometimes used by other sites)
.gov	U.S. government agency
.mil	U.S. military agency
.net	Network organizations

How does a **search engine work**?

Internet search engines are akin to computerized card catalogs at libraries. They provide a hyperlinked listing of locations on the World Wide Web according to the requested keyword or pattern of words submitted by the searcher. A search engine uses computer software called "spiders" or "bots" to automatically search out, inventory, and index Web pages. The spiders scan each Web page's content for words and the frequency of the words, then stores that information in a database. When the user submits words or terms, the search engine returns a list of sites from the database and ranks them according to the relevancy of the search terms.

What is the **most popular search engine**?

As of 2010, the most popular search engine was Google. The following chart shows the market share of the top five search engines.

Search Engine	Average Market Share
Google	71.6%
Yahoo!	14.6%
Bing	9.4%
Ask	2.5%
AOL	1.0%

When was **Google founded**?

In 1996 Stanford University graduate student Larry Page (1973–) and Sergey Brin (1973–) began collaborating on a search engine called BackRub. BackRub operated on the Stanford servers for more than a year until it began to take up too much bandwidth. In 1997, Page and Brin decided they needed a new name for the search engine and decided upon Google as a play on the mathematical term "googol". (A googol is the numeral 1 followed by 100 zeros.) Their goal was to organize the seemingly infinite amount of information on the Web. Google first incorporated as a company in September 1998 and the corporation went public in 2004.

How is **information** sent over the **Internet** kept **secure**?

Public-key cryptography is a means for authenticating information sent over the Internet. The system works by encrypting and decrypting information through the use of a combination of "keys." One key is a published "public key"; the second is a "private key," which is kept secret. An algorithm is used to decipher each of the keys. The method is for the sender to encrypt the information using the public key, and the recipient to decrypt the information using the secret private key.

The strength of the system depends on the size of the key: a 128-bit encryption is about $3 \times 1,026$ times stronger than 40-bit encryption. No matter how complex the encryption, as with any code, keeping the secret aspects secret is the important part to safeguarding the information.

Users can easily tell whether they are on a secure or non-secure Internet website by the prefix of the Web page address. Addresses that begin with "http://" are not secure while those that begin with "https://" are secure.

How much has **Internet usage changed** in recent years?

According to industry surveys, Internet usage has surged, increasing by nearly 400 percent between 2000 and 2009, far exceeding earlier estimates that there would be one billion users by 2005. Africa and the Middle East have shown the greatest rate of growth.

World Internet Usage*

Locale	Internet Users as of December 31, 2000	Internet Users as of December 31, 2009	Growth 2000–2009
Africa	4,514,400	86,217,900	1,809.8%
Asia	114,304,000	764,435,900	568.8%
Europe	105,096,093	425,773,571	305.1%
Middle East	202,687,005	58,309,546	1,675.1%
North America	108,096,800	259,561,000	140.1%
Latin America/Caribbean	18,068,919	186,922,050	934.5%
Oceania/Australia	7,620,480	21,110,490	177.0%
TOTAL	360,985,492	1,802,330,457	399.3%

*Source: Internet World Stats (www.internetworldstats.com).

What is **e-mail**?

Electronic mail, known more commonly as e-mail, uses communication facilities to transmit messages. A user can send a message to a single recipient or to many different recipients at one time. Different systems offer different options for sending, receiving, manipulating text, and addressing. For example, a message can be "registered," so that the sender is notified when the recipient looks at the message (though there is no way to tell if the recipient has actually read the message). Email messages may be forwarded to other recipients. Usually messages are stored in a simulated "mailbox" in the network server or host computer; some systems announce incoming mail if the recipient is logged onto the system. An organization (such as a corporation, university, or professional organization) can provide electronic mail facilities; national and international networks can provide them as well.

What is **spam**?

Spam, also called junk e-mail, is unsolicited e-mail. Spam is an annoyance to the recipient and may contain computer viruses or spyware. It often advertises products that are usually not of interest to the recipient and are oftentimes vulgar in content. Estimates suggest that as many as one billion spam messages are sent daily. Many e-mail programs have spam filters or blockers to detect spam messages and either delete them or send them to the "junk" mailbox.

What is a **hacker**?

A hacker is a skilled computer user. The term originally denoted a skilled programmer, particularly one skilled in machine code and with a good knowledge of the machine

---

Who sent the first e-mail?

In the early 1970s computer engineer Ray Tomlinson (1941–) noticed that people working at the same mainframe computer could leave one another messages. He imagined great utility of this communication system if messages could be sent to different mainframes. So he wrote a software program over the period of about a week that used file-transfer protocols and send-and-receive features. It enabled people to send messages from one mainframe to another over the Arpanet, the network that became the Internet. To make sure the messages went to the right system he adopted the @ symbol because it was the least ambiguous keyboard symbol and because it was brief.

---

and its operating system. The name arose from the fact that a good programmer could always hack an unsatisfactory system around until it worked.

The term later came to denote a user whose main interest is in defeating secure systems. The term has thus acquired a pejorative sense, with the meaning of one who deliberately and sometimes criminally interferes with data available through telephone lines. The activities of such hackers have led to considerable efforts to tighten security of transmitted data. The "hacker ethic" is that information-sharing is the proper way of human dealing, and, indeed, it is the responsibility of hackers to liberally impart their wisdom to the software world by distributing information. Many hackers are hired by companies to test their Web security.

What is **Wi-Fi**?

Wi-Fi is a wireless local area network. The Wi-Fi Alliance certifies that network devices comply with the appropriate standard, IEEE 802.11. A Wi-Fi hotspot is the geographic boundary covered by a Wi-Fi access point. There were estimates that at the end of 2009 there were more than a quarter million public hotspots.

When was the **term Web 2.0** first used?

The term Web 2.0 was first used in 2004 during a brainstorming conference between Tim O'Reilly (1954–) of O'Reilly Media and MediaLive International. Rather than believing as some had that the Web had "crashed" following the dot.com collapse, it was a turning point for the Web to become a platform for collaborative effort between Internet users, providers, and enterprises.

What is **Web 2.0**?

Web 2.0 is not a new version of the World Wide Web, but rather a collection of new technologies that changes the way users interact with the Web. When Tim Berners-

Lee (1955–) created the World Wide Web it was a repository of information with static content and users were generally unable to easily change or add to the content they were viewing. Newer technologies allow users to contribute to the Internet with blogs, wikis, and social networking sites.

A further distinction of Web 2.0 is "cloud computing" where data and applications ("apps") are stored on Web servers, rather than on individual computers, allowing users to access their documents, files, and data from any computer with a Web browser. Apps include many products, such as word processing and spreadsheets, that were traditionally found in software packages.

What are some **examples** of **user-generated content** on the World Wide Web?

Blogs (short for web logs) are akin to modern-day diaries (or logs) of thoughts and activities of the author. In the late 1990s software became available to create blogs using templates therefore making them accessible to a wide audience as a publishing tool. Blogs may be created by single individuals or by groups of contributors. Blog entries are organized in reverse chronological order with the most recent entries being seen first. Entries may include text, audio, images, video, and links to other sites. Blog authors may invite reader feedback via comments, which allows for dialogue between blog authors and readers. However, once posted, blog entries may not be edited.

Wikis, from the Hawaiian word *wikiwiki*, which means "fast", are Web pages that allow users to add and edit material in a collaborative fashion. The first wikis were developed in the mid–1990s by Ward Cunningham (1949–) as a way for users to quickly add content to Web pages. The advantage of this software was that the users did not need to know complicated languages to add material to the Web. One of the best known wikis is Wikipedia, an online, collaborative encyclopedia. Although entries to Wikipedia need to come from published sources and be based on fact, rather than the writer's opinion, there is no overall editorial authority on the site.

Podcasts are broadcast media that may be created by anyone and are available on demand. Unlike traditional broadcast media (radio and television), podcasts are easily created with a microphone, video camera, computer, and connection to the Web. Podcasting does not require sophisticated recording or transmitting equipment. Most podcasts are broadcast on a weekly, biweekly, or monthly schedule. While traditional broadcast media follow a set schedule, podcasts may be downloaded onto a computer or a portable device such as a MP3 player and listened to whenever it is convenient.

PHYSICS AND CHEMISTRY

ENERGY, MOTION, AND FORCE

What is **energy**?

Physicists define energy as the capacity to do work. Work is defined as the force required to move an object some distance. Examples of the different kinds of energy are heat energy, light energy, mechanical energy, potential energy, and kinetic energy. The law of the conservation of energy states that within an isolated system, energy may be transformed from one form to another, but it cannot be created nor can it be destroyed.

What are the **two forms** of **energy**?

The two forms of energy are kinetic energy and potential energy. Kinetic energy is the energy possessed by an object as a result of its motion while potential energy is the energy possessed by an object as a result of its position. As an example, a ball sitting on top of a fence has potential energy. When the ball falls off the fence it has kinetic energy. The potential energy is transformed into kinetic energy.

What is **inertia**?

Inertia is the tendency of all objects and matter in the universe to stay still, or, if moving, to continue moving in the same direction unless acted on by some outside force. This forms the first law of motion formulated by Isaac Newton (1642–1727). To move a body at rest, enough external force must be used to overcome the object's inertia; the larger the object is, the more force is required to move it. In his *Philosophae Naturalis Principia Mathematica*, published in 1687, Newton sets forth all three laws of motion. Newton's second law is that the force to move a body is equal to its mass times its acceleration (F = MA), and the third law states that for every action there is an equal and opposite reaction.

What is **superconductivity**?

Superconductivity is a condition in which many metals, alloys, organic compounds, and ceramics conduct electricity without resistance, usually at low temperatures. Heike Kamerlingh Onnes (1853–1926), a Dutch physicist, discovered superconductivity in 1911. He was awarded the Nobel Prize in Physics in 1913 for his low-temperature studies. The modern theory regarding the phenomenon was developed by three American physicists—John Bardeen (1908–1991), Leon N. Cooper (1930–), and John Robert Schrieffer (1931–). Known as the BCS theory after the three scientists, it postulates that superconductivity occurs in certain materials because the electrons in them, rather than remaining free to collide with imperfections and scatter, form pairs that can flow easily around imperfections and do not lose their energy. Bardeen, Cooper, and Schrieffer received the Nobel Prize in Physics for their work in 1972.

A further breakthrough in superconductivity was made in 1986 by J. Georg Bednorz (1950–) and K. Alex Müller (1927–). Bednorz and Müller discovered a ceramic material consisting of lanthanum, barium, copper, and oxygen, which became superconductive at 35°K (–238°C)—much higher than any other material. Bednorz and Müller won the Nobel Prize in Physics in 1987. This was a significant accomplishment since in most situations the Nobel Prize is awarded for discoveries made as many as 20 to 40 years earlier.

What are some **practical applications** of **superconductivity**?

A variety of uses have been proposed for superconductivity in fields as diverse as electronics, transportation, and power. Research continues to develop more powerful, more efficient electric motors and devices that measure extremely small magnetic fields for medical diagnosis. The field of electric power transmission has much to gain by developing superconducting materials since 15 percent of the electricity generated must be used to overcome the resistance of traditional copper wire. More powerful electromagnets will be utilized to build high-speed magnetically levitated trains, known as "maglevs."

What is **string theory**?

A relatively recent theory in particle physics, string theory conceives elementary particles not as points but as lines or loops. The idea of these "strings" is purely theoretical since no string has ever been detected experimentally. The ultimate expression of string theory may potentially require a new kind of geometry—perhaps one involving an infinite number of dimensions.

What is the role of friction in striking a match?

The head at the tip of a strike-anywhere match contains all the chemicals required to create a spark. A strike-anywhere match only needs to be rubbed against a surface with a high coefficient of friction, such as sandpaper, to create enough frictional heat to ignite the match. Safety matches differ from strike-anywhere matches since the chemicals necessary for ignition are divided between the match head and the treated strip found on the matchbox or matchbook. The friction between the match head and treated strip will ignite the match. Matches fail to ignite when wet because water reduces friction.

What is **friction**?

Friction is defined as the force that resists motion when the surface of one object slides over or comes in contact with the surface of another object. The three laws that govern the friction of an object at rest and the surface with which it is in contact state:

- Friction is proportional to the weight of an object.
- Friction is not determined by the surface area of the object.
- Friction is independent of the speed at which an object is moving along a surface provided the speed is not zero.

Although friction reduces the efficiency of machines and opposes movement, it is an essential force. Without friction it would be impossible to walk, drive a car, or even strike a match.

Why is a **lubricant**, such as oil, often used to counter the **force of friction**?

Lubricants, such as oil, are used to reduce friction. For example, in machines consisting of metal parts, the continuous rubbing of the parts together increases the temperature and creates heat. To prevent serious wear and damage to the machines, grease and oil are applied to reduce the friction.

Why do **golf balls** have **dimples**?

The dimples minimize the drag (a force that makes a body lose energy as it moves through a fluid or gas), allowing the ball to travel farther than a smooth ball would travel. The air, as it passes over a dimpled ball, tends to cling to the ball longer, reducing the eddies or wake effects that drain the ball's energy. A dimpled ball can travel up to 300 yards (275 meters), but a smooth ball only goes 70 yards (65 meters). A ball can have 300 to 500 dimples that can be 0.01 inch (0.25 millimeter) deep. Another effect

57

to get distance is to give the ball a backspin. With a backspin there is less air pressure on the top of the ball, so the ball stays aloft longer (much like an airplane).

Who successfully demonstrated that curve balls actually curve?

In 1959, Lyman Briggs (1874–1963) demonstrated that a ball can curve up to 17.5 inches (44.45 centimeters) over the 60 feet 6 inches (18.4 meters) it travels between a pitcher and a batter, ending the debate of whether curve balls actually curved or if the apparent change in course was merely an optical illusion. Briggs studied the effect of spin and speed on the trajectory and established the relationship between amount of curvature and the spin of the ball.

A rapidly spinning baseball experiences two lift forces that cause it to curve in flight. One is the Magnus force named after H.G. Magnus (1802–1870), the German physicist who discovered it, and the other is the wake deflection force. The Magnus force causes the curve ball to move sideways because the pressure forces on the ball's sides do not balance each other. The stitches on a baseball cause the pressure on one side of the ball to be less than on its opposite side. This forces the ball to move faster on one side than the other and forces the ball to "curve." The wake deflection force also causes the ball to curve to one side. It occurs because the air flowing around the ball in the direction of its rotation remains attached to the ball longer and the ball's wake is deflected.

Why does a boomerang return to its thrower?

Two well-known scientific principles dictate the characteristic flight of a boomerang: (1) the force of lift on a curved surface caused by air flowing over it; and (2) the unwillingness of a spinning gyroscope to move from its position.

When a person throws a boomerang properly, he or she causes it to spin vertically. As a result, the boomerang will generate lift, but it will be to one side rather than upwards. As the boomerang spins vertically and moves forward, air flows faster over the top arm at a particular moment than over the bottom arm. Accordingly, the top arm produces more lift than the bottom arm and the boomerang tries to twist itself, but because it is spinning fast it acts like a gyroscope and turns to the side in an arc. If the boomerang stays in the air long enough, it will turn a full circle and return to the thrower. Every boomerang has a built-in orbit diameter, which is not affected by a person throwing the boomerang harder or spinning it faster.

What is Maxwell's demon?

An imaginary creature who, by opening and shutting a tiny door between two volumes of gases, could, in principle, concentrate slower molecules in one (making it colder) and faster molecules in the other (making it hotter), thus breaking the second law of thermodynamics. Essentially this law states that heat does not naturally flow from a colder

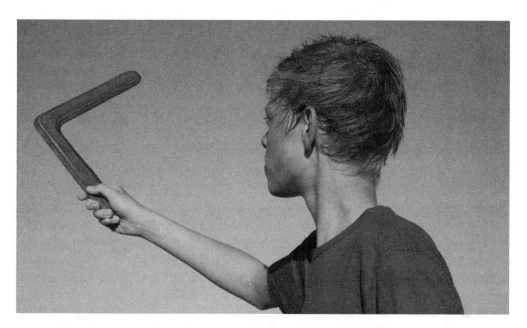

There is a lot of physics behind the way a boomerang will return to the person who throws it.

body to a hotter body; work must be expended to make it do so. This hypothesis was formulated in 1871 by James C. Maxwell (1831–1879), who is considered by many to be the greatest theoretical physicist of the nineteenth century. The demon would bring about an effective flow of molecular kinetic energy. This excess energy would be useful to perform work and the system would be a perpetual motion machine. About 1950, the French physicist Léon Brillouin (1889–1969) disproved Maxwell's hypothesis by demonstrating that the decrease in entropy resulting from the demon's actions would be exceeded by the increase in entropy in choosing between the fast and slow molecules.

Does **water** running down a **drain rotate** in a **different direction** in the **Northern** versus the **Southern Hemisphere**?

There is a widespread belief that water draining from a bathtub, sink, or toilet bowl in the Northern Hemisphere swirls counterclockwise while in the Southern Hemisphere the water drains clockwise due to the Coriolis effect. First described by the French mathematician and engineer Gaspard Gustav de Coriolis (1792–1843), the Coriolis effect is the apparent deflection of air masses and fluids caused by Earth's rotation. Although it does have an effect on fluids over great distances or long lengths of time, such as hurricanes, it is too weak to control fluids on a small-scale, such as bathtubs, sinks, or toilet bowls. These can drain in either direction in both hemispheres. The direction is determined by numerous factors including the shape of the container, the shape of the drain, the initial water-velocity and the tilt of the sink.

LIGHT, SOUND, AND OTHER WAVES

What is the **speed of light**?

Light travels at 186,282 miles (299,792 kilometers) per second or 12 million miles per minute.

What are the **primary colors** in light?

Color is determined by the wavelength of visible light (the distance between one crest of the light wave and the next). Those colors that blend to form "white light" are, from shortest wavelength to longest: red, orange, yellow, green, blue, indigo, and violet. All these monochromatic colors, except indigo, occupy large areas of the spectrum (the entire range of wavelengths produced when a beam of electromagnetic radiation is broken up). These colors can be seen when a light beam is refracted through a prism. Some consider the primary colors to be six monochromatic colors that occupy large areas of the spectrum: red, orange, yellow, green, blue, and violet. Many physicists recognize three primary colors: red, green, and blue. All other colors can be made from these by adding two primary colors in various proportions. Within the spectrum, scientists have discovered 55 distinct hues. Infrared and ultraviolet rays at each end of the spectrum are invisible to the human eye.

How do **polarized sunglasses** reduce glare?

Sunlight reflected from the horizontal surface of water, glass, and snow is partially polarized, with the direction of polarization chiefly in the horizontal plane. Such reflected light may be so intense as to cause glare. Polarized sunglasses contain filters that block (absorb) light that is polarized in a direction perpendicular to the transmission axes. The transmission axes of the lenses of polarized sunglasses are oriented vertically.

What were **Anders Ångström's contributions** to the development of **spectroscopy**?

Swedish physicist and astronomer Anders Jonas Ångström (1814–1874) was one of the founders of spectroscopy. His early work provided the foundation for spectrum analysis (analysis of the ranges of electromagnetic radiation emitted or absorbed). He investigated the sun spectra as well as that of the aurora borealis. In 1868, he established measurements for wavelengths of greater than 100 Frauenhofer. In 1907, the angstrom (Å, equal to 1×10^{-10} meters), a unit of wavelength measurement, was officially adopted.

Why was the **Michelson-Morley experiment** important?

This experiment on light waves, first carried out in 1881 by physicists Albert A. Michelson (1852–1931) and E.W. Morley (1838–1923) in the United States, is one of

Why does the color of clothing appear different in sunlight than it does in a store under fluorescent light?

White light is a blend of all the colors, and each color has a different wavelength. Although sunlight and fluorescent light both appear as "white light," they each contain slightly different mixtures of these varying wavelengths. When sunlight and fluorescent light (white light) are absorbed by a piece of clothing, only some of the wavelengths (composing white light) reflect off the clothing. When the retina of the eye perceives the "color" of the clothing, it is really perceiving these reflected wavelengths. The mixture of wavelengths determines the color perceived. This is why an article of clothing sometimes appears to be a different color in the store than it does on the street.

the most historically significant experiments in physics and led to the development of Einstein's theory of relativity. The original experiment, using the Michelson interferometer, attempted to detect the velocity of Earth with respect to the hypothetical "luminiferous ether," a medium in space proposed to carry light waves. The procedure measured the speed of light in the direction of Earth and the speed of light at right angles to Earth's motion. No difference was found. This result discredited the ether theory and ultimately led to the proposal by Albert Einstein (1879–1955) that the speed of light is a universal constant.

Is **light** a **wave** or a **particle**?

Scientists have debated for centuries whether light is a wave or a particle. Isaac Newton (1642–1727) was one of the early proponents of the particulate (or corpuscular) theory of light. According to this theory, light travels as a stream of particles that come from a source, such as the sun, travel to an object and are then reflected to an observer. One of the early proponents of the wave theory of light was Dutch physicist Christiaan Huygens (1629–1695). According to the wave theory of light, light travels through space in the form of a wave, similar to water waves. Albert Einstein's work in 1905 showed that light is a bundle of tiny particles, called photons. Scientists now believe that light has properties of both waves and particles, explained as wave-particle duality.

What were some of the **leading contributions** of **Albert Einstein**?

Albert Einstein was the principal founder of modern theoretical physics; his theory of relativity (speed of light is a constant and not relative to the observer or source of light), and the relationship of mass and energy ($E = mc^2$), fundamentally changed human understanding of the physical world.

During a single year in 1905, he produced three landmark papers. These papers dealt with the nature of particle movement known as Brownian motion, the quantum nature of electromagnetic radiation as demonstrated by the photoelectric effect, and the special theory of relativity. Although Einstein is probably best known for the last of these works, it was for his quantum explanation of the photoelectric effect that he was awarded the 1921 Nobel Prize in Physics.

His stature as a scientist, together with his strong humanitarian stance on major political and social issues, made him one of the outstanding men of the twentieth century.

What is the difference between **special** and **general relativity**?

Albert Einstein developed the theory of relativity in the early twentieth century. He published the theory of special relativity in 1905 and the theory of general relativity in 1916. Special relativity deals only with nonaccelerating (inertial) reference frames; general relativity deals with accelerating (noninertial) reference frames. Simply stated, according to the theory of special relativity, the laws of nature are the same for all observers whose frames of reference are moving with constant velocity with respect to each other. Published as an addenda to the special theory of relativity was the famous equation, $E = mc^2$, representing that mass and energy can be transformed into each other. In contrast, general relativity states that the laws of nature are the same for all observers even if they are accelerating with respect to each other.

How do **sound waves differ** from **light waves**?

Waves consist of a series of motions in regular succession carrying energy from one place to another without moving any matter. Periodic waves include ocean waves,

The X-1 in flight. This is the airplane Chuck Yeager flew when he broke the sound barrier in 1947.

sound waves, and electromagnetic waves. Visible light and radio waves are electromagnetic waves. Mechanical waves, such as ocean waves and sound waves, involve matter, but it is important to remember that there is no transport of matter. The water in an ocean wave does not move from one location to another merely the energy of the wave is transported. Light waves involve only energy without matter.

Does **sound travel faster** in **air** or **water**?

The speed of sound is not a constant; it varies depending on the medium in which it travels. The measurement of sound velocity in the medium of air must take into account many factors, including air temperature, pressure, and purity. At sea level and 32°F (0°C), scientists do not agree on a standard figure; estimates range from 740 to 741.5 miles (1,191.6 to 1,193.2 kilometers) per hour. As air temperature rises, sound velocity increases. Sound travels faster in water than in air and even faster in iron and steel. Sounds traveling a mile in air for five seconds will travel the same distance in one second underwater and one-third of a second in steel.

Who was the **first person** to **break** the **sound barrier**?

On October 14, 1947, Charles E. (Chuck) Yeager (1923–) was the first pilot to break the sound barrier. He flew a Bell X–1, attaining a speed of 750 miles (1,207 kilometers) per hour (Mach 1.06) and an altitude of 70,140 feet (21,379 meters) over the town of Victorville, California. The first woman to break the sound barrier was Jacqueline Cochran (c. 1906–1980). On May 18, 1953, she flew a North American F-86 Saber over Edwards Air Force Base in California, attaining the speed of 760 miles (1,223 kilometers) per hour.

63

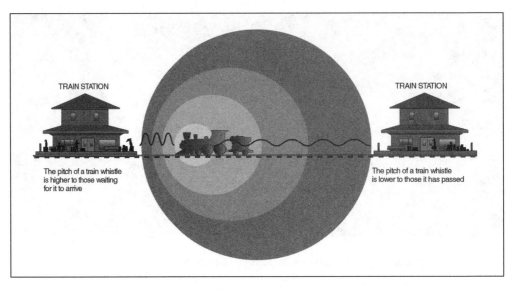

TRAIN STATION

The pitch of a train whistle is higher to those waiting for it to arrive

TRAIN STATION

The pitch of a train whistle is lower to those it has passed

Christian Doppler discovered that sound wavelengths are compressed when emitted by a moving object coming toward you and the wavelengths are increased as it moves away. Thus, a train whistle has a higher pitch as it approaches you and a lower pitch as it recedes. The Doppler Effect also holds true for light waves.

When is a **sonic boom heard**?

As long as an airborne object, such as a plane, is moving below the speed of sound (called Mach 1), the disturbed air remains well in front of the craft. But as the craft passes Mach 1 and is flying at supersonic speeds, a sharp air pressure rise occurs in front of the craft. In a sense, the air molecules are crowded together and collectively impact. What is heard is a claplike thunder called a sonic boom or a supersonic bang. There are many shocks coming from a supersonic aircraft, but these shocks usually combine to form two main shocks, one coming from the nose and one from the aft end of the aircraft. Each of the shocks moves at a different velocity. If the time difference between the two shock waves is greater than 0.10 seconds apart, two sonic booms will be heard. This usually occurs when an aircraft ascends or descends quickly. If the aircraft moves more slowly, the two booms will sound like only one boom to the listener.

What is the **Doppler effect**?

The Austrian physicist Christian Doppler (1803–1853) in 1842 explained the phenomenon of the apparent change in wavelength of radiation—such as sound or light—emitted either by a moving body (source) or by the moving receiver. The frequency of the wavelengths increases and the wavelength becomes shorter as the moving source approaches, producing high-pitched sounds and bluish light (called blue shift). Likewise, as the source recedes from the receiver the frequency of the wavelengths decreases, the sound is pitched lower, and light appears reddish (called red shift). This

> ## Do you really hear the ocean when you hold a seashell to your ear?
>
> **A**lthough the sounds heard in a seashell are reminiscent of hearing the ocean, in reality they are ambient, soft sounds that have been resonated and thereby amplified by the seashell's cavity. The extreme sensitivity of the human ear to sound is illustrated by the seashell resonance effect.

Doppler effect is commonly demonstrated by the whistle of an approaching train or the roar of a jet aircraft.

There are three differences between acoustical (sound) and optical (light) Doppler effects: The optical frequency change is not dependent on what is moving—the source or observer—nor is it affected by the medium through which the waves are moving, but acoustical frequency is affected by such conditions. Optical frequency changes are affected if the source or observer moves at right angles to the line connecting the source and observer. Observed acoustical changes are not affected in such a situation. Applications of the Doppler phenomenon include the Doppler radar and the measurement by astronomers of the motion and direction of celestial bodies.

What is the **sound frequency** of the **musical scale**?

Equal Tempered Scale

Note	Frequency (Hz)
C♭	261.63
C♯	277.18
D	293.67
D♯	311.13
E	329.63
F	349.23
F♯	369.99
G	392.00
G♯	415.31
A	440.00
A♯	466.16
B	493.88
Cn	523.25

Notes: ♯ indicates flat; ♭ indicates sharp; n indicates return to natural; Hz is Hertz.

The lowest frequency distinguishable as a note is about 20 hertz. The highest audible frequency is about 20,000 hertz. A hertz (Hz) is a unit of frequency that measures the number of the wave cycles per second frequency of a periodic phenomenon whose periodic time is one second (cycles per second).

What is a **decibel**?

A decibel is a measure of the relative loudness or intensity of sound. A 20 decibel sound is 10 times louder than a 10 decibel sound; 30 decibels is 100 times louder, etc. One decibel is the smallest difference between sounds detectable by the human ear.

Decibel	Level Equivalent
10	Light whisper
20	Quiet conversation
30	Normal conversation
40	Light traffic
50	Loud conversation
60	Noisy office
70	Normal traffic, quiet train
80	Rock music, subway
90	Heavy traffic, thunder
100	Jet plane at takeoff

What are the characteristics of **alpha, beta,** and **gamma radiation**?

Radiation is a term that describes all the ways energy is emitted by the atom as X rays, gamma rays, neutrons, or as charged particles. Most atoms, being stable, are nonradioactive; others are unstable and give off either particles or gamma radiation. Substances bombarded by radioactive particles can become radioactive and yield alpha particles, beta particles, and gamma rays.

Ernest Rutherford was the first to identify beta particles, which are high-speed electrons that can pass through solid matter.

Alpha particles, first identified by Antoine Henri Becquerel (1852–1908), have a positive electrical charge and consist of two protons and two neutrons. Because of their great mass, alpha particles can travel only a short distance, around 2 inches (5 centimeters) in air, and can be stopped by a sheet of paper.

Beta particles, identified by Ernest Rutherford (1871–1937), are extremely

66

high-speed electrons that move at the speed of light. They can travel far in air and can pass through solid matter several millimeters thick.

Gamma rays, identified by Marie (1867–1934) and Pierre Curie (1859–1906), are similar to X rays, but they usually have a shorter wavelength. These rays, which are bursts of photons, or very short-wave electromagnetic radiation, travel at the speed of light. They are much more penetrating than either the alpha or beta particles and can go through 7 inches (18 centimeters) of lead.

ELECTRICITY AND MAGNETISM

Who is the **founder** of the **science** of **magnetism**?

The English scientist William Gilbert (1544–1603) regarded Earth as a giant magnet and investigated its magnetic field in terms of dip and variation. He explored many other magnetic and electrostatic phenomena. The Gilbert (Gb), a unit of magnetism, is named for him.

John H. Van Vleck (1899–1980), an American physicist, made significant contributions to modern magnetic theory. He explained the magnetic, electrical, and optical properties of many elements and compounds with the ligand field theory, demonstrated the effect of temperature on paramagnetic materials (called Van Vleck paramagnetism), and developed a theory on the magnetic properties of atoms and their components.

What is a **Leyden jar**?

A Leyden jar, the earliest form of capacitor, is a device for storing an electrical charge. First described in 1745 by E. Georg van Kleist (c. 1700–1748), it was also used by Pieter van Musschenbroek (1692–1761), a professor of physics at the University of Leyden. The device came to be known as a Leyden jar and was the first device that could store large amounts of electric charge. The jars contained an inner wire electrode in contact with water, mercury, or wire. The outer electrode was a human hand holding the jar. An improved version coated the jar inside and outside with separate metal foils with the inner foil connected to a conducting rod and terminated in a conducting sphere. This eliminated the need for the liquid electrolyte. In use, the jar was normally charged from an electrostatic generator. The Leyden jar—which makes the hair stand up—is still used for classroom demonstrations of static electricity.

How does an **electromagnet work**?

Danish physicist Hans Christian Oersted (1777–1851) was the first to observe a connection between electricity and magnetism. He found that a current in a wire caused a

67

nearby compass needle to rotate thereby establishing that an electric current always produces a magnetic field around itself. When an electric current flows through a wire wrapped around a piece of iron, it creates a magnetic field in the iron. Electromagnets are used in a wide variety of applications, including doorbells, switches, and valves in heating and cooling equipment to telephones, large machinery used to lift heavy loads of scrap metal, and particle accelerators.

How do **permanent magnets** differ from **temporary magnets**?

Permanent magnets, also referred to as naturally occurring magnets, remain magnetized until they are demagnetized. Naturally occurring magnets are found in minerals such as magnetite and lodestone. These magnets are known as ferromagnets.

What **materials** are **used** to make the most **powerful permanent magnets**?

The most powerful permanent magnets are made from alloys of iron, boron, and neodymium. Magnetic strength is measured in units called tesla and gauss. One tesla equals 10,000 gauss. Most of the magnets used to decorate the refrigerator are 10 gauss. The most powerful permanent magnets produce magnetic fields of approximately 1.5 tesla.

Why is **static electricity** greater in **winter** than **summer**?

Static electricity results from an imbalance between negative and positive charges in an object. Most of the time the positive and negatives charges in an object are balanced, meaning the object is neutral; it is neither positively nor negatively charged. When the charges build up on an object, they must find a way to be released or transferred to restore the balance between negative and positive charges. During the winter, the air has very little water vapor in it and is dry. In the summer, the air contains more water vapor; the humidity is greater. Water is an electrical conductor allowing electrons to move from one object to another more easily. The humidity in the summer air allows extra electrons on charged objects to leak off into the air and attach to objects that have too few electrons. The dry air in the winter makes it more difficult for the extra electrons to leak off an object so static electricity, with its characteristic shock or spark, is more common.

How do dryer sheets and hair conditioner reduce or prevent static electricity?

Since materials that are electrical conductors allow free movements of charges, the goal of dryer sheets and hair conditioner is to turn electrical insulators, such as fabric and hair, into electrical conductors. When wet, fabrics and hair are usually negatively charged. Applying a cleansing agent that contains a positively charged detergent molecule to these wet fibers, such as dryer sheets and hair conditioner, clings to the wet fibers and remains in place giving fabrics and hair the soft, silky, non-clinging, static-free feeling.

How did the electrical **term ampere originate**?

It was named for André Marie Ampère (1775–1836), the physicist who formulated the basic laws of the science of electrodynamics. The ampere (A), often abbreviated as "amp," is the unit of electric current, defined at the constant current, that, maintained in two straight parallel infinite conductors placed one meter apart in a vacuum, would produce a force between the conductors of 2×10^{-7} newton per meter. For example, the amount of current flowing through a 100-watt light bulb is 1 amp; through a toaster, 10 amps; a TV set, 3 amps; a car battery, 50 amps (while cranking). A newton (N) is defined as a unit of force needed to accelerate one kilogram by one meter second^{-2}, or $1 N = 1 (Kg \times M)/s^{-2}$.

How did the electrical unit **volt originate**?

The unit of voltage is the volt, named after Alessandro Volta (1745–1827), the Italian scientist who built the first modern battery. (A battery, operating with a lead rod and vinegar, was also manufactured in ancient Egypt.) Voltage measures the force or "oomph" with which electrical charges are pushed through a material. Some common voltages are 1.5 volts for a flashlight battery; 12 volts for a car battery; 115 volts for ordinary household receptacles; and 230 volts for a heavy-duty household receptacle.

How did the electrical unit **watt originate**?

Named for the Scottish engineer and inventor James Watt (1736–1819), the watt is used to measure electric power. An electric device uses 1 watt when 1 volt of electric current drives 1 ampere of current through it.

What materials are the **best** and **worst conductors** of electricity?

Electrical conductivity is the ability of a material to transmit current or the movement of charged particles, most often protons. Materials that carry the flow of electrical current are called conductors. Metals, such as silver and aluminum, are some of the best conductors of electricity. Other good conductors of electricity are copper and gold. Materials that do not permit the flow of electrical current are called nonconductors or insulators. Wood, paper, and most plastics are examples of insulators. Resistance is defined as the extent to which a material prevents the flow of electricity. Materials with a low resistance have a high conductivity while those with a high resistance have a low conductivity. German physicist Georg Simon Ohm (1789–1854) was the first to describe the laws of electrical conductivity and resistance.

How do **lead-acid batteries** work?

Lead-acid batteries consist of positive and negative lead plates suspended in a diluted sulfuric acid solution called an electrolyte. Everything is contained in a chemically and electrically inert case. As the cell discharges, sulfur molecules from the electrolyte bond with the lead plates, releasing excess electrons. The flow of electrons is called electricity.

Who was **Nikola Tesla**?

Nikola Tesla (1856–1943) was a leading innovator in the field of electricity. Tesla held over one hundred patents, among which are patents for alternating current and the seminal patents for radio. Tesla's work for Westinghouse in the late 1880s led to the commercial production of electricity, including the Niagara Falls Power Project in 1895. After a bitter and prolonged public feud, Tesla's alternating current system was proven superior to Thomas Edison's direct current system. Tesla was responsible for many other innovations, including the Tesla coil, radio controlled boats, and neon and fluorescent lighting.

MATTER

What is an **atom**?

An atom is the smallest unit of an element, containing the unique chemical properties of that element. Atoms are very small—several million atoms could fit in the period at the end of this sentence.

Parts of an Atom

Subatomic Particle	Charge	Location
Proton	Positive	Nucleus
Neutron	Neutral	Nucleus
Electron	Negative	Orbits nucleus

Who is generally regarded as the **discoverer** of the **electron**, the **proton**, and the **neutron**?

The British physicist Sir Joseph John Thomson (1856–1940) in 1897 researched electrical conduction in gases, which led to the important discovery that cathode rays consist of negatively charged particles called electrons. The discovery of the electron inaugurated the electrical theory of the atom, and this, along with other work, entitled Thomson to be regarded as the founder of modern atomic physics.

Ernest Rutherford discovered the proton in 1919. He also predicted the existence of the neutron, later discovered by his colleague, James Chadwick (1891–1974). Chadwick was awarded the 1935 Nobel Prize in Physics for this discovery.

What is a **chemical bond**?

A chemical bond is an attraction between the electrons present in the outermost energy level or shell of a particular atom. This outermost energy level is known as the valence shell. Atoms with an unfilled outer shell are less stable and tend to share, accept, or donate electrons. When this happens, a chemical bond is formed.

What are the major **types** of **bonds**?

There are three major types of chemical bonds: covalent, ionic, and hydrogen. The type of bond that is established is determined by the electron structure. Ionic bonds are formed when electrons are exchanged between two atoms, and the resulting bond is relatively weak. Covalent bonds, the strongest type of bond, occur when electrons are shared between atoms. Hydrogen bonds are temporary, but they are important because they are crucial to the shape of a particular protein and have the ability to be rapidly formed and reformed. The following chart explains the types of bonds and their characteristics.

Three Types of Chemical Bonds

Type	Strength	Examples
Covalent	Strong	Bonds between hydrogen and oxygen in a molecule of water
Ionic	Moderate	Bond between Na^+ and Cl^- in salt
Hydrogen	Weak	Bonds between molecules of water

When was the first **atomic theory proposed**?

John Dalton (1766–1844), an English natural philosopher, chemist, physicist, and teacher, developed the concepts of an atomic theory in the early nineteenth century. He published *A New System of Chemical Philosophy* in 1808. His main concepts of atomic theory may be summarized as:

- All matter—solid, liquid and gas—consists of tiny, indivisible particles called atoms.
- All atoms of a given element have the same mass and are identical, but are different from the atoms of different elements.
- Chemical reactions involve the rearrangement of combinations of those atoms, not the destruction of atoms.
- When elements react to form compounds, they combine in simple, whole-number ratios.

Who proposed the modern theory of the atom?

The modern theory of atomic structure was first proposed by the Japanese physicist Hantaro Nagaoka (1865–1950) in 1904. In his model, electrons rotated in rings around a small central nucleus. In 1911, Ernest Rutherford discovered further evidence to prove that the nucleus of the atom is extremely small and dense and is surrounded by a much larger and less dense cloud of electrons. In 1913, the Danish physicist Niels Bohr (1885–1962) proposed a model that is known as the Bohr atom. It suggested that electrons orbit the nucleus in concentric quantum shells at certain well-specified distances from the nucleus corresponding to the electron's energy levels. These orbits are known as "Bohr orbits." Several years later, Erwin Schrödinger (1887–1961) proposed the Schrödinger wave equation, which provided a firm theoretical basis for the Bohr orbits.

Who invented the cyclotron?

The cyclotron was invented by Ernest Lawrence (1901–1958) at the University of California, Berkeley, in 1934, to study the nuclear structure of the atom. The cyclotron produced high energy particles that were accelerated outwards in a spiral rather than through an extremely long, linear accelerator.

A bust of Niels Bohr is displayed in Copenhagen, Denmark. Bohr proposed the model of the atom with which most people are familiar, showing electrons circling a nucleus of protons and neutrons.

What is the fourth state of matter?

Plasma, a mixture of free electrons and ions or atomic nuclei, is sometimes referred to as a "fourth state of matter." Plasmas occur in thermonuclear reactions as in the sun, in fluorescent lights, and in stars. When the temperature of gas is raised high enough, the collision of atoms becomes so violent that electrons are knocked loose from their nuclei. The result of a gas having loose, negatively charged electrons and heavier, positively charged nuclei is called a plasma.

All matter is made up of atoms. Animals and plants are organic matter; minerals and water are inorganic matter. Solid, liquid, and gas are the first three states of matter. Whether matter appears as a solid, liquid, or gas depends on how the molecules are held together in their chemical bonds. Solids have a rigid structure in the atoms of the molecules; in liquids the molecules are close together but not packed; in a gas, the molecules are widely spaced and move around, occasionally colliding but usually not interacting.

When were **liquid crystals discovered**?

Liquid crystals were observed by the Austrian botanist Friedrich Reinitzer (1857–1927) in 1888. He noticed that the solid organic compound cholesteryl benzoate became a cloudy liquid at 293°F (145°C) and a clear liquid at 354°F (179°C). The following year, the German physicist Otto Lehmann (1855–1922) used a microscope with a heating stage to determine that some molecules do not melt directly but first pass through a phase when they flow like a liquid but maintain the molecular structure and properties of a solid. He coined the phrase "liquid crystal" to describe this substance. Further experimentation showed that if an electrical charge is passed through a liquid crystal material, the liquid will line up according to the direction of the electrical field. Liquid crystals are used for electronic panel displays.

What is **antimatter**?

Antimatter is the exact opposite of normal matter. Antimatter was predicted in a series of equations derived by Paul Dirac (1902–1984). He was attempting to combine the theory of relativity with equations governing the behavior of electrons. In order to make his equations work, he had to predict the existence of a particle that would be similar to the electron, but opposite in charge. This particle, discovered in 1932, was the antimatter equivalent of the electron and called the positron (electrons with a positive charge). Other antimatter particles would not be discovered until 1955 when particle accelerators were finally able to confirm the existence of the antineutron and antiproton (protons with a negative charge). Antiatoms (pairings of positrons and antiprotons) are other examples of antimatter.

What is the difference between **nuclear fission** and **nuclear fusion**?

Nuclear fission is the splitting of an atomic nucleus into at least two fragments. Nuclear fusion is a nuclear reaction in which the nuclei of atoms of low atomic number, such as hydrogen and helium, fuse to form a heavier nucleus. Although in both nuclear fission and nuclear fusion substantial amounts of energy are produced, the

How did the quark get its name?

This theoretical particle, considered to be the fundamental unit of matter, was named by Murray Gell-Mann (1929–), an American theoretical physicist who was awarded the 1969 Nobel Prize in Physics. Its name was initially a playful tag that Gell-Mann invented, sounding something like "kwork." Later, Gell-Mann came across the line "Three quarks for Master Marks" in James Joyce's (1882–1941) *Finnegan's Wake*, and the tag became known as a quark. There are six kinds or "flavors" (up, down, strange, charm, bottom, and top) of quarks, and each "flavor" has three varieties or "colors" (red, blue, and green). All 18 types have different electric charges (a basic characteristic of all elementary particles). Three quarks form a proton (having one unit of positive electric charge) or a neutron (zero charge), and two quarks (a quark and an antiquark) form a meson. Like all known particles, a quark has its antimatter opposite, known as an antiquark (having the same mass but opposite charge).

amount of energy produced in fusion is far greater than the amount of energy produced in fission.

What was **Richard Feynman's contribution** to physics?

Richard Feynman (1918–1988) developed a theory of quantum electrodynamics that described the interaction of electrons, positrons, and photons, providing physicists a new way to work with electrons. He reconstructed quantum mechanics and electrodynamics in his own terms, formulating a matrix of measurable quantities visually represented by a series of graphs knows as the Feynman diagrams. Feynman was awarded the Nobel Prize in Physics in 1965.

What are **subatomic particles**?

Subatomic particles are particles that are smaller than atoms. Historically, subatomic particles were considered to be electrons, protons, and neutrons. However, the definition of subatomic particles has now been expanded to include elementary particles, which are so small that they do not appear to be made of anything more minute. The physical study of such particles became possible only during the twentieth century with the development of increasingly sophisticated apparatus. Many new particles have been discovered in the last half of the twentieth century.

A number of proposals have been made to organize the particles by their spin, their mass, or their common properties. One system is now commonly known as the Standard Model. This system recognizes two basic types of fundamental particles: quarks and leptons. Other force-carrying particles are called bosons. Photons, gluons, and weakons are bosons. Leptons include electrons, muons, taus, and three kinds of neutrinos. Quarks never occur alone in nature. They always combine to form particles called hadrons. According to the Standard Model, all other subatomic particles consist of some combination of quarks and their antiparticles. A proton consists of three quarks.

Who is generally regarded as the **founder** of **quantum mechanics**?

The German mathematical physicist Werner Karl Heisenberg (1901–1976) is regarded as the father of quantum mechanics (the theory of small-scale physical phenomena). His theory of uncertainty in 1927 overturned traditional classical mechanics and electromagnetic theory regarding energy and motion when applied to subatomic particles such as electrons and parts of atomic nuclei. The theory states that while it is impossible to specify precisely both the position and the simultaneous momentum (mass 3 velocity) of a particle, they can only be predicted. This means that the result of an action can be expressed only in terms of probability that a certain effect will occur. Heinsenberg was awarded the Nobel Prize in Physics in 1932.

Why is **liquid water** more **dense** than **ice**?

Pure liquid water is most dense at 39.2°F (3.98°C) and decreases in density as it freezes. The water molecules in ice are held in a relatively rigid geometric pattern by their hydrogen bonds, producing an open, porous structure. Liquid water has fewer bonds; therefore, more molecules can occupy the same space, making liquid water more dense than ice.

What substance, other than water, is **less dense as a solid** than as a liquid?

Only bismuth and water share this characteristic. Density (the mass per unit volume or mass/volume) refers to how compact or crowded a substance is. For instance, the density of water is 1 g/cm^3 (gram per cubic centimeter) or 1 kg/l (kilogram per liter); the density of a rock is 3.3 g/cm^3; pure iron is 7.9 g/cm^3; and Earth (as a whole) is 5.5 g/cm^3 (average). Water as a solid (i.e., ice) floats.

Who discovered the principle of **buoyancy**?

Buoyancy was first discovered by the Greek mathematician Archimedes (c. 287–212 B.C.E.). The famous story recounts how the king of Syracuse, Hieron II (c. 306–c. 215 B.C.E.), asked Archimdes to verify that his crown was made of pure gold without destroying the crown. When Archimedes entered his bath, he noticed that the water overflowed the tub. He realized that the volume water that flowed out of the bath had to be equal to the volume of his own body that was immersed in the bath. Shouting "Eureka," he ran through the streets of Syracuse announcing he had found a method to determine whether the king's crown was made of pure gold. He could measure the amount of water that was displaced by a block of pure gold of the same weight as the crown. If the crown was made of pure gold, it would displace the same amount of water as the block of gold. The principle of buoyancy, also known as Archimedes's principle, states that the buoyant force acting on an object placed in a fluid is equal to the weight of the fluid displaced by the object.

Who made the first organic compound to be synthesized from inorganic ingredients?

In 1828, Friedrich Wöhler (1800–1882) synthesized urea from ammonia and cyanic acid. This synthesis dealt a deathblow to the vital-force theory, which held that definite and fundamental differences existed between organic and inorganic compounds. The Swedish chemist Jöns Jakob Berzelius (1779–1848) had proposed that the two classes of compounds were produced from their elements by entirely different laws. Organic compounds were produced under the influence of a vital force and so were incapable of being prepared artificially. This distinction ended with Wöhler's synthesis.

CHEMICAL ELEMENTS

What were the two main aims of alchemy?

The main aim of alchemy, the early study of chemical reactions, was the transmutation (or transformation) of common elements into gold. Needless to say, all attempts to change a substance into gold were unsuccessful. A second aim of alchemy was to discover an elixir or universal remedy that would promote everlasting life. Again, this pursuit was unsuccessful.

What is a philosopher's stone?

A philosopher's stone is the name of a substance believed by medieval alchemists to have the power to change baser metals into gold or silver. It had, according to some, the power of prolonging life and of curing all injuries and diseases. The pursuit of it by alchemists led to the discovery of several chemical substances; however, the magical philosopher's stone has since proved fictitious.

Who are some of the founders of modern chemistry?

The history of chemistry in its modern form is often considered to begin with the British scientist Robert Boyle (1627–1691), although its roots can be traced back to the earliest recorded history. Best known for his discovery of Boyle's Law (volume of a gas is inversely proportional to its pressure at constant temperature), he was a pioneer in the use of experiments and the scientific method. A founder of the Royal Society, he worked to remove the mystique of alchemy from chemistry to make it a pure science.

The French chemist Antoine-Laurent Lavoisier (1743–1794) is regarded as another important founder of modern chemistry. Indeed, he is considered "the father of modern chemistry." His wideranging contributions include the discrediting of the phlogiston theory of combustion, which long had been a stumbling block to a true understanding of chemistry. He established modern terminology for chemical substances and did the first experiments in quantitative organic analysis. He is sometimes credited with having discovered or established the law of conservation of mass in chemical reactions.

In this 1702 engraving two angels hold the philosopher's stone. The furnace below them is being used for an alchemical transmutation.

John Dalton (1766–1844), an English chemist, who proposed an atomic theory of matter that became a basic theory of modern chemistry, is also an important figure in the development of the field. His theory, first proposed in 1803, states that each chemical element is composed of its own kind of atoms, all with the same relative weight.

Another important individual in the development of modern chemistry was Swedish chemist Jöns Jakob Berzelius (1779–1848). He devised chemical symbols, determined atomic weights, contributed to the atomic theory, and discovered several new elements. Between 1810 and 1816, he described the preparation, purification, and analysis of 2,000 chemical compounds. Then he determined atomic weights for 40 elements. He simplified chemical symbols, introducing a notation—letters with numbers—that replaced the pictorial symbols his predecessors used, and that is still used today. He discovered cerium (in 1803, with Wilhelm Hisinger), selenium (1818), silicon (1824), and thorium (1829).

What are the **four major divisions** of **chemistry**?

Chemistry has traditionally been divided into organic, inorganic, analytical, and physical chemistry. Organic chemistry is the study of compounds that contain carbon. More than 90 percent of all known chemicals are organic. Inorganic chemistry is the study of compounds of all elements except carbon. Analytical chemists determine the structure and composition of compounds and mixtures. They also develop and operate instruments and techniques for carrying out the analyses. Physical chemists use the principles of physics to understand chemical phenomena.

Who **proposed** the **phlogiston theory**?

Phlogiston was a name used in the eighteenth century to identify a supposed substance given off during the process of combustion. The phlogiston theory was developed in the early 1700s by the German chemist and physicist Georg Ernst Stahl (1660–1734).

In essence, Stahl held that combustible material such as coal or wood was rich in a material substance called "phlogiston." What remained after combustion was without phlogiston and could no longer burn. The rusting of metals also involved a transfer of phlogiston. This accepted theory explained a great deal previously unknown to chemists. For instance, metal smelting was consistent with the phlogiston theory, as was the fact that charcoal lost weight when burned. Thus the loss of phlogiston either decreased or increased weight.

The French chemist Antoine-Laurent Lavoisier demonstrated that the gain of weight when a metal turned to a calx was just equal to the loss of weight of the air in the vessel. Lavoisier also showed that part of the air (oxygen) was indispensable to combustion, and that no material would burn in the absence of oxygen. The transition from Stahl's phlogiston theory to Lavoisier's oxygen theory marks the birth of modern chemistry at the end of the eighteenth century.

When was **spontaneous combustion** first recognized?

Spontaneous combustion is the ignition of materials stored in bulk. This is due to internal heat buildup caused by oxidation (generally a reaction in which electrons are lost, specifically when oxygen is combined with a substance, or when hydrogen is removed from a compound). Because this oxidation heat cannot be dissipated into the surrounding air, the temperature of the material rises until the material reaches its ignition point and bursts into flame.

A Chinese text written before 290 C.E. recognized this phenomenon in a description of the ignition of stored oiled cloth. The first Western acknowledgment of spontaneous combustion was by J. P. F. Duhamel in 1757, when he discussed the gigantic conflagration of a stack of oil-soaked canvas sails drying in the July sun. Before spontaneous combustion was recognized, such events were usually blamed on arsonists.

Who **developed** the **periodic table**?

Dmitry Ivanovich Mendeleyev (1834–1907) was a Russian chemist whose name will always be linked with the development of the periodic table. He was the first chemist

Main-Group Elements

Transition Metals

Inner-Transition Metals

Atomic number 86	(222)	Atomic weight
Symbol	Rn	
Name	radon	

Period

Main-Group Elements

1 IA	2 IIA												13 IIIA	14 IVA	15 VA	16 VIA	17 VIIA	18 VIIIA
1 1.00794 H hydrogen																		2 4.002602 He helium
3 6.941 Li lithium	4 9.012182 Be beryllium												5 10.811 B boron	6 12.011 C carbon	7 14.00674 N nitrogen	8 15.9994 O oxygen	9 18.9984032 F fluorine	10 20.1797 Ne neon
11 22.989768 Na sodium	12 24.3050 Mg magnesium	3 IIIB	4 IVB	5 VB	6 VIB	7 VIIB	8	9 VIIIB	10	11 IB	12 IIB		13 26.981539 Al aluminum	14 28.0855 Si silicon	15 30.973762 P phosphorus	16 32.066 S sulfur	17 35.4527 Cl chlorine	18 39.948 Ar argon
19 39.0983 K potassium	20 40.078 Ca calcium	21 44.955910 Sc scandium	22 47.88 Ti titanium	23 50.9415 V vanadium	24 51.9961 Cr chromium	25 54.9305 Mn manganese	26 55.847 Fe iron	27 58.93320 Co cobalt	28 58.69 Ni nickel	29 63.546 Cu copper	30 65.39 Zn zinc		31 69.723 Ga gallium	32 72.61 Ge germanium	33 74.92159 As arsenic	34 78.96 Se selenium	35 79.904 Br bromine	36 83.80 Kr krypton
37 85.4678 Rb rubidium	38 87.62 Sr strontium	39 88.90585 Y yttrium	40 91.224 Zr zirconium	41 92.90638 Nb niobium	42 95.94 Mo molybdenum	43 (98) Tc technetium	44 101.07 Ru ruthenium	45 102.90550 Rh rhodium	46 106.42 Pd palladium	47 107.8682 Ag silver	48 112.411 Cd cadmium		49 114.82 In indium	50 118.710 Sn tin	51 121.75 Sb antimony	52 127.60 Te tellurium	53 126.90447 I iodine	54 131.29 Xe xenon
55 132.90543 Cs cesium	56 137.327 Ba barium	57 138.9055 *La lanthanum	72 178.49 Hf hafnium	73 180.9479 Ta tantalum	74 183.85 W tungsten	75 186.207 Re rhenium	76 190.2 Os osmium	77 192.22 Ir iridium	78 195.08 Pt platinum	79 196.96654 Au gold	80 200.59 Hg mercury		81 204.3833 Tl thallium	82 207.2 Pb lead	83 208.98037 Bi bismuth	84 (209) Po polonium	85 (210) At astatine	86 (222) Rn radon
87 (223) Fr francium	88 (226) Ra radium	89 (227) †Ac actinium	104 (261) Rf rutherfordium	105 (262) Db dubnium	106 (263) Sg seaborgium	107 (262) Bh bohrium	108 (265) Hs hassium	109 (266) Mt meitnerium	110 (269) Uun ununnilium	111 (272) Uuu unununium	112 (277) Uub ununbium							

*Lanthanides

58 140.115 Ce cerium	59 140.90765 Pr praseodymium	60 144.24 Nd neodymium	61 (145) Pm promethium	62 150.36 Sm samarium	63 151.965 Eu europium	64 157.25 Gd gadolinium	65 158.92534 Tb terbium	66 162.50 Dy dysprosium	67 164.93032 Ho holmium	68 167.26 Er erbium	69 168.93421 Tm thulium	70 173.04 Yb ytterbium	71 174.967 Lu lutetium

†Actinides

90 232.0381 Th thorium	91 231.0359 Pa protactinium	92 238.0289 U uranium	93 (237) Np neptunium	94 (244) Pu plutonium	95 (243) Am americium	96 (247) Cm curium	97 (247) Bk berkelium	98 (251) Cf californium	99 (252) Es einsteinium	100 (257) Fm fermium	101 (258) Md mendelevium	102 (259) No nobelium	103 (262) Lr lawrencium

The periodic table of the elements.

to really understand that all elements are related members of a single ordered system. He changed what had been a highly fragmented and speculative branch of chemistry into a true, logical science. His nomination for the 1906 Nobel Prize in Chemistry failed by one vote, but his name became recorded in perpetuity 50 years later when element 101 was called mendelevium.

According to Mendeleyev, the properties of the elements, as well as those of their compounds, are periodic functions of their atomic weights (in the 1920s, it was discovered that atomic number was the key rather than weight). Mendeleyev compiled the first true periodic table in 1869 listing all the 63 (then-known) elements. In order to make the table work, Mendeleyev had to leave gaps, and he predicted that further elements would eventually be discovered to fill them. Three were discovered in Mendeleyev's lifetime: gallium in 1875, scandium in 1879, and germanium in 1896.

There are 95 naturally occurring elements; of the remaining elements (elements 96 to 109), ten are undisputed. There are approximately more than 50 million chemical compounds, produced from the elements, registered with *Chemical Abstracts* as of 2009.

What were **Lothar Meyer's contributions** to the **periodic table**?

Lothar Meyer (1830–1895), a German chemist, prepared a periodic table that resembled closely Mendeleyev's periodic table. He did not publish his periodic table until after Mendeleyev's paper on the periodic table was published in 1869. It is believed his work was influential in causing some of the revisions Mendeleyev made in the second version of his periodic table, published in 1870. Specifically, Meyer focused on the periodicity of the physical properties of the elements, while Mendeleyev's focus was the chemical consequences of the periodic law.

What was the **first element** to be **discovered**?

Phosphorus was first discovered by German chemist Hennig Brand (c. 1630–c. 1710) in 1669 when he extracted a waxy white substance from urine that glowed in the dark. But Brand did not publish his findings. In 1680, phosphorus was rediscovered by the English chemist Robert Boyle.

What are the **alkali metals**?

These are the elements at the left of the periodic table: lithium (Li, element 3), potassium (K, element 19), rubidium (Rb, element 37), cesium (Cs, element 55), francium (Fr, element 87), and sodium (Na, element 11). The alkali metals are sometimes called the sodium family of elements, or Group I elements. Because of their great chemical reactivity (they easily form positive ions), none exists in nature in the elemental state.

What are the **alkaline Earth metals**?

These are beryllium (Be, element 4), magnesium (Mg, element 12), calcium (Ca, element 20), strontium (Sr, element 38), barium (Ba, element 56), and radium (Ra, element 88). The alkaline Earth metals are also called Group II elements. Like the alkali metals, they are never found as free elements in nature and are moderately reactive metals. Harder and less volatile than the alkali metals, these elements all burn in air.

What are the **transition elements**?

The transition elements are the 10 subgroups of elements between Group II and Group XIII, starting with period 4. They include gold (Au, element 79), silver (Ag, element 47), platinum (Pt, element 78), iron (Fe, element 26), copper (Cu, element 29), and other metals. All transition elements are metals. Compared to alkali and alkaline Earth metals, they are usually harder and more brittle and have higher melting points. Transition metals are also good conductors of heat and electricity. They have variable valences, and compounds of transition elements are often colored. Transition elements are so named because they comprise a gradual shift from the strongly electropositive elements of Groups I and II to the electronegative elements of Groups VI and VII.

What are the **transuranic chemical elements** and the names for elements 102 to 109?

Transuranic elements are those elements in the periodic system with atomic numbers greater than 92. Many of these elements are ephemeral, do not exist naturally outside the laboratory, and are not stable.

Elements 93 through 109

Element Number	Name	Symbol
93	Neptunium	Np
94	Plutonium	Pu
95	Americum	Am
96	Curium	Cm
97	Berkelium	Bk
98	Californium	Cf
99	Einsteinium	Es
100	Fermium	Fm
101	Mendelevium	Md
102	Nobelium	No
103	Lawrencium	Lr
104	Rutherfordium	Rf
105	Dubnium	Db

Element Number	Name	Symbol
106	Seaborgium	Sg
107	Bohrium	Bh
108	Hassium	Hs
109	Meitnerium	Mt
110	Darmstadium	Ds
111	Roentgenium	Rg
112	Copernicium	Cn

Elements 113 through 118 are unstable and have not been confirmed independently by researchers. The names for elements 113 through 118 are under review by the International Union of Pure and Applied Chemistry.

Which elements are the "noble metals"?

The noble metals are gold (Au, element 79), silver (Ag, element 47), mercury (Hg, element 80), and the platinum group, which includes platinum (Pt, element 78), palladium (Pd, element 46), iridium (Ir, element 77), rhodium (Rh, element 45), ruthenium (Ru, element 44), and osmium (Os, element 76). The term refers to those metals highly resistant to chemical reaction or oxidation (resistant to corrosion) and is contrasted to "base" metals, which are not so resistant. The term has its origins in ancient alchemy whose goals of transformation and perfection were pursued through the different properties of metals and chemicals. The term is not synonymous with "precious metals," although a metal, like platinum, may be both.

The platinum group metals have a variety of uses. In the United States more than 95 percent of all platinum group metals are used for industrial purposes. While platinum is a coveted material for jewelry making, it is also used in the catalytic converters of automobiles to control exhaust emissions, as are rhodium and palladium. Rhodium can also be alloyed with platinum and palladium for use in furnace windings, thermocouple elements, and in aircraft spark-plug electrodes. Osmium is used in the manufacture of pharmaceuticals and in alloys for instrument pivots and long-life phonograph needles.

What distinguishes **gold** and **silver** as elements?

Besides their use as precious metals, gold and silver have properties that distinguish them from other chemical elements. Gold is the most ductile and malleable metal—the thinnest gold leaf is 0.0001 millimeters thick. Silver is the most reflective of all metals; thus, it is used in mirrors.

What is **Harkin's rule**?

Atoms having even atomic numbers are more abundant in the universe than are atoms having odd atomic numbers. Chemical properties of an element are determined by its atomic number, which is the number of protons in the atom's nucleus.

What are some chemical **elements** whose **symbols** are **not derived** from their **English** names?

Modern Name	Symbol	Older Name
antimony	Sb	stibium
copper	Cu	cuprum
gold	Au	aurum
iron	Fe	ferrum
lead	Pb	plumbum
mercury	Hg	hydrargyrum
potassium	K	kalium
silver	Ag	argentum
sodium	Na	natrium
tin	Sn	stannum
tungsten	W	wolfram

Which **elements** are **liquid** at **room temperature**?

Mercury ("liquid silver," Hg, element 80) and bromine (Br, element 35) are liquid at room temperature 68° to 70°F (20° to 21°C). Gallium (Ga, element 31) with a melting point of 85.6°F (29.8°C), cesium (Cs, element 55) with a melting point of 83°F (28.4°C), and Francium (Fr, element 87) with a melting point of 80.6°F (27°C) are liquid at slightly above room temperature and pressure.

Which chemical **element** is the **most abundant** in the universe?

Hydrogen (H, element 1) makes up about 75 percent of the mass of the universe. It is estimated that more than 90 percent of all atoms in the universe are hydrogen atoms. Most of the rest are helium (He, element 2) atoms.

Which chemical **elements** are the **most abundant** on **Earth**?

Oxygen (O, element 8) is the most abundant element in Earth's crust, waters, and atmosphere. It comprises 49.5 percent of the total mass of these compounds. Silicon (Si, element 14) is the second most abundant element. Silicon dioxide and silicates make up about 87 percent of the materials in Earth's crust.

Why are the **rare gases** and **rare Earth elements** called "rare"?

Rare gases refers to the elements helium, neon, argon, krypton, and xenon. They are rare in that they are gases of very low density ("rarified") at ordinary temperatures and are found only scattered in minute quantities in the atmosphere and in some substances. In addition, rare gases have zero valence and normally will not combine with other elements to make compounds.

Rare Earth elements are elements numbered 58 through 71 in the periodic table plus yttrium (Y, element 39) and thorium (Th, element 90). They are called "rare Earths" because they are difficult to extract from monazite ore, where they occur. The term has nothing to do with scarcity or rarity in nature.

What is an **isotope**?

Elements are identified by the number of protons in an atom's nucleus. Atoms of an element that have different numbers of neutrons are isotopes of the same element. Isotopes of an element have the same atomic number but different mass numbers. Although the physical properties of atoms depend on mass, differences in atomic mass (mass numbers) have very little effect on chemical reactions. Common examples of isotopes are carbon–12 and carbon–14. Carbon–12 has six protons, six electrons, and six neutrons; carbon–14 has six protons, six electrons, and eight neutrons.

Which **elements** have the **most isotopes**?

The elements with the most isotopes, with 36 each, are xenon (Xe) with nine stable isotopes (identified from 1920 to 1922) and 27 radioactive isotopes (identified from 1939 to 1981), and cesium (Cs) with one stable isotope (identified in 1921) and 35 radioactive isotopes (identified from 1935 to 1983).

The element with the fewest number of isotopes is hydrogen (H), with three isotopes, including two stable ones—protium (identified in 1920) and deuterium, often called heavy water, (identified in 1931)—and one radioactive isotope—tritium (first identified in 1934, although not considered a radioactive isotope in 1939).

What is **heavy water**?

Heavy water, also called deuterium oxide (D_2O), is composed of oxygen and two hydrogen atoms in the form of deuterium, which has about twice the mass of normal hydro-

Are there more chemical compounds with even numbers of carbon atoms than with odd?

A team of chemists recently noted that the database of the Beilstein Information System, containing over 9.6 million organic compounds, includes significantly more substances with an even number of carbon atoms than with an odd number. Statistical analyses of smaller sets of organic compounds, such as the *Cambridge Crystallographic Database* or the *CRC Handbook of Chemistry and Physics*, led to the same results. A possible explanation for the observed asymmetry might be that organic compounds are ultimately derived from biological sources, and nature frequently utilized acetate, a C_2 building block, in its syntheses of organic compounds. It may therefore be that the manufacturers' and synthetic chemists' preferential use of relatively economical starting materials derived from natural sources has left permanent traces in chemical publications and databases.

gen. As a result, heavy water has a molecular weight of about 20, while ordinary water has a molecular weight of about 18. Approximately one part heavy water can be found in 6,500 parts of ordinary water, and it may be extracted by fractional distillation. It is used in thermonuclear weapons and nuclear reactors and as an isotopic tracer in studies of chemical and biochemical processes. Heavy water was discovered by Harold C. Urey (1893–1981) in 1931. He received the Nobel Prize in Chemistry in 1934 for his discovery of heavy hydrogen.

What does **half-life** mean?

Half-life is the time it takes for the number of radioactive nuclei originally present in a sample to decrease to one-half of their original number. Thus, if a sample has a half-life of one year, its radioactivity will be reduced to half its original amount at the end of a year and to one quarter at the end of two years. The half-life of a particular radionuclide is always the same, independent of temperature, chemical combination, or any other condition. Natural radiation was discovered in 1896 by the French physicist Antoine Henri Becquerel (1852–1908). His discovery initiated the science of nuclear physics.

Which **elements** have the **highest** and **lowest boiling points**?

Helium has the lowest boiling point of all the elements at −452.074°F (−268.93°C) followed by hydrogen −423.16°F (−252.87°C). According to the *CRC Handbook of Chemistry and Physics*, the highest boiling point for an element is that of rhenium

85

10,104.8°F (5,596°C) followed by tungsten 10,031°F (5,555°C). Other sources record tungsten with the highest boiling point followed by rhenium.

Which **element** has the **highest density**?

Either osmium or iridium is the element with the highest density; however, scientists have yet to gather enough conclusive data to choose between the two. When traditional methods of measurement are employed, osmium generally appears to be the densest element. Yet, when calculations are made based upon the space lattice, which may be a more reliable method given the nature of these elements, the density of iridium is 22.65 compared to 22.61 for osmium.

What is the **density** of **air**?

The density of dry air is 1.29 grams per liter at 32°F (0°C) at average sea level and a barometric pressure of 29.92 inches of mercury (760 millimeters).

The weight of one cubic foot of dry air at one atmosphere of barometric pressure is:

Temperature (Fahrenheit)	Weight per cubic foot (pounds)
50°	0.07788
60°	0.07640
70°	0.07495

Which **elements** are the **hardest** and **softest**?

Carbon is both the hardest and softest element occurring in two different forms as graphite and diamond. A single crystal of diamond scores the absolute maximum value on the Knoop hardness scale of 90. Based on the somewhat less informative abrasive hardness scale of Mohs, diamond has a hardness of 10. Graphite is an extremely soft material with a Mohs hardness of only 0.5 and a Knoop hardness of 0.12.

What are **isomers**?

Isomers are compounds with the same molecular formula but different structures due to the different arrangement of the atoms within the molecules. Structural isomers have atoms connected in different ways. Geometric isomers differ in their symmetry about a double bond. Optical isomers are mirror images of each other.

What are the **gas laws**?

The gas laws are physical laws concerning the behavior of gases. They include Boyle's law, which states that the volume of a given mass of gas at a constant temperature is

Carbon atoms can be arranged into molecules of graphite, which are soft, or diamonds, which are one of the hardest substances known.

inversely proportional to its pressure; and Charles's law, which states that the volume of a given mass of gas at constant pressure is directly proportional to its absolute temperature. These two laws can be combined to give the General or Universal gas law, which may be expressed as:

$$(\text{pressure} \times \text{volume})/\text{temperature} = \text{constant}$$

Avogadro's law states that equal volumes of all gases contain the same number of particles if they all have the same pressure and temperature.

The laws are not obeyed exactly by any real gas, but many common gases obey them under certain conditions, particularly at high temperatures and low pressures.

What is a **Lewis acid**?

Named after the American chemist Gilbert Newton Lewis (1875–1946), the Lewis theory defines an acid as a species that can accept an electron pair from another atom, and a base as a species that can donate an electron pair to complete the valence shell of another atom. Hydrogen ion (proton) is the simplest substance that will do this, but Lewis acids include many compounds—such as boron trifluoride (BF_3) and aluminum chloride ($AlCl_3$)—that can react with ammonia, for example, to form an addition compound or Lewis salt.

Which **chemical** is **used in greater quantities** than any other?

Sodium chloride (NaCl), or salt, has over 14,000 uses, and is probably used in greater quantities and for more applications than any other chemical.

87

What is the **sweetest chemical compound**?

The sweetest chemical compound is sucronic acid.

Sweetener	Relative sweetness
sucronic acid	200,000
saccharin	300
aspartame	180
cyclamate	30
sugar (sucrose)	1

TEMPERATURE, MEASUREMENT, AND METHODOLOGY

Who **invented** the **thermometer**?

The Greeks of Alexandria knew that air expanded as it was heated. Hero of Alexandria (first century C.E.) and Philo of Byzantium made simple "thermoscopes," but they were not real thermometers. In 1592, Galileo Galilei (1564–1642) made a kind of thermometer that also functioned as a barometer, and in 1612 his friend Santorio Santorio (1561–1636) adapted the air thermometer (a device in which a colored liquid was driven down by the expansion of air) to measure the body's temperature change during illness and recovery. Still, it was not until 1713 that Daniel Fahrenheit (1686–1736) began developing a thermometer with a fixed scale. He worked out his scale from two "fixed" points: the melting point of ice and the heat of the healthy human body. He realized that the melting point of ice was a constant temperature, whereas the freezing point of water varied. Fahrenheit put his thermometer into a mixture of ice, water, and salt (which he marked off as 0°) and using this as a starting point, marked off melting ice at 32° and blood heat at 96°. In 1835, it was discovered that normal blood measured 98.6°F. Sometimes Fahrenheit used spirit of wine as the liquid in the thermometer tube, but more often he used specially purified mercury. Later, the boiling point of water (212°F) became the upper fixed point.

What is the **Kelvin temperature scale**?

Temperature is the level of heat in a gas, liquid, or solid. The freezing and boiling points of water are used as standard reference levels in both the metric (centigrade or Celsius) and the English system (Fahrenheit). In the metric system, the difference between freezing and boiling is divided into 100 equal intervals called degree Celsius or degree centigrade (°C). In the English system, the intervals are divided into 180

What was unusual about the original Celsius temperature scale?

In 1742, the Swedish astronomer Anders Celsius (1701–1744) set the freezing point of water at 100°C and the boiling point of water at 0°C. It was Carolus Linnaeus (1707–1778) who reversed the scale, but a later textbook attributed the modified scale to Celsius and the name has remained.

units, with one unit called degree Fahrenheit (°F). But temperature can be measured from absolute zero (no heat, no motion); this principle defines thermodynamic temperature and establishes a method to measure it upward. This scale of temperature is called the Kelvin temperature scale, after its inventor, William Thomson, Lord Kelvin (1824–1907), who devised it in 1848. The Kelvin (K) has the same magnitude as the degree Celsius (the difference between freezing and boiling water is 100 degrees), but the two temperatures differ by 273.15 degrees (0K equals –273.15°C). Below is a comparison of the three temperatures:

Characteristic	K	°C	°F
Absolute zero	0	–273.15	–459.67
Freezing point of water	273.15	0	32
Normal human body temperature	310.15	37	98.6
100°F	290.95	17.8	100
Boiling point of water (at one atmosphere pressure)	373.15	100	212
Point of equality	233.15	–40.0	–40.0

To convert Celsius to Kelvin, add 273.15 to the temperature (K = C + 273.15). To convert Fahrenheit to Celsius, subtract 32 from the temperature and multiply the difference by 5; then divide the product by 9 (C = 5/9[F – 32]). To convert Celsius to Fahrenheit, multiply the temperature by 1.8, then add 32 (F = 9/5C + 32).

How are **Celsius** temperatures **converted** into **Fahrenheit** temperatures?

The formulas for converting Celsius temperatures into Fahrenheit (and the reverse) are as follows:

$$F = (C \times 9/5) + 32$$
$$C = (F - 32) \times 5/9$$

How is **"absolute zero"** defined?

Absolute zero is the theoretical temperature at which all substances have zero thermal energy. Originally conceived as the temperature at which an ideal gas at constant

> ## Does hot water freeze faster than cold?
>
> A bucket of hot water will not freeze faster than a bucket of cold water. However, a bucket of water that has been heated or boiled, then allowed to cool to the same temperature as the bucket of cold water, may freeze faster. Heating or boiling drives out some of the air bubbles in water; because air bubbles cut down thermal conductivity, they can inhibit freezing. For the same reason, previously heated water forms denser ice than unheated water, which is why hot-water pipes tend to burst before cold-water pipes.

pressure would contract to zero volume, absolute zero is of great significance in thermodynamics and is used as the fixed point for absolute temperature scales. Absolute zero is equivalent to 0K, –459.67°F, or –273.15°C.

The velocity of a substance's molecules determines its temperature; the faster the molecules move, the more volume they require, and the higher the temperature becomes. The lowest actual temperature ever reached was two-billionth of a degree above absolute zero (2×10^{-9}K) by a team at the Low Temperature Laboratory in the Helsinki University of Technology, Finland, in October 1989.

What is **STP**?

The abbreviation STP is often used for standard temperature and pressure. As a matter of convenience, scientists have chosen a specific temperature and pressure as standards for comparing gas volumes. The standard temperature is 0°C (273°K) and the standard pressure is 760 torr (one atmosphere).

What is the **pH scale**?

The pH scale is the measurement of the H1 concentration (hydrogen ions) in a solution. It is used to measure the acidity or alkalinity of a solution. The pH scale ranges from 0 to 14. A neutral solution has a pH of 7, one with a pH greater than 7 is basic, or alkaline; and one with a pH less than 7 is acidic. The lower the pH below 7, the more acidic the solution. Each whole-number drop in pH represents a tenfold increase in acidity.

pH Value	Examples of Solutions
0	Hydrochloric acid (HCl), battery acid
1	Stomach acid (1.0–3.0)
2	Lemon juice (2.3)
3	Vinegar, wine, soft drinks, beer, orange juice, some acid rain
4	Tomatoes, grapes, banana (4.6)

pH Value	Examples of Solutions
5	Black coffee, most shaving lotions, bread, normal rainwater
6	Urine (5–7), milk (6.6), saliva (6.2–7.4)
7	Pure water, blood (7.3–7.5)
8	Egg white (8.0), seawater (7.8–8.3)
9	Baking soda, phosphate detergents, Clorox™, Tums™
10	Soap solutions, milk of magnesia
11	Household ammonia (10.5–11.9), nonphosphate detergents
12	Washing soda (sodium carbonate)
13	Hair remover, oven cleaner
14	Sodium hydroxide (NaOH)

Who is known as the **founder** of **crystallography**?

The French priest and mineralogist René-Just Haüy (1743–1822) is called the father of crystallography. In 1781 Haüy had a fortunate accident when he dropped a piece of calcite and it broke into small fragments. He noticed that the fragments broke along straight planes that met at constant angles. Haüy hypothesized that each crystal was built from successive additions of what is now called a unit cell to form a simple geometric shape with constant angles. An identity or difference in crystalline form implied an identity or difference in chemical composition. This was the beginning of the science of crystallography.

By the early 1800s many physicists were experimenting with crystals; in particular, they were fascinated by their ability to bend light and separate it into its component colors. An important member of the emerging field of optical mineralogy was the British scientist David Brewster (1871–1868), who succeeded in classifying most known crystals according to their optical properties.

The work of French chemist Louis Pasteur (1822–1895) during the mid–1800s became the foundation for crystal polarimetry—a method by which light is polarized, or aligned to a single plane. Pierre Curie and his brother Jacques (1855–1941) discovered another phenomenon displayed by certain crystals called piezoelectricity. It is the creation of an electrical potential by squeezing certain crystals.

pH paper is treated with a chemical indicator that changes color depending on how acidic or basic the solution is in which it is dipped. You can purchase testing kits that include color charts to evaluate the results.

91

Perhaps the most important application of crystals is in the science of X-ray crystallography. Experiments in this field were first conducted by the German physicist Max von Laue (1879–1960). This work was perfected by William Henry Bragg (1862–1942) and William Lawrence Bragg (1890–1971), who were awarded the Nobel Prize in Physics in 1915 for their work. The synthesis of penicillin and insulin were made possible by the use of X-ray crystallography.

Who **invented chromatography**?

Chromatography was invented by the Russian botanist Mikhail Tswett (1872–1919) in the early 1900s. The technique was first used to separate different plant pigments from one another. Chromatography has developed into a widely used method to separate various components of a substance from one another. Three types of chromatography are high-performance liquid chromatography (HPLC), gas chromatography, and paper chromatography. Different chromatography techniques are used in forensic science and analytical laboratories.

What is **nuclear magnetic resonance**?

Nuclear magnetic resonance (NMR) is a process in which the nuclei of certain atoms absorb energy from an external magnetic field. Analytical chemists use NMR spectroscopy to identify unknown compounds, check for impurities, and study the shapes of molecules. They use the knowledge that different atoms will absorb electromagnetic energy at slightly different frequencies.

What is a **mole** in **chemistry**?

A mole (mol), a fundamental measuring unit for the amount of a substance, refers to either a gram atomic weight or a gram molecular weight of a substance. It is the quantity of a substance that contains $6.02 \times 1,023$ atoms, molecules, or formula units of that substance. This number is called Avogadro's number or constant after Amadeo Avogadro (1776–1856), who is considered to be one of the founders of physical science. Mole Day was organized by the National Mole Day Foundation to promote an awareness and enthusiasm for chemistry. It is celebrated each year on October 23.

ASTRONOMY AND SPACE

UNIVERSE

What was the **Big Bang**?

The Big Bang theory is the explanation most commonly accepted by astronomers for the origin of the universe. It proposes that the universe began as the result of an explosion—the Big Bang—15 to 20 billion years ago. Two observations form the basis of this cosmology. First, as Edwin Hubble (1889–1953) demonstrated, the universe is expanding uniformly, with objects at greater distances receding at greater velocities. Secondly, Earth is bathed in a glow of radiation that has the characteristics expected from a remnant of a hot primeval fireball. This radiation was discovered by Arno A. Penzias (1933–) and Robert W. Wilson (1936–) of Bell Telephone Laboratories. In time, the matter created by the Big Bang came together in huge clumps to form the galaxies. Smaller clumps within the galaxies formed stars. Parts of at least one clump became a group of planets—our solar system.

What is the **Big Crunch** theory?

According to the Big Crunch theory, at some point in the very distant future, all matter will reverse direction and crunch back into the single point from which it began. Two other theories predict the future of the universe: the Big Bore theory and the Plateau theory. The Big Bore theory, named because it has nothing exciting to describe, claims that all matter will continue to move away from all other matter and the universe will expand forever. According to the Plateau theory, expansion of the universe will slow to the point where it will nearly cease, at which time the universe will reach a plateau and remain essentially the same.

How **old** is the **universe**?

Recent data collected by the Hubble Space Telescope suggest that the universe may only be 8 billion years old. This contradicts the previous belief the universe was somewhere between 13 billion and 20 billion years old. The earlier figure was derived from the concept that the universe has been expanding at the same rate since its birth at the Big Bang. The rate of expansion is a ratio known as Hubble's constant. It is calculated by dividing the speed at which the galaxy is moving away from Earth by its distance from Earth. By inverting Hubble's Constant—that is, dividing the distance of a galaxy by its recessional speed—the age of the universe can be calculated. The estimates of both the velocity and distance of galaxies from Earth are subject to uncertainties, and not all scientists accept that the universe has always expanded at the same rate. Therefore, many still hold that the age of the universe is open to question.

How **old** is the **solar system**?

It is currently believed to be 4.5 billion years old. Earth and the rest of the solar system formed from an immense cloud of gas and dust. Gravity and rotational forces caused the cloud to flatten into a disc and much of the cloud's mass to drift into the center. This material became the sun. The leftover parts of the cloud formed small bodies called planetesimals. These planetesimals collided with each other, gradually forming larger and larger bodies, some of which became the planets. This process is thought to have taken about 25 million years.

How **large** is the **solar system**?

The size of the solar system can be visualized by imagining the sun (864,000 miles in diameter; 1,380,000 km) shrunk to a diameter of one inch (about the size of a ping-pong ball). Using the same size scale, Earth would be a speck 0.01 inches (0.25 mm) in diameter and about 9 feet (2.7 meters) away from the ping-pong ball sized sun. Our moon would have a diameter of 0.0025 inches (0.06 mm; the thickness of a human hair) and only a little over one-quarter inch (6.3 mm) from Earth. Jupiter, the largest planet in the solar system, appears as the size of a small pea (0.1 inches [2.5 mm] in diameter) and 46 feet (14 meters) from the sun.

What are **quasars**?

The name quasar originated as a contraction of "quasi-stellar" radio source. Quasars appear to be star like, but they have large redshifts in their spectra indicating that they are receding from Earth at great speeds, some at up to 90 percent the speed of light. Their exact nature is still unknown, but many believe quasars to be the cores of distant galaxies, the most distant objects yet seen. Quasars, also called quasi-stellar objects or QSOs, were first identified in 1963 by astronomers at the Palomar Observatory in California.

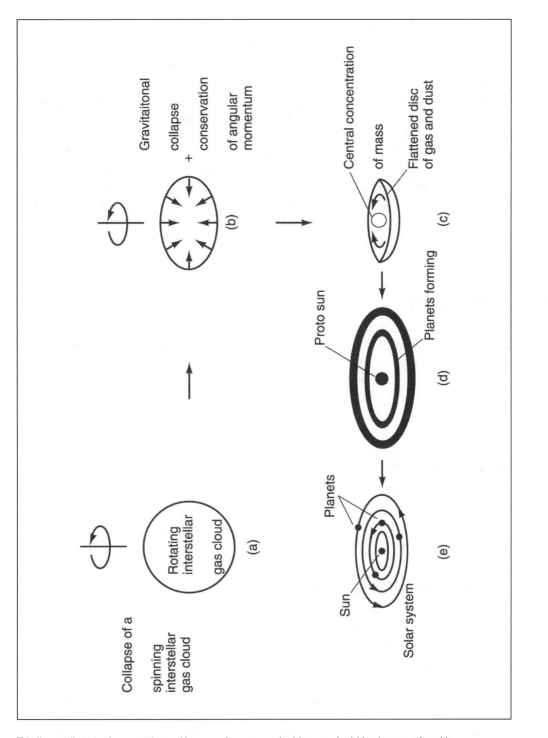

This diagram illustrates the current theory of how our solar system evolved from gas cloud (a) to its present form (e).

What is a **syzygy**?

A syzygy (*sizz*-eh-jee) is a configuration that occurs when three celestial bodies lie in a straight line, such as the sun, Earth, and the moon during a solar or lunar eclipse. The particular syzygy when a planet is on the opposite side of Earth from the sun is called an opposition.

STARS

What is a **supernova**?

A supernova is the death explosion of a massive star. Immediately after the explosion, the brightness of the star can outshine the entire galaxy, followed by a gradual fading. A supernova is a fairly rare event. The last supernova observed in our galaxy was in 1604. In February 1987, Supernova 1987A appeared in the Large Magellanic Cloud, a nearby galaxy.

What are the **four types** of **nebulae**?

The four types of nebulae are: emission, reflection, dark, and planetary. Primarily the birth place of stars, nebulae are clouds of gas and dust in space. Emission nebulae and reflection nebulae are bright nebulae. Emission nebulae are colorful and self-luminous. The Orion nebula, visible with the naked eye, is an example of an emission nebula. Reflection nebulae are cool clouds of dust and gas. They are illuminated by the light from nearby stars rather than by their own energy. Dark nebulae, also known as absorption nebulae, are not illuminated and appear as holes in the sky. The Horsehead nebula in the constellation Orion is an example of a dark nebula. Planetary nebulae are the remnants of the death of a star.

What is a **binary star**?

A binary star is a pair of stars revolving around a common center of gravity. About half of all stars are members of either binary star systems or multiple star systems, which contain more than two stars.

The bright star Sirius, about 8.6 light years away, is composed of two stars: one about 2.3 times the mass of the sun, the other a white dwarf star about 980 times the mass of Jupiter. Alpha Centauri, the nearest star to Earth after the sun, is actually three stars: Alpha Centauri A and Alpha Centauri B, two sunlike stars that orbit each other, and Alpha Centauri C, a low-mass red star that orbits around them.

What is a **black hole**?

When a star with a mass greater than about four times that of the sun collapses even the neutrons cannot stop the force of gravity. There is nothing to stop the contraction, and the star collapses forever. The material is so dense that nothing—not even light—can escape. The American physicist John Wheeler (1911–2008) gave this phenomenon the name "black hole" in 1967. Since no light escapes from a black hole, it cannot be observed directly. However, if a black hole existed near another star, it would draw matter from the other star into itself and, in effect, produce X rays. In the constellation of Cygnus, there is a strong X-ray source named Cygnus X-1. It is near a star, and the two revolve around each other. The unseen X-ray source has the gravitational pull of at least ten suns and is believed to be a black hole. Another type of black hole, a primordial black hole, may also exist dating from the time of the Big Bang, when regions of gas and dust were highly compressed. Recently, astronomers observed a brief pulse of X rays from Sagittarius A, a region near the center of the Milky Way galaxy. The origin of this pulse and its behavior led scientists to conclude that there is probably a black hole in the center of our galaxy.

There are four other possible black holes: a Schwarzschild black hole has no charge and no angular momentum; a Reissner-Nordstrom black hole has charge but

97

no angular momentum; a Kerr black hole has angular momentum but no charge; and a Kerr-Newman black hole has charge and angular momentum.

How far away is the **nearest black hole**?

The black hole nearest to Earth is V4641 Sagittarii. It is 1,600 light years (9,399,362, 677,500,000 miles) away from Earth.

What is a **pulsar**?

A pulsar is a rotating neutron star that gives off sharp regular pulses of radio waves at rates ranging from 0.001 to 4 seconds. Stars burn by fusing hydrogen into helium. When they use up their hydrogen, their interiors begin to contract. During this contraction, energy is released and the outer layers of the star are pushed out. These layers are large and cool; the star is now a red giant. A star with more than twice the mass of the sun will continue to expand, becoming a supergiant. At that point, it may blow up in an explosion called a supernova. After a supernova, the remaining material of the star's core may be so compressed that the electrons and protons become neutrons. A star 1.4 to 4 times the mass of the sun can be compressed into a neutron star only about 12 miles (20 kilometers) across. Neutron stars rotate very fast. The neutron star at the center of the Crab Nebula spins 30 times per second.

Some of these neutron stars emit radio signals from their magnetic poles in a direction that reaches Earth. These signals were first detected by Jocelyn Bell (1943–) of Cambridge University in 1967. Because of their regularity some people speculated that they were extraterrestrial beacons constructed by alien civilizations. This theory was eventually ruled out and the rotating neutron star came to be accepted as the explanation for these pulsating radio sources.

What does the **color** of a **star indicate**?

The color of a star gives an indication of its temperature and age. Stars are classified by their spectral type. From oldest to youngest and hottest to coolest, the types of stars are:

Type	Color	Temperature (°F)	Temperature (°C)
O	Blue	45,000–75,000	25,000–40,000
B	Blue	20,000–45,000	11,000–25,000
A	Blue-White	13,500–20,000	7,500–11,000
F	White	10,800–13,500	6,000–7,500
G	Yellow	9,000–10,800	5,000–6,000
K	Orange	6,300–9,000	3,500–5,000
M	Red	5,400–6,300	3,000–3,500

Each type is further subdivided on a scale of 0–9. The sun is a type G2 star.

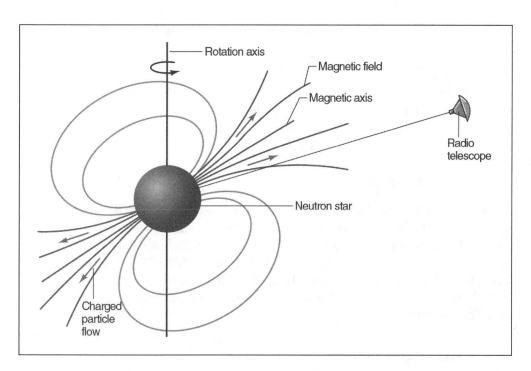

A rotating neutron star—or pulsar—emits radio waves that can be detected by radio telescopes on Earth.

What is the **most massive star**?

The Pistol Star is both the brightest and most massive star known. Located 25,000 light years away in the area of the constellation Sagittarius, this young (one- to three-million-year-old) star is as bright as ten million suns and may have weighed two hundred times the mass of the sun at one point in its young life.

Which **stars** are the **brightest**?

The brightness of a star is called its magnitude. Apparent magnitude is how bright a star appears to the naked eye. The lower the magnitude, the brighter the star. On a clear night, stars of about magnitude 16 can be seen with the naked eye. Large telescopes can detect objects as faint as 127. Very bright objects have negative magnitudes; the sun is –26.8.

Star	Constellation	Apparent Magnitude
Sirius	Canis Major	–1.47
Canopus	Carina	–0.72
Arcturus	Boötes	–0.06
Rigil Kentaurus	Centaurus	10.01

Star	Constellation	Apparent Magnitude
Vega	Lyra	10.04
Capella	Auriga	10.05
Rigel	Orion	10.14
Procyon	Canis Minor	10.37
Betelgeuse	Orion	10.41
Achernar	Eridanus	10.51

What is the **Milky Way**?

The Milky Way is a hazy band of light that can be seen encircling the night sky. This light comes from the stars that make up the Milky Way galaxy, the galaxy to which the sun and Earth belong. Galaxies are huge systems of stars separated from one another by largely empty space. Astronomers estimate that the Milky Way galaxy contains at least 100 billion stars and is about 100,000 light years in diameter. The galaxy is shaped like a compact disk with a central bulge, or nucleus, and spiral arms curving out from the center.

Which **galaxy** is **closest** to us?

The Andromeda galaxy is the galaxy closest to the Milky Way galaxy, where Earth is located. It is estimated to be 2.2 million light years away from Earth. Bigger than the Milky Way, Andromeda is a spiral-shaped galaxy that is also the brightest in Earth's sky.

What is the **Big Dipper**?

The Big Dipper is a group of seven stars that are part of the constellation Ursa Major. They appear to form a sort of bowl, composed of four stars, with a long handle, composed of three stars. The group is known as the Plough in Great Britain. The Big Dipper is almost always visible in the Northern Hemisphere. It serves as a convenient reference point when locating other stars; for example, an imaginary line drawn from the two end stars of the dipper leads to Polaris, the North Star.

Which **well-known star** is part of the **Little Dipper**?

The Little Dipper, part of the constellation Ursa Minor, is similar to the Big Dipper. It also has seven bright stars that form the shape of a ladle. Polaris, the North Star, is at the end of the handle of the Little Dipper.

Where is the **North Star**?

If an imaginary line is drawn from the North Pole into space, it will reach a star called Polaris, or the North Star, less than one degree away from the line. As Earth rotates on its axis, Polaris acts as a pivot-point around which all the stars visible in the Northern Hemisphere appear to move, while Polaris itself remains motionless. Identifying Polaris was important for navigation since in locating Polaris it was possible to identify north. In addition, the angle of Polaris above the horizon indicates latitude on Earth.

Was **Polaris** always the **North Star**?

Earth has had several North Stars. Earth slowly wobbles on its axis as it spins. This motion is called precession. Earth traces a circle in the sky over a period of 26,000 years. In Pharaonic times the North Star was Thuban; today it is Polaris; around 14,000 C.E. it will be Vega.

How **many constellations** are there and how were they named?

Constellations are groups of stars that seem to form some particular shape, such as that of a person, animal, or object. They only appear to form this shape and be close to each other from Earth; in actuality, the stars in a constellation are often very distant from each other. There are 88 recognized constellations whose boundaries were defined in the 1920s by the International Astronomical Union.

Various cultures in all parts of the world have had their own constellations. However, because modern science is predominantly a product of Western culture, many of the constellations represent characters from Greek and Roman mythology. When Europeans began to explore the Southern Hemisphere in the sixteenth and seventeeth centuries, they derived some of the new star patterns from the technological wonders of their time, such as the telescope.

Names of constellations are usually given in Latin. Individual stars in a constellation are usually designated with Greek letters in the order of brightness; the brightest star is alpha, the second brightest is beta, and so on. The genitive, or possessive, form of the constellation name is used, thus Alpha Orionis is the brightest star of the constellation Orion.

Constellation	Genitive	Abbreviation	Meaning
Andromeda	Andromedae	And	Chained Maiden
Antlia	Antliae	Ant	Air Pump
Apus	Apodis	Aps	Bird of Paradise
Aquarius	Aquarii	Aqr	Water Bearer
Aquila	Aquilae	Aql	Eagle
Ara	Arae	Ara	Altar
Aries	Arietis	Ari	Ram
Auriga	Aurigae	Aur	Charioteer
Boötes	Boötis	Boo	Herdsman
Caelum	Caeli	Cae	Chisel

101

Constellation	Genitive	Abbreviation	Meaning
Camelopardalis	Camelopardalis	Cam	Giraffe
Cancer	Cancri	Cnc	Crab
Canes Venatici	Canum Venaticorum	CVn	Hunting Dogs
Canis Major	Canis Majoris	CMa	Big Dog
Canis Minor	Canis Minoris	CMi	Little Dog
Capricornus	Capricorni	Cap	Goat
Carina	Carinae	Car	Ship's Keel
Cassiopeia	Cassiopeiae	Cas	Queen of Ethiopia
Centaurus	Centauri	Cen	Centaur
Cepheus	Cephei	Cep	King of Ethiopia
Cetus	Ceti	Cet	Whale
Chamaeleon	Chamaeleontis	Cha	Chameleon
Circinus	Circini	Cir	Compass
Columba	Columbae	Col	Dove
Coma Berenices	Comae Berenices	Com	Berenice's Hair
Corona Australis	Coronae Australis	CrA	Southern Crown
Corona Borealis	Coronae Borealis	CrB	Northern Crown
Corvus	Corvi	Crv	Crow
Crater	Crateris	Crt	Cup
Crux	Crucis	Cru	Southern Cross
Cygnus	Cygni	Cyg	Swan
Delphinus	Delphini	Del	Dolphin
Dorado	Doradus	Dor	Goldfish
Draco	Draconis	Dra	Dragon
Equuleus	Equulei	Equ	Little Horse
Eridanus	Eridani	Eri	River Eridanus
Fornax	Fornacis	For	Furnace
Gemini	Geminorum	Gem	Twins
Grus	Gruis	Gru	Crane
Hercules	Herculis	Her	Hercules
Horologium	Horologii	Hor	Clock
Hydra	Hydrae	Hya	Hydra, Greek monster
Hydrus	Hydri	Hyi	Sea Serpent
Indus	Indi	Ind	Indian
Lacerta	Lacertae	Lac	Lizard
Leo	Leonis	Leo	Lion
Leo Minor	Leonis Minoris	LMi	Little Lion
Lepus	Leporis	Lep	Hare

Constellation	Genitive	Abbreviation	Meaning
Libra	Librae	Lib	Scales
Lupus	Lupi	Lup	Wolf
Lynx	Lyncis	Lyn	Lynx
Lyra	Lyrae	Lyr	Lyre or Harp
Mensa	Mensae	Men	Table Mountain
Microscopium	Microscopii	Mic	Microscope
Monoceros	Monocerotis	Mon	Unicorn
Musca	Muscae	Mus	Fly
Norma	Normae	Nor	Carpenter's Square
Octans	Octantis	Oct	Octant
Ophiuchus	Ophiuchi	Oph	Serpent Bearer
Orion	Orionis	Ori	Orion, the Hunter
Pavo	Pavonis	Pav	Peacock
Pegasus	Pegasi	Peg	Winged Horse
Perseus	Persei	Per	Perseus, a Greek hero
Phoenix	Phoenicis	Phe	Phoenix
Pictor	Pictoris	Pic	Painter
Pisces	Piscium	Psc	Fish
Piscis Austrinus	Piscis Austrini	PsA	Southern Fish
Puppis	Puppis	Pup	Ship's Stern
Pyxis	Pyxidis	Pyx	Ship's Compass
Reticulum	Reticuli	Ret	Net
Sagitta	Sagittae	Sge	Arrow
Sagittarius	Sagittarii	Sgr	Archer
Scorpius	Scorpii	Sco	Scorpion
Sculptor	Sculptoris	Scl	Sculptor
Scutum	Scuti	Sct	Shield
Serpens	Serpentis	Ser	Serpent
Sextans	Sextantis	Sex	Sextant
Taurus	Tauri	Tau	Bull
Telescopium	Telescopii	Tel	Telescope
Triangulum	Trianguli	Tri	Triangle
Triangulum Australe	Trianguli Australis	TrA	Southern Australe Triangle
Tucana	Tucanae	Tuc	Toucan
Ursa Major	Ursae Majoris	UMa	Big Bear
Ursa Minor	Ursae Minoris	UMi	Little Bear
Vela	Velorum	Vel	Ship's Sail
Virgo	Virginis	Vir	Virgin
Volans	Volantis	Vol	Flying Fish
Vulpecula	Vulpeculae	Vul	Little Fox

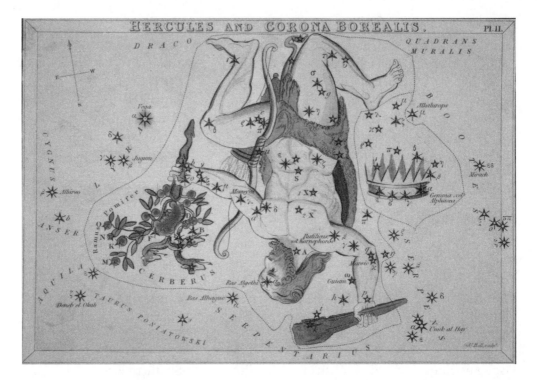

People in ancient times saw patterns in the stars in the night sky and named them afer animals and mythological creatures and gods such as the Roman demigod Hercules.

What is the **largest constellation**?

Hydra is the largest constellation, extending from Gemini to the south of Virgo. It has a recognizable long line of stars. The name "hydra" is derived from the watersnake monster killed by Hercules in ancient mythology.

Which **star** is the **closest** to **Earth**?

The sun, at a distance of 92,955,900 miles (149,598,000 kilometers), is the closest star to Earth. After the sun, the closest stars are the members of the triple star system known as Alpha Centauri (Alpha Centauri A, Alpha Centauri B, and Alpha Centauri C, sometimes called Proxima Centauri). They are 4.3 light years away.

How **hot** is the **sun**?

The center of the sun is about 27,000,000°F (15,000,000°C). The surface, or photosphere, of the sun is about 10,000°F (5,500°C). Magnetic anomalies in the photosphere cause cooler regions that appear to be darker than the surrounding surface. These sunspots are about 6,700°F (4,000°C). The sun's layer of lower atmosphere, the chromosphere, is only a few thousand miles thick. At the base, the chromosphere is about 7,800°F (4,300°C), but its temperature rises with altitude to the corona, the sun's outer layer of atmosphere, which has a temperature of about 1,800,000°F (1,000,000°C).

What is the **sun made of**?

The sun is an incandescent ball of gases. Its mass is 1.8×1027 tons or 1.8 octillion tons (a mass 330,000 times as great as Earth).

Element	Percent of Mass
Hydrogen	73.46
Helium	24.85
Oxygen	0.77
Carbon	0.29
Iron	0.16
Neon	0.12
Nitrogen	0.09
Silicon	0.07
Magnesium	0.05
Sulfur	0.04
Other	0.10

What is the **ecliptic**?

Ecliptic refers to the apparent yearly path of the sun through the sky with respect to the stars. In the spring, the ecliptic in the Northern Hemisphere is angled high in the evening sky. In fall, the ecliptic lies much closer to the horizon.

Why does the **color** of the **sun vary**?

Sunlight contains all the colors of the rainbow, which blend to form white light, making sunlight appear white. At times, some of the color wavelengths, especially blue, become scattered in Earth's atmosphere and the sunlight appears colored. When the sun is high in the sky, some of the blue rays are scattered in Earth's atmosphere. At such times, the sky looks blue and the sun appears to be yellow. At sunrise or sunset, when the light must follow a longer path through Earth's atmosphere, the sun looks red (red having the longest wavelengths).

How long does it take **light** from the **sun** to **reach Earth**?

Sunlight takes about 8 minutes and 20 seconds to reach Earth, traveling at 186,282 miles (299,792 kilometers) per second, although it varies with the position of Earth in its orbit. In January the light takes about 8 minutes 15 seconds to reach Earth; in July the trip takes about 8 minutes 25 seconds.

When will the sun die?

The sun is approximately 4.5 billion years old. About 5 billion years from now, the sun will have burned all of its hydrogen fuel into helium. As this process occurs, the sun will change from the yellow dwarf as we know it to a red giant. Its diameter will extend well beyond the orbit of Venus, and even possibly beyond the orbit of Earth. In either case, Earth will be burned to a cinder.

How long is a **solar cycle**?

The solar cycle is the periodic change in the number of sunspots. The cycle is taken as the interval between successive minima and is about 11.1 years long. During an entire cycle, solar flares, sunspots, and other magnetic phenomena move from intense activity to relative calm and back again. The solar cycle is one area of study to be carried out by up to ten ATLAS space missions designed to probe the chemistry and physics of the atmosphere. These studies of the solar cycle will yield a more detailed picture of Earth's atmosphere and its response to changes in the sun.

What is a **solar flare**?

A solar flare is a sudden, intense release of energy in the sun's outer atmosphere. The magnetic energy of a solar flare is released in the form of radiation at speeds of one million miles per hour (1.6 kilometers per hour). The amount of energy released is equivalent to millions of 100-megaton hydrogen bombs exploding at the same time. The frequency of solar flares is coordinated with the sun's 11-year cycle. The number of solar flares is greatest at the peak of the solar cycle. Solar flares may be visible from Earth. The first solar flare observed and recorded was in 1859 by Richard C. Carrington (1826–1875) who saw a sudden white light while observing sunspots.

What is the **sunspot cycle**?

It is the fluctuating number of sunspots on the sun during an 11-year period. The variation in the number of sunspots seems to correspond with the increase or decrease in the number of solar flares. An increased number of sunspots means an increased number of solar flares.

What is **solar wind**?

Solar wind is caused by the expansion of gases in the sun's outermost atmosphere, the corona. Because of the corona's extremely high temperature of 1,800,000°F (1,000,000°C), the gases heat up and their atoms start to collide. The atoms lose electrons and become electrically charged ions. These ions create the solar wind. Solar wind has a velocity of 310 miles (500 kilometers) per second, and its density is approximately 82 ions per cubic inch (5 ions per cubic centimeter). Because Earth is surrounded by strong magnetic forces, its magnetosphere, it is protected from the solar wind particles. In 1959, the Soviet spacecraft *Luna 2* acknowledged the existence of solar wind and made the first measurements of its properties.

When do **solar eclipses** happen?

A solar eclipse occurs when the moon passes between Earth and the sun and all three bodies are aligned in the same plane. When the moon completely blocks Earth's view of the sun and the umbra, or dark part of the moon's shadow, reaches Earth, a total eclipse occurs. A total eclipse happens only along a narrow path 100 to 200 miles (160 to 320 kilometers) wide called the track of totality. Just before totality, the only parts of the sun that are visible are a few points of light called Baily's beads shining through valleys on the moon's surface. Sometimes, a last bright flash of sunlight is seen—the diamond ring effect. During totality, which averages 2.5 minutes but may last up to 7.5 minutes, the sky is dark and stars and other planets are easily seen. The corona, the sun's outer atmosphere, is also visible.

If the moon does not appear large enough in the sky to completely cover the sun, it appears silhouetted against the sun with a ring of sunlight showing around it. This is an annular eclipse. Because the sun is not completely covered, its corona cannot be seen, and although the sky may darken it will not be dark enough to see the stars.

During a partial eclipse of the sun, the penumbra of the moon's shadow strikes Earth. A partial eclipse can also be seen on either side of the track of totality of an annular or total eclipse. The moon will cover part of the sun and the sky will not darken noticeably during a partial eclipse.

What is the **safest way to view** a solar eclipse?

Punch a pinhole in an index card and hold it 2 to 3 feet (0.6 to 0.9 meters) in front of another index card. The eclipse can be viewed safely through the hole. Encase the index card contraption in a box, using aluminum foil with a pinhole, and you'll see a sharper image of the eclipse. You may also purchase special glasses with aluminized Mylar lenses. Damage to the retina can occur if the eclipse is viewed with other devices such as photographic filters, exposed film, smoked glass, camera lenses, telescopes, or binoculars.

When and where will the **next ten total solar eclipses** occur?

The next total solar eclipse in the United States, August 21, 2017, will sweep a path 70 miles (113 kilometers) wide from Salem, Oregon, to Charleston, South Carolina. Below is a list of the upcoming solar eclipses across the world.

November 13, 2012	Australia, Pacific Ocean
March 20, 2015	North Atlantic Ocean
March 9, 2016	Indonesia, North Pacific Ocean
August 21, 2017	United States
July 2, 2019	South Pacific, Chile, Argentina
December 14, 2020	South Pacific, Chile, Argentina, South Atlantic
December 4, 2021	Antarctica
April 8, 2024	Mexico, central United States, eastern Canada
August 12, 2026	Arctic, Greenland, Iceland, Spain
August 2, 2027	Morocco, Spain, Algeria, Libya, Egypt, Saudi Arabia, Yemen, Somalia

PLANETS AND MOONS

How far are the **planets from the sun**?

The planets revolve around the sun in elliptical orbits, with the sun at one focus of the ellipse. Thus, a planet is at times closer to the sun than at other times. The distances given below are the average distance from the sun, starting with Mercury, the planet closest to the sun, and moving outward.

Planet	Average distance (miles)	Average distance (km)
Mercury	35,983,000	57,909,100
Venus	67,237,700	108,208,600
Earth	92,955,900	149,598,000
Mars	141,634,800	227,939,200
Jupiter	483,612,200	778,298,400
Saturn	888,184,000	1,427,010,000
Uranus	1,782,000,000	2,869,600,000
Neptune	2,794,000,000	4,496,700,000

Which are the **hottest** and **coldest planets**?

The hottest planet is Venus. It has a surface temperature of 900° Fahrenehit (480° Celsius). The carbon dioxide of Venus's atmosphere traps the heat producing very high surface temperatures. Neptune is the coldest planet with temperatures of –367° Fahrenheit (–224°Celsius) or 51.7°Kelvin.

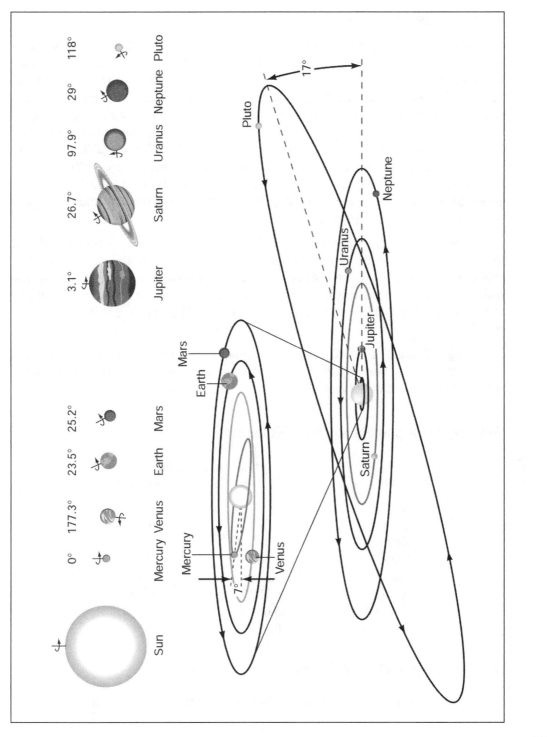

Our solar system.

Which **planets** have **rings**?

Jupiter, Saturn, Uranus, and Neptune all have rings. Jupiter's rings were discovered by *Voyager 1* in March 1979. The rings extend 80,240 miles (129,130 kilometers) from the center of the planet. They are about 4,300 miles (7,000 kilometers) in width and less than 20 miles (30 kilometers) thick. A faint inner ring is believed to extend to the edge of Jupiter's atmosphere. Saturn has the largest, most spectacular set of rings in the solar system. Saturn's ring system was first recognized by the Dutch astronomer Christiaan Huygens (1629–1695) in 1659. Its rings are 169,800 miles (273,200 kilometers) in diameter, but less than 10 miles (16 kilometers) thick. There are six different rings, the largest of which appears to be divided into thousands of ringlets. The rings appear to be composed of pieces of water ice ranging in size from tiny grains to blocks several tens of yards in diameter.

In 1977 when Uranus occulted (passed in front of) a star, scientists observed that the light from the star flickered or winked several times before the planet itself covered the star. The same flickering occurred in reverse order after the occultation. The reason for this was determined to be a ring around Uranus. Nine rings were initially identified, and *Voyager 2* observed two more in 1986. The rings are thin, narrow, and very dark.

Voyager 2 also discovered a series of at least four rings around Neptune in 1989. Some of the rings appear to have arcs, areas where there is a higher density of material than at other parts of the ring.

How long do the **planets** take to **orbit** the **sun**?

Planet	Period of Revolution	
	Earth Days	Earth Years
Mercury	88	0.24
Venus	224.7	0.62
Earth	365.26	1.00
Mars	687	1.88
Jupiter	4,332.6	11.86
Saturn	10,759.2	29.46
Uranus	30,685.4	84.01
Neptune	60,189	164.8

What are the **diameters** of the **planets**?

Planet	Diameter (Miles)*	Diameter (Kilometers)
Mercury	3,031	4,878
Venus	7,520	12,104

Planet	Diameter (Miles)*	Diameter (Kilometers)
Earth	7,926	12,756
Mars	4,221	6,794
Jupiter	88,846	142,984
Saturn	74,898	120,536
Uranus	31,763	51,118
Neptune	31,329	50,530

*All diameters are as measured at the planet's equator.

What are the **colors** of the **planets**?

Planet	Color
Mercury	Orange
Venus	Yellow
Earth	Blue, brown, green
Mars	Red
Jupiter	Yellow, red, brown, white
Saturn	Yellow
Uranus	Green
Neptune	Blue

What is the **gravitational force** on each of the **planets**, the **moon**, and the **sun** relative to **Earth**?

If the gravitational force on Earth is taken as 1, the comparative forces are:

Sun	27.9
Mercury	0.37
Venus	0.88
Moon	0.16
Mars	0.38
Jupiter	2.64
Saturn	1.15
Uranus	0.93
Neptune	1.22

Weight comparisons can be made by using this table. If a person weighed 100 pounds (45.36 kilograms) on Earth, then the weight of the person on the moon would be 16 pounds (7.26 kilograms) or 100 × 0.16.

Which planets are called **"inferior" planets** and which are **"superior" planets**?

An inferior planet is one whose orbit is nearer to the sun than Earth's orbit is. Mercury and Venus are the inferior planets. Superior planets are those whose orbits around the sun lie beyond that of Earth. Mars, Jupiter, Saturn, Uranus, and Neptune are the superior planets. The terms have nothing to do with the quality of an individual planet.

Is a **day** the same on all the **planets**?

No. A day, the period of time it takes for a planet to make one complete turn on its axis, varies from planet to planet. Venus and Uranus display retrograde motion; that is to say, they rotate in the opposite direction from the other planets. The table below lists the length of the day for each planet.

Length of day

Planet	Earth Days	Hours	Minutes
Mercury	58	15	30
Venus	243		32
Earth		23	56
Mars		24	37
Jupiter		9	50
Saturn		10	39
Uranus		17	14
Neptune		16	3

What are the **Jovian** and **terrestrial planets**?

Jupiter, Saturn, Uranus, and Neptune are the Jovian (the adjectival form for the word "Jupiter"), or Jupiter-like, planets. They are giant planets, composed primarily of light elements such as hydrogen and helium.

Mercury, Venus, Earth, and Mars are the terrestrial (derived from "terra," the Latin word for "earth"), or Earth-like, planets. They are small in size, have solid surfaces, and are composed of rocks and iron.

What is unique about the rotation of the planet Venus?

Unlike Earth and most of the other planets, Venus rotates in a retrograde, or opposite, direction with relation to its orbital motion around the sun. It rotates so slowly that only two sunrises and sunsets occur each Venusian year. Uranus's rotation is also retrograde.

Is it true that the **rotation speed** of **Earth varies**?

The rotation speed is at its maximum in late July and early August and at its minimum in April; the difference in the length of the day is about 0.0012 second. Since about 1900 Earth's rotation has been slowing at a rate of approximately 1.7 seconds per year. In the geologic past Earth's rotational period was much faster; days were shorter and there were more days in the year. About 350 million years ago, the year had 400 to 410 days; 280 million years ago, a year was 390 days long.

Is **Earth closer** to the **sun** in **winter** than in summer in the **Northern Hemisphere**?

Yes. However, Earth's axis, the line around which the planet rotates, is tipped 23.5° with respect to the plane of revolution around the sun. When Earth is closest to the sun (its perihelion, about January 3), the Northern Hemisphere is tilted away from the sun. This causes winter in the Northern Hemisphere while the Southern Hemisphere is having summer. When Earth is farthest from the sun (its aphelion, around July 4), the situation is reversed, with the Northern Hemisphere tilted toward the sun. At this time, it is summer in the Northern Hemisphere and winter in the Southern Hemisphere.

What is the **circumference** of **Earth**?

Earth is an oblate ellipsoid—a sphere slightly flattened at the poles and bulging at the equator. The distance around Earth at the equator is 24,902 miles (40,075 kilometers). The distance around Earth through the poles is 24,860 miles (40,008 kilometers).

What is the **precession** of the equinoxes?

The "precession of the equinoxes" is the 26,000-year circular movement of Earth's axis. It is caused by the bulging at the equator, which makes Earth's axis twist in such a way that the North and South Poles complete a circle every 26,000 years. Every year when the sun crosses the equator at the time of the equinox, it is in a slightly different position than the previous year. This movement proceeds eastward until a circle is completed.

113

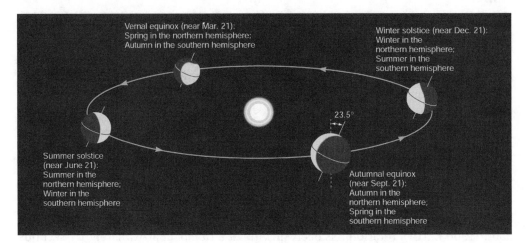

Vernal equinox (near Mar. 21):
Spring in the northern hemisphere;
Autumn in the southern hemisphere

Winter solstice (near Dec. 21):
Winter in the
northern hemisphere;
Summer in the
southern hemisphere

23.5°

Summer solstice
(near June 21):
Summer in the
northern hemisphere;
Winter in the
southern hemisphere

Autumnal equinox
(near Sept. 21):
Autumn in the
northern hemisphere;
Spring in the
southern hemisphere

The winter, vernal, summer, and autumnal equinoxes result from the earth's axis tilt and position around the sun.

Is **Pluto** a **planet**?

When Pluto was first discovered in 1930 by American astronomer Clyde Tombaugh (1906–1997), it was considered the ninth planet of our solar system. During the late 1990s and the beginning of the twenty-first century, astronomers began to discover more objects orbiting beyond Neptune in the area known as the transneptunian region. On the night of October 21, 2003, Mike Brown from Caltech, Chad Trujillo from the Gemini Observatory, and David Rabinowitz from Yale University discovered a new object more massive than Pluto with its own satellite. The Internaional Astronomical Union (IAU) began to debate the question of what constitutes a planet. In 2006, the IAU approved a new definition of a planet. This definition states that a planet is:

1. An object in orbit around the sun
2. An object with sufficient mass (large enough) to have its self-gravity pull itself into a round (or near spherical) shape
3. Has cleared the neighborhood around its orbit of other objects.

Pluto no longer meets the definition of a planet because of its size and it is in the transneptunian region, a zone of other similarly sized objects. Instead, Pluto is a dwarf planet.

How many **dwarf planets** are in the solar system?

A dwarf planet is an object in orbit around the sun that has sufficient mass (large enough) to have its own gravity pull itself into a round or near-round shape. There are currently five objects which are considered dwarf planets: Ceres, Pluto, Eris, Makemake, and Haumea. Scientists expect to discover additional dwarf planets or reclassify some large asteroids as dwarf planets.

What is a **small Solar System body**?

The International Astronomical Union defines a small Solar System body as all objects that orbit the sun and are too small for their own gravity to pull them into a nearly spherical shape. Simply stated, they are all objects that do not meet the definition of a planet or dwarf planet. Examples of small Solar System bodies are asteroids, near-Earth objects, Mars and Jupiter Trojan asteroids, most Centaurs, most Trans-Neptunian objects, and comets.

What are **Kuiper Belt Objects**?

Kuiper Belt Objects (KBOs) are, as their name implies, objects that originate from or orbit in the Kuiper Belt. Only one KBO was known for more than 60 years: Pluto. Many KBOs have been discovered since 1992, however, and the current estimate is that there are millions, if not billions, of KBOs.

KBOs are basically comets without tails: icy dirtballs that have collected together over billions of years. If they get large enough—such as Pluto did—they evolve as other massive planetlike bodies do, forming dense cores that have a different physical composition than the mantle or crust above it. Most short-period comets—those with relatively short orbital times of a few years to a few centuries—are thought to originate from the Kuiper Belt.

What are the largest **Kuiper Belt Objects** and how big are they?

The following table lists the largest KBOs in our solar system that are known of today.

Largest Kuiper Belt Objects

Name	Geometric Mean Diameter(km)
Eris	2,600
Pluto	2,390

115

Name	Geometric Mean Diameter(km)
Sedna	1,500
Quaoar	1,260
Charon	1,210
Orcus	940
Varuna	890
Ixion	820
Chaos	560
Huya	500

What does it mean when a **planet** is said to be **in opposition**?

A body in the solar system is in opposition when its longitude differs from the sun by 180°. In that position, it is exactly opposite the sun in the sky and it crosses the meridian at midnight.

How can an observer **distinguish planets** from **stars**?

In general, planets emit a constant light or shine, whereas stars appear to twinkle. The twinkling effect is caused by the combination of the distance between the stars and Earth and the refractive effect Earth's atmosphere has on a star's light. Planets are relatively closer to Earth than stars and their disklike shapes average out the twinkling effect, except when they're observed near Earth's horizon.

How **far** is the **moon** from Earth?

Since the moon's orbit is elliptical, its distance varies from about 221,463 miles (356,334 kilometers) at perigee (closest approach to Earth), to 251,968 miles (405,503 kilometers) at apogee (farthest point), with the average distance being 238,857 miles (384,392 kilometers).

How **many moons** does **each planet** have?

Planet	Number of Moons	Names of Some of the Moons
Mercury	0	
Venus	0	
Earth	1	The Moon (sometimes called Luna)
Mars	2	Phobos, Deimos
Jupiter	63	Metis, Adrastea, Amalthea, Thebe, Io, Europa, Ganymede, Callisto, Leda, Himalia, Lysithea, Elara, Ananke, Carme, Pasiphae, Sinope, Callirrhoe, Themisto, Megaclite, Taygete, Chaldene, Harpalyke, Kalkye

Planet	Number of Moons	Names of Some of the Moons
Saturn	61	Epimetheus, Janus, Mimas, Enceladus, Tethys, Telesto, Calypso, Dione, Helene, Rhea, Titan, Hyperion, Iapetus, Phoebe, Atlas, Prometheus, Pandora, Pan, Ymir, Paaliaq
Uranus	27	Cordelia, Ophelia, Bianca, Cressida, Desdemona, Juliet, Portia, Rosalind, Belinda, Puck, Miranda, Ariel, Umbriel, Titania, Oberon, Perdita, Cupid, Francisco, Stephano, Margaret
Neptune	13	Naiad, Thalassa, Despina, Galatea, Larissa, Proteus, Triton, Nereid, Halimede, Psamathe, Sao, Laomedeia, Neso

The four largest moons of Jupiter—Io, Europa, Calisto, and Ganymede—were discovered by Galileo in 1610.

Does the **moon** have an **atmosphere**?

The moon does have an atmosphere; however it is very slight, having a density of about 50 atoms per cubic centimeter. Recently, scientists have discovered frozen water on the moon.

What are the **diameter** and **circumference** of the **moon**?

The moon's diameter is 2,159 miles (3,475 kilometers) and its circumference is 6,790 miles (10,864 kilometers). The moon is 27 percent the size of Earth.

What are the **phases** of the **moon**?

The phases of the moon are changes in the moon's appearance during the month, which are caused by the moon's turning different portions of its illuminated hemisphere toward Earth. When the moon is between Earth and the sun, its daylight side is turned away from Earth, so it is not seen. This is called the new moon. As the moon continues its revolution around Earth, more and more of its surface becomes visible. This is called the waxing crescent phase. About a week after the new moon, half the moon is visible—the first quarter phase. During the next week, more than half of the moon is seen; this is called the waxing gibbous phase. Finally, about two weeks after the new moon, the moon and sun are on opposite sides of Earth.

The side of the moon facing the sun is also facing Earth, and all the moon's illuminated side is seen as a full moon. In the next two weeks the moon goes through the same phases, but in reverse from a waning gibbous to third or last quarter to waning crescent phase. Gradually, less and less of the moon is visible until a new moon occurs again.

What are moonquakes?

Similar to earthquakes, moonquakes are a result of the constant shifting of molten or partly molten material in the interior of the moon. These moonquakes are usually very weak. Other moonquakes may be caused by the impact of meteorites on the moon's surface. Still others occur at regular intervals during a lunar cycle, suggesting that gravitational forces from Earth have an effect on the moon similar to ocean tides.

Why does the **moon always** keep the same **face toward Earth**?

Only one side of the moon is seen because it always rotates in exactly the same length of time that it takes to revolve about Earth. This combination of motions (called "captured rotation") means that it always keeps the same side toward Earth.

What are the **names** of the **full moon** during each month?

Month	American Folk Name
January	Wolf Moon
February	Snow Moon
March	Sap Moon
April	Pink Moon
May	Flower Moon
June	Strawberry Moon
July	Buck Moon
August	Sturgeon Moon
September	Harvest Moon
October	Hunter Moon
November	Beaver Moon
December	Cold Moon

Is the moon really blue during a **blue moon**?

Although a bluish-looking moon can result from effects of Earth's atmosphere, the term "blue moon" does not refer to the color of the moon. For example, the phenomenon was widely observed in North America on September 26, 1950, due to Canadian forest fires that had scattered high-altitude dust, which refracted or absorbed certain wavelengths of light.

The popular definition of a "blue moon," is the second full moon in a single calendar month. Based on this definition, a blue moon occurs, on average, every 2.72 years.

Since 29.53 days pass between full moons (a synodial month), there is never a blue moon in February. On rare occasions, a blue moon can be seen twice in one year, but only in certain parts of the world. Blue moons will next occur:

- August 31, 2012
- July 31, 2015
- January 31, 2018
- March 31, 2018
- October 31, 2020
- August 31, 2023
- May 31, 2026
- December 31, 2028
- September 30, 2031
- July 31, 2034

However, this is a new definition of a blue moon based on an article that appeared in *Sky & Telescope* in 1946. The older definition of a blue moon, found in the *Maine Farmer's Almanac* dating back to 1819, describes a blue moon as the third full moon in a season of four full moons. To further complicate the calculation of a blue moon, astronomers define the start of a season based on the actual position of the sun throughout its annual orbit. The seasons will not be of equal length according to this definition. The *Maine Farmer's Almanac* preferred to define each season as being of equal length. Therefore, determining the occurrence of a blue moon based on the traditional definition will depend on whether the astronomical definition of a season or the *Maine Farmers Almanac* definition is used.

Upcoming Blue Moons (based on the traditional definition)

Astronomical Season	*Maine Farmer's Almanac* Season
August 21, 2013	August 21, 2013
May 21, 2016	May 21, 2016
May 18, 2019	February 19, 2019
August 22, 2021	November 19, 2021
August 19, 2024	August 19, 2024
May 20, 2027	February 20, 2027
August 24, 2029	November 21, 2029
August 21, 2032	August 21, 2032
May 22, 2035	May 22, 2035
May 18, 2038	February 19, 2038

The phases of a lunar eclipse.

Why do **lunar eclipses** happen?

A lunar eclipse occurs only during a full moon when the moon is on one side of Earth, the sun is on the opposite side, and all three bodies are aligned in the same plane. In this alignment Earth blocks the sun's rays to cast a shadow on the moon. In a total lunar eclipse the moon seems to disappear from the sky when the whole moon passes through the umbra, or total shadow, created by Earth. A total lunar eclipse may last up to one hour and 40 minutes. If only part of the moon enters the umbra, a partial eclipse occurs. A penumbral eclipse takes place if all or part of the moon passes through the penumbra (partial shadow or "shade") without touching the umbra. It is difficult to detect this type of eclipse from Earth. From the moon one could see that Earth blocked only part of the sun.

When and where will the **next ten total lunar eclipses** occur?

June 15, 2011	South America, Europe, Africa, Asia, Australia
December 10, 2011	Europe, East Africa, Asia, Australia, Pacific, North America
April 15, 2014	Australia, Pacific, Americas
October 8, 2014	Asia, Australia, Pacific, Americas
April 4, 2015	Asia, Australia, Pacific, Americas
September 28, 2015	East Pacific, Americas, Europe, Africa, West Asia

> ## What is the difference between a hunter's moon and a harvest moon?
>
> The harvest moon is the full moon nearest the autumnal equinox (on or about September 23). It is followed by a period of several successive days when the moon rises soon after sunset. In the Southern Hemisphere the harvest moon is the full moon closest to the vernal equinox (on or about March 21). This gives farmers extra hours of light for harvesting crops. The next full moon after the harvest moon is called the hunter's moon.

January 31, 2018	Europe, Africa, Asia, Australia
July 27, 2018	Asia, Australia, Pacific, western North America
January 21, 2019	South America, Europe, Africa, Asia, Australia
May 26, 2021	Asia, Australia, Pacific, Americas

What is the **moon's tail** that astronomers have discovered?

A glowing 15,000 mile (24,000 kilometer) long tail of sodium atoms streams from the moon. The faint, orange glow of sodium cannot be seen by the naked eye but it is detectable with the use of instruments. Astronomers are not certain of the source of these sodium atoms.

What is the **largest crater** on the **moon**?

The largest crater on the moon is Bailly. Its diameter is 184 miles (296 kilometers).

What are the **craters** on the moon that are **named** for the famous **Curie family**?

Curie—named for Pierre Curie (1859–1906), French chemist and Nobel prize winner.

Sklodowska—the family name of Marie Curie (1867–1934), French physical chemist and Nobel prize winner.

Joliot—named for physicist Frederic Joliot-Curie (1900–1958), Pierre and Marie's son-in-law and Nobel prize winner.

What is the **Genesis rock**?

The Genesis rock is a lunar rock brought to Earth by *Apollo 15*. It is approximately 4.15 billion years old, which is only 0.5 billion years younger than the generally accepted age of the moon.

COMETS AND METEORITES

Where are **asteroids found**?

The asteroids, also called the minor planets, are smaller than any of the eight major planets in the solar system and are not satellites of any major planet. The term asteroid means "starlike" because asteroids appear to be points of light when seen through a telescope.

Most asteroids are located between Mars and Jupiter, between 2.1 and 3.3 AUs (astronomical units) from the sun. Ceres, the largest and first to be discovered, was found by Giuseppe Piazzi on January 1, 1801, and has a diameter of 582 miles (936 kilometers). A second asteroid, Pallas, was discovered in 1802. Since then, astronomers have identified more than 18,000 asteroids and have established orbits for about 5,000 of them. Some of these have diameters of only 0.62 mile (1 kilometer). Originally, astronomers thought the asteroids were remnants of a planet that had been destroyed; now they believe asteroids to be material that never became a planet, possibly because it was affected by Jupiter's strong gravity.

Not all asteroids are in this main asteroid belt. Three groups of asteroids, the Near-Earth asteroids (NEAs), reside in the inner solar system. The Aten asteroids have orbits that lie primarily inside Earth's orbit. However, at their farthest point from the sun, these asteroids may cross Earth's orbit.

The Apollo asteroids cross Earth's orbit; some come even closer than the moon. The Amor asteroids cross the orbit of Mars, and some come close to Earth's orbit. The Trojan asteroids move in virtually the same orbit as Jupiter but at points 60° ahead or 60° behind the planet. In 1977 Charles Kowal discovered an object now known as Chiron orbiting between Saturn and Uranus. Originally cataloged as an asteroid, Chiron was later observed to have a coma (a gaseous halo), and it may be reclassified as a comet.

From where do **comets originate**?

According to a theory developed by Dutch astronomer Jan Oort (1900–1992), there is a large cloud, now called the Oort cloud, of gas, dust, and comets orbiting beyond Pluto out to perhaps 100,000 astronomical units (AU). Occasional stars passing close to this cloud disturb some of the comets from their orbits. Some fall inwards toward the sun.

Comets, sometimes called "dirty snowballs," are made up mostly of ice, with some dust mixed in. When a comet moves closer to the sun, the dust and ice of the core, or nucleus, heats up, producing a tail of material that trails along behind it. The tail is pushed out by the solar wind and almost always points away from the sun.

Most comets have highly elliptical orbits that carry them around the sun and then fling them back out to the outer reaches of the solar system, never to return. Occa-

What was the Tunguska Event?

On June 30, 1908, a violent explosion occurred in the atmosphere over the Podkamennaya Tunguska River in a remote part of central Siberia. The blast's consequences were similar to a hydrogen bomb going off, leveling thousands of square miles of forest. The shock of the explosion was heard more than 600 miles (960 kilometers) away. A number of theories have been proposed to account for this event.

Some people thought that a large meteorite or a piece of antimatter had fallen to Earth. But a meteorite, composed of rock and metal, would have created a crater and none was found at the impact site. There are no high radiation levels in the area that would have resulted from the collision of antimatter and matter. Two other theories include a mini-black hole striking Earth or the crash of an extraterrestrial spaceship. However, a mini-black hole would have passed through Earth and there is no record of a corresponding explosion on the other side of the world. As for the spaceship, no wreckage of such a craft was ever found.

The most likely cause of the explosion was the entry into the atmosphere of a piece of a comet, which would have produced a large fireball and blast wave. Since a comet is composed primarily of ice, the fragment would have melted during its passage through Earth's atmosphere, leaving no impact crater and no debris. Since the Tunguska Event coincided with Earth's passage through the orbit of Comet Encke, the explosion could have been caused by a piece of that comet.

sionally, however, a close passage by a comet near one of the planets can alter a comet's orbit, making it stay in the middle or inner solar system. Such a comet is called a short-period comet because it passes close to the sun at regular intervals. The most famous short-period comet is Comet Halley, which reaches perihelion (the point in its orbit that is closest to the sun) about every 76 years. Comet Encke, with an orbital period of 3.3 years, is another short-period comet.

When will **Halley's Comet return**?

Halley's Comet returns about every 76 years. It was most recently seen in 1986 and is predicted to appear again in 2061, then in 2134. Every appearance of what is now known as Comet Halley has been noted by astronomers since the year 239 B.C.E.

The comet is named for Edmund Halley (1656–1742), England's second Astronomer Royal. In 1682 he observed a bright comet and noted that it was moving in an orbit similar to comets seen in 1531 and 1607. He concluded that the three comets were actually one and the same and that the comet had an orbit of 76 years. In 1705

123

Halley published *A Synopsis of the Astronomy of Comets,* in which he predicted that the comet seen in 1531, 1607, and 1682 would return in 1758. On Christmas night, 1758, a German farmer and amateur astronomer named Johann Palitzsch spotted the comet in just the area of the sky that Halley had foretold.

Prior to Halley, comets appeared at irregular intervals and were often thought to be harbingers of disaster and signs of divine wrath. Halley proved that they are natural objects subject to the laws of gravity.

When was the comet **Hale-Bopp** first observed?

Comet Hale-Bopp was first observed by Alan Hale in New Mexico and Thomas Bopp in Arizona on July 22, 1995. Their discovery was announced by the International Astronomical Union on July 23, 1995. Hale-Bopp was closest to Earth in March 1997 when it was 122 million miles (196 million kilometers) away. It is a very large comet with a nucleus of approximately 25 miles (40 kilometers) in diameter, making it four times as large as Halley's comet. Although most comets have two tails, Hale-Bopp exhibited three tails. The two tails typical of most comets are the dust tail and the ion tail. The dust tail, consisting of dust and debris from the nucleus, streams behind the comet in its orbit. The ion tail, consisting of the comet's material interacting with the solar wind, faces away from the sun. Hale-Bopp's third tail was composed of neutral sodium atoms. Hale-Bopp was visible with the naked eye for nearly 19 months. It is not expected to return for 4,000 years.

How does a **meteorite differ** from a **meteoroid**?

A meteorite is a natural object of extraterrestrial origin that survives passage through Earth's atmosphere and hits Earth's surface. A meteorite is often confused with a meteoroid or a meteor. A meteoroid is a small object in outer space, generally less than 30 feet (10 meters) in diameter. A meteor (sometimes called a shooting star) is the flash of light seen when an object passes through Earth's atmosphere and burns as a result of heating caused by friction. A meteoroid becomes a meteor when it enters Earth's atmosphere; if any portion of a meteoroid lands on Earth, it is a meteorite.

There are three kinds of meteorites. Irons contain 85 percent to 95 percent iron; the rest of their mass is mostly nickel. Stony irons are relatively rare meteorites composed of about 50 percent iron and 50 percent silicates.

When do **meteor showers occur**?

There are a number of groups of meteoroids orbiting the sun just as Earth is. When Earth's orbit intercepts the path of one of these swarms of meteoroids, some of them enter Earth's atmosphere. When friction with the air causes a meteoroid to burn up, the streak, or shooting star, that is produced is called a meteor. Large numbers of meteors can produce a spectacular shower of light in the night sky. Meteor showers are named for the constellation that occupies the area of the sky from which they originate. Listed below are ten meteor showers and the dates during the year during which they can be seen.

Name of Shower	Dates
Quadrantids	January 1–6
Lyrids	April 19–24
Eta Aquarids	May 1–8
Perseids	July 25–August 18
Orionids	October 16–26
Taurids	October 20–November 20
Leonids	November 13–17
Phoenicids	December 4–5
Geminids	December 7–15
Ursids	December 17–24

How many **meteorites land** on **Earth** in a given **year**?

Approximately 26,000 meteorites, each weighing over 3.5 ounces (99.2 grams) land on Earth during a given year. Three thousand of these meteorites weigh more than 2.2 pounds (1 kg). This figure is compiled from the number of fireballs visually observed by the Canadian Camera Network. Of that number, only five or six falls are witnessed or cause property damage. The majority fall in the oceans, which cover over 70 percent of Earth's surface.

What are the **largest meteorites** that have been found in the world?

The famous Willamette (Oregon) iron, displayed at the American Museum of Natural History in New York, is the largest specimen found in the United States. It is 10 feet (3 meters) long and 5 feet (1.5 meters) high.

Name	Location	Weight (Tons/Tonnes)
Hoba West	Namibia	66.1/60
Ahnighito (The Tent)	Greenland	33.5/30.4
Bacuberito	Mexico	29.8/27
Mbosi	Tanzania	28.7/26
Agpalik	Greenland	22.2/20.1
Armanty	Outer Mongolia	22/20
Willamette	Oregon, USA	15.4/14
Chupaderos	Mexico	15.4/14
Campo del Cielo	Argentina	14.3/13

Name	Location	Weight (Tons/Tonnes)
Mundrabilla	Western Australia	13.2/12
Morito	Mexico	12.1/11

OBSERVATION AND MEASUREMENT

Who was the **first Astronomer Royal**?

The first Astronomer Royal was John Flamsteed (1646–1719). He was appointed Astronomer Royal in 1675 when the Royal Greenwich Observatory was founded. Until 1972, the Astronomer Royal also served as the Director of the Royal Greenwich Observatory.

Who is considered the **founder** of **systematic astronomy**?

The Greek scientist Hipparchus (c. 190–120 B.C.E.) is considered to be the father of systematic astronomy. He measured as accurately as possible the directions of objects in the sky. He compiled the first catalog of stars, containing about 850 entries, and designated each star's celestial coordinates, indicating its position in the sky. Hipparchus also divided the stars according to their apparent brightness or magnitudes.

What is a **light year**?

A light year is a measure of distance, not time. It is the distance that light, which travels in a vacuum at the rate of 186,282 miles (299,792 kilometers) per second, can travel in a year (365.25 days). This is equal to 5.87 trillion miles (9.46 trillion kilometers).

Besides the light year, what **other units** are used to **measure distances** in **astronomy**?

The astronomical unit (AU) is often used to measure distances within the solar system. One AU is equal to the average distance between Earth and the sun, or 92,955,630 miles (149,597,870 kilometers). The parsec is equal to 3.26 light years, or about 19.18 trillion miles (30.82 trillion kilometers).

How are new **celestial objects named**?

Many stars and planets have names that date back to antiquity. The International Astronomical Union (IAU), the professional astronomers organization, has attempted, in this century, to standardize names given to newly discovered celestial objects and their surface features.

Stars are generally called by their traditional names, most of which are of Greek, Roman, or Arabic origin. They are also identified by the constellation in which they appear, designated in order of brightness by Greek letters. Thus Sirius is also called alpha Canis Majoris, which means it is the brightest star in the constellation Canis Major. Other stars are called by catalog numbers, which include the star's coordinates. Several commercial star registries exist, and for a fee you can submit a star name to them. These names are not officially recognized by the IAU.

The IAU has made some recommendations for naming the surface features of the planets and their satellites. For example, features on Mercury are named for composers, poets, and writers; features of Venus for women; and features on Saturn's moon Mimas for people and places in Arthurian legend.

Comets are named for their discoverers. Newly discovered asteroids are first given a temporary designation consisting of the year of discovery plus two letters. The first letter indicates the half-month of discovery (A = first half of January, B = second half of January, etc.) and the second the order of discovery in that half-month. Thus asteroid 2002EM was the thirteenth (M) asteroid discovered in the first half of March (E) in 2002. After an asteroid's orbit is determined, it is given a permanent number and its discoverer is given the honor of naming it. Asteroids have been named after such diverse things as mythological figures (Ceres, Vesta), an airline (Swissair), and the Beatles (Lennon, McCartney, Harrison, Starr).

What is an **astrolabe**?

Invented by the Greeks or Alexandrians around 100 B.C.E., an astrolabe is a two-dimensional working model of the heavens, with sights for observations. It consists of two concentric, flat disks, one fixed, representing the observer on Earth, the other moving, which can be rotated to represent the appearance of the celestial sphere at a given moment. Given latitude, date, and time, the observer can read off the altitude and azimuth of the sun, the brightest stars, and the planets. By measuring the altitude of a particular body, one can find the time. The astrolabe can also be used to find times of sunrise, sunset, twilight, or the height of a tower or depth of a well. After 1600, it was replaced by the sextant and other more accurate instruments.

Who **invented** the **telescope**?

Hans Lippershey (c. 1570–1619), a German-Dutch lens grinder and spectacle maker, is generally credited with inventing the telescope in 1608 because he was the first scientist to apply for a patent. Two other inventors, Zacharias Janssen and Jacob Metius, also developed telescopes. Modern historians consider Lippershey and Janssen as the two likely candidates for the title of inventor of the telescope, with Lippershey possessing the strongest claim. Lippershey used his telescope for observing grounded objects from a distance.

In 1609, Galileo also developed his own refractor telescope for astronomical studies. Although small by today's standards, the telescope enabled Galileo to observe the Milky Way and to identify blemishes on the moon's surface as craters.

What are the differences between **reflecting** and **refracting telescopes**?

Reflecting telescopes capture light using a mirror while refracting telescopes capture light with a lens. The advantages of reflecting telescopes are: 1) they collect light with a mirror so there is no color fringing, and 2) since a mirror can be supported at the back there is no size limit. In an effort to alleviate the problem of color fringing always associated with lenses, Isaac Newton built a reflecting telescope in 1668 that collected light with mirrors.

For whom is the **Hubble Space Telescope named**?

Edwin Powell Hubble (1889–1953) was an American astronomer known for his studies of galaxies. His study of nebulae, or clouds—the faint, unresolved luminous patches in the sky—showed that some of them were large groups of many stars. Hubble classified galaxies by their shapes as being spiral, elliptical, or irregular.

Hubble's Law establishes a relationship between the velocity of recession of a galaxy and its distance. The speed at which a galaxy is moving away from our solar system (measured by its redshift, the shift of its light to longer wavelengths, presumed to be caused by the Doppler effect) is directly proportional to the galaxy's distance from it.

The Hubble Space Telescope was deployed by the space shuttle *Discovery* on April 25, 1990. The telescope, which would be free of distortions caused by Earth's atmosphere, was designed to see deeper into space than any telescope on land. However, on June 27, 1990, the National Aeronautics and Space Administration announced that the telescope had a defect in one of its mirrors that prevented it from properly focusing. Although other instruments, including one designed to make observations in ultraviolet light, were still operating, nearly 40 percent of the telescope's experiments had to be postponed until repairs were made. On December 2, 1993, astronauts were able to make the necessary repairs. Four of Hubble's six gyroscopes were replaced as well as two solar panels. Hubble's primary camera, which had a flawed mirror, was also replaced. Since that mission four other servicing missions have been conducted, dramatically improving the HST's capabilities.

What is the **Very Large Array** (VLA) and what information have we learned from it?

The Very Large Array (VLA) is one of the world's premier astronomical radio observatories. The VLA consists of 27 antennas arranged in a huge Y pattern up to 22 miles (36 kilometers) across—roughly one-and-a-half times the size of Washington, D.C. Each antenna is 81 feet (25 meters) in diameter; they are combined electronically to give the resolution of an antenna 22 miles (36 kilometers) across, with the sensitivity of a dish 422 feet (130 meters) in diameter. Each of the 27 radio telescopes in the VLA is the size of a house and can be moved on train tracks. In its twenty-second year of operation, the VLA has been one of the most productive observatories with more than 2,200 scientists using it for more than 10,000 separate observing projects. The VLA has been used to discover water on the planet Mercury, radio-bright coronae around ordinary stars, micro-quasars in our galaxy, gravitationally induced Einstein rings around distant galaxies, and radio counterparts to cosmologically distant gamma-ray bursts. The vast size of the VLA has allowed astronomers to study the details of super-fast cosmic jets, and even map the center of our galaxy.

EXPLORATION

Who are the **"fathers of space flight"**?

In 1903, Konstantin E. Tsiolkovsky (1857–1935), a Russian high school teacher, completed the first scientific paper on the use of rockets for space travel. Several years later, Robert H. Goddard (1882–1945) of the United States and Hermann Oberth (1894–1989) of Germany awakened wider scientific interest in space travel. These three men worked individually on many of the technical problems of rocketry and space travel. They are known, therefore, as the "fathers of space flight."

In 1919, Goddard wrote the paper, "A Method of Reaching Extreme Altitudes," which explained how rockets could be used to explore the upper atmosphere and described a way to send a rocket to the moon. During the 1920s Tsiolkovsky wrote a series of new studies that included detailed descriptions of multi-stage rockets. In 1923, Oberth wrote *The Rocket into Interplanetary Space*, which discussed the technical problems of space flight and also described what a spaceship would be like.

What is the difference between **zero-gravity** and **microgravity**?

Zero-gravity is the absence of gravity; a condition in which the effects of gravity are not felt; weightlessness. Microgravity is a condition of very low gravity, especially approaching weightlessness. On a spaceship, while in zero- or microgravity, objects would fall freely and float weightlessly. Both terms, however, are technically incorrect. The gravi-

An astronaut orbiting Earth actually experiences almost the same amount of gravitational pull as someone standing on the surface of the planet.

tation in orbit is only slightly less than the gravitation on Earth. A spacecraft and its contents continuously fall toward earth. It is the spacecraft's immense forward speed that appears to make Earth's surface curve away as the vehicle falls toward it. The continuous falling seems to eliminate the weight of everything inside the spacecraft. For this reason, the condition is sometimes referred to as weightlessness or zero-gravity.

What is meant by the phrase **"greening of the galaxy"**?

The expression means the spreading of human life, technology, and culture through interstellar space and eventually across the entire Milky Way galaxy, Earth's home galaxy.

What is a **"close encounter of the third kind"**?

UFO expert J. Allen Hynek (1910–1986) developed the following scale to describe encounters with extraterrestrial beings or vessels:

Close Encounter of the First Kind—sighting of a UFO at close range with no other physical evidence.

Close Encounter of the Second Kind—sighting of a UFO at close range, but with some kind of proof, such as a photograph, or an artifact from a UFO.

Close Encounter of the Third Kind—sighting of an actual extraterrestrial being.

Close Encounter of the Fourth Kind—abduction by an extraterrestrial spacecraft.

When was the Outer Space Treaty signed?

The United Nations Outer Space Treaty was signed on January 23, 1967. The treaty provides a framework for the exploration and sharing of outer space. It governs the outer space activities of nations that wish to exploit and make use of space, the moon, and other celestial bodies. It is based on a humanist and pacifist philosophy and on the principle of the nonappropriation of space and the freedom that all nations have to explore and use space. A very large number of countries have signed this agreement, including those from the Western alliance, the former Eastern bloc, and non-aligned countries.

Space law, or those rules governing the space activities of various countries, international organizations, and private industries, has been evolving since 1957, when the General Assembly of the United Nations created the Committee on the Peaceful Uses of Outer Space (COPUOS). One of its subcommittees was instrumental in drawing up the 1967 Outer Space Treaty.

Who was the **first man** in **space**?

Yuri Gagarin (1934–1968), a Soviet cosmonaut, became the first man in space when he made a full orbit of Earth in *Vostok I* on April 12, 1961. Gagarin's flight lasted only 1 hour and 48 minutes, but as the first man in space, he became an international hero. Partly because of this Soviet success, U.S. president John F. Kennedy (1917–1963) announced on May 25, 1961, that the United States would land a man on the moon before the end of the decade. The United States took its first step toward that goal when it launched the first American into orbit on February 20, 1962. Astronaut John H. Glenn Jr. (1921–) completed three orbits in *Friendship 7* and traveled about 81,000 miles (130,329 kilometers). Prior to this, on May 5, 1961, Alan B. Shepard Jr. (1923–1998) became the first American to pilot a spaceflight, aboard *Freedom 7*. This suborbital flight reached an altitude of 116.5 miles (187.45 kilometers).

What did NASA mean when it said *Voyager 1* and *2* would take a **"grand tour"** of the planets?

Once every 176 years the giant outer planets—Jupiter, Saturn, Uranus, and Neptune—align themselves in such a pattern that a spacecraft launched from Earth to Jupiter at just the right time might be able to visit the other three planets on the same mission. A technique called "gravity assist" used each planet's gravity as a power boost to point *Voyager* toward the next planet. The first opportune year for the "grand tour" was 1977.

What is the **message** attached to the *Voyager* spacecraft?

Voyager 1 (launched September 5, 1977) and *Voyager 2* (launched August 20, 1977) were unmanned space probes designed to explore the outer planets and then travel out of the solar system. A gold-coated copper phonograph record containing a message to any possible extraterrestrial civilization that they might encounter is attached to each spacecraft. The record contains both video and audio images of Earth and the civilization that sent this message to the stars.

The record begins with 118 pictures. These show Earth's position in the galaxy; a key to the mathematical notation used in other pictures; the sun; other planets in the solar system; human anatomy and reproduction; various types of terrain (seashore, desert, mountains); examples of vegetation and animal life; people of both sexes and of all ages and ethnic types engaged in a number of activities; structures (from grass huts to the Taj Mahal to the Sydney Opera House) showing diverse architectural styles; and means of transportation, including roads, bridges, cars, planes, and space vehicles.

The pictures are followed by greetings from Jimmy Carter, who was then president of the United States, and Kurt Waldheim, then secretary general of the United Nations. Brief messages in 54 languages, ranging from ancient Sumerian to English, are included, as is a "song" of the humpback whale.

The next section is a series of sounds common to Earth. These include thunder, rain, wind, fire, barking dogs, footsteps, laughter, human speech, the cry of an infant, and the sounds of a human heartbeat and human brainwaves.

The record concludes with approximately 90 minutes of music: "Earth's Greatest Hits." These musical selections were drawn from a broad spectrum of cultures and include such diverse pieces as a Pygmy girl's initiation song; bagpipe music from Azerbaijan; the Fifth Symphony, First Movement by Ludwig von Beethoven; and "Johnny B. Goode" by Chuck Berry.

It will be tens, or even hundreds of thousands of years before either Voyager comes close to another star, and perhaps the message will never be heard; but it is a sign of humanity's hope to encounter life elsewhere in the universe.

Which **astronauts** have **walked** on the **moon**?

Twelve astronauts have walked on the moon. Each Apollo flight had a crew of three. One crew member remained in orbit in the command service module (CSM) while the other two actually landed on the moon.

Apollo 11, July 16–24, 1969
- Neil A. Armstrong
- Edwin E. Aldrin Jr.
- Michael Collins (CSM pilot, did not walk on the moon)

Apollo 12, November 14–24, 1969

• Charles P. Conrad

• Alan L. Bean

• Richard F. Gordon Jr. (CSM pilot, did not walk on the moon)

Apollo 14, January 31–February 9, 1971

• Alan B. Shepard Jr.

• Edgar D. Mitchell

• Stuart A. Roosa (CSM pilot, did not walk on the moon)

Apollo 15, July 26–August 7, 1971

• David R. Scott

• James B. Irwin

• Alfred M. Worden (CSM pilot, did not walk on the moon)

Apollo 16, April 16–27, 1972

• John W. Young

• Charles M. Duke Jr.

• Thomas K. Mattingly, II (CSM pilot, did not walk on the moon)

Apollo 17, December 7–19, 1972

• Eugene A. Cernan

• Harrison H. Schmitt

• Ronald E. Evans (CSM pilot, did not walk on the moon)

Which **manned space flight** was the **longest**?

Dr. Valerij Polyakov manned a flight to the space station *Mir* on January 8, 1994. He returned aboard *Soyuz TM-20* on March 22, 1995, making the total time in space equal 438 days and 18 hours.

When and what was the **first animal** sent into **orbit**?

A small female dog named Laika, aboard the Soviet *Sputnik 2*, launched November 3, 1957, was the first animal sent into orbit. This event followed the successful Soviet launch on October 4, 1957, of *Sputnik 1*, the first man-made satellite ever placed in orbit. Laika was placed in a pressurized compartment within a capsule that weighed 1,103 pounds (500 kilograms). After a few days in orbit, she died, and *Sputnik 2* reentered Earth's atmosphere on April 14, 1958. Some sources list the dog as a Russian samoyed laika named "Kudyavka" or "Limonchik."

What were the **first monkeys** and **chimpanzees** in space?

On a United States *Jupiter* flight on December 12, 1958, a squirrel monkey named Old Reliable was sent into space, but not into orbit. The monkey drowned during recovery.

On another *Jupiter* flight, on May 28, 1959, two female monkeys were sent 300 miles (482.7 kilometers) high. Able was a 6-pound (2.7-kilogram) rhesus monkey and Baker was an 11-ounce (0.3-kilogram) squirrel monkey. Both were recovered alive.

A chimpanzee named Ham was used on a *Mercury* flight on January 31, 1961. Ham was launched to a height of 157 miles (253 kilometers) into space but did not go into orbit. His capsule reached a maximum speed of 5,857 miles (9,426 kilometers) per hour and landed 422 miles (679 kilometers) downrange in the Atlantic Ocean, where he was recovered unharmed.

On November 29, 1961, the United States placed a chimpanzee named Enos into orbit and recovered him alive after two complete orbits around Earth. Like the Soviets, who usually used dogs, the United States had to obtain information on the effects of space flight on living beings before they could actually launch a human into space.

Who was the **first woman** in **space**?

Valentina V. Tereshkova-Nikolaeva (1937–), a Soviet cosmonaut, was the first woman in space. She was aboard the *Vostok 6*, launched June 16, 1963. She spent three days circling Earth, completing 48 orbits. Although she had little cosmonaut training, she was an accomplished parachutist and was especially fit for the rigors of space travel.

The U.S. space program did not put a woman in space until 20 years later when, on June 18, 1983, Sally K. Ride (1951–) flew aboard the space shuttle *Challenger* mission STS-7. In 1987, she moved to the administrative side of NASA and was instrumental in issuing the "Ride Report," which recommended future missions and direction for NASA. She retired from NASA in August 1987 to become a research fellow at Stanford University after serving on the presidential commission that investigated the *Challenger* disaster in 1986. She was a physics professor at the University of California San Diego until 2001 when she founded Sally Ride Science. The company is dedicated to supporting girls' and boys' interests in science, math, and technology by showing science is fun with a variety of programs.

Sally K. Ride was the first American woman in space. She flew on the space shuttle *Challenger* in 1983.

What were the first words spoken by an astronaut after touchdown of the lunar module on the *Apollo 11* flight, and by an astronaut standing on the moon?

On July 20, 1969, at 4:17:43 P.M. Eastern Daylight Time (20:17:43 Greenwich Mean Time), Neil A. Armstrong (1930–) and Edwin E. Aldrin Jr. (1930–) landed the lunar module *Eagle* on the moon's Sea of Tranquility, and Armstrong radioed: "Houston, Tranquility Base here. The *Eagle* has landed." Several hours later, when Armstrong descended the lunar module ladder and made the small jump between the *Eagle* and the lunar surface, he announced: "That's one small step for man, one giant leap for mankind." The article "a" was missing in the live voice transmission, and was later inserted in the record to amend the message to "one small step for a man."

Who were the **first man** and **woman** to **walk in space**?

On March 18, 1965, the Soviet cosmonaut Alexei Leonov (1934–) became the first person to walk in space when he spent ten minutes outside his *Voskhod 2* spacecraft. The first woman to walk in space was Soviet cosmonaut Svetlana Savitskaya (1947–), who, during her second flight aboard the *Soyuz T-12* (July 17, 1984), performed 3.5 hours of extravehicular activity.

The first American to walk in space was Edward White II (1930–1967) from the spacecraft *Gemini 4* on June 3, 1965. White spent 22 minutes floating free attached to the *Gemini* by a lifeline. The photos of White floating in space are perhaps some of the most familiar of all space shots. Kathryn D. Sullivan (1951–) became the first American woman to walk in space when she spent 3.5 hours outside the *Challenger* orbiter during the space shuttle mission 41G on October 11, 1984.

American astronaut Bruce McCandless II (1937–) performed the first untethered space walk from the space shuttle *Challenger* on February 7, 1984, using an MMU (manual maneuvering unit) backpack.

What are some of the **accomplishments** of **female astronauts**?

- First American woman in space: Sally K. Ride—June 18, 1983, aboard *Challenger* STS-7.
- First American woman to walk in space: Kathryn D. Sullivan—October 11, 1984, aboard *Challenger* STS 41G.
- First woman to make three spaceflights: Shannon W. Lucid—June 17, 1985; October 18, 1989; and August 2, 1991.

135

- First African American woman in space: Mae Carol Jemison—September 12, 1992, aboard *Endeavour*.
- First American woman space shuttle pilot: Eileen M. Collins—February 3, 1995, aboard *Discovery*.

What **material** was used in the **U.S. flag** planted on the **moon** by astronauts Neil Armstrong and Edwin Aldrin Jr.?

The astronauts erected a 3-by-5 foot (0.9 by 1.5 meter) nylon U.S. flag, its top edge braced by a spring wire to keep it extended.

What was the **first meal** on the **moon**?

American astronauts Neil A. Armstrong and Edwin E. Aldrin Jr. ate four bacon squares, three sugar cookies, peaches, pineapple-grapefruit drink, and coffee before their historic moonwalk on July 20, 1969.

Who made the **first golf shot** on the moon?

Alan B. Shepard Jr., commander of *Apollo 14*, launched on January 31, 1971, made the first golf shot. He attached a six iron to the handle of the contingency sample return container, dropped a golf ball on the moon, and took a couple of one-handed swings. He missed with the first, but connected with the second. The ball, he reported, sailed for miles and miles.

Who was the **first African American** in space?

Guion S. Bluford Jr. (1942–), became the first African American to fly in space during the space shuttle *Challenger* mission STS-8 (August 30–September 5, 1983). Astronaut Bluford, who holds a Ph.D. in aerospace engineering, made a second shuttle flight aboard *Challenger* mission STS-61-A/Spacelab D1 (October 30–November 6, 1985). The first black man to fly in space was Cuban cosmonaut Arnaldo Tamayo-Mendez, who was aboard *Soyuz 38* and spent eight days aboard the Soviet space station *Salyut 6* during September 1980. Dr. Mae C. Jemison became the first African American woman in space on September 12, 1992 aboard the space shuttle *Endeavour* mission Spacelab-J.

Who were the **first married couple** to go into space together?

Astronauts Jan Davis and Mark Lee were the first married couple in space. They flew aboard the space shuttle *Endeavor* on an eight-day mission that began on September 12, 1992. Ordinarily, NASA bars married couples from flying together. An exception was made for Davis and Lee because they had no children and had begun training for the mission long before they got married.

Who was the first space tourist?

The first space tourist was Dennis Tito, an American businessman who paid an estimated $20 million to travel to space with a Russian crew to spend time on the International Space Station. He departed aboard the Russian *Soyuz* spacecraft on April 28, 2001 and returned to Earth on May 6, 2001.

Who has spent the **most time in space**?

The cosmonaut, Sergei Krikalev (1958–), has accumulated the most spaceflight time—803 days, 9 hours and 39 minutes in six flights. The cosmonaut, Valeri Polyakov (1942–), has spent 438 consecutive days in space from January 8, 1994 to March 22, 1995.

When was the **first U.S. satellite** launched?

Explorer 1, launched January 31, 1958, by the U.S. Army, was the first United States satellite launched into orbit. This 31-pound (14.06-kilogram) satellite carried instrumentation that led to the discovery of Earth's radiation belts, which would be named after University of Iowa scientist James A. Van Allen. It followed four months after the launching of the world's first satellite, the Soviet Union's *Sputnik 1*. On October 3, 1957, the Soviet Union placed the large 184-pound (83.5-kilogram) satellite into low Earth orbit. It carried instrumentation to study the density and temperature of the upper atmosphere, and its launch was the event that opened the space age.

What was the **mission** of the *Galileo* spacecraft?

Galileo, launched October 18, 1989, required almost six years to reach Jupiter after looping past Venus once and Earth twice. The *Galileo* spacecraft was designed to make a detailed study of Jupiter and its rings and moons over a period of years. On December 7, 1995, it released a probe to analyze the different layers of Jupiter's atmosphere. *Galileo* recorded a multitude of measurements of the planet, its four largest moons, and its mammoth magnetic field. The mission was originally scheduled to continue until the end of 1997, but, since it continued to operate successfully, missions exploring Jupiter's moons were added in 1997, 1999, and 2001. *Galileo* ended on September 21, 2003, when it passed into Jupiter's shadow and disintegrated in the planet's dense atmosphere.

Who was the **founder** of the **Soviet space program**?

Sergei P. Korolev (1907–1966) made enormous contributions to the development of Soviet manned space flight, and his name is linked with their most significant space

The *Galileo* probe made an extensive survey of Jupiter and the Galilean moons.

achievements. Trained as an aeronautical engineer, he directed the Moscow group studying the principles of rocket propulsion, and in 1946 took over the Soviet program to develop long-range ballistic rockets. Under Korolev, the Soviets used these rockets for space projects and launched the world's first satellite on October 4, 1957. Besides a vigorous, unmanned, interplanetary research program, Korolev's goal was to place men in space, and following tests with animals his manned space flight program was initiated when Yuri Gagarin was successfully launched into Earth's orbit.

How many **fatalities** have occurred during **space-related missions**?

The 14 astronauts and cosmonauts listed below died in space-related accidents.

Date	Astronaut/Cosmonaut	Mission
January 27, 1967	Roger Chaffee (U.S.)	Apollo 1
January 27, 1967	Edward White II (U.S.)	Apollo 1
January 27, 1967	Virgil "Gus" Grissom (U.S.)	Apollo 1
April 24, 1967	Vladimir Komarov (U.S.S.R.)	Soyuz 1
June 29, 1971	Viktor Patsayev (U.S.S.R.)	Soyuz 11
June 29, 1971	Vladislav Volkov (U.S.S.R.)	Soyuz 11
June 29, 1971	Georgi Dobrovolsky (U.S.S.R.)	Soyuz 11
January 28, 1986	Gregory Jarvis (U.S.)	STS 51L
January 28, 1986	Christa McAuliffe (U.S.)	STS 51L
January 28, 1986	Ronald McNair (U.S.)	STS 51L
January 28, 1986	Ellison Onizuka (U.S.)	STS 51L
January 28, 1986	Judith Resnik (U.S.)	STS 51L
January 28, 1986	Francis Scobee (U.S.)	STS 51L
January 28, 1986	Michael Smith (U.S.)	STS 51L
February 1, 2003	Rick D. Husband (U.S.)	STS 107
February 1, 2003	William C. McCool (U.S.)	STS 107
February 1, 2003	Michael P. Anderson (U.S.)	STS 107
February 1, 2003	Kalpana Chawla (U.S.)	STS 107
February 1, 2003	David M. Brown (U.S.)	STS 107

Date	Astronaut/Cosmonaut	Mission
February 1, 2003	Laurel Blair Salton Clark (U.S.)	STS 107
February 1, 2003	Ilan Ramon (Israel)	STS 107

Chaffee, Grissom, and White died in a cabin fire during a test firing of the *Apollo 1* rocket. Komarov was killed in *Soyuz 1* when the capsule's parachute failed. Dobrovolsky, Patsayev, and Volkov were killed during the *Soyuz II*'s re-entry when a valve accidently opened and released their capsule's atmosphere. Jarvis, McAuliffe, McNair, Onizuka, Resnik, Scobee, and Smith died when the space shuttle *Challenger* STS 51L exploded 73 seconds after liftoff. Husband, McCool, Anderson, Chawla, Brown, Clark, and Ramon died when the space shuttle *Columbia* disintegrated over Texas upon re-entry.

In addition, 19 other astronauts and cosmonauts have died of non-space related causes. Fourteen of these died in air crashes, four died of natural causes, and one died in an auto crash.

What was **one** of the **worst disasters** in the **U.S. space program** and what caused it?

Challenger mission STS 51L was launched on January 28, 1986, but exploded 73 seconds after liftoff. The entire crew of seven was killed, and the *Challenger* was completely destroyed. The investigation of the *Challenger* tragedy was performed by the Rogers Commission, established and named for its chairman, former secretary of state, William Rogers.

The consensus of the Rogers Commission (which studied the accident for several months) and participating investigative agencies was that the accident was caused by a failure in the joint between the two lower segments of the right solid rocket motor. The specific failure was the destruction of the seals that are intended to prevent hot gases from leaking through the joint during the propellant burn of the rocket motor. The evidence assembled by the commission indicated that no other element of the space shuttle system contributed to this failure.

Although the commission did not affix blame to any individuals, the public record made clear that the launch should not have been made that day. The weather was unusually cold at Cape Canaveral, and temperatures had dipped below freezing during the night. Test data had suggested that the seals (called O-rings) around the solid rocket booster joints lost much of their effectiveness in very cold weather.

What is the **composition** of the **tiles** on the underside of the **space shuttle** and how hot do they get?

The 20,000 tiles are composed of a low-density, high purity silica fiber insulator hardened by ceramic bonding. Bonded to a Nomex fiber felt pad, each tile is directly bond-

139

ed to the shuttle exterior. The maximum surface temperature can reach up to 922K to 978K (649°C to 704°C or 1,200°F to 1,300°F).

What are the **liquid fuels** used by the **space shuttles**?

Liquid hydrogen is used as a fuel, with liquid oxygen used to burn it. These two fuels are stored in chambers separately and then mixed to combust the two. Because oxygen must be kept below −183°C to remain a liquid, and hydrogen must be at −253°C to remain a liquid, they are both difficult to handle, but make useful rocket fuel.

What was the **cause** of the *Columbia* space shuttle **disaster**?

The *Columbia* space shuttle was launched on January 16, 2003 on mission STS-107. The mission was a research mission and the crew had many science experiments ranging from plant growth to a cancer drug study to studying the effects of microgravity on the cardiovascular system. The *Columbia* space shuttle was lost during its re-entry to Earth's atmosphere on February 1, 2003. The investigation of the disaster determined that a piece of foam insulation broke off shortly after liftoff and damaged the thermal protection system on the leading edge of the orbiter's left wing. As the space shuttle descended, super-hot gases entered the interior aluminum structure of the orbiter. The internal wing structure was weakened and eventually the atmospheric forces tore off the wing. The final communication between *Columbia* and NASA flight controllers occurred at an altitude of approximately 203,000 feet (61,900 meters) over Texas. Debris from *Columbia* was later found in Texas, Arkansas, and Louisiana.

How **long** was the *Mir* space station **in space**?

The first component, a 20.4 ton core module, of the *Mir* space station was launched in February 1986. The core module served as the living and working quarters, including the communications and command center, of the space station. Five more modules were launched and attached to ports on the core module over the next ten years. *Mir* had self-contained oxygen, power, and water generation capabilities allowing cosmonauts and astronauts to spend extended periods of time in space. Scientific investigations aboard *Mir* included space technology experiments, life science and biological research, astrophysics, and material processing tests. Much of the life science and biological research focused on the effects of microgravity on humans and flora and fauna.

Since some of the astronauts spent many months at a time in *Mir*, it was possible to study the physiological differences in long and short duration missions. Space technology studies investigated various materials for space use and the effects of the low Earth orbit environment on various materials. The *Mir* space station was de-orbited on March 23, 2001 after more than 86,000 orbits around Earth. The space station broke into several large pieces and thousands of smaller ones over the Pacific Ocean. There were no injuries.

Which **U.S. astronauts** spent time **aboard *Mir*?**

The collapse of the Soviet Union in 1991 allowed the United States and Russia to enter a new era of space collaboration. NASA made a financial commitment of $400 million to the Russian space program to continue the expeditions to the *Mir* space station. The space shuttle *Atlantis* was fitted with a special adaptor to allow it to dock with *Mir*. There were eleven space shuttle flights to *Mir* between February 1994 and June 1998. In a new spirit of cooperation, Russian cosmonauts began to fly on the shuttle, while American astronauts began to spend time on *Mir* conducting experiments with their Russian counterparts. The experiences gained by the U.S. astronauts on *Mir* were beneficial in planning for the International Space Station. A total of seven U.S. astronauts spent 980 cumulative days aboard *Mir*.

Astronaut	Dates	Total Number of Days
Shannon Lucid	March 22, 1996–September 26, 1996	188
C. Michael Foale	May 15, 1997–October 5, 1997	145
Andrew Thomas	January 22, 1998–June 12, 1998	141
Jerry Linenger	January 12, 1997–May 24, 1997	132
John Blaha	September 16, 1996–January 22, 1997	128
David Wolf	September 25, 1997–January 31, 1998	128
Norman E. Thagard	March 14, 1995–July 7, 1995	118

What is the **purpose** of the **International Space Station** (ISS)?

The International Space Station (ISS) orbits Earth at an altitude of about 250 miles (400 kilometers). The ISS travels at 17,500 miles per hour (32,410 kilometers per hour) in an orbit that extends from 52 degrees north latitude to 52 degrees south latitude. Its main purpose is to serve as a laboratory for materials, communications technology, and biological and medical research in an environment that lacks the force of gravity.

When was the **first module** of the **International Space Station** launched?

The first module of the International Space Station was launched by Russia in November 1998 following many years of planning. The space shuttle *Endeavour* brought the first U.S.-built module to the space station in December 1998. A second Russian module arrived in July 2000 followed by the first crew in December 2000. The members of the first crew to live in the space station were Sergei K. Krikalev and Yuri P. Gidzenko from Russia and William M. (Bill) Shepherd from the United States. Construction of the International Space Station is expected to be completed in 2011 and will, hopefully, be operational through 2015.

An artist's concept of the International Space Station.

Which **nations contribute** to the **International Space Station**?

Sixteen nations, representing five space agencies, contribute scientific, technological, and financial resources and knowledge to the International Space Station. The cooperating nations are: the United States (National Aeronautics and Space Administration or NASA), Russia (Russian Federal Space Agency), Canada (Canadian Space Agency or CSA), Japan (Japan Aerospace Exploration Agency or JAXA), Brazil, and eleven member countries of the European Space Agency (ESA). The eleven member countries of the ESA are: Belgium, Denmark, France, Germany, Italy, the Netherlands, Norway, Spain, Sweden, Switzerland, and the United Kingdom.

How **large** is the **International Space Station**?

When the ISS is completed, it will weigh 925,000 pounds (419,600 kilograms). It will measure 361 feet (110.03 meters) from end to end or the equivalent of a football field, including the end zones. There will be more than 33,000 cubic feet (935 cubic meters) of habitable space. The amount of habitable space will be almost equal to the room inside one and a half Boeing 747 jetliners.

What was the **music of the spheres**?

Inaudible to human beings, the "music of the spheres" is a theoretic harmony or music created by the movements of the planets and heavenly bodies. Its existence was

---

What is stardust and are we really made of it?

The heavy elements that comprise the earth—such as iron, silicon, oxygen, and carbon—originally formed in distant stars that exploded, sending their heavy elements into space. Earth and all life on it are, in fact, recycled stellar debris.

---

proclaimed by Pythagoras and accepted widely in the Greek world. It was later disproved by scientists.

An **asteroid came close** to hitting **Earth** sometime in **2002**. How much damage might it have done?

Asteroid 2002 EM7, a rock 230 feet (70 meters long), is estimated to have the capability of releasing the energy equivalent of a 4 megaton nuclear bomb.

What is widely considered to be **one** of the **earliest celestial observatories**?

Built in England between 2500 and 1700 B.C.E., Stonehenge is one of the earliest observatories or observatory-temples. It is widely believed that its primary function was to observe the mid-summer and mid-winter solstices.

Is anyone **looking** for **extraterrestrial life**?

A program called SETI (the Search for Extraterrestrial Intelligence) began in 1960, when American astronomer Frank Drake (1930–) spent three months at the National Radio Astronomy Observatory in Green Bank, West Virginia, searching for radio signals coming from the nearby stars Tau Ceti and Epsilon Eridani. Although no signals were detected and scientists interested in SETI have often been ridiculed, support for the idea of seeking out intelligent life in the universe has grown.

Project Sentinel, which used a radio dish at Harvard University's Oak Ridge Observatory in Massachusetts, could monitor 128,000 channels at a time. This project was upgraded in 1985 to META (Megachannel Extraterrestrial Assay), thanks in part to a donation by filmmaker Steven Spielberg. Project META is capable of receiving 8.4 million channels. NASA began a ten-year search in 1992 using radio telescopes in Arecibo, Puerto Rico, and Barstow, California.

Scientists are searching for radio signals that stand out from the random noises caused by natural objects. Such signals might repeat at regular intervals or contain mathematical sequences. There are millions of radio channels and a lot of sky to be examined. As of October 1995, Project BETA (Billion-channel Extraterrestrial Assay) has been scanning a quarter of a billion channels. This new design improves upon Project META 300-fold, making the challenge of scanning millions of radio channels seem less daunting. SETI has since developed other projects, some "piggybacking" on radio telescopes while engaged in regular uses. A program launched in 1999, SETI@HOME, uses the power of home computers while they are at rest.

EARTH

AIR

What is the **composition** of **Earth's atmosphere**?

Earth's atmosphere, apart from water vapor and pollutants, is composed of 78 percent nitrogen, 21 percent oxygen, and less than 1 percent each of argon and carbon dioxide. There are also traces of hydrogen, neon, helium, krypton, xenon, methane, and ozone. Earth's original atmosphere was probably composed of ammonia and methane; 20 million years ago the air started to contain a broader variety of elements. The atmosphere weighs approximately 5 million billion tons. It exerts an average of 14.7 pounds per square inch (PSI) of pressure on the surface of the planet.

How many **layers** does **Earth's atmosphere** contain?

The atmosphere, the "skin" of gas that surrounds Earth, consists of five layers that are differentiated by temperature:

The troposphere is the lowest level; it averages about 7 miles (11 kilometers) in thickness, varying from 5 miles (8 kilometers) at the poles to 10 miles (16 kilometers) at the equator. Most clouds and weather form in this layer. Temperature decreases with altitude in the troposphere.

The stratosphere ranges between 7 and 30 miles (11 to 48 kilometers) above Earth's surface. The ozone layer, important because it absorbs most of the sun's harmful ultraviolet radiation, is located in this band. Temperatures rise slightly with altitude to a maximum of about 32°F (0°C).

The mesosphere (above the stratosphere) extends from 30 to 55 miles (48 to 85 kilometers) above Earth. Temperatures decrease with altitude to −130°F (−90°C). 145

The thermosphere (also known as the heterosphere) is between 55 to 435 miles (85 to 700 kilometers). Temperatures in this layer range to 2,696°F (1,475°C).

The exosphere, beyond the thermosphere, applies to anything above 435 miles (700 kilometers). In this layer, temperature no longer has any meaning.

The ionosphere is a region of the atmosphere that overlaps the others, reaching from 30 to 250 miles (48 to 402 kilometers). In this region, the air becomes ionized (electrified) from the sun's ultraviolet rays. This area affects the transmission and reflection of radio waves. It is divided into three regions: the D region (at 35 to 55 miles [56 to 88 kilometers]), the E Region (Heaviside-Kennelly Layer, 55 to 95 miles [88 to 153 kilometers]), and the F Region (Appleton Layer, 95 to 250 miles [153 to 402 kilometers]).

What are the **Van Allen belts**?

The Van Allen belts (or zones) are two regions of highly charged particles above Earth's equator trapped by the magnetic field that surrounds Earth. Also called the magnetosphere, the first belt extends from a few hundred to about 2,000 miles (3,200 kilometers) above Earth's surface and the second is between 9,000 and 12,000 miles (14,500 to 19,000 kilometers). The particles, mainly protons and electrons, come from the solar wind and cosmic rays. The belts are named in honor of James Van Allen (1914–2006), the American physicist who discovered them in 1958 and 1959 with the aid of radiation counters carried aboard the artificial satellites *Explorer I* (1958) and *Pioneer 3* (1959).

In May 1998 there were a series of large, solar disturbances that caused a new Van Allen belt to form in the so-called "slot region" between the inner and outer Van Allen belts. The new belt eventually disappeared once the solar activity subsided. There were also a number of satellite upsets around the same time involving the *Galaxy IV* satellite, Iridium satellites, and others. This is not the first time that a temporary new belt has been observed to form in the same region, but it takes a prolonged period of solar storm activity to populate this region with particles.

Why is the **sky blue**?

The sunlight interacting with Earth's atmosphere makes the sky blue. In outer space the astronauts see blackness because outer space has no atmosphere. Sunlight consists of light waves of varying wavelengths, each of which is seen as a different color. The minute particles of matter and molecules of air in the atmosphere intercept and scatter the white light of the sun. A larger portion of the blue color in white light is scattered, more so than any other color because the blue wavelengths are the shortest. When the size of atmospheric particles are smaller than the wavelengths of the colors, selective scattering occurs—the particles only scatter one color and the atmosphere will appear to be that color. Blue wavelengths especially are affected, bouncing off the air particles to become visible. This is why the sun looks yellow (yellow equals white minus blue). At sunset, the sky changes color because as the sun drops to the horizon, sunlight has more atmosphere to pass through and loses more of its blue wavelengths. The orange and red, having the longer wavelengths and making up more of sunlight at this distance, are most likely to be scattered by the air particles.

PHYSICAL CHARACTERISTICS

What is the **mass** of the **earth**?

The mass of the earth is estimated to be 6 sextillion, 588 quintillion short tons (6.6 sextillion short tons), or 5.97×10^{24} kilograms, with the earth's mean density being 5.515 times that of water (the standard). This is calculated from using the parameters of an ellipsoid adopted by the International Astronomical Union in 1964 and recognized by the International Union of Geodesy and Geophysics in 1967.

What is the **interior** of the **earth** like?

The earth is divided into a number of layers. The topmost layer is the crust, which contains about 0.6 percent of the earth's volume. The depth of the crust varies from 3.5 to 5 miles (5 to 9 kilometers) beneath the oceans to 50 miles (80 kilometers) beneath some mountain ranges. The crust is formed primarily of rocks, such as granite and basalt.

Between the crust and the mantle is a boundary known as the Mohorovičić discontinuity (or Moho for short), named for Croatian seismologist Andrija Mohorovičić (1857–1936), who discovered it in 1909. Below the Moho is the mantle, extending down about 1,800 miles (2,900 kilometers). The mantle is composed mostly of oxygen, iron, silicon, and magnesium, and accounts for about 82 percent of the earth's volume. Although mostly solid, the upper part of the mantle, called the asthenosphere, is partially liquid.

The core-mantle boundary, also called the Gutenberg discontinuity for German-American seismologist Beno Gutenberg (1889–1960), separates the mantle from the core. Made up primarily of nickel and iron, the core contains about 17 percent

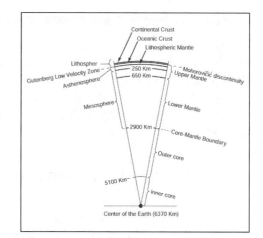

The layers of the earth's interior.

of the earth's volume. The outer core is liquid and extends from the base of the mantle to a depth of about 3,200 miles (5,155 kilometers). The solid inner core reaches from the bottom of the outer core to the center of the earth, about 3,956 miles (6,371 kilometers) deep. The temperature of the inner core is estimated to be about 7,000°F (3,850°C).

What causes **sinkholes**?

A sinkhole is a depression shaped like a well or funnel that occurs in a land surface. Most common in limestone regions, sinkholes are usually formed by the dissolving action of groundwater or the seepage of above-ground streams into the limestone below, causing cracks or fractures in subterranean rock. The collapse of cave roofs can also cause large sinkholes or landslides. The resulting depression may be several miles in diameter.

What is at the **center** of **Earth**?

Geophysicists have held since the 1940s that Earth's interior core is a partly crystallized sphere of iron and nickel that is gradually cooling and expanding. As it cools, this inner core releases energy to an outer core called the fluid core, which is composed of iron, nickel, and lighter elements, including sulfur and oxygen. Another model called the "nuclear earth model" holds that there is a small core, perhaps 5 miles (8 kilometers) wide, of uranium and plutonium surrounded by a nickel-silicon compound. The uranium and plutonium work as a natural nuclear reactor, generating radiating energy in the form of heat, which in turn drives charged particles to create Earth's magnetic field. The traditional model of Earth's core is still dominant; however, scientists have yet to disprove the nuclear earth model.

How does the **temperature** of the **earth change** as one goes deeper underground?

The earth's temperature increases with depth. Measurements made in deep mines and drill-holes indicate that the rate of temperature increase varies from place to place in the world, ranging from 59° to 167°F (15 to 75°C) per kilometer in depth. Actual temperature measurements cannot be made beyond the deepest drill-holes, which are a little more than 6.2 miles (10 kilometers) deep. Estimates suggest that the temperatures at Earth's center can reach values of 5,000°F (2,760°C) or higher.

What are the **highest** and **lowest points** on Earth?

The highest point on land is the top of Mt. Everest (in the Himalayas on the Nepal-Tibet border) at 29,035 feet [8,850 meters]) above sea level. This measurement, taken using satellite-based technology, includes the snow and ice layers which are estimated at between 30 feet (9 meters) to 60 feet (18 meters). The National Geographic Society accepted this height in November 1999. The U.S. National Imagery and Mapping Agency has also accepted 29,035 feet (8,850 meters) as the official height. There is hope that ground penetrating radar will one day be used to determine the snow pack depth. The height established by the Surveyor General of India in 1954 and accepted by the National Geographic Society was 29,028 feet (8,845 meters) plus or minus 10 feet (three meters) because of snow. Earlier measurements indicated the height of Mt. Everest was 29,002 feet (8,840 meters). Satellite measurements taken in 1987 indicated that Mt. Everest is 29,864 feet (9,102 meters) high, but this measurement was not adopted by the National Geographic Society.

The lowest point on land is the Dead Sea between Israel and Jordan, which is 1,312 feet (399 meters) below sea level. The lowest point on Earth's surface is thought to be in the Mariana Trench in the western Pacific Ocean, extending from southeast of Guam to northwest of the Mariana Islands. It has been measured as 36,198 feet (11,034 meters) below sea level.

What is the **tallest mountain**?

Mauna Kea, an inactive volcano on the island of Hawaii, measures 33,474 feet (10,203 meters) from its base to the top. However, it only rises 13,796 feet (4,205 meters) above sea level since its base is on the ocean floor.

Which **elements** are contained in **Earth's crust**?

The most abundant elements in Earth's crust are listed in the table below. In addition, nickel, copper, lead, zinc, tin, and silver account for less than 0.02 percent with all other elements comprising 0.48 percent.

Element	Percentage
Oxygen	47.0
Silicon	28.0
Aluminum	8.0
Iron	4.5
Calcium	3.5
Magnesium	2.5
Sodium	2.5
Potassium	2.5
Titanium	0.4
Hydrogen	0.2
Carbon	0.2
Phosphorus	0.1
Sulfur	0.1

What are the **highest** and **lowest elevations** in the **United States**?

Named in honor of U.S. president William McKinley (1843–1901), Mt. McKinley, Alaska, at 20,320 feet (6,194 meters), is the highest point in the United States and North America. Located in central Alaska, it is part of the Alaska Range. Its south peak measures 20,320 feet (6,194 meters) high and the north peak is 19,470 feet (5,931 meters) high. It boasts one of the world's largest unbroken precipices and is the main scenic attraction at Denali National Park. Denali means the "high one" or the "great one" and is a Native American name sometimes used for Mt. McKinley. Mt. Whitney, California, at 14,494 feet (4,421 meters), is the highest point in the continental United States. Death Valley, California, at 282 feet (86 meters) below sea level, is the lowest point in the United States and in the western hemisphere.

WATER

Does **ocean water circulate**?

Ocean water is in a constant state of movement. Horizontal movements are called currents while vertical movements are called upwelling and downwelling. Wind, tidal motion, and differences in density due to temperature or salinity are the main causes of ocean circulation. Temperature differentials arise from the equatorial water being warmer than water in the polar regions. In the Northern Hemisphere the currents circulate in a clockwise direction while in the Southern Hemisphere the currents circulate in a counterclockwise direction. In the equatorial regions the currents move in opposite directions—from left to right in the north and from right to left in the south. Currents moving north and south from equatorial regions carry warm water while those in polar regions carry cold water.

Major Cold Currents	Major Warm Currents
California	North Atlantic (Gulf Stream)
Humboldt	South Atlantic
Labrador	South Indian Ocean
Canaries	South Pacific
Benguela	North Pacific
Falkland	Monsoons
West Australian	Okhotsk

How much of **Earth's surface** is **land** and how much is **water?**

Approximately 30 percent of Earth's surface is land. This is about 57,259,000 square miles (148,300,000 square kilometers). The area of Earth's water surface is approximately 139,692,000 square miles (361,800,000 square kilometers), or 70 percent of the total surface area.

If Earth were a **uniform sphere**, how much **water** would **cover** the **surface**?

It is estimated that 97 percent of all water in the world or over one quadrillion acre-feet ($1,234 \times 10^{15}$ cubic meters) is contained within the oceans. If Earth were a uniform sphere, this volume of water would cover Earth to a depth of 800 feet (244 meters).

If you **melted** all the **ice** in the **world**, how high would the **oceans rise**?

If you melted all the ice in the world, some 5.5 million cubic miles (23 million cubic kilometers) in all, the oceans would rise 1.7 percent or about 180 feet (60 meters), which is enough, for example, for 20 stories of the Empire State Building to be underwater.

What **fraction** of an **iceberg** shows **above water**?

Only one seventh to one-tenth of an iceberg's mass shows above water.

What **color** is an **iceberg**?

Most icebergs are blue and white. However, one in a thousand icebergs in Antarctica is emerald green. They are only found in Antarctica because the Northern Hemisphere is not cold enough. These icebergs form when seawater freezes to the bottom of floating ice shelves. The ice looks green from the combination of yellow and blue. Yellow is from the yellowish-brown remains of dead plankton dissolved in seawater and trapped in the ice. Blue is present because although ices reflects virtually all the wavelengths of visible light, it absorbs slightly more red wavelengths than blue.

What is an **aquifer**?

Some rocks of the upper part of Earth's crust contain many small holes, or pores. When these holes are large or are joined together so that water can flow through them easily, the rock is considered to be permeable. A large body of permeable rock in which water is stored and flows through is called an aquifer (from the Latin for "water" and "to bear"). Sandstones and gravels are excellent examples of permeable rock.

The part of the iceberg you see above the ocean surface represents only 10 to 15 percent of the entire iceberg!

As water reservoirs, aquifers provide about 60 percent of American drinking water. The huge Ogallala Aquifer, underlying about two million acres of the Great Plains, is a major source of water for the central United States.

Water is purified as it is filtered through the rock, but it can be polluted by spills, dumps, acid rain, and other causes. In addition, recharging of water by rainfall often cannot keep up with the volume removed by heavy pumping. In some areas, the aquifer has been decreasing by 3.2 feet (1 meter) per year and only recharged at a rate of 1 mm (a little over 1/32 of an inch) per year. The Ogallala Aquifer's supply of water could be depleted by 80 percent by the year 2020.

What is the **chemical composition** of the **ocean**?

The ocean contains every known naturally occurring element plus various gases, chemical compounds, and minerals. Below is a sampling of the most abundant chemicals.

Constituent	Concentration (parts per million)
Chloride	18,980
Sodium	10,560
Sulfate	2,560
Magnesium	1,272
Calcium	400
Potassium	380
Bicarbonate	142
Bromide	65
Strontium	13
Boron	4.6
Fluoride	1.4

Why is the **sea blue**?

There is no single cause for the colors of the sea. What is seen depends in part on when and from where the sea is observed. Eminent authority can be found to support almost any explanation. Some explanations include absorption and scattering of light by pure water; suspended matter in sea water; the atmosphere; and color and brightness variations of the sky. For example, one theory is that when sunlight hits seawater, part of the white light, composed of different wavelengths of various colors, is absorbed, and some of the wavelengths are scattered after colliding with the water molecules. In clear water, red and infrared light are greatly absorbed but blue is least absorbed, so that the blue wavelengths are reflected out of the water. The blue effect requires a minimum depth of 10 feet (3 meters) of water.

What **causes waves** in the ocean?

The most common cause of surface waves is air movement (the wind). Waves within the ocean can be caused by tides, interactions among waves, submarine earthquakes or volcanic activity, and atmospheric disturbances. Wave size depends on wind speed, wind duration, and the distance of water over which the wind blows. The longer the distance the wind travels over water, or the harder it blows, the higher the waves. As the wind blows over water it tries to drag the surface of the water with it. The surface cannot move as fast as air, so it rises. When it rises, gravity pulls the water back, carrying the falling water's momentum below the surface. Water pressure from below pushes this swell back up again. The tug of war between gravity and water pressure constitutes wave motion. Capillary waves are caused by breezes of less than two knots. At 13 knots the waves grow taller and faster than they grow longer, and their steepness cause them to break, forming whitecaps. For a whitecap to form, the wave height must be one-seventh the distance between wave crests.

What are the **major oceans**?

Most geographers have recognized four major oceans—Pacific, Atlantic, Indian, and Arctic—for many years. In 2000, the International Hydrographic Organization, an intergovernmental organization, dedicated to safety in navigation and support of the marine environment, delimited a fifth ocean called the Southern Ocean. The Southern Ocean extends from the coast of Antarctica to 60° south latitude. It encompasses portions of the Atlantic, Indian, and Pacific Oceans.

How **deep** is the **ocean**?

The average depth of the ocean floor is 13,124 feet (4,000 meters). The average depth of the four major oceans is given below:

Ocean	Average depth Feet/Meters
Southern	14,450/4,404
Pacific	14,040/4,279
Indian	12,800/3,901
Atlantic	11,810/3,600
Arctic	4,300/1,311

153

There are great variations in depth because the ocean floor is often very rugged. The greatest depth variations occur in deep, narrow depressions known as trenches along the margins of the continental plates. The deepest measurements made—36,198 feet (11,034 meters), deeper than the height of the world's tallest mountains—were taken in Mariana Trench east of the Mariana Islands. In January 1960, the French oceanographer Jacques Piccard (1922–2008), together with the U.S. Navy Lieutenant David Walsh, took the bathyscaphe *Trieste* to the bottom of the Mariana Trench.

Ocean	Deepest Point	Feet/Meters
Pacific	Mariana Trench	35,840/10,924*
Atlantic	Puerto Rico Trench	28,232/8,605
Indian	Java Trench	23,376/7,125
Arctic	Eurasia Basin	17,881/5,450

*This is the the deepest point that the *Triest* reached, but there have been soundings as deep as 36,198 feet (11,034 meters).

How far can **sunlight penetrate** into the **ocean**?

The ocean is divided into three main zones based on depth and the level of light. The upper level, from 0 to 656 feet (200 meters) is called the epipelagic or in common terms, the sunlight zone. It is also known as the photic zone since this is where the most sunlight penetrates allowing for photosynthesis and the growth of plants in the ocean. The middle level is called the mesopelagic zone, or the dysphotic zone. It is often referred to as the twilight zone in common terms. It extends from 656 feet (200 meters) to 3,280 feet (1,000 meters). Some sunlight penetrates to these depths, but not enough to sustain plant growth. The bathypelagic zone extends from 3,280 feet (1,000 meters) to the depths of the ocean. It is the aphotic zone since sunlight cannot penetrate to these depths. In common terms, it is called the midnight zone.

What is a **tidal bore**?

A tidal bore is a large, turbulent, wall-like wave of water that moves inland or upriver as an incoming tidal current surges against the flow of a more narrow and shallow river, bay, or estuary. It can be 10 to 16 feet (3 to 5 meters) high and move rapidly (10 to 15 knots) upstream with, and faster than, the rising tide.

What are **rip tides** and why are they so dangerous?

At points along a coast where waves are high, a substantial buildup of water is created near the shore. This mass of water moves along the shore until it reaches an area of lower waves. At this point, it may burst through the low waves and move out from shore as a strong surface current moving at an abnormally rapid speed known as a rip

> ## Where are the world's highest tides?
>
> The Bay of Fundy (New Brunswick, Canada) has the world's highest tides. They average about 45 feet (14 meters) high in the northern part of the bay, far surpassing the world average of 2.5 feet (0.8 meter). The shape and size of the bay, as well as the increasing up-bay shallowness, cause these very high tides.

current. Swimmers who become exhausted in a rip current may drown unless they swim parallel to the shore.

Is there one **numerical value** for **sea level**?

Sea level is the average height of the sea surface. Scientists have calculated a mean sea level based on observations around the world. The mean average sea level takes into account all stages of ocean tides over a 19-year period. Numerous irregularities and slopes on the sea surface make it difficult to calculate an accurate measure of sea level.

What is the **difference** between an **ocean** and a **sea**?

There is no neatly defined distinction between ocean and sea. One definition says the ocean is a great body of interconnecting saltwater that covers 71 percent of Earth's surface. There are five major oceans—the Arctic, Atlantic, Indian, Pacific, and Southern—but some sources do not include the Arctic Ocean, calling it a marginal sea. The terms "ocean" and "sea" are often used interchangeably but a sea is generally considered to be smaller than an ocean. The name is often given to saltwater areas on the margins of an ocean, such as the Mediterranean Sea.

How much salt is in **brackish water**?

Brackish water has a saline (salt) content between that of freshwater and sea water. It is neither fresh nor salty, but somewhere in between. Brackish waters are usually regarded as those containing 0.5 to 30 parts per thousand salt, while the average saltiness of seawater is 35 parts per thousand.

How **salty** is **seawater**?

Seawater is, on average, 3.3 to 3.7 percent salt. The amount of salt varies from place to place. In areas where large quantities of freshwater are supplied by melting ice, rivers, or rainfall, such as the Arctic or Antarctic, the level of salinity is lower. Areas such as the Persian Gulf and the Red Sea have salt contents over 4.2 percent. If all the salt in the ocean were dried, it would form a mass of solid salt the size of Africa. Most of the

The salinity of the water in the Dead Sea is so extreme that people swimming in it can float on the surface with little effort at all.

ocean salt comes from processes of dissolving and leaking from the solid Earth over hundreds of millions of years. Some is the result of salty volcanic rock that flows up from a giant rift that runs through all the ocean's basins.

Is the **Dead Sea** really dead?

Because the Dead Sea, on the boundary between Israel and Jordan, is the lowest body of water on Earth's surface, any water that flows into it has no outflow. It is called "dead" because its extreme salinity makes impossible any animal or vegetable life except bacteria. Fish introduced into the sea by the Jordan River or by smaller streams die instantly. The only plant life consists primarily of halophytes (plants that grow in salty or alkaline soil). The concentration of salt increases toward the bottom of the lake. The water also has such a high density that bathers float on the surface easily.

What is the **bearing capacity** of **ice** on a **lake**?

The following chart indicates the maximum safe load. It applies only to clear lake ice that has not been heavily traveled. For early winter slush ice, ice thickness should be doubled for safety.

Ice thickness Inches/Centimeters	Examples	Maximum safe load Tons/Kilograms
2/5	One person on foot	
3/7.6	Group in single file	
7.5/19	Car or snowmobile	2/907.2
8/20.3	Light truck	2.5/1,361

Ice thickness Inches/Centimeters	Examples	Maximum safe load Tons/Kilograms
10/25.4	Medium truck	3.5/1,814.4
12/30.5	Heavy truck	9/7,257.6
15/38		10/9,072
20/50.8		25/22,680

Where is the **world's deepest lake**?

Lake Baikal, located in southeast Siberia, Russia, is approximately 5,371 feet (1,638 meters) deep at its maximum depth, Olkhon Crevice, making it the deepest lake in the world. Lake Tanganyika in Tanzania and Zaire is the second deepest lake, with a depth of 4,708 feet (1,435 meters).

Where are the **five largest lakes** in the world located?

Location	Area (miles²/km²)	Length (miles²/km²)	Depth (feet/meters)
Caspian Sea* (Eurasia)	143,244/370,922	760/1,225	3,363/1,025
Superior (North America)	31,700/82,103	350/560	1,330/406
Victoria (Africa)	26,828/69,464	250/360	270/85
Huron (North America)	23,000/59,570	206/330	750/229
Michigan (North America)	22,300/57,757	307/494	923/281

*Saltwater lake

What is a **yazoo**?

A yazoo is a tributary of a river that runs parallel to the river, being prevented from joining the river because the river has built up high banks. The name is derived from the Yazoo River, a tributary of the Mississippi River, which demonstrates this effect.

Which of the **Great Lakes** is the **largest**?

Lake	Surface Area (sq. miles/sq. km)	Maximum Depth (Feet/Meters)
Superior	31,700/82,103	1,333/406
Huron	23,010/59,600	750/229
Michigan	22,300/57,757	923/281
Erie	9,910/25,667	210/64
Ontario	7,540/9,529	802/244

157

Which lake is no longer one of the five largest lakes in the world?

The Aral Sea, located in Kazakhstan and Uzbekistan, was previously the fourth largest lake in the world with 26,000 square miles (67,340 square kilometers), but is now the sixth largest lake in the world. It is now divided into two distinct lakes with a total area of only 13,000 square miles (33,670 square kilometers) although the size varies with the seasons. One of the major reasons the lake has shrunk is the diversion of water for irrigation purposes.

Lake Superior is the largest of the Great Lakes. The North American Great Lakes form a single watershed with one common outlet to the sea—the St. Lawrence Seaway. The total volume of all five basins is 6,000 trillion gallons (22.7 trillion liters) equivalent to about 20 percent of the world's freshwater. Only Lake Michigan lies wholly within U.S. borders; the others share their boundaries with Canada. Some believe that Lake Huron and Lake Michigan are two lobes of one lake, since they are the same elevation and are connected by the 120-foot (36.5-meter) deep Strait of Mackinac, which is 3.6 to 5 miles (6 to 8 kilometers) wide. Gage records indicate that they both have similar water level regimes and mean long-term behavior, so that hydrologically they act as one lake. Historically they were considered two by the explorers who named them, but this is considered a misnomer by some.

What are the **longest rivers** in the world?

The two longest rivers in the world are the Nile in Africa and the Amazon in South America. However, which is the longest is a matter of some debate. The Amazon has several mouths that widen toward the South Atlantic, so the exact point where the river ends is uncertain. If the Pará estuary (the most distant mouth) is counted, its length is approximately 4,195 miles (6,750 kilometers). The length of the Nile as surveyed before the loss of a few miles of meanders due to the formation of Lake Nasser behind the Aswan Dam was 4,145 miles (6,670 kilometers). The table below lists the five longest river systems in the world (exact figures vary, depending on sources).

River	Length (miles/km)
Nile (Africa)	4,160/6,695
*Amazon (South America)	4,007/6,448
Chang jiang-Yangtze (Asia)	3,964/6,378
Mississippi-Missouri river system (North America)	3,710/5,970
Yenisei-Angara river system (Asia)	3,440/5,536

*Excluding Pará estuary

Why was Niagara Falls shut down for 30 hours in 1848?

The volume of Niagara waters depends on the height of Lake Erie at Buffalo, New York, a factor that varies with the direction and intensity of the wind. Changes of as much as 8 feet (2.5 meters) in the level of Lake Erie at the Niagara River source have been recorded. On March 29, 1848, a gale drove the floating ice in Lake Erie to the lake outlet, quickly blocking that narrow channel and shutting off a large proportion of the river's flow. Eyewitness accounts stated that the American falls were passable on foot, but for that day only.

What is the world's **highest waterfall**?

Angel Falls, named after the explorer and bush pilot Jimmy Angel, on the Carrao tributary in Venezuela is the highest waterfall in the world. It has a total height of 3,212 feet (979 meters) with its longest unbroken drop being 2,648 feet (807 meters).

It is difficult to determine the height of a waterfall because many are composed of several sections rather than one straight drop. The highest waterfall in the United States is Yosemite Falls on a tributary of the Merced River in Yosemite National Park, California, with a total drop of 2,425 feet (739 meters). There are three sections to the Yosemite Falls: Upper Yosemite is 1,430 feet (435 meters), Cascades (middle portion) is 675 feet (205 meters), and Lower Yosemite is 320 feet (97 meters).

When will **Niagara Falls disappear**?

The water dropping over Niagara Falls digs great plunge pools at the base, undermining the shale cliff and causing the hard limestone cap to cave in. Niagara has eaten itself 7 miles (11 kilometers) upstream since it formed 10,000 years ago. At this rate, it will disappear into Lake Erie in 22,800 years. The Niagara River connects Lake Erie with Lake Ontario, and marks the U.S.–Canada boundary (New York–Ontario).

LAND

Are there **tides** in the **solid part** of the **earth** as well as in its waters?

The solid Earth is distorted about 4.5 to 14 inches (11.4 to 35.6 centimeters) by the gravitational pull of the sun and moon. It is the same gravitational pull that creates the tides of the waters. When the moon's gravity pulls water on the side of Earth near to it, it pulls the solid body of the earth on the opposite side away from the water to create bulges on both sides, and causing high tides. These occur every 12.5 hours.

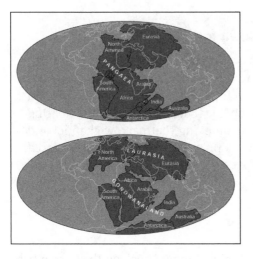

About 225 million years ago, there was one huge continent we now call Pangea (top). About 25 million years later, Pangea broke into Laurasia and Gondwanaland (bottom).

Low tides occur in those places from which the water is drained to flow into the two high-tide bulges. The sun causes tides on the earth that are about 33 to 46 percent as high as those due to the moon. During a new moon or a full moon when the sun and moon are in a straight line, the tides of the moon and the sun reinforce each other to make high tides higher; these are called spring tides. At the quarter moons, the sun and moon are out of step (at right angles), the tides are less extreme than usual; these are called neap tides. Smaller bodies of water, such as lakes, have no tides because the whole body of water is raised all at once, along with the land beneath it.

Do the **continents move**?

In 1912, a German geologist, Alfred Lothar Wegener (1880–1930), theorized that the continents had drifted or floated apart to their present locations and that once all the continents had been a single land mass near Antarctica called Pangaea (from the Greek word meaning *all-earth*). Pangaea then broke apart some 200 million years ago into two major continents called Laurasia and Gondwanaland. These two continents continued drifting and separating until the continents evolved into their present shapes and positions. Wegener's theory was discounted, but it has since been found that the continents do move sideways (not drift) at an estimated 0.75 inch (19 millimeters) annually because of the action of plate tectonics. American geologist William Maurice Ewing (1906–1974) and Harry Hammond Hess (1906–1969) proposed that Earth's crust is not a solid mass, but composed of eight major and seven minor plates that can move apart, slide by each other, collide, or override each other. Where these plates meet are major areas of mountain-building, earthquakes, and volcanoes.

How much of **Earth's surface** is **covered** with **ice**?

About ten percent of the world's land surface is glaciated, or permanently covered with ice. Approximately 5,712,800 square miles (14,800,000 square kilometers) are covered by ice in the form of ice sheets, ice caps, or glaciers. An ice sheet is a body of ice that blankets an area of land, completely covering its mountains and valleys. Ice sheets have an area of over 19,000 square miles (50,000 square kilometers); ice caps are smaller. Glaciers are larger masses of ice that flow, under the force of gravity, at a rate

Which is purer: glacier ice or regular ice?

Impurities found in the snow grains of glaciers have mostly moved to the boundaries of the grains and been flushed out. Glacier ice is like triple-distilled water and hence purer than regular ice.

of between 10 and 1,000 feet (3 to 300 meters) per year. Glaciers on steep slopes flow faster. The areas of glaciation in some parts of the world are:

Place	Square Miles/Square Km
South Polar regions (includes Antarctica)	5,340,000/13,830,000
North Polar regions (includes Greenland)	758,500/1,965,000
Alaska-Canada	22,700/58,800
Asia	14,600/37,800
South America	4,600/11,900
Europe	4,128/10,700
New Zealand	391/1,015
Africa	92/238

How much of **Earth's surface** is **permanently frozen**?

About one-fifth of Earth's land is permafrost, or ground that is permanently frozen. This classification is based entirely on temperature and disregards the composition of the land. It can include bedrock, sod, ice, sand, gravel, or any other type of material in which the temperature has been below freezing for over two years. Nearly all permafrost is thousands of years old. Permafrost is found in Canada, Russia, northern China, most of Greenland and Alaska, and Antarctica.

Where are the **northernmost** and **southernmost points** of land?

The most northern point of land is Cape Morris K. Jesup on the northeastern extremity of Greenland. It is at 83 degrees, 39 minutes north latitude and is 440 miles (708 kilometers) from the North Pole. However, the *Guinness Book of Records* reports that an islet of 100 feet (30 meters) across, called Oodaq, is more northerly at 83 degrees, 40 minutes north latitude and 438.9 miles (706 kilometers) from the North Pole. The southernmost point of land is the South Pole (since the South Pole, unlike the North Pole, is on land).

In the United States, the northernmost point of land is Point Barrow, Alaska (71 degrees, 23 minutes north latitude), and the southernmost point of land is Ka Lae or

161

South Cape (18 degrees, 55 minutes north latitude) on the island of Hawaii. In the 48 contiguous states, the northernmost point is Northwest Angle, Minnesota (49 degrees, 23 minutes north latitude); the southernmost point is Key West, Florida (24 degrees, 33 minutes north latitude).

How **thick** is the **ice** that **covers Antarctica**?

The average depth of the ice that covers Antarctica is 6,600 feet (2,000 meters) or more than a mile thick. In some areas, the ice is as thick as 2 miles (3 kilometers). Nearly 90 percent of the world's ice is found in Antarctica.

Who was the **first person** on **Antarctica**?

Historians are unsure who first set foot on Antarctica, the fifth largest continent, covering ten percent of Earth's surface with its area of 5.1 million square miles (13.2 million square kilometers). Between 1773 and 1775 British Captain James Cook (1728–1779) circumnavigated the continent. American explorer Nathaniel Palmer (1799–1877) discovered Palmer Peninsula in 1820, without realizing that this was a continent. That same year, Fabian Gottlieb von Bellingshausen (1779–1852) sighted the Antarctic continent. American sealer John Davis went ashore at Hughes Bay on February 7, 1821. In 1823, sealer James Weddell (1787–1834) traveled the farthest south (74 degrees south) that anyone had until that time and entered what is now called the Weddell Sea. In 1840, American Charles Wilkes (1798–1877), who followed the coast for 1,500 miles (2,414 kilometers), announced the existence of Antarctica as a continent. In 1841, Sir James Clark Ross (1800–1862) discovered Victoria Land, Ross Island, Mount Erebus, and the Ross Ice Shelf. In 1895, the whaler Henryk Bull landed on the Antarctic continent. Norwegian explorer Roald Amundsen (1872–1928) was the first to reach the South Pole on December 14, 1911. Thirty-four days later, Amundsen's rival Robert Falcon Scott (1868–1912) stood at the South Pole, the second to do so, but he and his companions died during their return trip.

When was the **Ice Age**?

Ice ages, or glacial periods, have occurred at irregular intervals for over 2.3 billion years. During an ice age, sheets of ice cover large portions of the continents. The exact reasons for the changes in Earth's climate are not known, although some think they are caused by changes in Earth's orbit around the sun.

The Great Ice Age occurred during the Pleistocene Epoch, which began about two million years ago and lasted until 11,000 years ago. At its height, about 27 percent of the world's present land area was covered by ice. In North America, the ice covered Canada and moved southward to New Jersey; in the Midwest, it reached as far south as St. Louis. Small glaciers and ice caps also covered the western mountains. Greenland was covered

Do loud noises cause avalanches?

Avalanches are not triggered by noise, but they are the only natural hazard that may be triggered by individuals walking, skiing, snowmobiling, and snowboarding on unstable snow. Many avalanches occur naturally without human provocation. Avalanches are great masses of snow that slide down mountainsides. The snow may be powdery, sliding over compacted older snow; slabs of snow that roll down the slope; or a combination of ice and snow from the slope, including rocks and other debris. Dry slab avalanches are the most dangerous, traveling at speeds of 60–80 miles (97–129 kilometers) per hour.

in ice as it is today. In Europe, ice moved down from Scandinavia into Germany and Poland; the British Isles and the Alps also had ice caps. Glaciers also covered the northern plains of Russia, the plateaus of Central Asia, Siberia, and the Kamchetka Peninsula.

The glaciers' effect on the United States can still be seen. The drainage of the Ohio River and the position of the Great Lakes were influenced by the glaciers. The rich soil of the Midwest is mostly glacial in origin. Rainfall in areas south of the glaciers formed large lakes in Utah, Nevada, and California. The Great Salt Lake in Utah is a remnant of one of these lakes. The large ice sheets locked up a lot of water; sea level fell about 450 feet (137 meters) below what it is today. As a result, some states, such as Florida, were much larger during the ice age.

The glaciers of the last ice age retreated about 11,000 years ago. Some believe that the ice age is not over yet; the glaciers follow a cycle of advance and retreat many times. There are still areas of the earth covered by ice, and this may be a time in between glacial advances.

What is a **moraine**?

A moraine is a mound, ridge, or any other distinct accumulation of unsorted, unstratified material or drift, deposited chiefly by direct action of glacier ice.

What is a **hoodoo**?

A hoodoo is a fanciful name for a grotesque rock pinnacle or pedestal, usually of sandstone, that is the result of weathering in a semi-arid region. An outstanding example of a hoodoo occurs in the Wasatch Formation at Bryce Canyon, Utah.

What are **sand dunes** and how are they formed?

Mounds of wind-blown sand in deserts and coastal areas are called dunes. Winds transport grains of sand until it accumulates around obstacles to form ridges and

163

mounds. Wind direction, the type of sand, and the amount of vegetation determine the type of dune. Dunes are named either for their shape (e.g., star dunes and parabolic dunes) or according to their alignment with the wind (e.g., longitudinal dunes and transverse dunes).

Which are the world's **largest deserts**?

Deserts are distinguished by two general characteristics. First, desert areas receive less than ten inches of precipitation per year. Second, due to the extreme dryness, there is little plant or animal life in most deserts. Many deserts form a band north and south of the equator at about 20 degrees latitude because moisture-bearing winds do not release their rain over these areas. As the moisture-bearing winds from the higher latitudes approach the equator, their temperatures increase and they rise higher and higher in the atmosphere. When the winds arrive over the equatorial areas and come in contact with the colder parts of Earth's atmosphere, they cool down and release all their water to create the tropical rain forests near the equator. However, other parts of the world, such as Antarctica, are also desert regions. The average annual precipitation in Antarctica is less than ten inches per year.

The Sahara Desert, the world's largest, is three times the size of the Mediterranean Sea. In the United States, the largest desert is the Mojave Desert in Southern California with an area of 15,000 square miles (38,900 square kilometers).

Desert	Location	Area (sq. miles/sq. km)	Location
Sahara	North Africa	3,500,000	9,065,000
Arabian	Arabian Peninsula	900,000	2,330,000
Australian*	Australia	600,000	1,554,000
Gobi	Mongolia and China	500,000	1,295,000
Libyan	Libya, SW Egypt, Sudan	450,000	1,165,500

*Includes the Simpson, the Great Victoria, the Sturt Stony, the Gibson, the Great Sandy, and the Tanami.

What is **quicksand**?

Quicksand is a mass of sand and mud that contains a large amount of water. A thin film separates individual grains of sand so the mixture has the characteristics of a liquid. Quicksand is found at the mouths of large rivers or other areas that have a constant source of water. Heavy objects, including humans, can sink when encountering quicksand and the mixture collapses. However, since the density of the sand/water mix is slightly greater than the density of the human body, most humans can actually float on quicksand.

Are all **craters** part of a **volcano**?

No, not all craters are of volcanic origin. A crater is a nearly circular area of deformed sedimentary rocks, with a central, ventlike depression. Some craters are caused by the collapse of the surface when underground salt or limestone dissolves. The withdrawal of groundwater and the melting of glacial ice can also cause the surface to collapse, forming a crater.

Craters are also caused by large meteorites, comets, and asteroids that hit Earth. These are called impact craters. A notable impact crater is Meteor Crater near Winslow, Arizona. It is 4,000 feet (1,219 meters) in diameter, 600 feet (183 meters) deep and is estimated to have been formed 30,000 to 50,000 years ago.

How are **caves formed**?

Water erosion creates most caves found along coastal areas. Waves crashing against the rock over many years wear away part of the rock forming a cave. Inland caves are also formed by water erosion—in particular, groundwater eroding limestone. As the limestone dissolves, underground passageways and caverns are formed.

What is the difference between **spelunking** and **speleology**?

Spelunking, or sport caving, is exploring caves as a hobby or for recreation. Speleology is the scientific study of caves and related phenomena.

How is **speleothem** defined?

Speleothem is a term given to those cave features that form after a cave itself has formed. They are secondary mineral deposits that are created by the solidification of fluids or from chemical solutions. These mineral deposits usually contain calcium carbonate ($CaCO_3$) or limestone, but gypsum or silica may also be found. Stalactites, stalagmites, soda straws, cave coral, boxwork, and cave pearls are all types of speleothems.

Why are the crystals in the **Cave of Crystals** so large?

The Cave of Crystals (Cuevo de los Cristales) was discovered in the Naica mine in Chihuahua, Mexico in 2000. Crystals in the cave are made of gypsum and measure 36 feet (11 meters) long and can weigh up to 55 tons—as long as school buses and as heavy as a small herd of elephants. The cave is located nearly 1,000 feet (300 meters) below the surface. The air temperature in the cave is 112°F (50°C) and the humidity is 90 to 100 percent. The crystals grew because the water temperature was 136°F (58°C)—the temperature at which the mineral anhydrite dissolves into gypsum. The crystallized form of gypsum is selenite. Conditions in the cave were constant, allowing the crystals to continue to form and grow to their large size until 1985, when miners used pumps to lower the water table, draining the cave. If the cave is allowed to fill with water again, the crystals will resume growing.

What is a **tufa**?

It is a general name for calcium carbonate ($CaCO_3$) deposits or spongy porous limestone found at springs in limestone areas, or in caves as massive stalactite or stalagmite deposits. Tufa, derived from the Italian word for "soft rock," is formed by the precipitation of calcite from the water of streams and springs.

Where is the **deepest cave** in the **world**?

The Krubera (Voronya) Cave in the Caucasus Mountains of Georgia is the deepest known cave in the world. Its greatest depth is 7,188 feet (2,191 meters).

Which is the **deepest cave** in the **United States**?

Lechuguilla Cave, in Carlsbad Caverns National Park, New Mexico, is the deepest cave in the United States. Its depth is 1,565 feet (477 meters). Unlike most caves in which carbon dioxide mixes with rainwater to produce carbonic acid, Carlsbad Caverns was shaped by sulfuric acid. The sulfuric acid was a result of a reaction between oxygen that was dissolved in groundwater and hydrogen sulfide that emanated from far below the cave's surface.

Where is the **longest cave system** in the world?

The Mammoth Cave system in Kentucky is the longest cave system in the world. The cave system consists of more than 367 miles (591 kilometers) of subterranean labyrinths. If the second and third longest caves in the world were combined, Mammoth Cave would still be the longest cave in the world.

How does a **stalactite** differ from a **stalagmite**?

A stalactite is a conical or cylindrical calcite formation hanging from a cave roof. It forms from the centuries-long buildup of mineral deposits resulting from the seepage of water from the limestone rock above the cave. This water containing calcium bicarbonate evaporates, losing some carbon dioxide, to deposit small quantities of calcium carbonate ($CaCO_3$; carbonate of lime), which eventually forms a stalactite.

A stalagmite is a stone formation that develops upward from the cave floor and resembles an icicle upside down. Formed from water containing calcite that drips from the limestone walls and roof of the cave, it sometimes joins a stalactite to form a column.

The spectacular "Big Room" in Carlsbad Caverns in New Mexico. These caverns also contain Lechuguilla Cave, the deepest cave in the United States.

What and where is the **Continental Divide** of North America?

The Continental Divide, also known as the Great Divide, is a continuous ridge of peaks in the Rocky Mountains that marks the watershed separating easterly flowing waters from westerly flowing waters in North America. To the east of the Continental Divide, water drains into Hudson Bay or the Mississippi River before reaching the Atlantic Ocean. To the west, water generally flows through the Columbia River or the Colorado River on its way to the Pacific Ocean.

How **long** is the **Grand Canyon**?

The Grand Canyon, cut out by the Colorado River over a period of 15 million years in the northwest corner of Arizona, is the largest land gorge in the world. It is 4 to 13 miles (6.4 to 21 kilometers) wide at its brim, 4,000 to 5,500 feet (1,219 to 1,676 meters) deep, and 217 miles (349 kilometers) long, extending from the mouth of the Little Colorado River to Grand Wash Cliffs (and 277 miles, 600 feet [445.88 kilometers] if Marble Canyon is included).

However, it is not the deepest canyon in the United States; that distinction belongs to Kings Canyon, which runs through the Sierra and Sequoia National Forests near East Fresno, California, with its deepest point being 8,200 feet (2,500 meters). Hell's Canyon of the Snake River between Idaho and Oregon is the deepest U.S. canyon in

167

low-relief territory. Also called the Grand Canyon of the Snake, it plunges 7,900 feet (2,408 meters) down from Devil Mountain to the Snake River.

What are the **LaBrea Tar Pits**?

The tar pits are located in an area of Los Angeles, California, formerly known as Rancho LaBrea. Heavy, sticky tar oozed out of the earth there, the scum from great petroleum reservoirs far underground. The pools were cruel traps for uncounted numbers of animals. Today, the tar pits are a part of Hancock Park, where many fossil remains are displayed along with life-sized reconstructions of these prehistoric species.

The tar pits were first recognized as a fossil site in 1875. However, scientists did not systematically excavate the area until 1901. By comparing Rancho LaBrea's fossil specimens with their nearest living relatives, paleontologists have a greater understanding of the climate, vegetation, and animal life in the area during the Ice Age. Perhaps the most impressive fossil bones recovered belong to such large extinct mammals as the imperial mammoth and the saber-toothed cat. Paleontologists have even found the remains of the western horse and the camel, which originated in North America, migrated to other parts of the world, and became extinct in North America at the end of the Ice Age.

VOLCANOES AND EARTHQUAKES

What is the most **famous volcano**?

The eruption of Mt. Vesuvius in Italy on August 79 C.E. is perhaps the most famous historical volcanic eruption. Vesuvius had been dormant for generations. When it erupted, entire cities were destroyed, including Pompeii, Stabiae, and Herculaneum. Pompeii and Stabiae were buried under ashes while Herculaneum was buried under a mud flow.

The dormant Mt. Fuji in Japan is an example of a composite cone volcano.

How many **kinds of volcanoes** are there?

Volcanoes are usually cone-shaped hills or mountains built around a vent connecting to reservoirs of molten rock, or magma, below Earth's surface. At times the molten rock is forced upwards by gas pressure until it breaks through weak

How do we determine when ancient volcanic eruptions occurred?

The most common method used to date ancient volcanic eruptions is carbon dating. Carbon dating relies on the rate of radioactive decay of carbon-14. It is used to date eruptions that took place more than two hundred years ago. Charcoal from trees burned during a volcanic eruption is almost pure carbon and is ideal for tracing the tiny amounts of carbon-14.

spots in Earth's crust. The magma erupts forth and lava flows or shoots into the air as clouds of lava fragments, ash, and dust. The accumulation of debris from eruptions cause the volcano to grow in size. There are four kinds of volcanoes:

- *Cinder cones* are built of lava fragments. They have slopes of 30 degrees to 40 degrees and seldom exceed 1,640 feet (500 meters) in height. Sunset Crater in Arizona and Paricutin in Mexico are examples of cinder cones.

- *Composite cones* are made of alternating layers of lava and ash. They are characterized by slopes of up to 30 degrees at the summit, tapering off to 5 degrees at the base. Mount Fuji in Japan and Mount St. Helens in Washington are composite cone volcanoes.

- *Shield volcanoes* are built primarily of lava flows. Their slopes are seldom more than 10 degrees at the summit and 2 degrees at the base. The Hawaiian Islands are clusters of shield volcanoes. Mauna Loa is the world's largest active volcano, rising 13,653 feet (4,161 meters) above sea level.

- *Lava domes* are made of viscous, pasty lava squeezed like toothpaste from a tube. Examples of lava domes are Lassen Peak and Mono Dome in California.

Where is the **Circle of Fire**?

The belt of volcanoes bordering the Pacific Ocean is often called the "Circle of Fire" or the "Ring of Fire." Earth's crust is composed of 15 pieces, called plates, which "float" on the partially molten layer below them. Most volcanoes, earthquakes, and mountain-building occur along the unstable plate boundaries. The Circle of Fire marks the boundary between the plate underlying the Pacific Ocean and the surrounding plates. It runs up the west coast of the Americas from Chile to Alaska (through the Andes Mountains, Central America, Mexico, California, the Cascade Mountains, and the Aleutian Islands) then down the east coast of Asia from Siberia to New Zealand (through Kamchatka, the Kurile Islands, Japan, the Philippines, Celebes, New Guinea, the Solomon Islands, New Caledonia, and New Zealand). Of the 850 active volcanoes in the world, over 75 percent of them are part of the Circle of Fire.

Which **island region** has the greatest **concentration** of **active volcanoes**?

In the early 1990s 1,133 seamounts (submerged mountains) and volcanic cones were discovered in and around Easter Island. Many of the volcanoes rise more than a mile above the ocean floor. Some are close to 7,000 feet (2,134 meters) tall, although their peaks are still 2,500 to 5,000 feet (760 to 1,500 meters) below the surface of the sea.

Which **volcanoes** in the contiguous **48 states** are considered **active** and have erupted in the past two hundred years?

Seven major volcanoes in the contiguous 48 states are considered active: two in California—Lassen Peak, and Mt. Shasta; four in Washington—Glacier Peak, Mt. Baker, Mt. Rainier, and Mt. St. Helens; and one in Oregon—Mt. Hood.

Which **volcanoes** have been the **most destructive**?

The five most destructive eruptions from volcanoes since 1700 are as follows:

Volcano	Date of Eruption	People Killed	Lethal Agent
Mt. Tambora, Indonesia	April 5, 1815	92,000	10,000 directly by the volcano; 82,000 from starvation afterwards
Karkatoa, Indonesia	Aug. 26, 1883	36,417	90% killed by a tsunami
Mt. Pelee, Martinque	Aug. 30, 1902	29,025	Pyroclastic flows
Nevada del Ruiz, Colombia	Nov. 13, 1985	23,000	Mud flow
Unzen, Japan	1792	14,300	70% killed by cone collapse; 30% by a tsunami

When did **Mount St. Helens erupt**?

Mount St. Helens, located in southwestern Washington state in the Cascades mountain range, erupted on May 18, 1980. Sixty-one people died as a result of the eruption. This was the first known eruption in the 48 contiguous United States to claim a human life. Geologists call Mount St. Helens a composite volcano (a steep-sided, often symmetrical cone constructed of alternating layers of lava flows, ash, and other volcanic debris). Composite volcanoes tend to erupt explosively. Mount St. Helens and the other active volcanoes in the Cascade Mountains are a part of the "Ring of Fire," the Pacific zone having frequent and destructive volcanic activity. Mt. St. Helens erupted again in October 2004. A steam plume billowed 10,000 feet (3,048 meters) into the air. The eruption continued for 2.5 years building a new lava dome. The new lava dome measured about 125 million cubic yards (95.6 million cubic meters), or a volume equal to nearly 200 large sports stadiums. The amount of lava that erupted would have been enough to pave seven highway lanes, 3 feet (0.9 meter) thick, from New York City to Portland, Oregon. There was no loss of life during this eruptive period.

Volcanoes have not only been active in Washington, but also in three other U.S. states: California, Alaska, and Hawaii. Lassen Peak is one of several volcanoes in the Cascade Range. It last erupted in 1921. Mount Katmai in Alaska had an eruption in 1912 in which the flood of hot ash formed the Valley of Ten Thousand Smokes 15 miles (24 kilometers) away. And Hawaii has its famed Mauna Loa, which is the world's largest active volcano, being 60 miles (97 kilometers) in width at its base.

How many different **kinds of faults** have been identified?

Faults are fractures in Earth's crust. Faults are classified as normal, reverse, or strike-slip. Normal faults occur when the end of one plate slides vertically down the end of another. Reverse faults occur when one plate slides vertically up the end of another. A strike-slip fault occurs when two plate ends slide past each other horizontally. In an oblique fault there is plate movement vertically and horizontally at the same time.

Where is the **San Andreas Fault**?

Perhaps the most famous fault in the world, the San Andreas Fault extends from Mexico north through most of California. The San Andreas Fault is not a single fault, but rather a system of faults. The northern half of the fault, near San Francisco, has reverse faults and is mostly mountainous. The southern half, near Los Angeles, has mostly normal faults. Land development has made it difficult to see the fault except in a few locations, notably near Lake San Andreas south of San Francisco. The fault was named in honor of Andrew Lawson (1861–1952), a geologist who studied the 1906 San Francisco earthquake.

How does a **seismograph** work?

A seismograph records earthquake waves. When an earthquake occurs, three types of waves are generated. The first two, the P and S waves, are propagated within the earth, while the third, consisting of Love and Rayleigh waves, is propagated along the planet's surface. The P wave travels about 3.5 miles (5.6 kilometers) per second and is the first wave to reach the surface. The S wave travels at a velocity of a little more than half of the P waves. If the velocities of the different modes of wave propagation are known, the distance

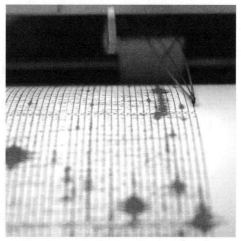

Seismographs detect earthquake waves and record their strength on rolls of moving paper.

171

between the earthquake and an observation station may be deduced by measuring the time interval between the arrival of the faster and slower waves.

When the ground shakes, the suspended weight of the seismograph, because of its inertia, scarcely moves, but the shaking motion is transmitted to the marker, which leaves a record on the drum.

What is the **Richter scale**?

On a machine called a seismograph, the Richter scale measures the magnitude of an earthquake, i.e., the size of the ground waves generated at the earthquake's source. The scale was devised by American geologist Charles W. Richter (1900–1985) in 1935. Every increase of one number means a tenfold increase in magnitude.

Richter Scale

Magnitude	Possible Effects
1	Detectable only by instruments
2	Barely detectable, even near the epicenter
3	Felt indoors
4	Felt by most people; slight damage
5	Felt by all; damage minor to moderate
6	Moderately destructive
7	Major damage
8	Total and major damage

What is the **modified Mercalli Scale**?

The modified Mercalli Scale is a means of measuring the intesity of an earthquake. Unlike the Richter scale, which uses mathematical calculation to measure seismic waves, the modified Mercalli Scale uses the effects of an earthquake on the people and structures in a given area to determine its intensity. It was invented by Guiseppe Mercalli (1850–1914) in 1902 and modified by Harry Wood (1879–1958) and Frank Neumann in the 1930s to take into consideration such modern inventions as the automobile and the skyscraper.

The Modified Mercalli Scale

I. Only felt by a few under especially favorable circumstances.

II. Felt only by a few sleeping persons, particularly on upper floors of buildings. Some suspended objects may swing.

III. Felt quite noticeably indoors, especially on upper floors of buildings, but may not be recognized as an earthquake. Standing automobiles may rock slightly. Vibration like passing of truck.

IV. During the day felt indoors by many, outdoors by few. At night some awakened. Dishes, windows, doors disturbed; walls make creaking sound. Sensation like heavy truck striking building. Standing automobiles rocked noticeably.

V. Felt by nearly everyone, many awakened. Some dishes, windows, and so on broken; cracked plaster in a few places; unstable objects overturned. Disturbances of trees, poles, and other tall objects sometimes noticed. Pendulum clocks may stop.

VI. Felt by all; many frightened and run outdoors. Some heavy furniture moved; a few instances of fallen plaster and damaged chimneys. Damage slight.

VII. Everybody runs outdoors. Damage negligible in buildings of good design and construction; slight to moderate in well-built ordinary structures; considerable in poorly built or badly designed structures; some chimneys broken. Noticed by persons driving cars.

VIII. Damage slight in specially designed structures; considerable in ordinary substantial buildings with partial collapse; great in poorly built structures. Panel walls thrown out of frame structures. Fall of chimneys, factory stacks, columns, monuments, walls. Heavy furniture overturned. Sand and mud ejected in small amounts. Changes in well water. Persons driving cars disturbed.

IX. Damage considerable in specially designed structures; well-designed frame structures thrown out of plumb; great in substantial buildings, with partial collapse. Building shifted off foundations. Ground cracked conspicuously. Underground pipes broken.

X. Some well-built wooden structures destroyed; most masonry and frame structures destroyed with foundations; ground badly cracked. Rails bent. Landslides considerable from river banks and steel slopes. Shifted sand and mud. Water splashed, slopped over banks.

XI. Few, if any, (masonry) structures remain standing. Bridges destroyed. Broad fissures in ground. Underground pipelines completely out of service. Earth slumps and landslips in soft ground. Rails bent greatly.

XII. Damage total. Waves seen on ground surface. Lines of sight and level distorted. Objects thrown into the air.

Who **invented** the ancient **Chinese earthquake detector**?

The detector, invented by Zhang Heng (78–139 C.E.) around 132 C.E., was a copper-domed urn with dragons' heads circling the outside, each containing a bronze ball. Inside the dome was suspended a pendulum that would swing when the ground shook and knock a ball from the mouth of a dragon into the waiting open mouth of a bronze toad below. The ball made a loud noise and signaled the occurrence of an earthquake. Knowing which ball had been released, one could determine the direction of the earthquake's epicenter (the point on Earth's surface directly above the quake's point of origin).

Of what magnitude was the earthquake that hit San Francisco on April 18, 1906?

The historic 1906 San Francisco earthquake took a mighty toll on the city and surrounding area. Over 700 people were killed; the newly constructed $6 million city hall was ruined; the Sonoma Wine Company collapsed, destroying 15 million gallons (57 million liters) of wine. The quake registered 8.3 on the Richter scale and lasted 75 seconds total. Many poorly constructed buildings built on landfills were flattened and the quake destroyed almost all of the gas and water mains. Fires broke out shortly after the quake, and when they were finally eliminated, 3,000 acres of the city, the equivalent of 520 blocks, were charred. Damage was estimated to be $500 million, and many insurance agencies went bankrupt after paying out the claims.

Another large earthquake hit San Francisco on October 17, 1989. It measured 7.1 on the Richter scale, killed 67 people, and caused billions of dollars worth of damage.

When did the **most severe earthquake** in **American history** occur?

The New Madrid series of earthquakes (a series of quakes starting on December 16, 1811, and lasting until March 1812) is considered to be the most severe earthquake event in U.S. history. It shook more than two-thirds of the United States and was felt in Canada. It changed the level of land by as much as 20 feet (6 meters), altered the course of the Mississippi River, and created new lakes, such as Lake St. Francis west of the Mississippi and Reelfoot Lake in Tennessee. Because the area was so sparsely populated, no known loss of life occurred. Scientists agree that at least three, and possibly five, of the quakes had surface wave magnitudes of 8.0 or greater. The largest was probably a magnitude of 8.8, which is larger than any quake yet experienced in California.

What is a **tsunami**?

A tsunami is a giant wave set in motion by a large earthquake occurring under or near the ocean that causes the ocean floor to shift vertically. This vertical shift pushes the water ahead of it, starting a tsunami. These are very long waves (100 to 200 miles [161 to 322 kilometers]) with high speeds (500 mph [805 kph]) that, when approaching shallow water, can grow into a 100-foot (30.5-meter) high wave as its wavelength is reduced abruptly. Ocean earthquakes below a magnitude of 6.5 on the Richter scale, and those that shift the sea floor only horizontally, do not produce these destructive waves. The highest recorded tsunami was 1,719 feet (524 meters) high along Lituya

Bay, Alaska, on July 9, 1958. Caused by a giant landslip, it moved at 100 miles (161 kilometers) per hour. This wave would have swamped the Petronas Towers in Kuala Lumpur, Malaysia, which are 1,483 feet (452 meters) high.

Where do most tsunamis occur?

Tsunamis occur most frequently in the Pacific Ocean, although they also occur in the Caribbean and Mediterranean Seas and the Atlantic and Indian Oceans. The Sumatra tsunami, which followed a strong earthquake (9.0 on the Richter scale) in the Indian Ocean on December 26, 2004, is perhaps the most devastating on record. The tsunami had waves up to 100 feet (30 meters) in height. Damage and deaths were recorded in 11 countries with a total of more than 230,000 people being killed. Indonesia (168,000 people killed), Thailand, Malaysia, Bangladesh, India, Sri Lanka, the Maldives, and the eastern coast of Africa in Somalia, Kenya, and Tanzania all recorded damage and deaths from this event.

OBSERVATION AND MEASUREMENT

What is the Gaia hypothesis?

British scientists James Lovelock (1919–) and Lynn Margulis (1938–) proposed the Gaia hypothesis in the 1970s. According to the theory, all living and non-living organisms on Earth form a single unity that is self-regulated by the organisms themselves. Therefore, the whole planet can be considered a huge single organism. Evidence for this theory is the stability of atmospheric conditions over eons.

What is magnetic declination?

It is the angle between magnetic north and true north at a given point on Earth's surface. It varies at different points on Earth's surface and at different times of the year.

Which direction does a compass needle point at the North Pole?

At the north magnetic pole, the compass needle would be attracted by the ground and point straight down.

The Gaia hypothesis speculates that our entire planet could be considered as one huge living organism.

175

What is the **Piri Re'is map**?

In 1929 a map was found in Constantinople that caused great excitement. Painted on parchment and dated in the Muslim year 919 (1541 according to the Christian calendar), it was signed by an admiral of the Turkish navy known as Piri Re'is. This map appears to be one of the earliest maps of America, and it shows South America and Africa in their correct relative longitudes. The mapmaker also indicated that he had used a map drawn by Christopher Columbus (1451–1506) for the western part. It was an exciting statement because for several centuries geographers had been trying to find a "lost map of Columbus" supposedly drawn by him in the West Indies.

What is a **Foucault pendulum**?

An instrument devised by Jean Foucault (1819–1868) in 1851 to prove that Earth rotates on an axis, the pendulum consisted of a heavy ball suspended by a very long fine wire. Sand beneath the pendulum recorded the plane of rotation of the pendulum over time.

A reconstruction of Foucault's experiment is located in Portland, Oregon, at its Convention Center. It swings from a cable 90 feet (27.4 meters) long, making it the longest pendulum in the world.

What are the major **eras, periods**, and **epochs** in geologic time?

Modern dating techniques have given a range of dates as to when the various geologic time periods have started; they are listed below:

The Geologic Time Scale
(in millions of years ago)

Eon	Era	Period	Sub-Period	Epoch
Precambrian 4,500–543	Hadean 4,500–3,800			
	Archaean 3,800–2,500			
	Proterozoic 2,500–543	Paleoproterozoic 2,500–1,600		
		Mesoproterozoic 1,600–900		
		Neoproterozoic 900–543		
Phanerozoic 543 to present	Paleozoic 543–248	Cambrian 543–490		
		Ordovician 490–443		

Eon	Era	Period	Sub-Period	Epoch
		Silurian 443–417		
		Devonian 417–354		
		Carboniferous 354–290	Mississippian 354–323	
			Pennsylvanian 323–290	
		Permian 290–248		
	Mesozoic 248–65	Triassic 248–206		
		Jurassic 206–144		
		Cretaceous 144–65		
	Cenozoic 65 to present	Tertiary 65–1.8	Paleogene 65–23.8	Paleocene 65–54.8
				Eocene 54.8–33.7
				Oligocene 33.7–23.8
			Neogene 23.8–1.8	Miocene 23.8–5.3
				Pliocene 5.3–1.8
		Quaternary 1.8 to present		Pleistocene 1.8 to 10,000 years ago
				Holocene 10,000 years ago to present

What is the **prime meridian**?

The north-south lines on a map run from the North Pole to the South Pole and are called "meridians." The word "meridian" means "noon": When it is noon on one place on the line, it is noon at any other point as well. The lines are used to measure longitudes, or how far east or west a particular place might be, and they are 69 miles (111 kilometers) apart at the equator. The east-west lines are called parallels, and, unlike meridians, are all parallel to each other. They measure latitude, or how far north or south a particular place might be. There are 180 lines circling the earth, one for each

177

degree of latitude. The degrees of both latitude and longitude are divided into 60 minutes, which are then further divided into 60 seconds each.

The prime meridian is the meridian of 0 degrees longitude, used as the origin for measurement of longitude. The meridian of Greenwich, England, is used almost universally for this purpose.

What is **Mercator's projection** for maps?

The Mercator projection is a modification of a standard cylindrical projection, a technique used by cartographers to transfer the spherical proportions of the earth to the flat surface of a map. For correct proportions, the parallels, or lines of latitude, are spaced at increasing distances toward the poles, resulting in severe exaggeration of size in the polar regions. Greenland, for example, appears five times larger than it actually is. Created by Flemish cartographer Gerardus Mercator (1512–1594) in 1569, this projection is useful primarily because compass directions appear as straight lines, making it ideal for navigation.

When were **relief maps** first used?

The Chinese were the first to use relief maps, in which the contours of the terrain were represented in models. Relief maps in China go back at least to the third century B.C.E. Some early maps were modeled in rice or carved in wood. It is likely that the idea of making relief maps was transmitted from the Chinese to the Arabs and then to Europe. The earliest known relief map in Europe was a map showing part of Austria, made in 1510 by Paul Dox.

Who was the **first person** to **map** the **Gulf Stream**?

In his travels to and from France as a diplomat, Benjamin Franklin (1706–1790) noticed a difference in speed in the two directions of travel between France and America. He was the first to study ships' reports seriously to determine the cause of the

> ## From what distance are satellite photographs taken?
>
> U.S. Department of Defense satellites orbit at various distances above Earth. Some satellites are in low orbit, 100 to 300 miles (160 to 483 kilometers) above the surface, while others are positioned at intermediate altitudes from 500 to 1,000 miles (804 to 1,609 kilometers) high. Some have an altitude of 22,300 miles (35,880 kilometers).

speed variation. As a result, he found that there was a current of warm water coming from the Gulf of Mexico that crossed the North Atlantic Ocean in the direction of Europe. In 1770, Franklin mapped it.

Franklin thought the current started in the Gulf of Mexico. However, the Gulf Stream actually originates in the western Caribbean Sea and moves through the Gulf of Mexico, the Straits of Florida, then north along the east coast of the United States to Cape Hatteras in North Carolina, where it becomes northeast. The Gulf Stream eventually breaks up near Newfoundland, Canada, to form smaller currents or eddies. Some of these eddies blow toward the British Isles and Norway, causing the climate of these regions to be milder than other areas of northwestern Europe.

What are **Landsat maps**?

They are images of Earth taken at an altitude of 567 miles (912 kilometers) by an orbiting Landsat satellite, or ERTS (Earth Resources Technology Satellite). The Landsats were originally launched in the 1970s. Rather than cameras, the Landsats use multispectral scanners, which detect visible green and blue wavelengths, and four infrared and near-infrared wavelengths. These scanners can detect differences between soil, rock, water, and vegetation; types of vegetation; states of vegetation (e.g., healthy/unhealthy or underwatered/well-watered); and mineral content. The differences are especially accurate when multiple wavelengths are compared using multispectral scanners. Even visible light images have proved useful—some of the earliest Landsat images showed that some small Pacific islands were up to 10 miles (16 kilometers) away from their charted positions.

The results are displayed in "false-color" maps, where the scanner data is represented in shades of easily distinguishable colors—usually, infrared is shown as red, red as green, and green as blue. The maps are used by farmers, oil companies, geologists, foresters, foreign governments, and others interested in land management. Each image covers an area approximately 115 square miles (185 square kilometers). Maps are offered for sale by the United States Geological Survey.

Other systems that produce similar images include the French SPOT satellites, Russian Salyut and Mir manned space stations, and NASA's Airborne Imaging Spectrometer, which senses 128 infrared bands. NASA's Jet Propulsion Laboratories are developing instruments that will sense 224 bands in infrared, which will be able to detect specific minerals absorbed by plants.

Why are **topographic maps** important?

Topographic maps are distinguished by the contour lines that portray the shape and elevation of the land. They display the three-dimensional ups and downs of the terrain on a two-dimensional surface. Topographic maps usually portray natural features, such as mountains, valleys, plains, lakes, rivers, and vegetation, and man-made features, such as roads, boundaries, transmission towers, and even major buildings. Topographic maps are used by professionals for engineering, environment management, conservation of natural resources, energy exploration, public works design, and general commercial and residential planning. They are also used by individuals for outdoor recreational activities such as camping, hiking, biking, and fishing.

What do the **7.5 minute quadrangle** maps represent?

One of the most popular maps of the United States Geologic Survey (USGS) is the 7.5-minute quadrangle map. Produced at a scale of 1:24,000, each of these maps covers a four-sided area of 7.5 minutes latitude and 7.5 minutes longitude. The United States has been divided into precisely measured quadrangles. Each quadrangle in a state has a unique name. The maps of adjacent quadrangles may be combined to form a large map. There are more than 54,000 USGS topographic maps that cover the 48 contiguous states and Hawaii.

What is **Global Positioning System** (GPS) and how does it work?

The Global Positioning System (GPS) has three parts: the space part, the user part, and the control part. The space part consists of 24 satellites in orbit 11,000 nautical miles (20,300 kilometers) above Earth. The user part consists of a GPS receiver, which may be hand-held or mounted in a vehicle. The control part consists of five ground stations worldwide to assure the satellites are working properly. Using a GPS receiver an individual can determine his or her location on or above Earth to within about 300 feet (90 meters).

CLIMATE AND WEATHER

TEMPERATURE

What is **El Niño**?

El Niño is the unusual warming of the surface waters of large parts of the tropical Pacific Ocean. Occurring around Christmastime, it is named after the Christ child. El Niño occurs erratically every three to seven years. It brings heavy rains and flooding to Peru, Ecuador, and Southern California, and a milder winter with less snow to the northeastern United States. Studies reveal that El Niño is not an isolated occurrence, but is instead part of a pattern of change in the global circulation of the oceans and atmosphere. The 1982–1983 El Niño was one of the most severe climate events of the twentieth century in both its geographical extent as well as in the degree of warming (14°F or 8°C).

What is **La Niña**?

La Niña is the opposite of El Niño. It refers to a period of cold surface temperatures of the tropical Pacific Ocean. La Niña winters are especially harsh with heavy snowfall in the northeastern United States and rain in the Pacific Northwest.

Is the world actually **getting warmer**?

Global surface temperatures increased at a rate of near 0.11°F/decade (0.06°C/decade) during most of the twentieth century, but the rate increased to approximately 0.29°F/decade (0.16°C/decade) during the past 30 years. There have been two sustained periods of warming, one beginning around 1910 and ending around 1945, and the most recent beginning about 1976. The decade 2000-2009 is the warmest on

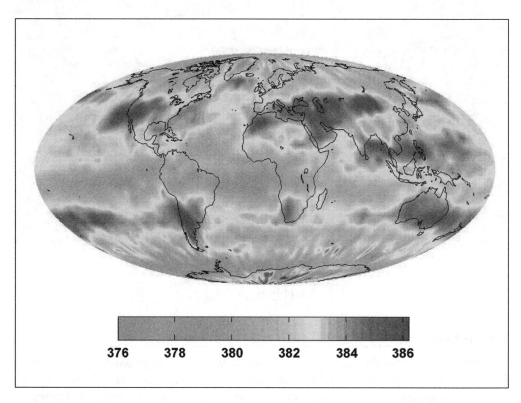

| 376 | 378 | 380 | 382 | 384 | 386 |

An image taken by NASA's Atmospheric Infrared Sounder (AIRS) in July 2008 shows levels of carbon dioxide in Earth's atmosphere (the numbers indicate parts per billion). Most people in the scientific community worry that man-made air pollutants like carbon dioxide are causing global warming.

record with an average global surface temperature of 0.96°F (0.54°C) above the twentieth century average. The following chart indicates the top ten warmest years:

Year	°F/°C above twentieth century average
2005	1.11/0.62
1998	1.08/0.60
2003	1.04/0.58
2002	1.03/0.57
2006	1.01/0.56
2009	1.01/0.56
2007	0.99/0.55
2004	0.97/0.54
2001	0.94/0.25
2008	0.86/0.48
1997	0.86/0.48

Which is **colder**—the **North Pole** or the **South Pole**?

The South Pole is considerably colder than the North Pole. The average temperature at the South Pole is –56°F (–49°C), which is approximately 35° lower than the average temperature at the North Pole. The North Pole is surrounded by an ocean. The water of the ocean retains heat so temperatures at the North Pole are warmer than in the South Pole. The South Pole is located on the continent of Antarctica, which is a large snow- and ice-covered landmass surrounded by ocean. In addition, the elevation of the South Pole is 9,300 feet (2,900 meters), while the North Pole is at sea level, resulting in colder temperatures at the South Pole. Finally, the combination of snow, ice, and extremely dry air reflects nearly 75 percent of the incoming solar radiation. Any heat that is radiated back to the atmosphere is lost instead of being absorbed by water vapor in the atmosphere.

What are the **highest** and **lowest recorded temperatures** on Earth?

The highest temperature in the world was recorded as 136°F (58°C) at Al Aziziyah (el-Aziia), Libya, on September 13, 1922; the highest temperature recorded in the United States was 134°F (56.7°C) in Death Valley, California, on July 10, 1913. The temperatures of 140°F (60°C) at Delta, Mexico, in August 1953 and 136.4°F (58°C) at San Luis, Mexico, on August 11, 1933, are not internationally accepted. The lowest temperature was –128.6°F (–89.6°C) at Vostok Station in Antarctica on July 21, 1983. The record cold temperature for an inhabited area was –90.4°F (–68°C) at Oymyakon, Siberia (population 4,000), on February 6, 1933. This temperature tied with the readings at Verkhoyansk, Siberia, on January 3, 1885, and February 5 and 7, 1892. The lowest temperature reading in the United States was –79.8°F (–62.1°C) on January 23, 1971, in Prospect Creek, Alaska; for the contiguous 48 states, the coldest temperature was –69.7°F (–56.5°C) at Rogers Pass, Montana, on January 20, 1954.

What is a **heat wave**?

A heat wave is a period of two days in a row when apparent temperatures on the National Weather Service heat index exceed 105°F to 110°F (40°C to 43°C). Heat waves can be extremely dangerous. According to the National Weather Service, 175–200 Americans die from heat in a normal summer. Between 1936 and 1975, as many as 15,000 Americans died from problems related to heat. In 1980, 1,250 people died during a brutal heat wave in the Midwest. In 1995, more than 500 people died in the city of Chicago from heat-related problems. A majority of these individuals were the elderly living in high-rise apartment buildings without proper air conditioning. Large concentrations of buildings, parking lots, and roads create an "urban heat island" in cities with higher temperatures than the surrounding open, rural areas.

What is the **heat index**?

The index is a measure of what hot weather feels like to the average person for various temperatures and relative humidities. Heat exhaustion and sunstroke are inclined to happen when the heat index reaches 105°F (40°C). The chart below provides the heat index for some temperatures and relative humidities.

Relative Humidity	Air Temperature (°F)										
	70	75	80	85	90	95	100	105	110	115	120
	Feels like (°F)										
0%	64	69	73	78	83	87	91	95	99	103	107
10%	65	70	75	80	85	90	95	100	105	111	116
20%	66	72	77	82	87	93	99	105	112	120	130
30%	67	73	78	84	90	96	104	113	123	135	148
40%	68	74	79	86	93	101	110	123	137	151	
50%	69	75	81	88	96	107	120	135	150		
60%	70	76	82	90	100	114	132	149			
70%	70	77	85	93	106	124	144				
80%	71	78	86	97	113	136					
90%	71	79	88	102	122						
100%	72	80	91	108							

Which place has the **maximum** amount of **sunshine** in the **United States**?

Yuma, Arizona, has an annual average of 90 percent of sunny days, or over 4,000 sunny hours per year. St. Petersburg, Florida, had 768 consecutive sunny days from February 9, 1967, to March 17, 1969. On the other extreme, the South Pole has no sunshine for 182 days annually, and the North Pole has none for 176 days.

How can the **temperature** be determined from the frequency of **cricket chirps**?

Listen for the chirping of either katydids or crickets, then count the number of chirps you hear in one minute. For the following equations "C" equals the number of chirps per minute:

For katydids, Fahrenheit temperature = 60 + (C–19)/3

For crickets, Fahrenheit temperature = 50 + (C–50)/4

Why are the hot, humid days of summer called **"dog days"**?

This period of extremely hot, humid, sultry weather that traditionally occurs in the Northern Hemisphere in July and August received its name from the dog star Sirius of

Why was 1816 known as the year without a summer?

The eruption of Mount Tambora, a volcano in Indonesia, in 1815 threw billions of cubic yards of dust over 15 miles (24 kilometers) into the atmosphere. Because the dust penetrated the stratosphere, wind currents spread it throughout the world. As a consequence of this volcanic activity, in 1816 normal weather patterns were greatly altered. Some parts of Europe and the British Isles experienced average temperatures 2.9 to 5.8°F (1.6 to 3.2°C) below normal. In New England heavy snow fell between June 6 and June 11 and frost occurred in every month of 1816. Crop failures occcured in Western Europe and Canada as well as in New England. By 1817 the excess dust had settled and the climate returned to more normal conditions.

the constellation Canis Major. At this time of year, Sirius, the brightest visible star, rises in the east at the same time as the sun. Ancient Egyptians believed that the heat of this brilliant star added to the sun's heat to create this hot weather. Sirius was blamed for the withering droughts, sickness, and discomfort that occurred during this time. Traditional dog days start on July 3 and end on August 11.

What is **Indian summer**?

The term Indian summer dates back to at least 1778. The term may relate to the way the Native Americans availed themselves of the nice weather to increase their winter food supplies. It refers to a period of pleasant, dry, warm days in the middle to late autumn, usually after the first killing frost.

AIR PHENOMENA

What is a **bishop's ring**?

It is a ring around the sun, usually with a reddish outer edge. It is probably due to dust particles in the air, since it is observed after all great volcanic eruptions.

When does the **green flash** phenomenon occur?

On rare occasions, the sun may look bright green for a moment, as the last tip of the sun is setting. This green flash occurs because the red rays of light are hidden below the horizon and the blue are scattered in the atmosphere. The green rays are seldom seen because of dust and pollution in the lower atmosphere. It may best be seen when the air

An aurora seen over Alaska. Auroras are caused by solar winds interacting with the earth's magnetosphere.

is cloudless and when a distant, well-defined horizon exists, as on an ocean.

How often does an aurora appear?

Because it depends on solar winds (electrical particles generated by the sun) and sunspot activity, the frequency of an aurora cannot be determined. Auroras usually appear two days after a solar flare (a violent eruption of particles on the sun's surface) and reach their peak two years into the eleven-year sunspot cycle. The auroras, occurring in the polar regions, are broad displays of usually colored light at night. The northern polar aurora is called Aurora Borealis or Northern Lights and the southern polar aurora is called the Aurora Australis.

When and by whom were clouds first classified?

The French naturalist Jean Lamarck (1744–1829) proposed the first system for classifying clouds in 1802. His work, however, did not receive wide acclaim. A year later the Englishman Luke Howard (1772–1864) developed a cloud classification system that has been generally accepted and is still used today. Clouds are distinguished by their general appearance ("heap clouds" and "layer clouds") and by their height above the ground. Latin names and prefixes are used to describe these characteristics. The shape names are cirrus (curly or fibrous), stratus (layered), and cumulus (lumpy or piled). The prefixes denoting height are cirro (high clouds with bases above 20,000 feet [6,096 meters]) and alto (mid-level clouds from 6,500 to 20,000 feet [2,000 to 6,096 meters]). There is no prefix for low clouds. Nimbo or nimbus is also added as a name or prefix to indicate that the cloud produces precipitation.

What are the four major cloud groups and their types?

1. High Clouds—composed almost entirely of ice crystals. The bases of these clouds start at 20,000 feet (6,096 meters) and reach 40,000 feet (12,192 meters).
   - *Cirrus* (from Latin, "lock of hair")—thin, feather-like crystal clouds in patches or narrow bands. The large ice crystals that often trail downward in well-defined wisps are called "mares tails."
   - *Cirrostratus*—thin, white cloud layer that resembles a veil or sheet. This layer can be striated or fibrous. Because of the ice content, these clouds are associated with the halos that surround the sun or moon.

- *Cirrocumulus*—thin clouds that appear as small white flakes or cotton patches and may contain super-cooled water.

2. Middle Clouds—composed primarily of water. The height of the cloud bases range from 6,500 to 20,000 feet (2,000 to 6,096 meters).
   - *Altostratus*—appears as a bluish or grayish veil or layer of clouds that can gradually merge into altocumulus clouds. The sun may be dimly visible through it, but flat, thick sheets of this cloud type can obscure the sun.
   - *Altocumulus*—a white or gray layer or patches of solid clouds with rounded shapes.

3. Low Clouds—composed almost entirely of water that may at times be super-cooled; at subfreezing temperatures, snow and ice crystals may be present as well. The bases of these clouds start near Earth's surface and climb to 6,500 feet (2,000 meters) in the middle latitudes.
   - *Stratus*—gray, uniform, sheet-like clouds with a relatively low base, or they can be patchy, shapeless, low gray clouds. Thin enough for the sun to shine through, these clouds bring drizzle and snow.
   - *Stratocumulus*—globular rounded masses that form at the top of the layer.
   - *Nimbostratus*—seen as a gray or dark, relatively shapeless, massive cloud layer containing rain, snow, and ice pellets.

4. Clouds with Vertical Development—contain super-cooled water above the freezing level and grow to great heights. The cloud bases range from 1,600 feet (488 meters) to 20,000 feet (6,096 meters).
   - *Cumulus*—detached, fair-weather clouds with relatively flat bases and dome-shaped tops. These usually do not have extensive vertical development and do not produce precipitation.
   - *Cumulonimbus*—unstable, large, vertical clouds with dense boiling tops that bring showers, hail, thunder, and lightning.

Are there **clouds** that form in **specific areas** or **times** of day?

Lenticular clouds form only over mountain peaks. They look like a stack of different layers of cloud matter. Noctilucent clouds are the highest in Earth's atmosphere. They form only between sunset and sunrise between the latitudes of 50° and 70° north and south of the equator.

What is **cloud seeding**?

Cloud seeding is an attempt to modify the weather to produce precipitation. A cloud contains water droplets and ice crystals. Precipitation occurs when the ice crystals grow large enough to fall as rain, snow, hail, or other precipitation. Unless the conditions are right for ice crystals to grow, precipitation does not occur. Cloud seeding is

Although they look light and fluffy floating in the sky, clouds actually consist of water droplets and are heavy. A one cubic kilometer (1 km3 meaning 1 km × 1 km × 1km or 0.6 miles × 0.6 miles × 0.6 miles) cumulus cloud has an estimated weight of 2.211 billion pounds (1.003 billion kilograms). Clouds float because they are suspended on dry air. An equal volume (1 km³) of dry air weighs 2.220 billion pounds (1.007 billion kilograms). Dry air has a density of 1.007 kilograms/cubic meter while the density of the moist cloud is 1.003 kilograms/cubic meter. Clouds float because they are less dense than the drier air.

an attempt to convert the super-cooled droplets of liquid water in a cloud to ice crystals. Dry ice (solid carbon dioxide) and silver iodide are the substances most often used to seed clouds to transform the water droplets to ice crystals. Seeding is most efficient when the seeding agent is dropped from an airplane on the top of a cloud.

What is **lightning**?

Lightning is an electrical discharge occurring in the atmosphere accompanied by a vivid flash of light. During a thunderstorm, a positive charge builds in the upper part of a cloud and a negative charge builds in the lower part of the cloud. The difference between the positive and negative charges increases, generating an electrical field, until the electrical charge jumps from one area to another. Lightning may travel from cloud to ground or cloud to air or cloud to cloud or stay within a cloud. The main types of lightning are:

- Streak lightning: A single or multiple zigzagging line from cloud to ground.
- Forked lightning: Lightning that forms two branches simultaneously.
- Sheet lightning: A shapeless flash covering a broad area.
- Ribbon lightning: Streak lightning blown sideways by the wind to make it appear like parallel successive strokes.
- Bead or chain lightning: A stroke interrupted or broken into evenly spaced segments or beads.
- Ball lightning: A rare form of lightning in which a persistent and moving luminous white or colored sphere is seen. It can last from a few seconds to several minutes, and it travels at about a walking pace. Spheres have been reported to vanish harmlessly, or to pass into or out of rooms—leaving, in some cases, signs of their passage, such as a hole in a window pane. Sphere dimensions vary but are most commonly from 4 to 8 inches (10 to 20 centimeters) in diameter.

• Heat lightning: Lightning seen along the horizon during hot weather and believed to be a reflection of lightning occurring beyond the horizon.

How **common** are **lightning strikes**?

Thunderstorms are a very common natural phenomenon. At any given moment, there are 1,800 thunderstorms in progress somewhere on the earth. Lightning detection systems in the United States monitor an average of 25 million flashes of lightning from cloud to ground every year.

What is the **color** of **lightning**?

The atmospheric conditions determine the color of lightning. Blue lightning within a cloud indicates the presence of hail. Red lightning within a cloud indicates the presence of rain. Yellow or orange lightning indicates a large concentration of dust in the air. White lightning is a sign of low humidity in the air.

How **hot** is **lightning**?

The temperature of the air around a bolt of lightning is about 54,000°F (30,000°C), which is six times hotter than the surface of the sun, yet many times people survive being struck by a bolt of lightning. American park ranger Roy Sullivan (1912–1983), for example, was hit by lightning seven times between 1942 and 1977. In cloud-to-ground lightning, its energy seeks the shortest route to Earth, which could be through a person's shoulder, down the side of the body, through the leg, and to the ground. As long as the lightning does not pass across the heart or spinal column, the victim usually does not die.

How many **volts** are in **lightning**?

A stroke of lightning discharges from 10 to 100 million volts of electricity. An average lightning stroke has 30,000 amperes.

How **long** is a **lightning stroke**?

The visible length of the streak of lightning depends on the terrain and can vary greatly. In mountainous areas where clouds are low, the flash can be as short as 300 yards (273 meters); whereas in flat terrain, where clouds are high, the bolt can measure as long as 4 miles (6.5 kilometers). The usual length is about 1 mile (1.6 kilometers), but streaks of lightning up to 20 miles (32 kilometers) have been recorded. The stroke channel is very narrow—perhaps as little as half an inch (1.27 centimeters). It is surrounded by a "corona envelope" or a glowing discharge that can be as wide as 10 to 20 feet (3 to 6 meters) in diameter. The speed of lightning can vary from 100 to 1,000 miles (161 to 1,610 kilometers) per second for the downward leader track; the return stroke is 87,000 miles (140,070 kilometers) per second (almost half the speed of light).

How is the **distance** from a **lightning** flash **calculated**?

Count the number of seconds between seeing a flash of lightning and hearing the sound of the thunder. Divide the number by five to determine the number of miles away that the lightning flashed.

Does lightning ever **strike twice** in the **same place**?

Lightning can and often does strike twice in the same place. Since lightning bolts head for the highest and most conductive point, that point often receives multiple strikes of lightning in the course of a storm. In fact, tall buildings, such as the Empire State Building in New York, can be struck several times during the same storm. During one storm, lightning struck the Empire State Building 12 times. Designed as a lightning rod for the surrounding area, the Empire State Building is struck by lightning about one hundred times per year during multiple storms.

Why are **lightning rods** important?

The lightning rod was invented by Benjamin Franklin (1706–1790) around 1750 following his experiment with the kite and the key. In 1752, Franklin tied a metal key to the end of kite string and flew it during a thunderstorm. Franklin suspected that lightning was a natural form of electricity. He knew that if lightning was electricity, it would be attracted to the metal key. When sparks jumped from the metal key, he understood that electrical current had traveled from the electrified air above down the kite string to the key. This experiment confirmed that lightning is an electrical phenomenon. A lightning rod is often placed on the top of buildings to attract lightning bolts. They are designed to provide a safe path to ground the electricity so that it does not damage the building. In recent years, lightning rods have become even more important because the metal pipes that used to be installed for indoor plumbing and could serve as lightning rods are being replaced by nonconductive PVC pipes.

How many fatalities occur each year in the United States **due to lightning**?

During the years 1980 to 2009, an average of 57 people were killed each year by lightning. The ten-year average (2000 to 2009) decreased to only 41 people per year. An estimated 300 people are injured by lightning strikes in the United States annually.

What is **Saint Elmo's fire**?

Saint Elmo's fire has been described as a corona from electric discharge produced on high grounded metal objects, chimney tops, and ship masts. Since it often occurs during

What are fulgurites?

Fulgurites (from the Latin word *fulgur*, meaning "lightning") are petrified lightning, created when lightning strikes an area of dry sand. The intense heat of the lightning melts the sand surrounding the stroke into a rough, glassy tube forming a fused record of its path. These tubes may be 0.5 to 2 inches (1.5 to 5 centimeters) in diameter, and up to 10 feet (3 meters) in length. They are extremely brittle and break easily. The inside walls of the tube are glassy and lustrous while the outside is rough, with sand particles adhering to it. Fulgurites are usually tan or black in color, but translucent white ones also have been found.

thunderstorms, the electrical source may be lightning. Another description refers to this phenomenon as weak static electricity formed when an electrified cloud touches a high exposed point. Molecules of gas in the air around this point become ionized and glow. The name originated with sailors who were among the first to witness the display of spearlike or tufted flames on the tops of their ships' masts. Saint Elmo (which is a corruption of Saint Ermo) is the patron saint of sailors, so they named the fire after him.

How do **rainbows occur**?

A rainbow is a spectrum of light formed when sunlight interacts with droplets. Upon entering a water droplet, the white light is refracted, and dispersed, that is, spread apart into its individual wavelengths, just as in a prism. The light inside the droplet then reflects against the back of the water droplet before it refracts and disperses as it exits the droplet. The angle between entering and leaving is 40° for blue light, 42° for red.

What **conditions** must be met in order to see a **rainbow**?

There are two main conditions for witnessing a rainbow. The first is that the observer must be between the sun and the water droplets. The water droplets can either be rain, mist from a waterfall, or the spray of a garden hose. The second condition is that the angle between the sun, the water droplets, and the observer's eyes must be between 40° and 42°. Therefore, rainbows are most easily seen when the sun is close to the horizon so the rays striking the droplets are close to horizontal.

Is there such a thing as a **completely circular rainbow**?

All rainbows would be completely round except that the ground gets in the way of completing the circle. However, if viewed from a high altitude, such as an airplane, circular rainbows can been seen when the angle between the sun, the water droplets, and the plane is between the 40° and 42°. In this case, the rainbow is horizontal, meaning that it is parallel to the ground and therefore not blocked by the ground. This is quite a sight!

What is the **order of colors** in a **rainbow**?

Red, orange, yellow, green, blue, indigo, and violet are the colors of the rainbow, but these are not necessarily the sequence of colors that an observer might see. Rainbows are formed when raindrops reflect sunlight. As sunlight enters the drops, the different wavelengths of the colors that compose sunlight are refracted at different lengths to produce a spectrum of color. Each observer sees a different set of raindrops at a slightly different angle. Drops at different angles from the observer send different wavelengths (i.e., different colors) to the observer's eyes. Since the color sequence of the rainbow is the result of refraction, the color order depends on how the viewer sees this refraction from the viewer's angle of perception.

WIND

Where do **haboobs occur?**

Haboobs, derived from the Arabic word *habb* meaning "to blow," are violent dust storms with strong winds of sand and dust. They are most common in the Sahara region of Africa and the deserts of southwestern United States, Australia, and Asia.

What is **wind shear**?

Wind shear refers to rapid changes in wind speed and/or direction over short distances and is usually associated with thunderstorms. It is especially dangerous to aircraft.

What is the effect of a **microburst** on **aircraft**?

Microbursts are downbursts with a diameter of 2.5 miles (4 kilometers) or less. Often associated with thunderstorms, they can generate winds of hurricane force that change direction abruptly. Headwinds can become tailwinds in a matter of seconds, resulting in a loss of airspeed and loss of altitude. After microbursts caused several major air catastrophes in the 1970s and 1980s, the Federal Aviation Administration (FAA) installed warning and radar systems at airports to alert pilots to wind shear and microburst conditions.

What conditions result in **microclimates**?

Microclimates are small-scale regions where the average weather conditions are measurably different from the larger, surrounding region. Differences in temperature, precipitation, wind, or cloud cover can produce microclimates. Frequent causes of microclimates are differences in elevation, mountains that alter wind patterns, shorelines, and man-made structures, such as buildings, that can alter the wind patterns. Microclimates can be found anywhere from a protected corner in the backyard to larger areas, such as several miles from the shore near an inland lake

What is the **Coriolis effect**?

The nienteenth-century French engineer Gaspard C. Coriolis (1792–1843) discovered that the rotation of Earth deflects streams of air. Because Earth spins to the east, all moving objects in the Northern Hemisphere tend to turn somewhat to the right of a straight path, while those in the Southern Hemisphere turn slightly left. The Coriolis effect explains the lack of northerly and southerly winds in the tropics and polar regions; the northeast and southeast trade winds and the polar easterlies all owe their westward deflection to the Coriolis effect.

These three images taken over time clearly show a microburst in action. These sudden downbursts of air can be very dangerous to aircraft.

When was **jet stream discovered**?

A jet stream is a flat and narrow tube of air that moves more rapidly than the surrounding air. Discovered by World War II bomber pilots flying over Japan and the Mediterranean Sea, jet streams have become important with the advent of airplanes capable of cruising at over 30,000 feet (9,144 meters). The currents of air flow from west to east and are usually a few miles deep, up to 100 miles (160 kilometers) wide, and well over 1,000 miles (1,600 kilometers) in length. The air current must flow at over 57.5 miles (92 kilometers) per hour.

There are two polar jet streams, one in each hemisphere. They move between 30 and 70 degrees latitude, occur at altitudes of 25,000 to 35,000 feet (7,620 to 10,668 meters), and achieve maximum speeds of over 230 miles (368 kilometers) per hour. The subtropical jet streams (again one per hemisphere) wander between 20 and 50 degrees latitude. They are found at altitudes of 30,000 to 45,000 feet (9,144 to 13,715 meters) and have speeds of over 345 miles (552 kilometers) per hour.

What is a Siberian express?

This term describes storms that are severely cold and cyclonic; they descend from northern Canada and Alaska to other parts of the United States.

Why are the **horse latitudes** called by that name?

The horse latitudes are two high pressure belts characterized by low winds about 30 degrees north and south of the equator. Dreaded by early sailors, these areas have undependable winds with periods of calm. In the Northern Hemisphere, particularly near Bermuda, sailing ships carrying horses from Spain to the New World were often becalmed. When water supplies ran low, these animals were the first to be rationed water. Dying from thirst or tossed overboard, the animals were sacrificed to conserve water for the men. Explorers and sailors reported that the seas were "strewn with bodies of horses," which may be why the areas are called the horse latitudes. The term might also be rooted in complaints by sailors who were paid in advance and received no overtime when the ships slowly traversed this area. During this time they were said to be "working off a dead horse."

What are **halcyon days**?

This term is often used to refer to a time of peace or prosperity. Among sailors, it is the two-week period of calm weather before and after the shortest day of the year, approximately December 21. The phrase is taken from halcyon, the name the ancient Greeks gave to the kingfisher. According to legend, the halcyon built its nest on the surface of the ocean and was able to quiet the winds while its eggs were hatching.

What is an **Alberta clipper**?

An Alberta clipper is a little gyrating storm that develops on the Pacific front, usually over the Rocky Mountains of Alberta, Canada. This quick-moving storm moves southeast into the Great Plains, leaving a trail of cold air.

What is a **Chinook**?

A Chinook is a wind that is generally warm and originates from the eastern slope of the Rocky Mountains. It often moves from the southwest in a downslope manner, causing a noticeable rise in temperature that helps to warm the plains just east of the Rocky Mountains.

The Chinook is classified as a katabatic wind. A katabatic wind develops because of cold, heavy air spilling down sloping terrain, moving the lighter, warmer air in front of it. The air is dried and heated as it streams down the slope. At times the falling air becomes warmer than the air it restores. Some katabatic winds have been interestingly named, like Taku, a frigid wind in Alaska, or Santa Ana, a warmer wind from the Sierras.

What is the **Beaufort scale**?

The Beaufort scale was devised in 1805 by a British Admiral, Sir Francis Beaufort (1774–1857), to help mariners in handling ships. It uses a series of numbers from 0 to 17 to indicate wind speeds and applies to both land and sea.

Beaufort number	Name	Wind speed (mph/kph)
0	Calm	< 1 / < 1.5
1	Light air	1–3 / 1.5–4.8
2	Light breeze	4–7 / 6.4–11.3
3	Gentle breeze	8–12 / 12.9–19.3
4	Moderate breeze	13–18 / 21–29
5	Fresh breeze	19–24 / 30.6–38.6
6	Strong breeze	25–31 / 40.2–50
7	Moderate gale	32–38 / 51.5–61.1
8	Fresh gale	39–46 / 62.8–74
9	Strong gale	47–54 / 75.6–86.9
10	Whole gale	55–63 / 88.5–101.4
11	Storm	64–73 / 103–117.5
12–17	Hurricane	74+ / 119.1+

Is **Chicago** the **windiest city**?

Dodge City, Kansas, with an average wind speed of 13.9 miles per hours (22.4 kilometers per hour), ranks number one as the windiest city. By comparison, the average wind speed in Chicago is only 10.3 miles per hour (16.6 kilometers per hour). The highest surface wind ever recorded was on Mount Washington, New Hampshire, at an elevation of 6,288 feet (1.9 kilometers). On April 12, 1934, its wind speed was 231 miles (371.7 kilometers) per hour and its average wind speed was 35 miles (56.3 kilometers) per hour. The windiest place on Earth is Antarctica where winds can gust up to 200 miles per hour (322 kilometers per hour), which is twice as hard as in an average hurricane.

Who is associated with **developing** the **concept** of **wind chill**?

The Antarctic explorer Paul A. Siple (1908–1968) coined the term wind chill in his 1939 dissertation, "Adaptation of the Explorer to the Climate of Antarctica." Siple was the youngest member of Admiral Richard Byrd's Antarctica expedition in 1928–1930, and later made other trips to the Antarctic as part of Byrd's staff and for the U.S.

A bust of Admiral Richard Byrd was erected at McMurdo Station, Antarctica.

Department of the Interior assigned to the U.S. Antarctic Expedition. He also served in many other endeavors related to the study of cold climates.

What is meant by the **wind chill factor**?

The wind chill factor, or wind chill index, is a number that expresses the cooling effect of moving air at different temperatures. It indicates in a general way how many calories of heat are carried away from the surface of the body. The National Weather Service began reporting the equivalent wind chill temperature along with the actual air temperature in 1973. For years it was believed that the index overestimated the wind's cooling effect on skin. In 2001–2002 a new wind chill index was instituted, and additional corrections are expected in the next few years. Below is the official wind chill chart from the National Weather Service as of 2009.

Is there a **formula** for **computing** the **wind chill**?

Here is how you calculate the New Wind Chill Index:

$$T_{wc} 5 35.74 1 0.6215T - 35.75(V^{0.16}) 1 0.4275T(V^{0.16})$$

T_{wc} is the wind chill in degrees F, V is the wind speed in miles per hour, and T is the temperature in degrees F.

How does a **cyclone differ** from a **hurricane** or a **tornado**?

All three wind phenomena are rotating winds that spiral in toward a low-pressure center as well as upward. Their differences lie in their size, wind velocity, rate of travel, and duration. Generally, the faster the winds spin, the shorter (in time) and smaller (in size) the event becomes.

A cyclone has rotating winds from 10 to 60 miles per hour (16 to 97 kilometers per hour), can be up to 1,000 miles (1,600 kilometers) in diameter, travels about 25 miles per hour (40 kilometers per hour), and lasts from one to several weeks.

A hurricane (or typhoon, as it is called in the Pacific Ocean area) has winds that vary from 75 to 200 miles per hour (120 to 320 kilometers per hour), moves between 10 to 20 miles per hour (16 to 32 kilometers per hour), can have a diameter up to 600 miles (960 kilometers), and can exist from several days to more than a week. A tornado can reach a rotating speed of 300 miles per hour (400 kilometers per hour), travels between 25 to 40 miles per hour (40 to 64 kilometers per hour), and generally lasts only minutes, although some have lasted for five to six hours. Its diameter can range from 300 yards (274 meters) to 1 mile (1.6 kilometers) and its average path length is 5 miles (8 kilometers), with a maximum of 300 miles (483 kilometers).

A tornado touches down in Oklahoma.

Typhoons, hurricanes, and cyclones tend to breed in low-altitude belts over the oceans, generally from 5 degrees to 15 degrees latitude north or south. A tornado generally forms several thousand feet above Earth's surface, usually during warm, humid weather; many times it is in conjunction with a thunderstorm. Although a tornado can occur in many places, they mostly appear on the continental plains of North America (i.e., from the Plains States eastward to western New York and the southeastern Atlantic states).

What is the **Fujita** and **Pearson Tornado Scale**?

The Fujita and Pearson Tornado Scale was developed in 1971 by University of Chicago professor T. Theodore Fujita (1920–1998) and Allen Pearson (1925–), who was then the director of the National Severe Storms Forecast Center. It ranked tornadoes by their wind speed, path, length, and width. Tornadoes are not assessed based on actual wind speed and damage, but rather the scale determines wind speed based on damage. Sometimes known simply as the Fujita scale, the rankings ranged from F0 (very weak) to F6 (inconceivable).

F0—Light damage: damage to trees, billboards, and chimneys.

F1—Moderate damage: mobile homes pushed off their foundations and cars pushed off roads.

F2—Considerable damage: roofs torn off, mobile homes demolished, and large trees uprooted.

F3—Severe damage: even well-constructed homes torn apart, trees uprooted, and cars lifted off the ground.

Wind Chill Chart

Wind speed (mph) / Temperature (°F)

Wind speed (mph)	40	35	30	25	20	15	10	5	0	-5	-10	-15	-20	-25	-30	-35	-40	-45
5	36	31	25	19	13	7	1	-5	-11	-16	-22	-28	-34	-40	-46	-52	-57	-63
10	34	27	21	15	9	3	-4	-10	-16	-22	-28	-35	-41	-47	-53	-59	-66	-72
15	32	25	19	13	6	0	-7	-13	-19	-26	-32	-39	-45	-51	-58	-64	-71	-77
20	30	24	17	11	4	-2	-9	-15	-22	-29	-35	-42	-48	-55	-61	-68	-74	-81
25	29	23	16	9	3	-4	-11	-17	-24	-31	-37	-44	-51	-58	-64	-71	-78	-84
30	28	22	15	8	1	-5	-12	-19	-26	-33	-39	-46	-53	-60	-67	-73	-80	-87
35	28	21	14	7	0	-7	-14	-21	-27	-34	-41	-48	-55	-62	-69	-76	-82	-89
40	27	20	13	6	-1	-8	-15	-22	-29	-36	-43	-50	-57	-64	-71	-78	-84	-91
45	26	19	12	5	-2	-9	-16	-23	-30	-37	-44	-51	-58	-65	-72	-79	-86	-93
50	26	19	12	4	-3	-10	-17	-24	-31	-38	-45	-52	-60	-67	-74	-81	-88	-95
55	25	18	11	4	-3	-11	-18	-25	-32	-39	-46	-54	-61	-68	-75	-82	-89	-97
60	25	17	10	3	-4	-11	-19	-26	-33	-40	-48	-55	-62	-69	-76	-84	-91	-98

Wind Chill Chart

Wind speed (kph)	Temperature (°C)													
	0	-1	-2	-3	-4	-5	-10	-15	-20	-25	-30	-35	-40	-45
6	-2	-3	-4	-5	-7	-8	-14	-19	-25	-31	-37	-42	-48	-54
8	-3	-4	-5	-6	-7	-9	-14	-20	-26	-32	-38	-44	-50	-56
10	-3	-5	-6	-7	-8	-9	-15	-21	-27	-33	-39	-45	-51	-57
15	-4	-6	-7	-8	-9	-11	-17	-23	-29	-35	-41	-48	-54	-60
20	-5	-7	-8	-9	-10	-12	-18	-24	-30	-37	-43	-49	-56	-62
25	-6	-7	-8	-10	-11	-12	-19	-25	-32	-38	-44	-51	-57	-64
30	-6	-8	-9	-10	-12	-13	-20	-26	-33	-39	-46	-52	-59	-65
35	-7	-8	-10	-11	-12	-14	-20	-27	-33	-40	-47	-53	-60	-66
40	-7	-9	-10	-11	-13	-14	-21	-27	-34	-41	-48	-54	-61	-68
45	-8	-9	-10	-12	-13	-15	-21	-28	-35	-42	-48	-55	-62	-69
50	-8	-10	-11	-12	-14	-15	-22	-29	-35	-42	-49	-56	-63	-69
55	-8	-10	-11	-13	-14	-15	-22	-29	-36	-43	-50	-57	-63	-70
60	-9	-10	-12	-13	-14	-16	-23	-30	-36	-43	-50	-57	-64	-71
65	-9	-10	-12	-13	-15	-16	-23	-30	-37	-44	-51	-58	-65	-72
70	-9	-11	-12	-14	-15	-16	-23	-30	-37	-44	-51	-58	-65	-72
75	-10	-11	-12	-14	-15	-17	-24	-31	-38	-45	-52	-59	-66	-73
80	-10	-11	-13	-14	-16	-17	-24	-31	-38	-45	-52	-60	-67	-74
85	-10	-11	-13	-14	-16	-17	-24	-31	-39	-46	-53	-60	-67	-74
90	-10	-12	-13	-15	-16	-18	-25	-32	-39	-46	-53	-61	-68	-75
95	-10	-12	-13	-15	-16	-18	-25	-32	-39	-47	-54	-61	-68	-75
100	-11	-12	-14	-15	-16	-18	-25	-32	-40	-47	-54	-61	-69	-76
105	-11	-12	-14	-15	-17	-18	-25	-33	-40	-47	-55	-62	-69	-76
110	-11	-12	-14	-15	-17	-18	-26	-33	-40	-48	-55	-62	-70	-77

F4—Devastating damage: houses leveled, cars thrown, and objects become flying missiles.

F5—Incredible damage: structures lifted off foundations and carried away; cars become missiles. Less than two percent of tornadoes are in this category.

F6—An F6 tornado has never been recorded, but we surmise the damage would be devasting.

Fujita and Pearson Tornado Scale

Scale	Speed (mph)	Path length (miles)	Path width
0	< 72	< 1.0	< 17 yards
1	73–112	1.0–3.1	18–55 yards
2	113–157	3.2–9.9	56–175 yards
3	158–206	10.0–31.0	176–556 yards
4	207–260	32.0–99.0	0.34–0.9 miles
5	261–318	100–315	1.0–3.1 miles
6	319–380	316–999	3.2–9.9 miles

How does the **Enhanced Fujita Scale** differ from the original Fujita Scale?

The National Weather Service adapted the Enhanced Fujita (EF) Scale on February 1, 2007 to rate tornadoes. The enhanced scale has six categories, EF0 to EF5, representing increasing levels of damage. It was revised to better estimate wind speeds by considering different types of construction and low-populated areas with few structures. The enhanced scale offers more detailed descriptions of potential damages by using 28 Damage Indicators based on different building structures and vegetation.

Enhanced Fujita Scale

Scale	Wind Speed (mph/kph)	Damages
EF0	65–85/105–137	Tree branches break off, trees with shallow roots fall over; house siding and gutters damaged; some roof shingles peel off or other minor roof damage.
EF1	86–110/137–177	Mobile homes overturned; doors, windows, and glass broken; severe damage to roofs.
EF2	111–135/178–217	Large tree trunks split and big trees fall over; mobile homes destroyed, and homes on foundations are shifted; cars lifted off the ground; roofs torn off; some lighter objects thrown at the speed of missiles.
EF3	136–165/218–265	Trees broken and debarked; mobile homes completely destroyed and houses on foundations lose stories, and buildings with weaker foundations are lifted and

		blown distances; commercial buildings such as shopping malls are severely damaged; heavy cars are thrown and trains are tipped over.
EF4	166–200/266–322	Frame houses leveled; cars thrown long distances; larger objects become dangerous projectiles.
EF5	> 200/> 323	Homes are completely destroyed and even steel-reinforced buildings are severely damaged; objects the size of cars are thrown distances of 300 feet (90 meters) or more. Total devastation.

Have **wind speeds** during **tornadoes** been accurately **measured**?

No, tornado wind speeds have been scientifically estimated using Doppler radar and video observations, but there have been no successful attempts to physically measure wind speeds using an anemometer. Many severe tornadoes will destroy an anemometer before it records the wind speed during a tornado. Furthermore, they may occur in many random locations without equipment in place to measure wind speed.

How **long** do most **tornadoes last**?

Most tornadoes have short lifespans, lasting less than ten minutes though they may be as short as only several seconds. Some may last an hour or longer. Tornadoes during the early to mid-1900s were often reported to be longer lived, though many climatologists believe these may have been tornado series instead of one single event. An average tornado will travel 5 miles (8 kilometers) during its lifespan. The tornado of March 18, 1925 traveled 219 miles (352 kilometers) through Missouri, Illinois, and Indiana at an average speed of 60 to 73 miles per hour (97 to 117 kilometers per hour).

What was the **greatest natural disaster** in **U.S.** history?

The greatest natural disaster occurred when a hurricane struck Galveston, Texas, on September 8, 1900, and killed over 8,000 people.

What is the **biggest** known **tornado**?

The tornado on May 22, 2004 in Hallam, Nebraska holds the record for the peak width of any tornado at nearly 2.5 miles (4 kilometers) across.

Do **tornadoes** in the **Northern** and **Southern Hemispheres** rotate in the **same direction**?

In general, tornadoes in the Northern Hemisphere rotate counterclockwise (cyclonically) while those in the Southern Hemisphere rotate clockwise (anticyclonically).

Occasionally, anticyclonic tornadoes have been observed in the Northern Hemisphere. Typically, anticyclonic tornadoes in the Northern Hemisphere are weaker twisters associated with weak storm cells or sometimes appearing as waterspouts. In 1998, a tornado spinning anticyclonically was observed near Sunnyvale, California. Rarer, but still possible, is a supercell, which generates both cyclonic and anticyclonic tornadoes.

When do most **tornadoes happen**?

Tornadoes can happen at any time of the year and at any time during the day or night. Tornadoes are most likely to happen in the United States during March through August. There is a general northward shift in the "tornado season" from late winter through mid-summer. Regionally, tornado activity increases in the central Gulf Coast as early as February, then shifts to the southeastern Atlantic states in March and April before moving to the southern Plains in May and early June. The greatest amount of tornado activity is in June and July in the northern Plains and upper Midwest. Statistically, most tornadoes strike between 4:00 P.M. and 9:00 P.M..

How many **tornadoes** occur in the **United States** each year?

There are about 1,000 tornadoes in the United States each year. Compiling an actual average is difficult since reporting methods have changed over the last several decades so the officially recorded tornado climatologies are believed to be incomplete. Some tornadoes, especially ones that cause little or no damage in remote areas, may not be reported. The three-year average for 2007 to 2009 is 1,316 tornadoes per year. The largest single outbreak of tornadoes occurred on April 3 and 4, 1974; 148 tornadoes were recorded in this "Super Outbreak" in the Great Plains and Midwestern states. Six of these tornadoes had winds greater than 260 miles (420 kilometers) per hour and some of them were the strongest ever recorded.

How many **fatalities** occur each **year** in the **United States** due to tornadoes?

During the past 30 years, 1979 to 2008, an average of 57 people were killed each year by tornadoes; nearly identical to the average number of people killed by lightning during the same time period. The 10 year average, 1999 to 2008, for deaths due to tornadoes was 63 people per year.

Year	Total Fatalities
2008	124
2007	81
2006	66
2005	38
2004	35
2003	54
2002	55
2001	40
2000	41
1999	94
1998	130
1997	67
1996	25
1995	30

How are **hurricanes classified**?

The Saffir/Simpson Hurricane Damage-Potential scale assigns numbers 1 through 5 to measure the disaster potential of a hurricane's winds and its accompanying storm surge. The purpose of the scale, developed in 1971 by Herbert Saffir (1917–2007) and Robert Simpson (1912–), is to help disaster agencies gauge the potential significance of these storms in terms of assistance.

Saffir/Simpson Hurricane Scale Ranges

Scale number (category)	Barometric pressure (in inches)	Winds (miles per hour)	Surge (in feet)	Damage
1	> 28.94	74–95	4–5	Minimal
2	28.50–28.91	96–110	6–8	Moderate
3	27.91–28.47	111–130	9–12	Extensive
4	27.17–27.88	131–155	13–18	Extreme
5	< 27.17	> 155	> 18	Catastrophic

Damage categories:

Minimal—No real damage to building structures. Some tree, shrubbery, and mobile home damage. Coastal road flooding and minor pier damage.

Moderate—Some roof, window, and door damage. Considerable damage to vegetation, mobile homes, and piers. Coastal and low-lying escape routes flood two to four hours before center of storm arrives. Small craft can break moorings in unprotected areas.

Extensive—Some structural damage to small or residential buildings. Mobile homes destroyed. Flooding near coast destroys structures and floods of homes 5 feet (1.5 meters) above sea level as far inland as 6 miles (9.5 kilometers).

Extreme—Extensive roof, window, and door damage. Major damage to lower floors of structures near the shore, and some roof failure on small residences. Complete beach erosion. Flooding of terrain 10 feet (3 meters) above sea level as far as 6 miles (9.5 kilometers) inland requiring massive residential evacuation.

Catastrophic—Complete roof failure to many buildings; some complete building failure, with small utility buildings blown away. Major damage to lower floors of all structures 19 feet (5.75 meters) above sea level located within 500 yards (547 meters) of the shoreline. Massive evacuation of residential areas on low ground 5 to 10 miles (8 to 16 kilometers) from shoreline may be required.

What is the **origin** of the **term "hurricane"**?

The term "hurricane" is derived from "Hurican," the Carib god of evil, which was derived from the Mayan god "Hurakan." Hurakan was one of the Mayan creator gods, who blew his breath across the chaotic water and brought forth dry land.

How do **hurricanes** get their **names**?

Since 1950, hurricane names have been officially selected from library sources and are decided on during the international meetings of the World Meteorological Organization (WMO). The names are chosen to reflect the cultures and languages found in the Atlantic, Caribbean, and Hawaiian regions. When a tropical storm with rotary action and wind speeds above 39 miles (63 kilometers) per hour develops, the National Hurricane Center near Miami, Florida, selects a name from one of the six listings for Region 4 (Atlantic and Caribbean area). Letters Q, U, X, Y, and Z are not included because of the scarcity of names beginning with those letters.

2011	2012	2013	2014	2015	2016
Arlene	Alberto	Andrea	Arthur	Ana	Alex
Bret	Beryl	Barry	Bertha	Bill	Bonnie
Cindy	Chris	Chantal	Cristobal	Claudette	Colin
Don	Debbie	Dorian	Dolly	Danny	Danielle
Emily	Ernesto	Erin	Edouard	Erika	Earl
Franklin	Florence	Fernand	Fay	Fred	Fiona

2011	2012	2013	2014	2015	2016
Gert	Gordon	Gabrielle	Gonzalo	Grace	Gaston
Harvey	Helene	Humberto	Hanna	Henri	Hermine
Irene	Isaac	Ingrid	Isaias	Ida	Igor
Jose	Joyce	Jerry	Josephine	Joaquin	Julia
Katia	Kirk	Karen	Kyle	Kate	Karl
Lee	Leslie	Lorenzo	Laura	Larry	Lisa
Maria	Michael	Melissa	Marco	Mindy	Matthew
Nate	Nadine	Nestor	Nana	Nicholas	Nicole
Ophelia	Oscar	Olga	Omar	Odette	Otto
Philippe	Patty	Pablo	Paulette	Peter	Paula
Rina	Rafael	Rebekah	Rene	Rose	Richard
Sean	Sandy	Sebastien	Sally	Sam	Shary
Tammy	Tony	Tanya	Teddy	Teresa	Tomas
Vince	Valerie	Van	Vicky	Victor	Virginie
Whitney	William	Wendy	Wilfred	Wanda	Walter

Which hurricane names have been retired?

Once a hurricane has done a great deal of damage and caused loss of life, its name is retired from the six-year list cycle. Countries affected by such hurricanes will petition the WMO to have the name retired and replaced with a new name.

Atlantic Storm Names Retired into Hurricane History

Hurricane Name	Year	Area Affected
Agnes	1972	Florida, Northeast U.S.
Alicia	1983	North Texas
Allen	1980	Antilles, Mexico, South Texas
Allison*	2001	Gulf of Mexico, Texas, Louisiana
Andrew	1992	Bahamas, South Florida, Louisiana
Anita	1977	Mexico
Audrey	1957	Louisiana, North Texas
Betsy	1965	Bahamas, Southeast Florida, Southeast Louisiana
Beulah	1967	Antilles, Mexico, South Texas
Bob	1991	North Carolina, Northeast U.S.
Camille	1969	Louisiana, Mississippi, Alabama
Carla	1961	Texas
Carmen	1974	Mexico, Central Louisiana
Carol	1954	Northeast U.S.

Hurricane Name	Year	Area Affected
Celia	1970	South Texas
Charley	2004	Florida, Cuba, Captiva Island
Cesar	1996	Honduras
Cleo	1964	Lesser Antilles, Haiti, Cuba, Southeast Florida
Connie	1955	North Carolina
David	1979	Lesser Antilles, Hispañola, Florida, Eastern U.S.
Dean	2007	Caribbean, Yucatan Peninsula, Mexico
Dennis	2005	Cuba, Haiti, Jamaica, Florida
Diana	1990	Mexico
Diane	1955	Mid-Atlantic U.S., Northeast U.S.
Donna	1960	Bahamas, Florida, Eastern U.S.
Dora	1964	Northeast Florida
Edna	1954	Maine, New Brunswick
Elena	1985	Mississippi, Alabama, Western Florida
Eloise	1975	Antilles, Northwest Florida, Alabama
Fabian	2003	Bermuda
Felix	2007	Nicaragua, Honduras
Fifi	1974	Yucatan Peninsula, Louisiana
Flora	1963	Haiti, Cuba
Floyd	1999	North Carolina, Eastern U.S.
Fran	1996	North Carolina
Frances	2004	Bahamas, Florida
Frederic	1979	Alabama, Mississippi
Georges	1998	Antigua, St. Kitts and Nevis, Puerto Rico, Domenican Republic, Haiti, Cuba
Gilbert	1988	Lesser Antilles, Jamaica, Yucatan Peninsula, Mexico
Gloria	1985	North Carolina, Northeast U.S.
Gustav	2008	Haiti, Dominican Republic, Louisiana
Hattie	1961	Belize, Guatemala
Hazel	1954	Antilles, North and South Carolina
Hilda	1964	Louisiana
Hortense	1996	Puerto Rico, Dominican Republic
Hugo	1989	Antilles, South Carolina
Ike	2008	Haiti, Cuba, Texas
Inez	1966	Lesser Antilles, Hispañola, Cuba, Florida Keys, Mexico
Ione	1955	North Carolina
Iris	2001	Belize
Isabel	2003	North Carolina, Virginia, Eastern U.S.
Isidore	2002	Cuba, Yucatan Peninsula, Louisiana
Ivan	2004	Barbados, Grenada, Grand Cayman, Alabama, Gulf Coast
Janet	1955	Lesser Antilles, Belize, Mexico

Hurricane Name	Year	Area Affected
Jeanne	2004	Dominican Republic, Haiti, Florida
Joan	1988	Curacao, Venezuela, Colombia, Nicaragua (crossed into the Pacific and became Miriam)
Juan	2003	Nova Scotia, Prince Edward Island (Atlantic Canada)
Katrina	2005	Louisiana, Mississippi, Florida, Alabama, Georgia
Keith	2000	Belize, southern Mexico
Klaus	1990	Martinique
Lenny	1999	Antilles
Lili	2002	Cuba, Louisiana, Windward Islands, Cayman Islands
Luis	1995	Leeward Islands, Lesser Antilles, Bermuda
Marilyn	1995	Bermuda
Michelle	2001	Cuba, Jamaica
Mitch	1998	Central America, Nicaragua, Honduras
Noel	2007	Haiti, Cuba, Bahamas, Puerto Rico
Opal	1995	Florida Panhandle
Paloma	2008	Cayman Islands, Cuba, Jamaica
Rita	2006	Louisiana, Texas
Roxanne	1995	Yucatan Peninsula
Stan	2005	Mexico, Central America
Wilma	2005	Cuba, Yucatan Peninsula, Mexico, Florida

*Allison is the only name that has been retired as a tropical storm. It never became a hurricane.

Which **U.S. hurricanes** have caused the **most deaths**?

The ten deadliest United States hurricanes are listed below.

Hurricane	Year	Deaths
Texas (Galveston)	1900	8,000
Florida (Lake Okeechobee)	1928	2,500
Louisiana, Mississippi (Katrina)	2005	1,833
Louisiana (Cheniere Caminanda)	1893	1,100–1,400
South Carolina, Georgia (Sea Islands)	1893	1,000–2,000
Georgia, South Carolina	1881	700
Louisiana/Texas (Audrey)	1957	416
Florida (Keys)	1935	408
Louisiana (Last Island)	1856	400
Florida (Miami), Mississippi, Alabama, Pensacola	1926	372

Which **hurricane** in the **United States** was the **most destructive**?

The economic loss caused by Hurricane Katrina is estimated at $100 to $150 billion making it the costliest natural disaster in United States. The death toll for Katrina was

An image taken from space shows Hurricane Katrina heading for the Gulf Coast.

less than 2,000, which is significant although nowhere near the death toll of at least 8,000 during the Galveston hurricane in 1900. Katrina made landfall in Plaquemines Parish, Louisiana as a category 3 hurricane with winds of 125 miles per hour (201 kilometers per hour) on August 29, 2005. The coastal areas of Louisiana (including New Orleans), Mississippi, and Alabama suffered extensive damage.

PRECIPITATION

What is a **"white-out"**?

An official definition for "white-out" does not exist. It is a colloquial term that can describe any condition during snowfall that severely restricts visibility. That may mean a blizzard, or snow squall, etc. If you get some sunlight in the mix, that makes the situation even worse—it's like driving in fog with your headlights on high-beam. The light gets backscattered right into your eyes and you can't see.

What is the **dew point**?

The dew point is the temperature at which air is full of moisture and cannot store any more. When the relative humidity is 100 percent, the dew point is either the same as or lower than the air temperature. If a fine film of air contacts a surface and is chilled to below the dew point, then actual dew is formed. This is why dew often forms at night or early morning: as the temperature of the air falls, the amount of water vapor the air can hold also decreases. Excess water vapor then condenses as very small drops on whatever it touches. Fog and clouds develop when sizable volumes of air are cooled to temperatures below the dew point.

What is the **shape** of a **raindrop**?

Although a raindrop has been illustrated as being pear-shaped or tear-shaped, high-speed photographs reveal that a large raindrop has a spherical shape with a hole not quite through it (giving it a doughnut-like appearance). Water surface tension pulls the drop into this shape. As a drop larger than 0.08 inch (2 millimeters) in diameter

> ## How fast does rain fall?
>
> The speed of rainfall varies with drop size and wind speed. A typical raindrop in still air falls about 600 feet (182 meters) per minute or about 7 miles (11 kilometers) per hour. Large raindrops can reach speeds of 16 to 20 miles (26 to 32 kilometers) per hour. Very small raindrops may fall slowly at no more than 1 mile (1.6 kilometer) per hour.

falls, it will become distorted. Air pressure flattens its bottom and its sides bulge. If it becomes larger than 0.25 inch (6.4 millimeters) across, it will keep spreading cross-wise as it falls and will bulge more at its sides, while at the same time, its middle will thin into a bow-tie shape. Eventually in its path downward, it will divide into two smaller spherical drops.

What is the **greatest amount** of **rainfall** ever measured?

The table below summarizes the places that experienced the greatest amount of rainfall and just how much was measured.

Date	Place	Time duration	Inches/Centimeters
November 26, 1970	Barst, Guadeloupe, West Indies	1 minute	1.50/3.80
March 15–16, 1952	Cilaos, La Réunion, Indian Ocean	24 hours	73.62/187.00
July 25–26, 1979	Alvin, Texas	24 hours	19.00/48.30
July 1861	Cherrapunji, Meghalaya, India	calendar month	366.14/930.00
August 1, 1860 to July 31, 1861	Cherrapunji, Meghalaya, India	12 months	1,041.78/2,646.00
December 1981–December 1982	Kukui, Maui, Hawaii	12 months	739.00/1,877.00

Where is the **rainiest place** on Earth?

The wettest place in the world is Mawsynram, India, with an average annual rainfall of 467.4 inches (1,187 centimeters) per year. Second place is Tutunendo, Colombia, with an average annual rainfall of 463.4 inches (1,177 centimeters) per year. It is estimated that Lloro, Columbia receives 523.6 inches (1,330 centimeters) of rain per year, but this amount has not been verified. The place that has the most rainy days per year is Mount Wai-'ale'ale on Kauai, Hawaii. It has up to 350 rainy days annually.

How far away can thunder be heard?

Thunder is the crash and rumble associated with lightning. It is caused by the explosive expansion and contraction of air heated by the stroke of lightning. This results in sound waves that can be heard easily 6 to 7 miles (9.7 to 11.3 kilometers) away. Occasionally such rumbles can be heard as far away as 20 miles (32.2 kilometers). The sound of great claps of thunder is produced when intense heat and the ionizing effect of repeated lightning occurs in a previously heated air path. This creates a shockwave that moves at the speed of sound.

In contrast, the longest rainless period in the world was from October 1903 to January 1918 at Arica, Chile—a period of 14 years. In the United States the longest dry spell was 767 days at Bagdad, California, from October 3, 1912, to November 8, 1914.

When do **thunderstorms occur**?

In the United States thunderstorms usually occur in the summertime, especially from May through August. Thunderstorms tend to occur in late spring and summer when large amounts of tropical maritime air move across the United States. Storms usually develop when the surface air is heated the most from the sun (2 P.M. to 4 P.M.). Thunderstorms are relatively rare in the New England area, North Dakota, Montana, and other northern states (latitude 60 degrees), where the air is often too cold. These storms are also rare along the Pacific Ocean because the summers there are too dry for these storms to occur. Florida, the Gulf states, and the southeastern states tend to have the most storms, averaging 70 to 90 annually. The mountainous southwest averages 50 to 70 storms annually. In the world, thunderstorms are most plentiful in the areas between latitude 35 degrees north and 35 degrees south; in these areas there can be as many as 3,200 storms within a 12-hour nighttime period. As many as 1,800 storms can occur at once throughout the world.

How **loud** is **thunder**?

A clap of thunder can be as loud as 120 decibels, which is comparable to the noise at a rock concert, a chainsaw, or a pneumatic drill.

How **large** can **hailstones** become?

The average hailstone is about 0.25 inch (0.64 centimeter) in diameter. However, hailstones weighing up to 7.5 pounds (3.4 kilograms) are reported to have fallen in Hyderabad state in India in 1939, although scientists think these huge hailstones may be several stones that partly melted and stuck together. On April 14, 1986, hailstones weighing 2.5 pounds (1 kilogram) were reported to have fallen in the Gopalgang district of Bangladesh.

The largest hailstone ever recorded in the United States fell in Coffeyville, Kansas, on September 3, 1970. It measured 5.57 inches (14.15 centimeters) in diameter and weighed 1.67 pounds (0.75 kilogram).

Hail is precipitation consisting of balls of ice. Hailstones usually are made of concentric, or onion-like, layers of ice alternating with partially melted and refrozen snow, structured around a tiny central core. It is formed in cumulonimbus, or thunderclouds, when freezing water and ice cling to small particles in the air, such as dust. The winds in the cloud blow the particles through zones of different temperatures, causing them to accumulate additional layers of ice and melting snow and to increase in size.

Hail Size Estimates

Description	Size
Pea size	0.25 inch
Penny/Dime	0.75 inch
Nickel	0.88 inch
Quarter	1.00 inch
Half Dollar or Susan B. Anthony Dollar	1.25 inches
Ping Pong Ball	1.50 inches
Golfball	1.75 inches
Hen Egg	2.00 inches
Tennis Ball	2.50 inches
Baseball	2.75 inches
Grapefruit	4.00 inches

What is the **difference** between **freezing rain** and **sleet**?

Freezing rain is rain that falls as a liquid but turns to ice on contact with a freezing object to form a smooth ice coating called glaze. Usually freezing rain only lasts a short time, because it either turns to rain or to snow. Sleet is frozen or partially frozen rain in the form of ice pellets. Sleet forms when rain falls from a warm layer of air, passes through a freezing air layer near Earth's surface, and forms hard, clear, tiny ice pellets that can hit the ground so fast that they bounce off with a sharp click.

How does **snow form**?

Snow is not frozen rain. Snow forms by sublimation of water vapor—the turning of water vapor directly into ice, without going through the liquid stage. High above the ground, chilled water vapor turns to ice when its temperature reaches the dew point. The result of this sublimation is a crystal of ice, usually hexagonal. Snow begins in the form of these tiny hexagonal ice crystals in the high clouds; the young crystals are the seeds from which snowflakes will grow. As water vapor is pumped up into the air by updrafts, more water is deposited on the ice crystals, causing them to grow. Soon some of the larger crystals fall to the ground as snowflakes.

Are all **snowflakes** shaped **alike**?

Some snowflakes may have strikingly similar shapes, but these twins are probably not molecularly identical. In 1986, cloud physicist Nancy Knight believed she found a uniquely cloned pair of crystals on an oil coated slide that had been hanging from an airplane. This pair may have been the result of breaking off from a star crystal, or were attached side by side, thereby experiencing the same weather conditions simultaneously. Unfortunately the smaller aspects of each of the snow crystals could not be studied because the photograph was unable to capture possible molecular differences. So, even if the human eye may see twin flakes, on a minuscule level these flakes are different.

When does **frost form**?

A frost is a crystalline deposit of small thin ice crystals formed on objects that are at freezing or below freezing temperatures. This phenomenon occurs when atmospheric water vapor condenses directly into ice without first becoming a liquid; this process is called sublimation. Usually frost appears on clear, calm nights, especially during early autumn when the air above the earth is quite moist. Permafrost is permanently frozen ground that never thaws out completely.

How **much water** is in an **inch** of **snow**?

An average figure is 10 inches (25 centimeters) of snow is equal to 1 inch (2.5 centimeters) of water. Heavy, wet snow has a high water content; 4 to 5 inches (10 to 12 centimeters) may contain 1 inch (2.5 centimeters) of water. A dry, powdery snow might require 15 inches (38 centimeters) of snow to equal 1 inch (2.5 centimeters) of water.

What is the **record** for the **greatest snowfall** in the United States?

The record for the most snow in a single storm is 189 inches (480 centimeters) at Mount Shasta Ski Bowl in California from February 13–19, 1959. For the most snow in a 24-hour day, the record goes to Silver Lake, Colorado, on April 14–15, 1921, with 76 inches (193 centimeters) of snow. The year record goes to Paradise, Mount Rainier, in Washington, with 1,224.5 inches (3,110 centimeters) from February 19, 1971, to February 18, 1972. The highest average annual snowfall is 241 inches (612 centimeters) for Blue Canyon, California. In March 1911, Tamarack, California, had the deepest snow accumulation—over 37.5 feet (11.4 meters).

WEATHER PREDICTION

What is the **difference** between **weather** and **climate**?

Weather is the current condition of the atmosphere. Climate is the long-term average weather for a particular place or region. Often a 30-year average, although also calculated for ten-year averages, climate describes the average weather pattern of a locale. Climatic elements include precipitation, temperature, humidity, sunshine, wind velocity, and other measures of weather such as fog, frost, and other storms.

When was the **National Weather Service founded**?

The National Weather Service is part of the National Oceanic and Atmospheric Administration (NOAA). It was started on February 9, 1870 when President Ulysses S. Grant (1822–1885) signed a joint resolution of Congress authorizing the secretary of war to establish a national weather service. Its first official name was the National Weather Bureau. It was renamed the U.S. Weather Bureau in 1891 and became the National Weather Service in 1967. Its mission is to provide weather, hydrologic, and climate forecasts and warning for the citizens of the United States, its territories, adjacent waters, and ocean areas in order to protect life and property and for the enhancement of the national economy. Weather observations are now made hourly and daily by government agencies, volunteer/citizen observers, ships, planes, automatic weather stations, and Earth-orbiting satellites.

Who was **Cleveland Abbe**?

Cleveland Abbe (1838–1916), nicknamed the "Father of the Weather Bureau," was appointed director of the Cincinnati observatory in 1868. *The Weather Bulletin of the*

Cleveland Abbe

Cincinnati Observatory began including a three-day weather forecast on September 22, 1869 prepared by forecaster Cleveland Abbe. In January 1871, Abbe was appointed as a civilian assistant in the National Weather Bureau where he continued to prepare weather reports and forecasts.

What is the **difference** between a National Weather Service **statement, advisory, watch**, and **warning**?

The National Weather Service will issue a *statement* as a "first alert" of the possibility of severe weather. An *advisory* is issued when weather conditions are not life threatening, but individuals need to be alert to weather conditions. A weather *watch* is issued when conditions are more favorable than usual for dangerous weather conditions, e.g., tornadoes and violent thunderstorms. A watch is a recommendation for planning, preparation, and increased awareness (i.e., to be alert for changing weather, listen for further information, and think about what to do if the danger materializes). A *warning* is issued when a particular weather hazard is either imminent or has been reported. A warning indicates the need to take action to protect life and property. The type of hazard is reflected in the type of warning (e.g., tornado warning, blizzard warning).

Are there specific **criteria** to prompt the National Weather Service to **issue** an **advisory, watch**, or **warning**?

Certain advisories, watches, or warnings are issued when specific conditions are met while others will vary by location. Examples of when some common National Weather Service advisories, watches, and warnings issued are:

- Hurricane watch: Issued when a hurricane poses a possible threat to specific coastal areas, generally within 36 hours.
- Hurricane warning: Issued when sustained winds of 74 miles per hour (119 kilometers per hour) or higher associated with a hurricane are expected in a specified coastal area in 24 hours or less. A hurricane warning may remain in effect even when winds are below hurricane force when dangerously high water and/or exceptionally high waves continue to persist.

What were Benjamin Franklin's contributions to meteorology?

Benjamin Franklin, who is said to have discovered electricity by flying a kite in a storm, made the important discovery that low pressure systems caused the atmosphere to circulate in a rotating pattern. He made this discovery in 1743, after unsuccessfully attempting to view an eclipse on October 21. There was a storm in Philadelphia at the time, but he later learned that the skies were clear in Boston that day. He then found out that the storm that had been in Philadelphia traveled to Boston the next day. From this information, he surmised that the storm was traveling in a clockwise manner from southwest to northeast. Franklin thus concluded that the low pressure system was causing the storm to move in this manner.

- Severe thunderstorm warning: Issued when either a severe thunderstorm is indicated by radar or a spotter reports a thunderstorm producing hail 0.75 inch (2 centimeters) in diameter and/or winds equal to or exceeding 58 miles per hour (93 kilometers per hour). Lightning frequency is not a criterion for issuing a severe thunderstorm warning. It is advisable to seek safe shelter immediately when a severe thunderstorm warning is issued.

- Tornado watch: Issued when conditions are favorable for the development of tornadoes in the area. They normally are issued well in advance of the actual occurrence of severe weather for a duration of four to eight hours.

- Tornado warning: Issued when a tornado is indicated by radar or sighted by spotters. People in the affected area should seek safe shelter immediately. They are usually issued for around 30 minutes.

- Wind advisory: Issued when there are sustained winds of 25 to 39 miles per hour (40 to 63 kilometers per hour) and/or gusts to 57 miles per hour (92 kilometers per hour).

- Winter weather advisory: Issued when a low pressure system produces a combination of winter weather (snow, freezing rain, sleet) that presents a hazard, but does not meet warning criteria.

- Winter storm watch: Issued when there is the potential for heavy snow or significant ice accumulations usually at least 24 to 36 hours in advance. The criteria for this watch can vary from place to place.

- Winter storm warning: Issued when a winter storm is producing or is forecast to produce heavy snow or significant ice accumulations. The criteria for this warning can vary from place to place.

What is **Doppler radar**?

Doppler radar measures frequency differences between signals bouncing off objects moving away from or toward it. By measuring the difference between the transmitted and received frequencies, Doppler radar calculates the speed of the air in which the rain, snow, ice crystals, and even insects are moving. It can then be used to predict speed and direction of wind and amount of precipitation associated with a storm. The National Weather Service has installed a series of NEXRAD (Next Generation Radar) Doppler Radar systems throughout the country. They are especially helpful in measuring the speed of tornadoes and other violent thunderstorms.

When did **modern weather forecasting** begin?

On May 14, 1692, a weekly newspaper, *A Collection for the Improvement of Husbandry and Trade*, gave a seven-day table with pressure and wind readings for the comparable dates of the previous year. Readers were expected to make up their own forecasts from the data. Other journals soon followed with their own weather features. In 1771, a new journal called the *Monthly Weather Paper* was completely devoted to weather prediction. The first daily newspaper weather report was published on August 31, 1848 in the *Daily News* in London. The first daily weather forecast was published in the *Times* of London in 1860. The first broadcast of weather forecasts was done by the University of Wisconsin's station 9XM at Madison, Wisconsin, on January 3, 1921.

When was the **first weather satellite** launched?

The first weather satellite, the Television and Infrared Observation Satellite (TIROS I), was launched by NASA on April 1, 1960. Although the images were not of the same resolution as we have now, they were able to reveal the organization and structure of clouds and storms. One of its accomplishments was to see a previously undetected tropical storm near Australia. The information was conveyed to the people so they could prepare for the approaching storm. It operated for 77 days until mid-June 1960 when an electrical fire caused it to cease operating.

Barometers measure atmospheric pressure, which in turn can help predict the weather. Higher pressure systems bring fair weather, while low pressure systems bring clouds and precipitation.

Who **invented** the **barometer**?

A barometer is a device that measures air pressure. It was invented in 1644 by Evangelista Torricelli (1608–1647). As a student of Galileo Galilei (1564–1642) for a short time, Torricelli was inspired by Galileo's observation that piston pumps can only lift water up 33 feet (about 10 meters), after which point it is impossible to pump the water any higher. Torricelli proposed that air had weight, and therefore, exerted pressure. He tested his theory by filling a dish with mercury, a liquid that is 13.6 times denser than water. (A liquid that is denser than water allowed him to use a smaller quantity and a glass tube, which was easier to manipulate). He then took a glass tube 4 feet (1.2 meters) long glass tube that was open on one end, filled it with mercury, and turned it upside down with the open end beneath the surface of the mercury in the dish. Some, but not all of the mercury, flowed from the tube into the dish; 30 inches (760 millimeters) remained. The only force that was able to support the mercury in the tube was the weight of the air exerting pressure on the mercury in the dish.

The word "barometer" which means "weight measure" (from the Greek *báros*, meaning "weight" and *métron*, meaning "meter") was not coined until 1665 by Robert Boyle (1627–1691). Boyle also changed the design for the barometer by using a U-shaped tube, thus eliminating the need for a mercury reservoir. The English physicist Robert Hooke (1635–1703) further improved on the barometer by creating an easy-to-read dial display.

217

What is **barometric pressure** and what does it mean?

Barometric, or atmospheric, pressure is the force exerted on a surface by the weight of the air above that surface, as measured by an instrument called a barometer. Pressure is greater at lower levels because the air's molecules are squeezed under the weight of the air above. So while the average air pressure at sea level is 14.7 pounds per square inch (1,013.53 hecto Pascals), at 1,000 feet (304 meters) above sea level, the pressure drops to 14.1 pounds per square inch (972.1 hecto Pascals), and at 18,000 feet (5,486 meters) the pressure is 7.3 pounds (503.32 hecto Pascals), about half of the figure at sea level. Changes in air pressure bring weather changes. High pressure areas bring clear skies and fair weather; low pressure areas bring wet or stormy weather. Areas of very low pressure have serious storms, such as hurricanes.

How **accurate** is *The Old Farmer's Almanac* in predicting weather?

The Farmer's Almanac published in Lewiston, Maine, and *The Old Farmer's Almanac* published in Dublin, New Hampshire, both make predictions about the weather for the coming year when they are published. Each book claims to have secret formulas to predict the weather. The *Old Farmer's Almanac* claims it has an accuracy rate of 80 percent. Prior to modern forecasting techniques using radar, satellites and computer simulation, many people relied on the almanacs for long-range weather forecasts. However, many of the predictions in the almanacs are very general in nature, such as a five-day period in November may be predicted to be sunny/cool without a range of expected temperatures or whether it will be sunny each of the five days.

Can **weather** be **predicted** from the **stripes** on a **wooly-bear caterpillar**?

It is an old superstition that the severity of the coming winter can be predicted by the width of the brown bands or stripes around the wooly-bear caterpillar in the autumn.

If the brown bands are wide, says the superstition, the winter will be mild, but if the brown bands are narrow, a rough winter is foretold. Studies at the American Museum of Natural History in New York failed to show any connection between the weather and the caterpillar's stripes.

Are there **trees** that **predict** the **weather** and **tell time**?

Observing the leaves of a tree may be an old-fashioned method of predicting the weather, but farmers have noted that when maple leaves curl and turn bottom up in a blowing wind, rain is sure to follow. Woodsmen claim they can tell how rough a winter is going to be by the density of lichens on a nut tree. Trees can also be extraordinary timekeepers: Griffonia, in tropical west Africa, has 2-inch (5-centimeter) inflated pods that burst with a hearty noise, indicating that it is time for farmers of the Accra Plains to plant crops; Trichilia is a 60-foot (18-meter) tree that flowers in February and again in August, signaling that it is time, just before the second rains arrive, for the second planting of corn. In the Fiji Islands, planting yams is cued by the flowering of the coral tree.

Is a **halo** around the **sun** or **moon** a sign of **rain** or **snow** approaching?

The presence of a ring around the sun or, more commonly, the moon in the night sky, indicates very high ice crystals composing cirrostratus clouds. The brighter the ring, the greater the odds of precipitation and the sooner it may be expected. Rain or snow will not always fall, but two times out of three, precipitation will start to fall within 12 to 18 hours. These cirroform clouds are a forerunner of an approaching warm front and an associated low pressure system.

MINERALS, METALS, AND OTHER MATERIALS

ROCKS AND MINERALS

What are the **three main groups** of **rocks**?

Rocks can be conveniently placed into one of three groups—igneous, sedimentary, and metamorphic.

- *Igneous rocks*, such as granite, pegmatite, rhyolite, obsidian, gabbro, and basalt, are formed by the solidification of molten magma that emerges through Earth's crust via volcanic activity. The nature and properties of the crystals vary greatly, depending in part on the composition of the original magma and partly on the conditions under which the magma solidified. There are thousands of different igneous rock types. For example, granite is formed by slow cooling of molten material (within the earth). It has large crystals of quartz, feldspars, and mica.

- *Sedimentary rocks*, such as brecchia, sandstone, shale, limestone, chert, and coals, are produced by the accumulation of sediments. These are fine rock particles or fragments, skeletons of microscopic organisms, or minerals leached from rocks that have accumulated from weathering. These sediments are then redeposited underwater and later compressed in layers over time. The most common sedimentary rock is sandstone, which is predominantly quartz crystals.

- *Metamorphic rocks*, such as marble, slate, schist, gneiss, quartzite, and hornsfel, are formed by the alteration of igneous and sedimentary rocks through heat and/or pressure. One example of these physical and chemical changes is the formation of marble from thermal changes in limestone.

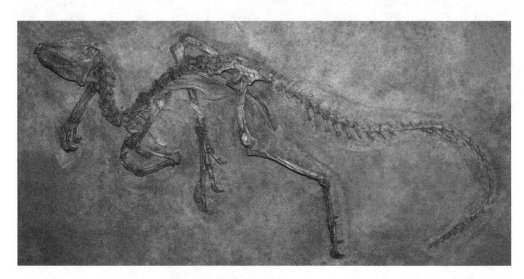

A dinosaur fossil. It's unusual to find complete fossils such as this one, making such finds very valuable to paleontologists.

What is **petrology** and what does a **petrologist** do?

Petrology is the science of rocks. A petrologist is a person who studies the mineralogy of rocks and the record of the geological past contained within rocks. From rocks, a petrologist can learn about past climates and geography, past and present composition of the earth, and the conditions that prevail within the interior of the earth.

How are **fossils formed**?

Fossils are the remains of animals or plants that were preserved in rock before the beginning of recorded history. It is unusual for complete organisms to be preserved; fossils usually represent the hard parts of animals, such as bones or shells, and leaves, seeds, or woody parts of plants.

Some fossils are simply the bones, teeth, or shells themselves, which can be preserved for a relatively short period of time. Another type of fossil is the imprint of a buried plant or animal that decomposes, leaving a film of carbon that retains the form of the organism.

Some buried material is replaced by silica and other materials that permeate the organism and replace the original material in a process called petrification. Some woods are replaced by agate or opal so completely that even the cellular structure is duplicated. The best examples of this can be found in Petrified Forest National Park in Arizona.

Molds and casts are other very common fossils. A mold is made from an imprint, such as a dinosaur footprint, in soft mud or silt. This impression may harden, then be covered with other materials. The original footprint will have formed a mold and the sediments filling it will be a cast of the footprint.

What are Indian Dollars?

They are six-sided, disk-shaped, twin crystals of aragonite ($CaCO^3$), which have altered to calcite but retained their outer form. They occur in large numbers in northern Colorado, where they are known as "Indian Dollars." In New Mexico they are called "Aztec Money" and in western Kansas they are called "Pioneer Dollars."

How **old** are **fossils**?

The oldest known fossils are of bacteria that left their impressions approximately 3.5 billion years ago. The oldest animal fossils are of invertebrates that lived approximately 700 million years ago. The largest number of fossils come from the Cambrian period of 505–590 million years ago, when living organisms began to develop skeletons and hard parts. Since these parts tended to last longer than ordinary tissue, they were more likely to be preserved in clay and become fossilized.

What is a **tektite**?

Tektites are silica-rich glass objects (rocks) found scattered in selected regions of Earth's surface. They are generally black, oblong, teardrop, or dumbbell-shaped, and several centimeters in length. They are formed from molten rock resulting when a meteorite, asteroid, or comet fragment impacts Earth's surface. The molten rock is hurled high into the atmosphere, where it rapidly cools into its unique shape and physical characteristics. Their mode of formation is considered indisputable evidence of such impacts. Tektites range from 0.7 to 35 million years in age.

What is **cinnabar**?

Cinnabar is the main ore of the mineral mercury. Its cinnamon to scarlet-red color makes it a colorful mineral. It is produced primarily in the United States (California, Oregon, Texas, and Arkansas), Spain, Italy, and Mexico. It is often used as a pigment.

How does a **rock differ** from a **mineral**?

Mineralogists use the term "mineral" for a substance that has all four of the following features: it must be found in nature; it must be made up of substances that were never alive (organic); it has the same chemical makeup wherever it is found; and its atoms are arranged in a regular pattern to form solid crystals.

Although "rocks" are sometimes described as an aggregate or combination of one or more minerals, geologists extend the definition to include clay, loose sand, and certain limestones.

From what type of **stone** was **Mount Rushmore** National Memorial carved?

Granite. The monument, in the Black Hills of southwestern South Dakota, depicts the 60-foot-high (18-meter-high) faces of four United States presidents: George Washington (1732–1799), Thomas Jefferson (1743–1826), Abraham Lincoln (1809–1865), and Theodore Roosevelt (1858–1919). Sculptor Gutzon Borglum (1867–1941) designed the monument, but died before the completion of the project; his son, Lincoln, finished it. From 1927 to 1941, 360 people, mostly construction workers, drillers, and miners, "carved" the figures using dynamite.

What is the **composition** of the **Rock of Gibraltar**?

It is composed of gray limestone, with a dark shale overlay on parts of its western slopes. Located on a peninsula at the southern extremity of Spain, the Rock of Gibraltar is a mountain at the east end of the Strait of Gibraltar, the narrow passage between the Atlantic Ocean and the Mediterranean Sea. "The Rock" is 1,398 feet (425 meters) tall at its highest point.

What is the **Mohs scale**?

The Mohs scale is a standard of ten minerals by which the hardness of a mineral is rated. It was introduced in 1812 by the German mineralogist Friedrich Mohs (1773–1839). The minerals are arranged from softest to hardest. Harder minerals, with higher numbers, can scratch those with a lower number.

The Mohs Scale

Hardness	Mineral	Comment
1	Talc	Hardness 1–2 can be scratched by a fingernail
2	Gypsum	Hardness 2–3 can be scratched by a copper coin
3	Calcite	Hardness 3–6 can be scratched by a steel pocket knife
4	Fluorite	
5	Apatite	
6	Orthoclase	Hardness 6–7 will not scratch glass
7	Quartz	
8	Topaz	Hardness 8–10 will scratch glass
9	Corundum	
10	Diamond	

Who was the **first person** to attempt a **color standardization scheme** for minerals?

The German mineralogist Abraham Gottlob Werner (c. 1750–1817) devised a method of describing minerals by their external characteristics, including color. He worked out an arrangement of colors and color names, illustrated by an actual set of minerals.

> ## If diamond is the hardest substance, what is the next hardest substance?
>
> Cubic boron nitride, which is the hardest ceramic, is the second hardest substance in the world.

What is meant by the term **strategic minerals**?

Strategic minerals are minerals essential to national defense—the supply of which a country uses but cannot produce itself. A third to a half of the 80 minerals used by industry could be classed as strategic minerals. Wealthy countries, such as the United States, stockpile these minerals to avoid any crippling effect on their economy or military strength if political circumstances were to cut off their supplies. The United States, for instance, stockpiles bauxite (10.5 million metric dry tons), manganese (1.7 million metric tons), chromium (1.4 million metric tons), tin (59,993 metric tons), cobalt (189 metric tons), tantalum (635 tons), palladium (1.25 million troy ounces), and platinum-group metals (platinum—4,704 kilograms, palladium—16,715 kilograms and iridium—784 kilograms).

What is **pitchblende**?

Pitchblende is a massive variety of uraninite, or uranium oxide, found in metallic veins. It is a radioactive material and the most important ore of uranium. In 1898, Marie (1867–1934) and Pierre (1859–1906) Curie discovered that pitchblende contained radium, a rare element that has since been used in medicine and the sciences.

What is **galena**?

Galena is a lead sulphide (PbS) and the most common ore of lead, containing 86.6 percent lead. Lead-gray in color, with a brilliant metallic luster, galena has a specific gravity of 7.5 and a hardness of 2.5 on the Mohs scale, and usually occurs as cubes or a modification of an octahedral form. Mined in Australia, it is also found in Canada, China, Mexico, Peru, and the United States (Missouri, Kansas, Oklahoma, Colorado, Montana, and Idaho).

What is **stibnite**?

Stibnite is a lead-gray mineral (Sb_2S_3) with a metallic luster. It is the most important ore of antimony, and is also known as antimony glance. One of the few minerals that fuse easily in a match flame (977°F or 525°C), stibnite has a hardness of two on the Mohs scale and a specific gravity of 4.5 to 4.6. It is commonly found in hydrothermal

225

veins or hot springs deposits. Stibnite is mined in Germany, Romania, France, Bolivia, Peru, and Mexico. The Yellow Pine mine at Stibnite, Idaho, is the largest producer in the United States, but California and Nevada also have deposits.

What are **Cape May diamonds**?

They are pure quartz crystals of many sizes and colors, found in the vicinity of the Coast Guard station in Cape May, New Jersey. When polished and faceted, these crystals have the appearance of real diamonds. Prior to the development of modern gem examination equipment, many people were fooled by these quartz crystals. The possibility of finding a Cape May diamond on one's own, and the availability of already-polished faceted stones, has been a long-standing tourist attraction in the Cape May area.

Are there any **diamond mines** in the **United States**?

The United States has no commercial diamond mines. The only significant diamond deposit in North America is Crater of Diamonds State Park near Murfreesboro, Arkansas. It is on government-owned land and has never been systematically developed. For a small fee, tourists can dig there and try to find diamonds. The largest crystal found there weighed 40.23 carats and was named the "Uncle Sam" diamond.

Diamonds crystallize directly from rock melts rich in magnesium and saturated with carbon dioxide gas that has been subjected to high pressures and temperatures exceeding 2,559°F (1,400°C). These rock melts originally came from deep in Earth's mantle at depths of 93 miles (150 kilometers).

Diamonds are minerals composed entirely of the element carbon, with an isometric crystalline structure. The hardest natural substance, gem diamonds have a density of 3.53, though black diamonds (black carbon cokelike aggregates of microscopic crystals) may have a density as low as 3.15. Diamonds have the highest thermal conductivity of any known substance. This property enables diamonds to be used in cutting tools, because they do not become hot.

How can a **genuine diamond** be **identified**?

There are several tests that can be performed without the aid of tools. A knowledgeable person can recognize the surface lustre, straightness and flatness of facets, and high light reflectivity. Diamonds become warm in a warm room and cool if the surroundings are cool. A simple test that can be done is exposing the stones to warmth and cold and then touching them to one's lips to determine their appropriate temperature. This is especially effective when the results of this test are compared to the results of the test done on a diamond known to be genuine. Another test is to pick up the stone with a moistened fingertip. If this can be done, then the stone is likely to be a diamond. The majority of other stones cannot be picked up in this way.

A skilled jeweler can assess the value of a diamond based on cut, clarity, color, and carat size.

The water test is another simple test. A drop of water is placed on a table. A perfectly clean diamond has the ability to almost "magnetize" water and will keep the water from spreading. An instrument called a diamond probe can detect even the most sophisticated fakes. Gemologists always use this as part of their inspection.

How is the **value** of a **diamond determined**?

Demand, beauty, durability, rarity, freedom from defects, and perfection of cutting generally determine the value of a gemstone. But the major factor in establishing the price of gem diamonds is the control over output and price as exercised by the Central Selling Organization's (CSO) Diamond Trading Corporation Ltd. The CSO is a subsidiary of DeBeers Consolidated Mines Ltd.

What are the **four "C"s** of **diamonds**?

The four Cs are cut, color, clarity, and carat. Cut refers to the proportions, finish, symmetry, and polish of the diamond. These factors determine the brilliance of a diamond. Color describes the amount of color the diamond contains. Color ranges from colorless to yellow with tints of yellow, gray, or brown. Colors can also range from intense yellow to the more rare blue, green, pink, and red. Clarity describes the cleanness or purity of a diamond as determined by the number and size of imperfections. Carat is the weight of the diamond.

227

How are **diamonds weighed**?

The basic unit is a carat, which is set at 200 milligrams (0.00704 ounces or 1/142 of an avoirdupois ounce). A well-cut, round diamond of one carat measures almost exactly 0.25 inch (6.3 millimeters) in diameter. Another unit commonly used is the point, which is one hundredth of a carat. A stone of one carat weighs 100 points. "Carat" as a unit of weight should not be confused with the term "karat" used to indicate purity of the gold into which gems are mounted.

Which **diamond** is the **world's largest**?

The Cullinan Diamond, weighing 3,106 carats, is the world's largest. It was discovered on January 25, 1905, at the Premier Diamond Mine, Transvaal, South Africa. Named for Sir Thomas M. Cullinan (1862–1936), chairman of the Premier Diamond Company, it was cut into nine major stones and 96 smaller brilliants. The total weight of the cut stones was 1,063 carats, only 35 percent of the original weight.

Cullinan I, also known as the "Great Star of Africa" or the "First Star of Africa," is a pear-shaped diamond weighing 530.2 carats. It is 2.12 inches (5.4 centimeters) long, 1.75 inches (4.4 centimeters) wide, and 1 inch (2.5 centimeters) thick at its deepest point. It was presented to Britain's King Edward VII (1841–1910) in 1907, and was set in the British monarch's sceptre with the cross. It is still the largest cut diamond in the world.

Cullinan II, also known as the "Second Star of Africa," is an oblong stone that weighs 317.4 carats. It is set in the British Imperial State Crown.

What are the **common cuts** of gemstones?

Modern gem cutting uses faceted cutting for most transparent gems. In faceted cutting, numerous facets—geometrically disposed to bring out the beauty of light and color to the best advantage—are cut. The four most common cuts are the brilliant, the rose, the baguette, and the step or trap cut. The step or trap cut is also known as the emerald cut and is used for emeralds. The brilliant and rose cuts are often used for diamonds.

What is the **difference** between **cubic zirconium** and **diamonds**?

Cubic zirconium is a gemstone material that is an imitation of diamonds. The word "imitation" is key. The U.S. Federal Trade Commission defines imitation materials as resembling the natural material in appearance only. Cubic zirconia may be cut the same way as diamonds. It is very dense and solid, weighing 1.7 times more than a diamond of the same millimeter size.

Besides the Cullinan diamonds, what are the **largest precious stones**?

The largest ruby is a 8,500 carat stone that is 5.5 inches (14 centimeters) tall, carved to resemble the Liberty Bell. The largest star ruby is the 6,465 carat "Eminent Star"

What is cubic zirconium?

Cubic zirconium was discovered in 1937 by two German mineralogists, M. V. Stackelberg and K. Chudoba. It became popular with jewelry designers in the 1970s after Soviet scientists under the direction of V. V. Osika learned how to "grow" the mineral in the Lebedev Physical Institute laboratory. Most of the cubic zirconium on the market is chemically composed of zirconium oxide and yttrium oxide. The two compounds are melted together at a very high temperature (almost 5,000°F [2,760°C]) using the skull melt method. This method uses a radio-frequency generator to heat the zirconium oxide. A careful cooling of the mixture produces the flawless crystals that become cubic zirconia gemstones.

from India that has a six-line star. The largest cut emerald was found in Carnaiba, Brazil, in August 1974. It is 86,136 carats. A 2,302 carat sapphire from Anakie, Queensland, Australia, was carved into a 1,318 carat head of Abraham Lincoln, making it the largest carved sapphire. "The Lone Star," at 9,719.5 carats, is the largest star sapphire. The largest natural pearl is the "Pearl of Lao-tze," also called the "Pearl of Allah." Found in May 1934 in the shell of a giant clam at Palawan, Philippines, the pearl weighs 14 pounds, 1 ounce (6.4 kilograms).

How does the **emerald** get its **color**?

Emerald is a variety of green beryl ($Be_3Al_2Si_6O_{18}$) that is colored by a trace of chromium (Cr), which replaces the aluminum (Al) in the beryl structure. Other green beryls exist; but if no chromium is present, they are, technically speaking, not emeralds.

Which **two gems** contain the mineral **corundum**?

Both rubies and sapphires contain the mineral corundum (Al_2O_3). Chromium (Cr) ions replace small amounts of aluminum in rubies giving them their characteristic red color. In sapphires, iron (Fe) and titanium (Ti) ions replace some of the aluminum producing the characteristic blue color.

How is the star in **star sapphires** produced?

Sapphires are composed of gem-quality corundum (Al_2O_3). Color appears in sapphires when small amounts of iron and titanium are present. Star stapphires contain needles of the mineral rutile that will display as a six-ray star figure when cut in the unfaceted cabochon (dome or convex) form. The most highly prized star sapphires are blue. Black or white star sapphires are less valuble. Since a ruby is simply the red variety of corundum, star rubies also exist.

What is a **tiger's eye**?

Tiger's eye is a semiprecious quartz gem that has a vertical luminescent band like that of a cat's eye. To achieve the effect of a cat's eye, veins of parallel blue asbestos fibers are first altered to iron oxides and then replaced by silica. The gem has a rich yellow to yellow-brown or brown color.

The parallel yellow and brown bands of the tiger's eye lend it the appearance of its namesake.

METALS

What is **coltan**?

Coltan is the shortened name for the metallic ore columbite-tantalite. When refined it becomes a heat-resistant powder, tantalum, which can hold a high electrical charge. These properties make it a vital element in creating capacitors, the electronic elements that control current flow inside miniature circuit boards. Tantalum capacitors are used in almost all cell phones, laptop computers, pagers, and other electrical devices.

Which **metallic element** is the **most abundant**?

Aluminum is the most abundant metallic element on the surface of Earth and the moon; it comprises more than eight percent of Earth's crust. It is never free in nature, combining with oxygen, sand, iron, titanium, and other substances; its ores are mainly bauxites (aluminum hydroxide). Nearly all rocks, particularly igneous rocks, contain aluminum as aluminosilicate minerals. Napoleon III (1808–1883) recognized that the physical characteristic of its lightness could revolutionize the arms industry, so he granted a large subsidy to French chemist Sainte-Claire Deville (1818–1881) to develop a method to make its commercial use feasible. In 1854, Deville obtained the first pure aluminum metal through the process of reduction of aluminum chloride. In 1886, American Charles Martin Hall (1863–1914) and Frenchman Paul Heroult (1863–1914) independently discovered an electrolytic process to produce aluminum from bauxite.

Which **industries** use the most **aluminum** in the United States?

Aluminum is important in all segments of the world economy. The use of aluminum exceeds that of any other metal except iron measured in both quantity and value. In 2009, 33 percent of the aluminum used in the United States was in the transportation industry for the manufacture of automobiles, trucks, railcars, marine vessels, and airplanes. Alloys have high tensile strengths and are of considerable industrial importance to the aerospace industry. Because of aluminum's resistance to corrosion, low density, and excellent heat-conducting property, 26 percent of the aluminum was used by the packaging industry for drink and food containers and covers, and foil pouches and wraps in 2009. In 2009, the building construction industry used 14 percent of aluminum alloys in such items as gutters, panels, siding, window frames, and roofing. It is a good conductor of electricity and is widely used in power and telephone cables, light bulbs, and electrical equipment. Examples of the numerous other products containing aluminum and aluminum alloys are cookware, golf clubs, air conditioners, lawn furniture, license plates, paints, refrigerators, rocket fuel, and zippers.

Why are **alchemical symbols** for metals and **astrological symbols** for planets **identical**?

The ancient Greeks and Romans knew seven metals and also knew seven "planets" (the five nearer planets plus the sun and the moon). They related each planet to a specific metal. Alchemy, originating in about the third century B.C.E., focused on changing base metals, such as lead, into gold. Although at times alchemy bordered on mysticism, it contained centuries of chemical experience, which provided the foundation for the development of modern chemistry.

English name	Chemical symbol	Latin name	Alchemical symbol
Gold	Au	aurum	A (Sun)
Silver	Ag	argentum	B (Moon)
Copper	Cu	cuprum	F (Venus)
Iron	Fe	ferrum	C (Mars)
Mercury	Hg	hydrargyrum	D (Mercury)
Tin	Sn	stannum	E (Jupiter)
Lead	Pb	plumbum	G (Saturn)

What are the **precious metals**?

This is a general term for expensive metals that are used for making coins, jewelry, and ornaments. The name is limited to gold, silver, and platinum. Expense or rarity does not make a metal precious, but rather it is a value set by law that states that the object made of these metals has a certain intrinsic value. The term is not synonymous with "noble metals," although a metal (such as platinum) may be both noble and precious.

What is **24 karat gold**?

The term "karat" refers to the percentage of gold versus the percentage of an alloy in a piece of jewelry or a decorative object. Gold is too soft to be usable in its purest form and has to be mixed with other metals. One karat is equal to one-24th part fine gold. Thus, 24 karat gold is 100 percent pure and 18 karat gold is 18/24 or 75 percent pure.

How far can a troy ounce of gold, if formed into a thin wire, be stretched before it breaks?

Ductility is the characteristic of a substance to lend itself to shaping and stretching. A troy ounce of gold (31.1035 grams) can be drawn into a fine wire that is 50 miles (80 kilometers) long.

Karatage	Percentage of fine gold
24	100
22	91.75
18	75
14	58.5
12	50.25
10	42
9	37.8
8	33.75

Is **white gold** really gold?

White gold is the name of a class of jeweler's white alloys used as substitutes for platinum. Different grades vary widely in composition, but usual alloys consist of between 20 percent and 50 percent nickel, with the balance gold. A superior class of white gold is made of 90 percent gold and 10 percent palladium. Other elements used include copper and zinc. The main use of these alloys is to give the gold a white color.

How **thick** is **gold leaf**?

Gold leaf is pure gold that is hammered or rolled into sheets or leaves so extremely thin that it can take 300,000 units to make a stack one inch high. The thickness of a single gold leaf is typically 0.0000035 inch (3.5 millionths of an inch), although this may vary widely according to which manufacturer makes it. Also called gold foil, it is used for architectural coverings and for hot-embossed printing on leather.

What are the chief **gold-producing countries**?

China is the leading gold-producing nation in the world followed by Australia, South Africa, and the United States. In the United States, Nevada is the leading gold producer. Commercial usage in 2009 was estimated as follows: jewelry and arts, 72 percent; industrial (mainly electronic), 7 percent; dental and other industrial uses, 21 percent.

Estimated world mine production of the top six countries in 2009 was:

Country	Gold Production
China	300 metric tons (300,000 kg)
Australia	220 metric tons (220,000 kg)
South Africa	210 metric tons (210,000 kg)
United States	210 metric tons (210,000 kg)
Russia	185 metric tons (185,000 kg)
Peru	180 metric tons (180,000 kg)

What is **fool's gold**?

Iron pyrite (FeS_2) is a mineral popularly known as "fool's gold." Because of its metallic luster and pale brass yellow color, it is often mistaken for gold. Real gold is much heavier, softer, not brittle, and not grooved.

What is **sterling silver**?

Sterling silver is a high-grade alloy that contains a minimum of 925 parts in 1,000 of silver (92.5 percent silver and 7.5 percent of another metal—usually copper).

What is **German silver**?

Nickel silver, sometimes known as German silver or nickel brass, is a silver-white alloy composed of 52 percent to 80 percent copper, 10 percent to 35 percent zinc, and 5 percent to 35 percent nickel. It may also contain a small percent of lead and tin. There are other forms of nickel silver, but the term "German silver" is the name used in the silverware trade.

Which **metal** is the main component of **pewter**?

Tin—at least 90 percent. Antimony, copper, and zinc may be added in place of lead to harden and strengthen pewter. Pewter may still contain lead, but high lead content will both tarnish the piece and dissolve into food and drink to become toxic. The alloy used today in fine quality pieces contain 91–95 percent minimum tin, 8 percent maximum antimony, 2.5 percent maximum copper, and 0–5 percent maximum bismuth, as determined by the European Standard for pewter.

Where were the **first successful ironworks** in **America**?

Although iron ore in this country was first discovered in North Carolina in 1585, and the manufacture of iron was first undertaken (but never accomplished) in Virginia in 1619, the first successful ironworks in America was established by Thomas Dexter and Robert Bridges near the Saugus River in Lynn, Massachusetts. As the original promoters of the enterprise, they hired John Winthrop Jr. from England to begin production. By 1645, a blast furnace had begun operations, and by 1648 a forge was working there.

Metallurgists in several countries developed stainless steel, a group of iron-based alloys combined with chromium in order to be resistant to rusting and corrosion. Chromium was used in small amounts in 1872 to strengthen the steel of the Eads Bridge over the Mississippi River, but it wasn't until the 1900s that a truly rust-resistant alloy was developed. Metallurgists in several countries developed stainless steel between 1903 and 1912. An American, Elwood Haynes (1857–1925), developed several alloy steels and in 1911 produced stainless steel. Harry Brearley (1871–1948) of Great Britain receives most of the credit for its development. In 1913, he discovered that adding chromium to low carbon steel improved its resistance to corrosion. Frederick Becket (1875–1942), a Canadian-American metallurgist, and German scientists Philip Monnartz and W. Borchers were among the early developers.

What is **high speed steel**?

High speed steel is a general name for high alloy steels that retain their hardness at very high temperatures and are used for metal-cutting tools. All high speed steels are based on either tungsten or molybdenum (or both) as the primary heat-resisting alloying element. These steels require a special heat so that their unique properties can be fully realized. The manufacturing process consists of heating the steel to a temperature of 2,150°F to 2,400°F (1,175°C to 1,315°C) to obtain solution of a substantial percentage of the alloy carbides, quenching to room temperature, tempering at 1,000°F to 1,150°F (535°C to 620°C), and again cooling to room temperature.

What **material** is used to make a **tuning fork**?

A tuning fork, an instrument that when struck emits a fixed pitch, is generally made of steel. Some tuning forks are made of aluminum, magnesium-alloy, fused quartz, or other elastic materials.

Which **countries** have **uranium** deposits?

Uranium, a radioactive metallic element, is the only natural material capable of sustaining nuclear fission. But only one isotope, uranium-235, which occurs in one molecule out of 40 of natural uranium, can undergo fission under neutron bombardment. Mined in various parts of the world, it must then be converted during purification to uranium dioxide (UO_2). Uranium deposits occur throughout the world, with approximately 75 percent of all known uranium deposits in Australia. Other countries with significant deposits are Kazakhstan (17 percent of the world's total), Canada, United States, South

Africa, Namibia, Brazil, Niger, and Russia. Canada is the largest exporter of uranium.

What is **technetium**?

Technetium (Tc, element 43) is a radioactive metallic element that does not occur naturally either in its pure form or as compounds; it is produced during nuclear fission. A fission product of molybdenum (Mo, element 42), Tc can also occur as a fission product of uranium (U, element 92). It was the first element to be made artificially in 1937 when it was isolated and extracted by Carlo Perrier (1886–1948) and Emilio Segrè (1905–1989).

U.S. coins such as dimes were once made out of 100 percent gold, silver, or copper. Making such coins today would be far too expensive, so dimes, for instance, are now a mix of copper and nickel.

Technetium has found significant application in diagnostic imaging and nuclear medicine. Ingested soluble technetium compounds tend to concentrate in the liver and are valuable in labeling and in radiological examination of that organ. Also, by technetium labeling of blood serum components, diseases involving the circulatory system can be explored.

What is the **composition** of **U.S. coins** currently in circulation?

During colonial times, coins were composed of gold, silver and copper. The U.S. Mint produced gold coins until 1933, during the Great Depression. The silver in quarters and dimes was replaced in 1966. Today, nickels, dimes, quarters, and half-dollars are composed of copper and nickel. They have a copper core and an outer layer composed of a 75 percent copper, 25 percent nickel alloy. Pennies, once copper coins, are now composed of copper-plated zinc.

NATURAL SUBSTANCES

What is **obsidian**?

Obsidian is a volcanic glass that usually forms in the upper parts of lava flows. Embryonic crystal growths, known as crystallites, make the glass an opaque, jet-black color. Red or brown obsidian could result if iron oxide dust is present. There are some well-known formations in existence, including the Obsidian Cliffs in Yellowstone Park and Mount Hekla in Iceland.

Is **lodestone** a **magnet**?

Lodestone is a naturally occurring variety of magnetic iron oxide or magnetite. Lodestone is frequently called a natural magnet because it attracts iron objects and possesses polarity. It was used by early mariners to find magnetic north. Other names for lodestone are loadstone, leading stone, and Hercules stone.

What is **red dog**?

Red dog is the residue from burned coal dumps. The dumps are composed of waste products incidental to coal mining. Under pressure in these waste dumps, the waste frequently ignites from spontaneous combustion, producing a red-colored ash, which is used for driveways, parking lots, and roads.

What are some **uses** for **coal** other than as an energy resource?

In the past, many of the aromatic compounds, such as benzene, toluene, and xylene were made from coal. These compounds are now chiefly by-products of petroleum. Naphthalene and phenanthrene are still obtained from coal tar. Coal tar, a by-product of coal, is used in roofing.

What is **diatomite**?

Diatomite (also called diatomaceous earth) is a white- or cream-colored, friable, porous rock composed of the fossil remains of diatoms (small water plants with silica cell walls). These fossils build up on the ocean bottoms to form diatomite, and in some places these areas have become dry land or diatomacceous earth. Chemically inert and having a rough texture and other unusual physical properties, it is suitable for many scientific and industrial purposes, including use as a filtering agent; building material; heat, cold, and sound insulator; catalyst carrier; filler absorbent; abrasive; and ingredient in pharmaceutical preparations. Dynamite is made from it by soaking it in the liquid explosive nitroglycerin.

What is **fly ash**?

Fly ash is the very fine portion or ash residue that results from the combustion of coal. The fly ash portion is usually removed electrostatically from the coal combustion gases before they are released to the atmosphere. About 31 percent of the 57 million metric tons produced annually in the United States are beneficially used; the remainder must be disposed of in ponds or landfills.

In **coal mining** what is meant by **damp**?

Damp is a poisonous or explosive gas in a mine. The most common type of damp is firedamp, also known as methane. White damp is carbon monoxide. Blackdamp (or chokedamp) is a mixture of nitrogen and carbon dioxide formed by mine fires and explosions of firedamp in mines. Blackdamp extinguishes fire and suffocates its victims.

What is **fuller's earth**?

It is a naturally occurring white or brown clay containing aluminum magnesium silicate. Fuller's earth acts as a catalyst and was named for a process known as fulling—a process used to clean grease from wool and cloth. It is currently used for lightening the color of oils and fats, as a pigment extender, as a filter, as an absorbent (for example, in litter boxes to absorb animal waste), and in floor sweeping compounds.

How is **charcoal made**?

Commercial production of charcoal uses wood processing residues, such as sawdust, shavings, milled wood, and bark, as a raw material. Depending on the material, the residues are placed in kilns or furnaces and heated at low oxygen concentrations. A Herreshoff furnace can produce at least a ton of charcoal per hour.

How **much wood** is used to make a **ton** of **paper**?

In the United States, the wood used for the manufacture of paper is mainly from small diameter bolts and pulpwood. It is usually measured by the cord or by weight. Although the fiber used in making paper is overwhelmingly wood fiber, a large percentage of other ingredients is needed. One ton of a typical paper requires two cords of wood, but also requires 55,000 gallons (208,000 liters) of water, 102 pounds (46 kilograms) of sulfur, 350 pounds (159 kilograms) of lime, 289 pounds (131 kilograms) of clay, 1.2 tons of coal, 112 kilowatt hours of power, 20 pounds (9 kilograms) of dye and pigments, and 108 pounds (49 kilograms) of starch, as well as other ingredients.

What **products** come from **tropical forests**?

Products from Tropical Forests

Woods	Houseplants	Spices	Foods	Oils, Medicines, Alkaloids	Fibers	Gums
Balsa	*Anthurium*	Allspice	Avocado	Camphor oil	Bamboo	Strophanthus
Mahogany	Croton	Black pepper	Banana	Cascarilla oil	Jute/Kenaf	Chicle latex
Rosewood	*Dieffenbachia*	Cardamom	Coconut	Coconut oil	Kapok	Copaiba
Sandalwood	*Dracaena*	Cayenne	Grapefruit	Eucalyptus oil	Raffia	Copal
Teak	Fiddle-leaf fig	Chili	Lemon	Oil of star anise	Ramie	Gutta percha
	Mother-in-law's tongue	Cinnamon	Lime	Palm oil	Rattan	Rubber latex
	Parlor ivy	Cloves	Mango	Patchouli oil		Tung oil

Houseplants	Spices	Foods	Oils, Medicines, Alkaloids
Philodendron	Ginger	Orange	Rosewood oil
Rubber tree plant	Mace	Papaya	Tolu balsam oil
Schefflera	Nutmeg	Passion fruit	Annatto
Silver vase bromeliad	Paprika	Pineapple	Curare
Spathiphyllum	Sesame seeds	Plantain	Diosgenin
Swiss cheese plant	Turmeric	Tangerine	Quinine
Zebra plant	Vanilla bean	Brazil nuts	Reserpine
		Cashew nuts	Cane sugar
		Chocolate	Strychnine
		Coffee	Ylang-Ylang
		Cucumber	
		Hearts of palm	
		Macadamia nuts	
		Manioc/tapioca	
		Okra	
		Peanuts	
		Peppers	
		Cola beans	
		Tea	

What does **one acre** of **trees yield** when cut and processed?

There are about 660 trees on one acre in a forest. When cut, one acre of trees may yield approximately 105,000 board feet of lumber or more than 30 tons of paper or 16 cords of firewood.

What does **one cord** of **wood yield** when processed?

A cord of wood may produce:

- 12 dining room table sets (seating eight each)
- 250 copies of the Sunday *New York Times*

- 2700 copies of an average (36-page) daily newspaper
- 1,000 pounds (454 kilograms) to 2,000 pounds (907 kilograms) of paper, depending on the quality and grade of the paper
- 61,370 standard (#10) envelopes

Which **woods** are used for **telephone poles**?

The principal woods used for telephone poles are southern pine, Douglas fir, western red cedar, and lodgepole pine. Ponderosa pine, red pine, jack pine, northern white cedar, other cedars, and western larch are also used.

Which **woods** are used for **railroad ties**?

Many species of wood are used for ties. The more common are oaks, gums, Douglas fir, mixed hardwoods, hemlock, southern pine, sycamore, and mixed softwoods.

Does any **type** of **wood sink** in **water**?

Ironwood is a name applied to many hard, heavy woods. Some ironwoods are so dense that their specific gravity exceeds 1.0 and they are therefore unable to float in water. North American ironwoods include the American hornbeam, the mesquite, the desert ironwood, and leadwood (*Krugiodendron ferreum*), which has a specific gravity of 1.34–1.42, making it the heaviest in the United States.

The heaviest wood is black ironwood (*Olea laurifolia*), also called South African ironwood. Found in the West Indies, it has a specific gravity of 1.49 and weighs up to 93 pounds (42.18 kilograms) per foot. The lightest wood is *Aeschynomene hispida*, found in Cuba, with a specific gravity of 0.044 and a weight of 2.5 pounds (1.13 kilograms) per foot. Balsa wood (*Ochroma pyramidale*) varies between 2.5 and 24 pounds (1 to 10 kilograms) per foot.

What is **amber**?

Amber is the fossil resin of trees. The two major deposits of amber are in the Dominican Republic and Baltic. Amber came from a coniferous tree that is now extinct. Amber is usually yellow or orange in color, semitransparent or opaque with a glossy surface. It is used by both artisans and scientists.

How is **petrified wood** formed?

Petrified wood is formed when water containing dissolved minerals such as calcium carbonate ($CaCO_3$) and silicate infiltrates wood or other structures. The process takes thousands of years. The foreign material either replaces or encloses the organic matter and often retains all of the structural details of the original plant material. Botanists find

239

The Petrified Forest in Arizona is famous for its vast tracts of ancient trees that have had their organic matter replaced with minerals.

these types of fossils to be very important since they allow for the study of the internal structure of extinct plants. After a time, wood seems to have turned to stone because the original form and structure are retained. The wood itself does not turn to stone.

What is **rosin**?

Rosin is the resin produced after the distillation of turpentine, obtained from several varieties of pine trees, especially the longleaf pine (*Pinus palustris*) and the slash pine (*Pinus caribaea*). Rosin has many industrial uses, including the preparation of inks, adhesives, paints, sealants, and chemicals. Rosin is also used by athletes and musicians to make smooth surfaces less slippery.

What are **naval stores**?

Naval stores are products of coniferous trees as pine and spruce. These products include pitch, tar, resin, turpentine, pine oil, and terpenes. The term "naval stores" originated in the seventeenth century when these materials were used for building and maintaining wooden sailing ships.

Why are **essential oils** called "essential"?

Called essential oils because of their ease of solubility in alcohol to form essences, essential oils are used in flavorings, perfumes, disinfectants, medicine, and other

> ## What is excelsior?
>
> **E**xcelsior is a trade name dating from the mid-nineteenth century for the curly, fine wood shavings used as packing material when shipping breakable items. It is also used as a cushioning and stuffing material. Poplar, aspen, basswood or cottonwood are woods that are often made into excelsior.

products. They are naturally occurring volatile aromatic oils found in uncombined forms within various parts of plants, such as leaves and pods. These oils contain as one of their main ingredients a substance belonging to the terpene group. Examples of essential oils include bergamot, eucalyptus, ginger, pine, spearmint, and wintergreen oils. Extracted by distillation or enfleurage (extraction using fat) and mechanical pressing, these oils can now be made synthetically.

What is **gutta percha**?

Gutta percha is a rubberlike gum obtained from the milky sap of trees of the *Sapotaceae* family, found in Indonesia and Malaysia. Once of great economic value, gutta percha is now being replaced by plastics in many items, although it is still used in some electrical insulation and dental work. The English natural historian John Tradescant (c. 1570–1638) introduced gutta percha to Europe in the 1620s, and its inherent qualities gave it a slow but growing place in world trade. By the end of World War II, however, many manufacturers switched from gutta percha to plastics, which are more versatile and cheaper to produce.

What is **ambergris**?

Ambergris, a highly odorous, waxy substance found floating in tropical seas, is a secretion from the sperm whale (*Physeter catodon*). The whale secretes ambergris to protect its stomach from the sharp bone of the cuttlefish, a squid-like sea mollusk, which it ingests. Ambergris is used in perfumery as a fixative to extend the life of a perfume and as a flavoring for food and beverages. Today ambergris is synthesized and used by the perfume trade, which has voluntarily refused to purchase natural ambergris to protect sperm whales from exploitation.

From where do **frankincense** and **myrrh originate**?

Frankincense is an aromatic gum resin obtained by tapping the trunks of trees belonging to the genus *Boswellia*. The milky resin hardens when exposed to the air and forms irregular lumps—the form in which it is usually marketed. Also called

241

olibanum, frankincense is used in pharmaceuticals, as a perfume, as a fixative, and in fumigants and incense. Myrrh comes from a tree of the genus *Commiphora*, a native of Arabia and Northeast Africa. It too is a resin obtained from the tree trunk and is used in pharmaceuticals, perfumes, and toothpaste.

Where does **isinglass** come from?

Isinglass is the purest form of animal gelatin. It is manufactured from the swimming bladder of sturgeon and other fishes. It is used in the clarification of wine and beer as well as in the making of some cements, jams, jellies, and soups.

What are **natural fibers**?

Natural fibers come from plants or animals. Examples of fabrics of animal origin are wool and silk. Wool is made from the fibers of animal coats including sheep, goats, rabbits, and alpacas. Cotton, linen, hemp, ramie, and jute are all fabrics that have fibers of plant origin.

Fiber	Origin
Wool	Sheep
Mohair	Goat
Angora	Rabbit
Camel hair	Camel
Cashmere	Kashmir goat
Alpaca fleece	Alpaca
Silk	Silkworm cocoons
Cotton	Cotton plant seed pods
Linen	Flax
Hemp	*Cannabis Sativa* stem
Ramie	*Boehmeria nivea* (Chinese grass)
Jute	Glossy fiber

How is **silk made**?

Silk fiber is a continuous protein filament produced by a silkworm to form its cocoon. The principal species used in commercial silkmaking is the mulberry silkworm (the larva of the silk moth *Bombyx mori*) belonging to the order Lepidoptera. The raw silk fiber has three elements—two filaments excreted from both of the silkworm's glands and a soluble silk gum called sericin, which cements the filaments together. It is from these filaments that the caterpillar constructs a cocoon around itself.

The process of silkmaking starts with raising silkworms on diets of mulberry leaves for five weeks until they spin their cocoons. Then the cocoons are treated with heat to kill the silkworms inside (otherwise when the moths emerged, they would break the long silk filaments). After the cocoons are soaked in hot water, the filaments of five to ten cocoons are unwounds in the reeling process, and twisted into a single thicker filament; still too fine for weaving, these twisted filaments are twisted again into a thread that can be woven.

What is **cashmere**?

Kashmir goats, which live high in the plateaus of the area from northern China to Mongolia, are covered in a coarse outer hair that helps protect them from the cold, harsh weather. As insulation, these goats have a softer, finer layer of hair or down under the coarse outer hair. This fine hair is shed annually and processed to make cashmere. Each goat produces enough cashmere to make one sweater every four years.

MAN–MADE PRODUCTS

What are the major **distinguishing characteristics** of **ceramics**?

Ceramics are crystalline compounds of metallic and nonmetallic elements. Ceramics are the most rigid of all materials with an almost total absence of ductility. They have the highest known melting points of materials with some being as high as 7,000°F (3,870°C) and many that melt at temperatures of 3,500°F (1,927°C). Glass, brick, cement and plaster, dinnerware, artware, and porcelain enamel are all examples of ceramics.

What are **composite materials**?

Composite materials, or simply composites, consist of two parts; the reinforcing phase and the binder or matrix. Composites may be natural substances, such as wood and bone, or man-made substances. A composite product is different from each of its components and is often superior to each individual component. The binder or matrix of a composite is the material that supports the reinforcing phase. The reinforcing phase is usually in the form of particles, fibers, or flat sheets. Reinforced concrete is an example of a composite material. The steel rods embedded in the concrete (the matrix) are the reinforcing phase adding strength and flexibility to the concrete. High-performance composites are composites that perform better than traditional structural materials, such as steel. Most high-performance composites have fibers in the reinforcing phase and a polymer matrix. The fibers may be glass, boron, silicon carbide, aluminum oxide, or a type of polymer. The fibers are often interwoven to form bundles. The purpose of the matrix, usually a polymer, in a high-performance composite, is to hold the fibers together and protect them.

243

Why is **Styrofoam** a **good insulator**?

Styrofoam insulates well because the foam form increases the length of path for heat flow through the material. It also reduces the effective cross-sectional area across which the heat can flow.

How is **dry ice** made?

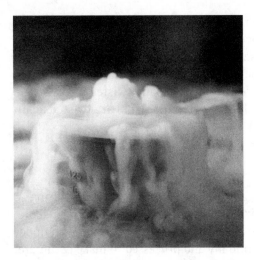

Dry ice is the solid form of carbon dioxide. A useful refrigerant, it is also often used in theatrical performances to create fog effects.

Dry ice a solid form of carbon dioxide (CO_2) used primarily to refrigerate perishables that are being transported from one location to another. The carbon dioxide, which at normal temperatures is a gas, is stored and shipped as a liquid in tanks that are pressurized at 1,073 pounds per square inch. To make dry ice, the carbon dioxide liquid is withdrawn from the tank and allowed to evaporate at a normal pressure in a porous bag. This rapid evaporation consumes so much heat that part of the liquid CO_2 freezes to a temperature of −109°F (−78°C). The frozen liquid is then compressed by machines into blocks of "dry ice," which will melt into a gas again when set out at room temperature.

It was first made commercially in 1925 by the Prest-Air Devices Company of Long Island City, New York, through the efforts of Thomas Benton Slate. It was used by Schrafft's of New York in July 1925 to keep ice cream from melting. The first large sale of dry ice was made later in that year to Breyer Ice Cream Company of New York. Although used mostly as a refrigerant or coolant, other uses include medical procedures such as freezing warts, blast cleaning, freeze branding animals, and creating special effects for live performances and films.

Why is **sulfuric acid** important?

Sometimes called "oil of vitriol," or vitriolic acid, sulfuric acid (H_2SO_4) has become one of the most important of all chemicals. It was little used until it became essential for the manufacture of soda in the eighteenth century. It is prepared industrially by the reaction of water with sulfur trioxide, which in turn is made by chemical combination of sulfur dioxide and oxygen by one of two processes (the contact process or the chamber process). Many manufactured articles in common use depend in some way on sulfuric acid for their production. Ninety percent of the sulfuric acid manufactured in the United States is used in the production of fertilizers and other inorganic chemicals.

What is **aqua regia**?

Aqua regia, also known as nitrohydrochloric acid, is a mixture of one part concentrated nitric acid and three parts concentrated hydrochloric acid. The chemical reaction between the acids makes it possible to dissolve all metals except silver. The reaction of metals with nitrohydrochloric acid typically involves oxidation of the metals to a metallic ion and the reduction of the nitric acid to nitric oxide. The term comes from Latin and means "royal water." It was named by the alchemists for its ability to dissolve gold and platinum, which were called the "noble metals."

Who developed the process for **making ammonia**?

Known since ancient times, ammonia (NH_3) has been commercially important for more than one hundred years. The first breakthrough in the large-scale synthesis of ammonia resulted from the work of Fritz Haber (1863–1934). In 1913, Haber found that ammonia could be produced by combining nitrogen and hydrogen ($N_2 + 3H_2 \leftrightarrow 2NH_3$) with a catalyst (iron oxide with small quantities of cerium and chromium) at 131°F (55°C) under a pressure of about 200 atmospheres. The process was adapted for industrial-quality production by Karl Bosch (1874–1940). Thereafter, many improved ammonia-synthesis systems, based on the Haber-Bosch process, were commercialized using various operating conditions and synthesis loop-designs. One of the five top inorganic chemicals produced in the United States, it is used in refrigerants, detergents, and other cleaning preparations, explosives, fabrics, and fertilizers. Most ammonia production in the United States is used for fertilizers. It has been shown to produce cancer of the skin in humans in doses of 1,000 milligram per kilogram (2.2 pounds) of body weight.

What does the symbol **H_2O_2** stand for?

It is hydrogen peroxide, a syrupy liquid compound used as a strong bleaching, oxidizing, and disinfecting agent. It is usually made either in anthrahydroquinone autoxidation processes or electrolytically. The primary use of hydrogen peroxide is in bleaching wood pulp. A more familiar use is as a three percent solution as an antiseptic and germicide. Undiluted, it can cause burns to human skin and mucous membranes, is a fire and explosion risk, and can be highly toxic.

What **percentage of salt** consumed in the **United States** is used for **de-icing roads**?

In 2009, an estimated 43 percent of the salt (sodium chloride) consumed in the United States was used to de-ice roads. Although calcium chloride may also be effective in de-icing roads, it is not as economical.

245

What is buckminsterfullerene?

It is a large molecule in the shape of a soccer ball, containing 60 carbon atoms, whose structure is the shape of a truncated icosahedron (a hollow, spherical object with 32 faces, 12 of them pentagons and the rest hexagons). This molecule was named buckminsterfullerene because of the structure's resemblance to the geodesic domes designed by American architect R. Buckminster Fuller (1895–1983). The molecule was formed by vaporizing material from a graphite surface with a laser. Large molecules containing only carbon atoms have been known to exist around certain types of carbon-rich stars. Similar molecules are also thought to be present in soot formed during the incomplete combustion of organic materials.

Chemist Richard Smalley (1943–2005) identified buckminsterfullerene in 1985 and speculated that it may be fairly common throughout the universe. Since that time, other stable, large, even-numbered carbon clusters have been produced. This new class of molecules has been called "fullerenes" since they all seem to have the structure of a geodesic dome. They are also popularly known as "bucky balls." Buckminsterfullerene (C_{60}) seems to function as an insulator, conductor, semi-conductor, and superconductor in various compounds. Although no practical application has yet to be developed for it or the other fullerenes, research is expected to result in new types of materials, lubricants, coatings, catalysts, electro-optical devices, and medical applications.

What is the **lightest known solid**?

The lightest solid is silica aerogels, made of tiny spheres of bonded silicon and oxygen atoms linked together into long strands separated with air pockets. They appear almost like frozen wisps of smoke. They also have the lowest conductivity, lowest solid density, highest porosity, highest surface area, and the highest dielectric constant, giving them the potential of being used in many applications. Understandably, their use is not currently widespread due to the expense to create them, and the difficulty in insulating capabilities will allow their use in place of fiberglass and polyurethane foam, significantly reducing global energy consumption and greenhouse gas emissions.

Is **glass** a **solid** or a **liquid**?

Even at room temperature, glass appears to be a solid in the ordinary sense of the word. However, it actually is a fluid with an extremely high viscosity, which refers to the internal friction of fluids. Viscosity is a property of fluids by which the flow motion is gradually damped (slowed) and dissipated by heat. Viscosity is a familiar phenomenon in daily life. An opened bottle of wine can be poured: the wine flows easily under the influence of gravity. Maple syrup, on the other hand, cannot be poured so easily; under the action of gravity, it flows sluggishly. The syrup has a higher viscosity than the wine.

Glass is usually composed of mixed oxides based around the silicon dioxide (SiO_2) unit. A very good electrical insulator, and generally inert to chemicals, commercial glass is manufactured by the fusing of sand (silica, SiO_2), limestone ($CaCO_2$), and soda (sodium carbonate, Na_2CO3) at temperatures around 2,552°F to 2,732°F (1,400°C to 1,500°C). On cooling, the melt becomes very viscous, and at about 932°F (500°C, known as glass transition temperature), the melt "solidifies" to form soda glass. Small amounts of metal oxides are used to color glass, and its physical properties can be changed by the addition of substances like lead oxide (to increase softness, density, and refractive ability for cutglass and lead crystal), and borax (to significantly lower thermal expansion for cookware and laboratory equipment). Other materials can be used to form glasses if rapidly cooled from the liquid or gaseous phase to prevent an ordered crystalline structure from forming.

Glass objects might have been made as early as 2500 B.C.E. in Egypt and Mesopotamia, and glass blowing developed about 100 B.C.E. in Phoenicia.

What is **crown glass**?

In the early 1800s, window glass was called crown glass. It was made by blowing a bubble, then spinning it until flat. This left a sheet of glass with a bump, or crown, in the center. This blowing method of window-pane making required great skill and was very costly. Still, the finished crown glass produced a distortion through which everything looked curiously wavy, and the glass itself was also faulty and uneven. By the end of the nineteenth century, flat glass was mass-produced and was a common material. The cylinder method replaced the old method, and used compressed air to produce glass that could be slit lengthwise, reheated, and allowed to flatten on an iron table under its own weight. New furnaces and better polishing machines made the production of plate-glass a real industry. Today, almost all flat glass is produced by a float-glass process, which reheats the newly formed ribbon of glass and allows it to cool without touching a solid surface. This produces inexpensive glass that is flat and free from distortion.

What are the **advantages** of **tempered glass**?

Tempered glass is glass that is heat-treated. Glass is first heated and then the surfaces are cooled rapidly. The edges cool first, leaving the center relatively hot compared to the surfaces. As the center cools, it forces the surfaces and edges into compression. Tempered glass is approximately four times as strong as annealed glass. It is able to resist temperature differences of 200°F–300°F (90°C–150°C). Since it resists breakage it is favored in many applications for its safety characteristics such as automobiles, doors, tub and shower enclosures, and skylights.

When were **glass blocks invented**?

Dating back to 1847, glass blocks were originally used as telegraph insulators. They were much smaller and thicker than structural glass blocks and were used mostly in the southeastern Untied States until eventually replaced with porcelain and other types of insulating materials. Glass building bricks were invented in Europe in the early 1900s as thin blocks of glass supported by a grid. Structural glass blocks have been manufactured in the United States since Pittsburgh-based Pittsburgh Corning began producing them in 1938. Blocks made at that time measured approximately 8 inches (20 centimeters) square by nearly 5 inches (13 centimeters) in depth, and cast a greenish tint as light transmitted through it. Today's glass blocks can be a square foot in size, much more uniformly shaped, and available in many different sizes, textures, and colors.

How is **bulletproof glass** made?

Bulletproof glass is composed of two sheets of plate glass with a sheet of transparent resin in between, molded together under heat and pressure. When subjected to a severe blow, it will crack without shattering. Today's bulletproof glass is a development of laminated or safety glass, invented by the French chemist Edouard Benedictus (1878–1930). It is basically a multiple lamination of glass and plastic layers.

Who invented **thermopane glass**?

Thermopane insulated window glass was invented by C.D. Haven in the United States in 1930. It is two sheets of glass that are bonded together in such a manner that they enclose a captive air space in between. Often this space is filled with an inert gas that increases the insulating quality of the window. Glass, in general, is also one of the best transparent materials because it allows the short wavelengths of solar radiation to pass through it, but prohibits nearly all of the long waves of reflected radiation from passing back through it.

What is the **float glass process**?

Manufacture of high-quality flat glass, needed for large areas and industrial uses, depends on the float glass process, invented by Alastair Pilkington (1920–1995) in 1952. The float process departs from all other glass processes where the molten glass flows from the melting chamber into the float chamber, which is a molten tin pool approximately 160 feet (49 meters) long and 12 feet (3.5 meters) wide. During its passage over this molten tin, the hot glass assumes the perfect flatness of the tin surface and develops excellent thickness uniformity. The finished product is as flat and smooth as plate glass without having been ground and polished.

> ## How is the glass used in movie stunts made?
>
> The "glass" might be made of candy (sugar boiled down to a translucent pane) or plastic. This looks like glass and will shatter like glass, but will not cut a performer.

Who developed fiberglass?

Coarse glass fibers were used for decoration by the ancient Egyptians. Other developments were made in Roman times. Parisian craftsman Ignace Dubus-Bonnel was granted a patent for the spinning and weaving of drawn glass strands in 1836. In 1893, the Libbey Glass Company exhibited lampshades at the World's Columbian Exposition in Chicago that were made of coarse glass thread woven together with silk. However, this was not a true woven glass. Between 1931 and 1939, the Owens Illinois Glass Company and the Corning Glass Works developed practical methods of making fiberglass commercially.

Once the technical problem of drawing out the glass threads to a fraction of their original thinness was solved—basically an endless strand of continuous glass filament as thin as 1/5000 of an inch (0.005 millimeters)—the industry began to produce glass fiber for thermal insulation and air filters, among other uses. When glass fibers were combined with plastics during World War II, a new material was formed. Glass fibers did for plastics what steel did for concrete—gave strength and flexibility. Glass-fiber-reinforced plastics (GFRP) became very important in modern engineering. Fiberglass combined with epoxy resins and thermosetting polyesters are now used extensively in boat and ship construction, sporting goods, automobile bodies, and circuit boards in electronics.

When was cement first used?

Cements are finely ground powders that, when mixed with water, set to a hard mass. The cement used by the Egyptians was calcined gypsum, and both the Greeks and Romans used a cement of calcined limestone. Roman concrete (a mixture of cement, sand, and some other fine aggregate) was made of broken brick embedded in a pozzolanic lime mortar. This mortar consisted of lime putty mixed with brick dust or volcanic ash. Hardening was produced by a prolonged chemical reaction between these components in the presence of moisture. With the decline of the Roman empire, concrete fell into disuse.

The first step toward its reintroduction was in 1756, when English engineer John Smeaton (1724–1792) found that when lime containing a certain amount of clay was burned, it would set under water. This cement resembled what had been made by the Romans. Further investigations by James Parker in the same decade led to the commercial production of natural hydraulic cement. In 1824, Englishman Joseph Aspdin (1799–1855) obtained a patent for what he called "portland cement," a material produced from a synthetic mixture of limestone and clay. He called it "portland" because it resembled a building stone that was quarried on the Isle of Portland off the coast of Dorset. The manufacture of this cement spread rapidly to Europe and the United States by 1870. Today, concrete is often reinforced or prestressed, increasing its load-bearing capabilities.

How were **early macadam roads** different from **modern paved roads**?

Macadam roads developed originally in England and France and are named after the Scottish road builder and engineer John Louden MacAdam (1756–1836). The term "macadam" originally designated road surface or base in which clean, broken, or crushed ledge stone was mechanically locked together by rolling with a heavy weight and bonded together by stone dust screenings that were worked into the spaces and then "set" with water. With the beginning of the use of bituminous material (tar or asphalt), the terms "plain macadam," "ordinary macadam," or "waterbound macadam" were used to distinguish the original type from the newer bituminous macadam. Waterbound macadam surfaces are almost never built now in the United States, mainly because they are expensive and the vacuum effect of vehicles loosens them. Many miles of bituminous macadam roads are still in service, but their principal disadvantages are their high crowns and narrowness. Today's roads that carry very heavy traffic are usually surfaced with very durable portland cement.

What is **Belgian block**?

Belgian block is a road-building material, first used in Brussels, Belgium, and introduced into New York about 1850. Its shape is a truncated pyramid with a base of about 5 to 6 inches (13 to 15 centimeters) square and a depth of 7 to 8 inches (18 to 20.5 centimeters). The bottom of the block is not more than 1 inch (2.5 centimeters) different from the top. The original blocks were cut from trap-rock from the Palisades of New Jersey.

Belgian blocks replaced cobblestones mainly because their regular shape allowed them to remain in place better than cobblestones. They were not universally adopted, however, because they would wear round and create joints or openings that would then form ruts and hollows. Although they provided a smooth surface compared to the uneven cobblestones, they still made for a rough and noisy ride.

What is **solder**?

Solder is an alloy of two or more metals used for joining other metals together. One example of solder is half-and-half composed of equal parts of lead and tin. Other metals used in solder are aluminum, cadmium, zinc, nickel, gold, silver, palladium, bismuth, copper, and antimony. Various melting points to suit the work are obtained by varying the proportions of the metals.

Solder is an ancient joining method, mentioned in the Bible (Isaiah 41:7). There is evidence of its use in Mesopotamia some 5,000 years ago, and later in Egypt, Greece, and Rome. Use of numerous types of solder is currently wide and varied, and future use looks bright as well. As long as circuitry based on electrical and magnetic impulses and composed of a combination of conductors, semiconductors, and insulators continues to be in use solder will remain indispensable.

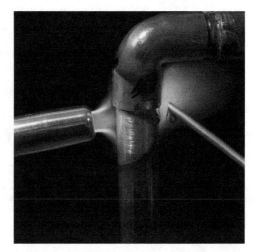

Solder is a metal alloy that is melted and then used for joining metals together. It is very useful in everything from plumbing to circuit boards.

What is **slag**?

Slag is a non-metallic by-product of iron production that is drawn from the surface of pig iron in the blast furnace. Slag can also be produced in smelting copper, lead, and other metals. Slag from steel blast furnaces contains lime, iron oxide, and silica. The slag from copper and lead-smelting furnaces contains iron silicate and oxides of other metals in small amounts. Slag is used in cements, concrete, and roofing materials as well as ballast for roads and railways.

What is **creosote**?

Creosote is a yellowish, poisonous, oily liquid obtained from the distillation of coal or wood tar (coal tar constitutes the major part of the liquid condensate obtained from the "dry" distillation or carbonization of coal to coke). Crude creosote oil, also called dead oil or pitchoil, is obtained by distilling coal tar and is used as a wood preservative. Railroad ties, poles, fence posts, marine pilings, and lumber for outdoor use are impregnated with creosote in large cylindrical vessels. This treatment can greatly extend the useful life of wood that is exposed to the weather. Creosote that is distilled from wood tar is used in pharmaceuticals. Other uses of creosote include disinfectants and solvents. In 1986, the United States Environmental Protection Agency (EPA) began restricting the use of creosote as a wood preservative because of its poisonous and carcinogenic nature.

What is **carbon black**?

Carbon black is finely divided carbon produced by incomplete combustion of methane or other hydrocarbon gases (by letting the flame impinge on a cool surface). This forms a very fine pigment containing up to 95 percent carbon, which gives a very intense black color that is widely used in paints, inks, and protective coatings, and as a colorant for paper and plastics. It is also used in large amounts by the tire industry in the production of vulcanized rubber.

How is **parchment paper** made?

Most parchment paper is now vegetable parchment. It is made from a base paper of cotton rags or alpha cellulose, known as waterleaf, which contains no sizing or filling materials. The waterleaf is treated with sulfuric acid, converting a part of the cellulose into a gelatin-like amyloid. When the sulfuric acid is washed off, the amyloid film hardens on the paper. The strength of the paper is increased and will not disintegrate even when fully wet. Parchment paper can withstand heat and items will not stick to it.

How is **sandpaper** made?

Sandpaper is a coated abrasive that consists of a flexible-type backing (paper) upon which a film of adhesive holds and supports a coating of abrasive grains. Various types of resins and hide glues are used as adhesives. The first record of a coated abrasive is in thirteenth-century China, when crushed seashells were bound to parchment using natural gums. The first known article on coated abrasives was published in 1808 and described how calcined, ground pumice was mixed with varnish and spread on paper with a brush. Most abrasive papers are now made with aluminum oxide or silicon carbide, although the term sandpapering is still used. Quartz grains are also used for wood polishing. The paper used is heavy, tough, and flexible, and the grains are bonded with a strong glue.

Why is **acid-free paper** important?

Acid-free paper is important for the preservation of printed materials on paper. Acidic papers deteriorate quickly. Papers treated with an alkaline reserve, most frequently chalk, neutralizes the acids extending the life span of the paper.

Why is **titanium dioxide** the most widely used **white pigment**?

Titanium dioxide (TiO_2), also known as titania, titanic anhydride, titanium oxide, or titanium white, has become the predominant white pigment in the world because of its high refractive index, lack of absorption of visible light, ability to be produced in the right size range, and its stability. It is the whitest known pigment, unrivalled for color, opacity, stain resistance, and durability; it is also non-toxic. The main consuming industries are paint, printing inks, plastics, and ceramics. Titanium dioxide is also used in the manufacture of floor coverings, paper, rubber, and welding rods.

When and where was **gunpowder invented**?

The explosive mixture of saltpeter (potassium nitrate), sulfur, and charcoal called gunpowder was known in China at least by 850 C.E., and probably was discovered by Chinese alchemists searching for components to make artificial gold. Early mixtures had too little saltpeter (50 percent) to be truly explosive; 75 percent minimum is needed to get a detonation. The first use of the mixture was in making fireworks. Later, the Chinese used it in incendiary-like weapons. Eventually it is thought that the Chinese found the correct proportions to utilize its explosive effects in rockets and "bamboo bullets."

However, some authorities still maintain that the "Chinese gunpowder" really had only pyrotechnic qualities, and "true" gunpowder was a European invention. Roger Bacon (1214–1292) had a formula for it and so might have the German monk Berthold Schwartz who is believed to have lived in the fourteenth century. The year 1353 is often given as the date of his invention of gunpowder. Its first European use depended on the development of firearms in the fourteenth century. Not until the seventeenth century was gunpowder used in peacetime, for mining and civil engineering applications.

How are **colored fireworks** made?

Fireworks existed in ancient China in the ninth century where saltpeter (potassium nitrate), sulfur, and charcoal were mixed to produce the dazzling effects. Magnesium burns with a brilliant white light and is widely used in making flares and fireworks. Various other colors can be produced by adding certain substances to the flame. Strontium compounds color the flame scarlet, barium compounds produce yellowish-green, copper produces a blue-green, lithium creates purple, and sodium results in yellow. Iron and aluminum granules give gold and white sparks, respectively.

What is the **chemical formula** for **TNT**?

TNT is the abbreviation for 2,4,6-trinitrotoluene [$C_7H_5N3O_6$ or $C_6H_2(CH_3)(NO_2)_3$]. TNT is a powerful, highly explosive compound widely used in conventional bombs. Discovered by Joseph Wilbrand (1811–1894) in 1863, it is made by treating toluene with nitric acid and sulfuric acid. This yellow crystalline solid with a low melting point has low shock sensitivity and even burns without exploding. This makes it safe to handle and cast; but once detonated, it explodes violently.

Who **invented dynamite**?

Dynamite was not an accidental discovery but the result of a methodical search by the Swedish technologist Alfred Nobel (1833–1896). Nitroglycerine had been discovered in 1849 by the Italian organic chemist Ascanio Sobriero (1812–1888), but it was so sensitive and difficult to control that it was useless. Nobel sought to turn nitroglycerine into a manageable solid by absorbing it into a porous substance. From 1866 to 1867 he tried an unusual mineral, kieselguhr, and created a doughlike explosive that was controllable. He also invented a detonating cap incorporating mercury fulminate with which nitroglycerine could be detonated at will.

When was **plastic invented**?

In the mid–1850s, Alexander Parkes (1813–1890) experimented with nitrocellulose (or guncotton). Mixed with camphor, it made a hard but flexible transparent material, which he called "Parkesine." He teamed up with a manufacturer to produce it, but there was no demand for it, and the firm went bankrupt. An American, John Wesley Hyatt (1837–1920), acquired the patent in 1868 with the idea of producing artificial ivory for billiard balls. Improving the formula and with an efficient manufacturing process, he marketed the material, intended for use in making a few household articles, under the name "celluloid." It soon found use in the manufacture of novelty and fancy goods—buttons, letter openers, boxes, hatpins, combs, and the like were products often made of celluloid. The material also became the medium for cinematography: celluloid strips coated with a light-sensitive "film" were ideal for shooting and showing moving pictures.

Celluloid was the only plastic material until 1904, when Belgian scientist Leo Hendrik Baekeland (1863–1944) succeeded in producing a synthetic shellac from formaldehyde and phenol. Called "bakelite," it was the first of the thermosetting plastics (i.e., synthetic materials that, having once been subjected to heat and pressure, became extremely hard and resistant to high temperatures). Bakelite and other, more versatile plastics, eventually eclipsed celluloid, and by the 1940s, celluloid's markets had shrunk so that it was no longer of commercial importance. Today, ping pong balls are almost the only product still made with celluloid.

Why can't metal objects be placed in a microwave?

Metal objects should not be used in a microwave (except as directed by the manufacturer) because the microwaves are reflected off these materials and do not allow the heat to penetrate the food. The oven could be damaged by an arc between the metal utensils and the cavity interior or door assembly if the cooking load is not large enough to absorb the microwave energy.

What are **man-made fibers**?

Man-made fibers are totally made by chemical means or may be fibers of regenerated cellulose. According to U.S. law, the fibers must be labeled in accordance with generic groups as follows:

Man-made Fiber	Derived From
Acetate	Cellulose acetate
Acrylic	Acrylic resins
Metallic	Any type of fabric made with metallic yarns. Made by twisting thin metal foil around cotton, silk, linen, or rayon yarn
Modacrylic	Acrylic resins
Nylon	Synthetic polyamides extracted from coal and petroleum
Rayon	Trees, cotton, and woody plants
Saran	Vinylidene chloride
Spandex	Polyurethane
Triacetate	Regenerated cellulose
Vinyl	Polyvinyl chloride

What is **Kevlar**?

The registered trademark Kevlar refers to synthetic fiber called liquid crystalline polymers. Discovered by Stephanie Kwolek (1923–), Kevlar is a thin, very strong fiber. It is best known for its use in bulletproof garments.

Who **invented Teflon**?

In 1938, the American engineer Roy J. Plunkett (1910–1994) at DuPont de Nemours discovered the polymer of tetraluorethylene (PTFE) by accident. This fluorocarbon is marketed under the name of Fluon in Great Britain and Teflon in the United States. Patented in 1939 and first exploited commercially in 1954, PTFE is resistant to all acids and has exceptional stability and excellent electrical insulating properties. It is used in making piping for corrosive materials, in insulating devices for radio transmitters, in pump gaskets, and in computer microchips. In addition, its nonstick properties make PTFE an ideal material for surface coatings. In 1956, French engineer Marc Gregoire discovered a process whereby he could fix a thin layer of Teflon on an aluminum surface. He then patented the process of applying it to cookware, and the nostick frying pan was created.

Who **invented Velcro**?

The inspiration for Velcro came about when Swiss engineer George deMaestral (1907–1990) took a closer look at those annoying cockleburs that stick so well to things like dog fur and your favorite pair of socks. He recreated what he discovered into a new type of fastener.

When were **microfibers invented**?

Microfibers are very fine fibers with a diameter of less than one denier. "Denier" is the term used to describe the diameter or fineness of a fiber. It is the weight in grams of a 9,000-meter (9,842 yards) length of fiber. Many microfibers are 0.5 to 0.6 denier. As a comparison, microfibers are 100 times finer than human hair and half the diameter of a fine silk fiber. The first microfiber fabric was Ultrasuede™ created by Miyoshi Okamoto in 1970 at Toray Industries in Japan. His colleague, Toyohiko Hikota, perfected the process to create a soft, supple, stain-resistant fabric.

Who made the first successful **synthetic gemstone**?

In 1902, Auguste Victor Louis Verneuil (1856–1913) synthesized the first man-made gemstone—a ruby. Verneuil perfected a "flame-fusion" method of producing crystals of ruby and other corundums within a short time period.

ENERGY

FOSSIL FUELS

Why are **coal, oil**, and **natural gas** called **fossil fuels**?

Coal, oil, and natural gas are composed of the remains of organisms that lived as long ago as 500 million years. These microscopic organisms (such as phytoplankton) became incorporated into the bottom sediments and then were converted, with time, to oil and gas. Coal is the remains of plants and trees (changing into peat and then lignite) that were buried and subjected to pressure, temperature, and chemical processes for millions of years. Fossil fuels are nonrenewable sources of energy. There is a finite supply of the resources for fossil fuels. Eventually these resources will diminish to the point of being too expensive or too environmentally damaging to retrieve. Fossil fuels provide more than 85 percent of all the energy consumed in the United States; including two-thirds of the electricity and nearly all of the transportation fuels.

How and when was **coal formed**?

Coal is formed from the remains of plants that have undergone a series of far-reaching changes, turning into a substance called peat, which subsequently was buried. Through millions of years, Earth's crust buckled and folded, subjecting the peat deposits to very high pressure and changing the deposits into coal. The Carboniferous, or coal-bearing, period occurred about 250 million years ago. Geologists in the United States sometimes divide this period into the Mississippian and the Pennsylvanian periods. Most of the high-grade coal deposits are to be found in the strata of the Pennsylvanian period.

A compressor in a pillar mine tunnel.

What **types of coal** are there?

The first stage in the formation of coal converts peat into lignite, a dark-brown type of coal. Lignite is then converted into subbituminous coal as pressure from overlying materials increases. Under still greater pressure, a harder coal called bituminous, or soft, coal is produced. Intense pressure changes bituminous coal into anthracite, the hardest of all coals.

What is **cannel coal**?

Cannel coal is a type of coal that possesses some of the properties of petroleum. Valued primarily for its quick-firing qualities, it burns with a long, luminous flame. It is made up of coal-like material mixed with clay and shale, and it may also look like black shale, being compact and dull black in color.

How is underground **coal mined**?

There are two basic types of underground mining methods: room and pillar and longwall. In room and pillar mines, coal is removed by cutting rooms, or large tunnels, in the solid coal, leaving pillars of coal for roof support. Longwall mining takes successive slices over the entire length of a long working face. In the United States, about two-thirds of the coal recovered by underground mining is by room and pillar method; the other third is recovered by longwall mining. Coal seams in the United

What is a miner's canary?

Miner's canary refers to the birds used by miners to test the purity of the air in the mines. At least three birds were taken by exploring parties, and the distress of any one bird was taken as an indication of carbon monoxide danger. Some miners used mice rather than birds. This method of safety was used prior to the more sophisticated equipment used today.

States range in thickness from a thin film to 50 feet (15 meters) or more. The thickest coal beds are in the western states, ranging from 10 feet (3 meters) in Utah and New Mexico to 50 feet (15 meters) in Wyoming. Other places, such as Great Britian, use the longwall method.

Which **country** is the **largest producer** of **coal**?

China is the world's largest producer of coal producing an estimated 3,209,677 thousand short tons in 2009.

Which **countries** have the **greatest reserves** of **coal**?

Coal is one of the most abundant fossil fuels. Over half of all the global reserves of coal are found in the United States (29 percent), Russia (19 percent), and China (14 percent). The world's total recoverable reserves of coal were estimated at 826,001 million tons at the end of 2009. At the current rate of production (2010), coal reserves are estimated to last for 119 years.

How much of the **world's energy** use is fueled by **coal**?

Coal provides over 23 percent of global primary energy needs and generates around 39 percent of the world's electricity. In the United States, coal supplies more than half of the electricity consumed. Nearly 70 percent of total global steel production is dependent on coal.

Where are the **largest oil** and **gas fields** in the world and in the United States?

The Ghawar field, discovered in 1948 in Saudi Arabia, is the largest in the world; it measures 150 × 22 miles (241 × 35 kilometers). The largest oil field in the United States is the Permian Basin, which covers approximately 100,000 square miles (160,934 square kilometers) in southeast New Mexico and western Texas.

Which country is the **largest supplier** of **oil** to the **United States**?

Canada is the largest supplier of crude oil to the United States. In 2008, the United States imported 912,263 thousand barrels of oil from Canada. The next largest single country supplier of crude oil to the United States was Saudi Arabia with 559,750 thousand barrels of crude oil.

Which U.S. **state produces** the most **crude oil**?

Texas is the largest producer of crude oil in the United States. In 2008, Texas produced 398,014 thousand barrels of oil. Alaska is the second largest producer of crude oil in the United States producing 249,874 thousand barrels of oil in 2008.

When was the **first oil well** in the **United States** drilled?

The Drake well at Titusville, Pennsylvania, was completed on August 28, 1859 (some sources list the date as August 27). The driller, William "Uncle Billy" Smith (d. 1890), went down 69.5 feet (21 meters) to find oil for Edwin L. Drake (1819–1880), the well's operator. Within 15 years, Pennsylvania oil field production reached over ten million 360-pound (163-kilogram) barrels a year.

When was **offshore drilling** for oil **first done**?

The first successful offshore oil well was built off the coast at Summerland, Santa Barbara County, California, in 1896.

Why is **Pennsylvania crude oil** so highly valued?

The waxy, sweet paraffinic oils found in Pennsylvania first became prominent because high quality lubricating oils and greases could be made from them. Similar grade crude oil is also found in West Virginia, eastern Ohio, and southern New York. Different types of crude oil vary in thickness and color, ranging from a thin, clear oil to a thick, tar-like substance.

What is the process known as **hydrocarbon cracking**?

Cracking is a process that uses heat to decompose complex substances. Hydrocarbon cracking is the decomposition by heat, with or without catalysts, of petroleum or heavy petroleum fractions (groupings) to give materials of lower boiling points. Thermal cracking, developed by William Burton (1865–1954) in 1913, uses heat and pressure to break some of the large heavy hydrocarbon molecules into smaller gasoline-grade ones. The cracked hydrocarbons are then sent to a flash chamber where the various fractions are separated. Thermal cracking not only doubles the gasoline yield, but has improved gasoline quality, producing gasoline components with good anti-knock characteristics (no premature fuel ignition).

Why was **lead added** to **gasoline** and why is lead-free gasoline used in cars?

Tetraethyl lead was used for more than 40 years to improve the combustion characteristics of gasoline. It reduces or eliminates "knocking" (pinging caused by premature ignition) in large high-performance engines and in smaller high-compression engines. It provides lubrication to the extremely close-fitting engine parts, where oil has a tendency to wash away or burn off. However, lead will ruin and effectively destroy the catalyst presently used in emission control devices installed in new cars. Therefore, only lead-free gasoline must be used.

When did the use of **lead-free fuel** become **mandatory** in the United States?

The sale of leaded gasoline for motor vehicles ended in 1996. All vehicles manufactured after July 1974 for sale in the United States were required to use unleaded gasoline.

What is a **reformulated gasoline**?

Oil companies are being required to offer new gasolines that burn more cleanly and have less impact on the environment. Typically, reformulated gasoline (RFG) contains: lower concentrations of benzene, aromatics, and olefins; less sulfur; a lower Reid vapor pressure (RVP); and some percentage of an oxygenate (non-aromatic component), such as methyl tertiary butyl ether (MTBE). MTBE is a high-octane gasoline blending component produced by the reaction of isobutylene and methanol. It was developed to meet the ozone ambient air quality standards, but its unique characteristics as a water pollutant pose a challenge to the Environmental Protection Agency (EPA) in meeting the requirements of the Clean Air Act, the Safe Drinking Water Act, and the Underground Storage Tank Program. The Clean Air Act called for reformulated gasoline to be sold in the cities with the worst smog pollution beginning January 1, 1995. Reformulated gasoline is now used in 17 cities and the District of Columbia.

What kinds of **additives** are in **gasoline** and why?

Additive	Function
Antiknock compounds	Increase octane number
Scavengers	Remove combustion products of antiknock compounds
Combustion chamber modifiers	Suppress surface ignition and spark plug deposit fouling
Antioxidants	Provide storage stability
Metal deactivators	Supplement storage stability
Antirust agents	Prevent rusting in gasoline-handling systems
Anti-icing agents	Suppress carburetor and fuel system freezing
Detergents	Control carburetor and induction system cleanliness
Upper-cylinder lubricants	Lubricate upper cylinder areas and control intake system deposits
Dyes	Indicate presence of antiknock compounds and identify makes and grades of gasoline

What do the **octane numbers** of gasoline **mean**?

The octane number is a measure of the gasoline's ability to resist engine knock. Two test fuels, normal heptane and isooctane, are blended for test results to determine octane number. Normal heptane has an octane number of zero and isooctane a value of 100. Gasolines are then compared with these test blends to find one that makes the same knock as the test fuel. The octane rating of the gasoline under testing is the percentage by volume of isooctane required to produce the same knock. For example, if the test blend has 85 percent isooctane, the gasoline has an octane rating of 85. The octane rating that appears on gasoline pumps is an average of research octane determined in laboratory tests with engines running at low speeds, and motor octane, determined at higher speeds.

When did **gasoline stations open**?

The first service station (or garage) was opened in Bordeaux, France, in December 1895 by A. Barol. It provided overnight parking, repair service, and refills of oil and "motor spirit." In April 1897 a parking and refueling establishment—Brighton Cycle and Motor Co.—opened in Brighton, England.

The pump that would be used to eventually dispense gasoline was devised by Sylanus Bowser of Fort Wayne, Indiana, in September 1885; it originally dispensed kerosene. Twenty years later Bowser manufactured the first self-regulating gasoline pump. In 1912, a Standard Oil of Louisiana superstation opened in Memphis, Tennessee, featuring 13 pumps, a ladies' restroom, and a maid who served ice water to waiting customers. On December 1, 1913, in Pittsburgh, Pennsylvania, the Gulf Refining Company opened the first drive-in station as a 24-hour-a-day operation. Only 30 gallons (114 liters) of gasoline were sold the first day.

Alternative sources of energy are becoming more available in recent years, such as biodiesel as fuel for automobiles, as concerns about the environment increase.

What are the **advantages** and **disadvantages** of the **alternative fuels** to gasoline to power automobiles?

The emissions of gasoline are a major source of air pollution in most U.S. urban areas. Researchers are investigating alternative fuels, which are cleaner, yet also economical. Alternative fuels include ethanol, methanol, biodiesel, compressed natural gas, electricity, and hydrogen. Currently, none of the alternatives delivers as much energy content as gasoline, so more of each of these fuels must be consumed to equal the distance that the energy of gasoline propels the automobile. Some predict that in the future families may have different vehicles that use different fuels for different types of driving. For example, electric or hybrids for short trips, an ethanol/electric hybrid for longer trips, and a biodiesel vehicle for hauling large loads.

Alternative	Advantages	Disadvantages
Biodiesel from vegetable oils	Cleaner than petrodiesel; reduced emissions	Cost of pure biodiesel; blends (B30) turn into solids in low temperatures
Electricity from batteries	No vehicle emissions, good for stop-and-go driving	Short-lived bulky batteries; limited trip range
Ethanol from corn,	Relatively clean fuel	Costs; corrosive damage biomass, etc.
Hydrogen from electrolysis, etc.	Plentiful supply; non-toxic emissions	High cost; highly flammable

263

Alternative	Advantages	Disadvantages
Methanol from methanol gas, coal, biomass, wood	Cleaner combustion; less volatile	Corrosive; some irritant emissions
Natural gas from hydrocarbons and petroleum deposits	Cheaper on energy basis; relatively clean	Cost to adapt vehicle; bulky storage; sluggish performance
Propane(liquefied petroleum gas)	Cleaner; good supply from domestic sources; less expensive	Vehicles must be retrofitted less readily available

How do the **costs** of various **fuels compare**?

Fuel prices are reported in the units in which they are typically sold, for example, dollars per gallon. However, since the energy content per gallon of each fuel is different, the price paid per unit of energy is often different from the price paid per gallon. The following chart shows the average price per gallon, the average price in gasoline gallon equivalents, and the average price in dollars per million BTU as of 2009.

Average Fuel Prices (2009)

Fuel	Ave. Price/Gal.	Ave. Price in Gas Equivalents	Ave. Price in Dollars/Million BTU
Gasoline	$2.64	$2.64	$22.90
Compressed Natural Gas	$1.86*	$1.86	$16.14
Ethanol (E85)	$2.27	$3.21	$27.80
Propane	$2.69	$3.72	$32.22
Biodiesel (B20)	$2.88	$2.63	$22.70
Biodiesel (B99-B100)	$3.19	$3.14	$27.21

*(per gas gallon equivalent)

Interest in alternative fuels increases when the actual price differential per gallon increases even if the savings is not as great on an energy-equivalent basis.

How is **gasohol made**?

Gasohol, a mixture of 90 percent unleaded gasoline and 10 percent ethyl alcohol (ethanol), has gained some acceptance as a fuel for motor vehicles. It is comparable in performance to 100 percent unleaded gasoline with the added benefit of superior anti-knock properties (no premature fuel ignition). No engine modifications are needed for the use of gasohol and all auto manufacturers approve the use of gasohol (blends of 10 percent ethanol) in gasoline vehicles.

Since corn is the most abundant U.S. grain crop, it is predominantly used in producing ethanol. However, the fuel can be made from other organic raw materials, such as oats, barley, wheat, milo, sugar beets, or sugar cane. Potatoes, cassava (a starchy plant), and cellulose (if broken up into fermentable sugars) are possible other sources. The corn starch is processed through grinding and cooking. The process requires the conversion of a starch into a sugar, which in turn is converted into alcohol by reaction with yeast. The alcohol is distilled and any water is removed until it is 200 proof (100 percent alcohol).

One acre of corn yields 250 gallons (946 liters) of ethanol; an acre of sugar beets yields 350 gallons (1,325 liters), while an acre of sugar can produce 630 gallons (2,385 liters). In the future, motor fuel could conceivably be produced almost exclusively from garbage, but currently its conversion remains an expensive process.

How does **ethanol differ** from **gasohol**?

Ethanol (E85) is a mixture of 85 percent ethyl alcohol and 15 percent gasoline. It can only be used in flexible fuel vehicles (FFVs). Flexible fuel vehicles are designed to run on gasoline, ethanol (E85), or any mixture of the two.

What are the main **components** found in motor **vehicle exhaust**?

The main components of exhaust gas are nitrogen, carbon dioxide, and water. Smaller amounts of nitrogen oxides, carbon monoxide, hydrocarbons, aldehydes, and other products of incomplete combustion are also present. The most important air pollutants, in order of amount produced, are carbon monoxide, nitrogen oxides, and hydrocarbons.

How much **carbon dioxide** is **produced** per **gallon** of **gasoline**?

A gallon of gasoline produces 19.4 pounds (8.8 kilograms) of carbon dioxide. The average passenger vehicle produces 5.2 metric tons (11,464 pounds) of carbon dioxide annually.

What is the **cleanest fossil fuel**?

Natural gas, composed primarily of methane, is the cleanest of all fossil fuels, producing only carbon dioxide and water vapor when it is burned. Coal and oil produce higher levels of harmful emissions, including nitrogen oxides and sulfur dioxide when burned. Coal creates the most carbon dioxide when it is burned. Natural gas emits 25 to 50 percent less carbon dioxide than either oil or coal for each unit of energy produced.

What are synthetic fuels?

Synthetic fuels, commonly called synfuels, are gaseous and liquid fuels produced synthetically from coal and oil shale. Basically, coal is converted to gaseous or liquid forms in coal-based synfuels. These are easier to transport and burn more cleanly than coal itself. Synthetic natural gas is also produced from coal. Disadvantages of these processes are that they require a large amount of water and the new fuels have 30–40 percent less fuel content than pure coal. Synfuels from oil shale are produced by extracting the oils from the rocky base. Gasoline and kerosene can be produced from oil shale. Synfuel may also be obtained from biomass from human and animal waste. The waste is converted to methane by the action of anaerobic bacteria in a digester.

Fossil Fuel Emission Levels
(Pounds per Billion BTU of Energy Input)

Pollutant	Natural Gas	Oil	Coal
Carbon Dioxide	117,000	164,000	208,000
Carbon Monoxide	40	33	208
Nitrogen Oxides	92	448	457
Sulfur Dioxide	1	1,122	2,591

Which **countries** have the **highest reserves** of **natural gas**?

Worldwide reserves of natural gas are estimated at 6,254 trillion cubic feet with 41 percent of the total located in the Middle East and 32 percent of the total located in Europe and the former U.S.S.R.

Country	Natural Gas Reserves (Trillion Cubic Feet)
Russia	1,680
Iran	892
Saudi Arabia	258
United States	238

What is **cogeneration**?

Cogeneration is an energy production process involving the simultaneous generation of thermal (steam or hot water) and electric energy by using a single primary heat source. By producing two kinds of useful fuels in the same facility the net energy yield from the primary fuel increases from 30–35 percent to 80–90 percent. Cogeneration can result in significant cost savings and can reduce any possible environmental effects conventional energy production may produce. Cogeneration facilities have been installed at a variety of sites, including oil refineries, chemical plants, paper mills, utility complexes, and mining operations.

RENEWABLE AND ALTERNATIVE ENERGY

How do **renewable energy resources differ** from fossil fuels?

The main sources of renewable energy are biofuels, such as wood, hydropower, geothermal, solar, and wind. Unlike fossil fuels, renewable energy resources are being replenished continuously and will never run out. Some sources of renewable energy, such as solar power and wind power, are perpetual, meaning the wind will blow and the sun will shine no matter how much energy is used. Sources of renewable energy that rely on agriculture, such as woods, are renewable as long as they are not depleted and exploited too rapidly.

What are the **three types** of **primary energy** that **flow continuously** on or to the surface of Earth?

- *Geothermal energy* is heat contained beneath Earth's crust and brought to the surface in the form of steam or hot water. The five main sources of this geothermal reservoir are: dry, super-heated steam from steam fields below Earth's surface; mixed hot water, wet steam, etc., from geysers; dry rocks (into which cold water is pumped to create steam); pressurized water fields of hot water and natural gas beneath ocean beds; and magma (molten rock in or near volcanoes and 5 to 30 miles [8 to 48 kilometers] below Earth's crust). Most buildings in Iceland are heated by geothermal energy, and a few communities in the United States, such as Boise, Idaho, use geothermal home heating. Electric power production, industrial processing, and space heating, are fed from geothermal sources. The California Geysers project is the world's largest geothermal electric generating complex, with 200 steam wells that provide some 1,300 megawatts of power.

- *Solar radiation* utilization depends on the weather, number of cloudy days, and the ability to store energy for night use. The process of collecting and storing is difficult and expensive. A solar thermal facility (LUZ International Solar Thermal Plant) in the Mojave Desert currently produces 274 megawatts and is used to supplement the power needs of Los Angeles utilities companies. Japan has four million solar panels on roofs, and two-thirds of the houses in Israel have them; 90 percent of Cyprus homes do as well. Solar photo voltaic cells can generate electric current when exposed to the sun. Virtually every spacecraft and satellite since 1958 utilizes this kind of resource, too.

- *Tidal and wave energy* contain enormous amounts of energy to be harnessed. The first tidal-powered mill was built in England in 1100; another in Woodbridge, England, built in 1170, has functioned for over 800 years. The Rance River Power Station in France, in operation since 1966, was the first large tidal electric generator plant, producing 160 megawatts. A tidal station works like a hydropower dam, with its turbines spinning as the tide flows through them. Unfortunately, the tidal period of 13.5 hours causes problems in integrating the peak use with the peak generation ability. Ocean wave energy can also be made to drive electrical generators.

How much of the **U.S. energy supply** is from **renewable sources**?

Renewable energy resources accounted for only eight percent of total energy consumption in the United States in 2009. Hydropower is the single largest source of renewable energy in the United States, accounting for 35 percent of total renewable energy.

What are the **advantages** of **solar power**?

Solar energy is a clean, abundant, and safe energy source. More energy falls from the sun on Earth in one hour than is used by everyone in the world in one year. Over a two-week period, Earth gets as much energy from the sun as is stored in all known reserves of coal, oil, and natural gas. Solar energy can be used to heat water and spaces for homes and businesses or can be converted into electricity. Solar energy accounts for only about one percent of the total renewable energy resources.

Solar panels convert solar radiation into clean, renewable energy. However, the disadvantage is that the panels obviously do not work at night or on overcast days.

What is the difference between **passive** solar energy systems and **active solar energy systems**?

Passive solar energy systems use the architectural design, the natural materials, or absorptive structures of the building as an energy saving system. The building itself serves as a solar collector and storage device. An example would be thick-walled stone and adobe dwellings that slowly collect heat during the day and gradually release it at night. Passive systems require little or no investment of external equipment.

Active solar energy systems require a separate collector, a storage device, and controls linked to pumps or fans that draw heat from storage when it is available. Active solar systems generally pump a heat-absorbing fluid medium (air, water, or an antifreeze solution) through a collector. Collectors, such as insulated water tanks, vary in size, depending on the number of sunless days in a locale. Another heat storage system uses eutectic (phase-changing) chemicals to store a large amount of energy in a small volume.

How is **solar energy converted** into electricity?

Solar energy is converted into electricity using photovoltaic (PV) cells or concentrating solar power plants. Photovoltaic cells convert sunlight directly into electricity. Individual PV cells are combined in modules of about 40 cells to form a solar panel; 10 to 20 solar panels are used to power a typical home. The panels are usually mounted on the home facing south or mounted onto a tracking device that follows the sun for the maximum exposure to sunlight. Power plants and other industrial locations combine more solar panels to generate electricity.

Concentrating solar power plants collect the heat (energy) from the sun to heat a fluid, which produces steam that drives a generator to produce electricity. The three main types of concentrating solar power systems are parabolic trough, solar dish, and solar power tower, which describe the different types of collectors. Parabolic troughs collectors have a long, rectangular, U-shaped reflector or mirror focused on the sun with a tube (receiver) along its length. A solar dish looks very much like a large satellite dish that concentrates the sunlight into a thermal receiver that absorbs and collects the heat and transfers it to the engine generator. The engine produces mechanical power, which is used to run a generator converting mechanical power into electrical power. A solar tower uses a field of flat, sun-tracking mirrors, called heliostats, to collect and concentrate the sunlight onto a tower-mounted heat exchanger (receiver). A fluid is heated in the receiver to generate steam, which is used in a generator to produce electricity.

How does a **solar cell generate electricity**?

A solar cell, also called a photovoltaic (PV) cell, consists of several layers of silicon-based material. When photons, particles of solar energy from sunlight, strike a photovoltaic cell, they are reflected, pass through, or are absorbed. Absorbed photons provide energy to generate electricity. The top p-layer absorbs light energy. This energy frees electrons at the junction layer between the p-layer and the n-layer. The freed electrons collect at the bottom n-layer. The loss of electrons from the top layer produces "holes" in the layer that are then filled by other electrons. When a connection, or circuit, is completed between the p-layer and n-layer the flow of electrons creates an electric current. The photovoltaic effect, including the naming of the p-layer and n-layer, was discovered by Russell Ohl (1898–1987), a researcher at Bell Labs, in 1940.

When were **photovoltaic cells** developed?

A group of researchers at Bell Labs, Calvin Fuller (1902–1994), Daryl Chapin (1906–1995), and Gerald Pearson (1905–1987), developed the first practical silicon solar cell in 1954. The earliest PV cells were used to power U.S. space satellites. The use of PV cells was then expanded to power small items such as calculators and watches.

Where is the **largest solar generating plant** located?

The largest solar generating plant in the world is the Solar Energy Generating System (SEGS) located in California's Mojave Desert. Consisting of nine power plants, SEGS VIII and IX located in Harper Lake are individually and collectively the largest solar generating power plants in the world.

Who **invented** the **fuel cell**?

The earliest fuel cell, known as a "gas battery," was invented by Sir William Grove (1811–1896) in 1839. Grove's fuel cell incorporated separate test tubes of hydrogen and oxygen, which he placed over strips of platinum. It was later modified by Francis Thomas Bacon (1904–1992) with nickel replacing the platinum. A fuel cell is equivalent to a generator—it converts a fuel's chemical energy directly into electricity.

What is **biomass energy**?

The catch-all term biomass includes all the living organisms in an area. Wood, crops and crop waste, and wastes of plant, mineral, and animal matter are part of the biomass. Much of it is in garbage, which can be burned for heat energy or allowed to decay and produce methane gas. However, some crops are grown specifically for energy, including sugar cane, sorghum, ocean kelp, water hyacinth, and various species of trees. It has been estimated that 90 percent of U.S. waste products could be burned to provide as much energy as 100 million tons of coal (20 percent will not burn, but can be recycled). The use of biomass energy is significantly higher in developing countries where electricity and motor vehicles are scarcer.

Which plant has been investigated as a source of petroleum?

A number of plant species have been investigated as potential sources of petroleum. The shrub called the gopher plant (*Euphorbia lathyrus*) produces significant quantities of a milk-like sap—called latex—that is an emulsion of hydrocarbons in water. Another candidate is *Pittosporum resiniferum*, a native of the Philippines. The fruit of this plant, called a petroleum nut, is quite large and the oil harvested from it is frequently used for illumination. Various experiments are in progress to use vegetable and seed oils as diesel substitutes, particularly in farm machinery.

In Western Europe, there are over two hundred power plants that burn rubbish to produce electricity. France, Denmark, and Switzerland recover 50, 60, and 80 percent of their municipal waste respectively. Biomass can be converted into biofuels such as biogas or methane, methanol, and ethanol. However, the process has been more costly than the conventional fossil fuel processes. Rubbish buried in the ground can provide methane gas through an aerobic decomposition. One ton of refuse can produce 8,000 cubic feet (227 cubic meters) of methane.

Which **woods** have the **best heating** quality in a wood-burning stove?

Wood accounts for 28 percent of the total of renewable energy resources in the United States. Woods that have high heat value, meaning that one cord equals 200 to 250 gallons (757 to 946 liters) of fuel oil or 250 to 300 cubic feet (7 to 8.5 cubic meters) of natural gas, are hickory, beech, oak, yellow birch, ash, hornbeam, sugar maple, and apple.

Woods that have medium heat value, meaning that one cord equals 150 to 200 gallons (567 to 757 liters) of fuel oil or 200 to 250 cubic feet (5.5 to 7 cubic meters) of natural gas, are white birch, douglas fir, red maple, eastern larch, big leaf maple, and elm.

Woods that have a low heat value, meaning that one cord equals 100 to 150 gallons (378 to 567 liters) of fuel oil or 150 to 200 cubic feet (4 to 5.5 cubic meters) of natural gas, are aspen, red alder, white pine, redwood, western hemlock, eastern hemlock, Sitka spruce, cottonwood, western red cedar, and lodgepole pine.

How much **wood** is in a **cord**?

A cord of wood is a pile of logs 4 feet (1.2 meters) wide, 4 feet (1.2 meters) high, and 8 feet (2.4 meters) long. It may contain from 77 to 96 cubic feet of wood. The larger the unsplit logs the larger the gaps, with fewer cubic feet of wood actually in the cord. Burning one full cord of wood produces the same amount of energy as one ton of coal.

How much **energy** does one **wind turbine** generate?

The amount of energy one wind turbine generates depends on the size of the turbine and the speed of the wind through the rotor. Wind turbines have power ratings that range from 250 watts to 5 megawatts (MW). The average U.S. household consumes approximately 10,000 kilowatts of electricity per year. One 10 kW wind turbine can generate approximately 10,000 kW per year or enough to power a typical household. A larger wind turbine will generate electricity for more homes.

Where are the **largest wind farms**?

Wind turbines harness the wind's energy to generate electricity. The turbines, consisting of two or three blades, are mounted on tall towers to capture the wind. Wind turbines can stand-alone for water pumping, communications, or for use by individual homeowners or farmers to supply electric power. Wind farms are clusters of wind turbines that generate electricity. The largest wind farms in the United States are in Texas, which has a total installed capacity of 7,118 megawatts. The following chart indicates the top five states with their capacity for electricity generated by wind:

Texas	7,118 megawatts
Iowa	2,791 megawatts
California	2,517 megawatts
Minnesota	1,754 megawatts
Washington	1,447 megawatts

Which **states generate** the **highest** percentage of **electricity** from **wind**?

The United States was able to generate more than 25,000 megawatts (MW) of electricity from the wind in 2008. This supplied an estimated 73 billion kilowatt-hours of electricity which was enough to power about seven million average American homes. Minnesota and Iowa both got over seven percent of their electricity from wind energy.

State	Percentage of Power Needs from Wind
Minnesota	7.48
Iowa	7.10
Colorado	5.91
North Dakota	4.80
New Mexico	4.41

Which **countries** in the world generate the **most electricity** using **wind** energy?

Denmark receives approximately 20 percent of its electricity from wind energy. Spain and Portugal receive over 11 percent of their electricity from wind energy.

This geothermal plant in Iceland takes advantage of the energy from underground heat sources. Geothermal energy is clean and reliable.

When was the **first geothermal power plant** built?

The first geothermal power station was built in 1904 in Larderello, Italy, by Prince Piero Conti. The Prince attached a generator to a natural steam-driven engine to produce enough power to illuminate five light bulbs. In 1913 the first geothermal power plant with a capacity of 250 kilowatts was installed.

Which **states** have **geothermal power plants**?

Geothermal power plants in the United States produced 14.86 billion kilowatt-hours of electricity or 0.4 percent of the total U.S. electricity generation. Six states—California, Nevada, Hawaii, Idaho, Montana, and Utah—all have geothermal power plants. The 34 geothermal power plants in California produce almost 90 percent of all geothermal generated electricity in the United States.

What are the **advantages** and **disadvantages** of **geothermal power**?

Geothermal energy is a renewable resource that is mostly emission and pollution-free. Costs to operate a geothermal power plant are relatively inexpensive. It has the potential to produce power consistently, unlike wind and solar power. Currently, drilling for geothermal energy is expensive. Since rocks lose heat over time, new sites and wells have to be drilled.

273

Where was the **first hydroelectric power plant** in the United States built?

Appleton, Wisconsin, is the site of the first hydroelectric generating plant in the United States. It was built in 1882.

What is the most **commonly used renewable energy** source?

Hydroelectric power is the most common renewable energy source. Based on consumption, hydroelectric power accounts for 34 percent of the total amount of renewable energy resources. The capacity for hydroelectric power is about 95,000 megawatts. Hydropower facilities in the United States can generate enough power to supply 28 million households or the equivalent of nearly 500 million barrels of oil. In 2008 it accounted for 6 percent of total electricity generation and 67 percent of generation from renewables.

NUCLEAR ENERGY

What is the **life** of a **nuclear power plant**?

While there is some controversy about the subject, the working life of a nuclear power plant as defined by the industry and its regulators is approximately 40 years, which is about the same as that of other types of power stations.

What are the **two types** of **nuclear reactors** in the United States?

Nuclear power plants contain a reactor unit that releases heat to turn water into steam which then drives the turbine-generators that generate electricity. There are two types of reactors in the United States: boiling-water reactors and pressurized-water reactors. In a boiling-water reactor, the water heated by the reactor core turns directly into steam in the reactor vessel and is then used to power the turbine-generator. In a pressurized-water reactor, the water heated by the reactor core is kept under pressure and does not turn to steam. The hot, radioactive, liquid water then flows through and heats thousands of tubes in a steam generator. Outside the hot tubes in the steam generator is nonradioactive water which boils and turns to steam. The radioactive water flows back into the reactor core where it is reheated and then returns to the steam generator.

When did the **first nuclear power plant** generate electricity for **residential use**?

The first nuclear power plant to generate electricity for residential use was a small facility in Obninsk, in the former Soviet Union, on June 27, 1954. The capacity of the power plant was only 5 megawatts. Due to the status of international relations during the Cold War, this accomplishment was not reported widely in the West. Two years later, in August 1956, the Calder Hall Unit 1 power plant became operational in Great Britain. Its initial capacity was 50 megawatts of electricity. Other units were added and it continued to operate for nearly 50 years until March 2003. Calder Hall was the longest operating reactor in history.

When did the **first commercial nuclear power plant** in the United States become operational?

The first commercial electricity-generating nuclear power plant in the United States was located in Shippingport, Pennsylvania. It became operational on December 2, 1957, supplying energy to the Pittsburgh area. It continued to supply power until 1982. Although the Shippingport power plant was the first large-scale nuclear power plant in the United States, it was not the first civilian power plant. A few months earlier, in July 1957, the Santa Susana Sodium Reactor Experimental in California began providing power.

Where is the **oldest operational nuclear power plant** in the United States?

The oldest operational plant in the United States is the Oyster Creek, New Jersey, plant. It became operational on December 1, 1969. The original license for the Oyster Creek power plant was scheduled to expire on April 9, 2009. On April 8, 2009, the U.S. Nuclear Regulatory Commission approved a 20-year license extension until April 9, 2029. The Nine Mile Point, Unit 1 also became operational on December 1, 1969. Its original license was to expire on August 22, 2009, but on October 31, 2006, the U.S. Nuclear Regulatory Commission granted a 20-year extension to its license, extending it to August 22, 2029. Since the license for Oyster Creek was issued before the license for Nine Mile Point, Oyster Creek is officially the oldest operating U.S. reactor.

How many **nuclear power plants** are there **worldwide**?

As of 2008, 440 reactors were operational with 44 more under construction.

Country	Number of Units Under construction	Number of Reactors
Argentina	2	1
Armenia	1	
Belgium	7	
Brazil	2	
Bulgaria	2	2
Canada	18	
China	11	11
Czech Republic	6	

Country	Number of Units Under construction	Number of Reactors
Finland	4	1
France	59	1
Germany	17	
Hungary	4	
India	17	6
Iran		1
Japan	55	2
Korea, South	20	5
Lithuania	1	
Mexico	2	
Netherlands	1	
Pakistan	2	1
Romania	2	
Russia	31	8
South Africa	2	
Slovakia	6	
Slovenia	1	
Spain	8	
Sweden	10	
Switzerland	5	
Taiwan	6	2
Ukraine	15	2
United Kingdom	19	
United States	104	1

As of 2008, the United States had 104 reactors in operation, with one under construction. The plants generated 100,683 megawatts of electricity or 19.7 percent of domestic electricity in 2008.

Which **countries** produce the **most commercial nuclear power**?

The United States produced the most commercial nuclear power in 2008. However, the percentage of electricity power by nuclear energy is greater for other countries.

World Nuclear Power

Country	Number of Reactors	Total Megawatts Electricity	Percent of Nation's Total Generated Electricity
United States	104	100,683	19.7
France	59	63,260	76.2

The Chernobyl nuclear plant in the Ukraine is the site of the worst nuclear disaster in history. Interestingly, though radiation levels are still high, the abandoned region is filled with wildlife and the Ukrainian government has opened the area up to tourism.

Country	Number of Reactors	Total Megawatts Electricity	Percent of Nation's Total Generated Electricity
Japan	55	47,278	
Russia	31	21,743	16.9
Germany	17	20,470	
Ukraine	15	13,107	47.4
Belgium	7	5,824	53.8
Slovakia	4	1,711	56.4
Lithuania	1	1,185	72.9

How much **energy** does a **typical nuclear power plant** generate?

In 2008, the "average" nuclear power plant generated about 12.4 billion kilowatt-hours (kWh). The smallest nuclear plant has a single reactor with 476 MW of generation capacity and the largest has three reactors with a total of 3,825 MW of capacity.

What have been some notable **nuclear accidents**?

Date	Site	Description of Incident
October 7, 1957	Windscale plutonium production reactor (near Liverpool, England)	Fire in the reactor; released radioactive material; blamed for 39 cancer deaths

Date	Site	Description of Incident
January 3, 1961	Idaho Falls, Idaho	Explosion in the reactor; radiation contained; 3 workers killed
October 5, 1966	Enrico Fermi demonstration breeder reactor (near Detroit, Michigan)	Partial core meltdown; radiation contained
January 21, 1969	Lucens Vad, Switzerland	Coolant malfunction; released radiation into a cavern; cavern sealed
March 22, 1975	Brown's Ferry Reactor (Decatur, Alabama)	Fire causes cooling water levels to become dangerously low
March 28, 1979	Three Mile Island, Pennsylvania	Partial core meltdown; minimal radiation released
April 25, 1981	Tsunga, Japan	Workers exposed to radiation during repairs to nuclear plant
January 6, 1986	Kerr-McGee nuclear plant (Gore, Oklahoma)	Cylinder of nuclear material burst; 100 workers hospitalized; 1 death
April 26, 1986	Chernobyl, Ukraine (USSR in 1986)	Fire and explosions; radioactive material spread over much of Europe; at least 31 dead in the immediate aftermath; tens of thousands of cancer deaths and increased birth defects
September 30, 1999	Uranium-reprocessing facility (Tokaimura, Japan)	Container with uranium is overloaded; workers and nearby residents exposed to extremely high radiation levels

What actually happened at **Three Mile Island**?

The worst commercial nuclear accident in the United States occurred at Three Mile Island. The Three Mile Island nuclear power plant in Pennsylvania experienced a partial meltdown of its reactor core and radiation leakage. On March 28, 1979, just after 4:00 A.M., a water pump in the secondary cooling system of the Unit 2 pressurized water reactor failed. A relief valve jammed open, flooding the containment vessel with radioactive water. A backup system for pumping water was down for maintenance.

Temperatures inside the reactor core rose, fuel rods ruptured, and a partial (52 percent) meltdown occurred, because the radioactive uranium core was almost entirely uncovered by coolant for 40 minutes. The thick steel-reinforced containment building prevented nearly all the radiation from escaping—the amount of radiation released into the atmosphere was one-millionth of that at Chernobyl. However, if the coolant had not been replaced, the molten fuel would have penetrated the reactor containment vessel, where it would have come into contact with the water, causing a steam explosion, breaching the reactor dome, and leading to radioactive contamination of the area similar to the Chernobyl accident.

What is the **Rasmussen report**?

Dr. Norman Rasmussen (1927–2003) of the Massachusetts Institute of Technology (MIT) conducted a study of nuclear reactor safety for the U.S. Atomic Energy Commission. The 1975 study cost $4 million and took three years to complete. It concluded that the odds against a worst-case accident occurring were astronomically large—ten million to one. The worst-case accident projected about three thousand early deaths and $14 billion in property damage due to contamination. Cancers occurring later due to the event might number 1,500 per year. The study concluded that the safety features engineered into a plant are very likely to prevent serious consequences from a meltdown. Other groups criticized the Rasmussen report and declared that the estimates of risk were too low. After the Chernobyl disaster in 1986, some scientists estimated that a major nuclear accident might in fact happen every decade.

What is a **meltdown**, and what does it have to do with the **"China Syndrome"**?

A meltdown is a type of accident in a nuclear reactor in which the fuel core melts, resulting in the release of dangerous amounts of radiation. In most cases the large containment structure that houses a reactor would prevent the radioactivity from escaping. However, there is a small possibility that the molten core could become hot enough to burn through the floor of the containment structure and go deep into the earth. Nuclear engineers call this type of situation the "China Syndrome." The phrase derives from a discussion on the theoretical problems that could result from a meltdown, when a scientist commented that the molten core could bore a hole through the earth, coming out—if one happened to be standing in North America—in China. Although the scientist was grossly exaggerating, some took him seriously. In fact, the core would only bore a hole about 30 feet (10 meters) into the earth, but even this distance would have grave repercussions. All reactors are equipped with emergency systems to prevent such an accident from occurring.

What caused the **Chernobyl accident**?

The worst nuclear power accident in history, which occurred at the Chernobyl nuclear power plant in the Ukraine, will affect, in one form or another, 20 percent of the republic's population (2.2 million people). On April 26, 1986, at 1:23:40 A.M., during unauthorized experiments by the operators in which safety systems were deliberately circumvented in order to learn more about the plant's operation, one of the four reactors rapidly overheated and its water coolant "flashed" into steam. The hydrogen formed from the steam reacted with the graphite moderator to cause two major explosions and a fire. The explosions blew apart the 1,000 ton (907 metric ton) lid of the reactor, and released radioactive debris high into the atmosphere. It is estimated that 50 tons of fuel went into the atmosphere. An additional 70 tons of fuel and 700 tons of radioactive reactor graphite settled in the vicinity of the damaged unit.

Human error and design features (such as a positive void coefficient type of reactor, use of graphite in construction, and lack of a containment building) are generally cited as the causes of the accident. Thirty-one people died from trying to stop the fires. More than 240 others sustained severe radiation sickness. Eventually 150,000 people living near the reactor were relocated; some of whom may never be allowed to return home. Fallout from the explosions, containing radioactive isotope cesium-137, was carried by the winds westward across Europe.

The problems created by the Chernobyl disaster are overwhelming and continue today. Particularly troubling is the fact that by 1990–1991, a five-fold increase had occurred in the rate of thyroid cancers in children in Belarus. A significant rise in general morbidity has also taken place among children in the heaviest-hit areas of Gomel and Mogilev.

What was the **distribution** of **radioactive fallout** after the 1986 **Chernobyl** accident?

Radioactive fallout, containing the isotope cesium-137, and nuclear contamination covered an enormous area, including Byelorussia, Latvia, Lithuania, the central portion of the then Soviet Union, the Scandinavian countries, the Ukraine, Poland, Austria, Czechoslovakia, Germany, Switzerland, northern Italy, eastern France, Romania, Bulgaria, Greece, Yugoslavia, the Netherlands, and the United Kingdom. The fallout, extremely uneven because of the shifting wind patterns, extended 1,200 to 1,300 miles

(1,930 to 2,090 kilometers) from the point of the accident. Estimates of the effects of this fallout range from 28,000 to 100,000 deaths from cancer and genetic defects within the next 50 years. In particular, livestock in high rainfall areas received unacceptable dosages of radiation.

How is **nuclear waste stored**?

Nuclear wastes consist either of fission products formed from the splitting of uranium, cesium, strontium, or krypton, or from transuranic elements formed when uranium atoms absorb free neutrons. Wastes from transuranic elements are less radioactive than fission products; however, these elements remain radioactive far longer—hundreds of thousands of years. The types of waste are irradiated fuel (spent fuel) in the form of 12-foot (4-meter) long rods, high-level radioactive waste in the form of liquid or sludge, and low-level waste (non-transuranic or legally high-level) in the form of reactor hardware, piping, toxic resins, water from fuel pool, and other items that have become contaminated with radioactivity.

Currently, most spent nuclear fuel in the United States is safely stored in specially designed pools at individual reactor sites around the country. If pool capacity is reached, licensees may move toward use of above-ground dry storage casks. The three low-level radioactive waste disposal sites are Barnwell located in South Carolina, Hanford located in Washington, and Envirocare located in Utah. Each site accepts low-level radioactive waste from specific regions of the country.

Most high-level nuclear waste has been stored in double-walled, stainless-steel tanks surrounded by 3 feet (1 meter) of concrete. The current best storage method, developed by the French in 1978, is to incorporate the waste into a special molten glass mixture, then enclose it in a steel container and bury it in a special pit. The Nuclear Waste Policy Act of 1982 specified that high-level radioactive waste would be disposed of underground in a deep geologic repository. Yucca Mountain, Nevada, was chosen as the single site to be developed for disposal of high-level radioactive waste. However, the Yucca Mountain site continues to be controversial due to dormant volcanoes in the vicinity and known earthquake fault lines. As of 2010, the Energy Department withdrew its application for a nuclear-waste repository at Yucca Mountain.

What is **vitrification**?

Vitrification is a technique that transforms radioactive liquid waste into a solid. The nuclear material is mixed with sand or clay. The mixture is then heated to the melting point of the sand or to the point the clay forms a ceramic. Although the resulting composite material is still radioactive, it is much easier and safer to handle, transport, and store the radioactive liquid waste as a solid.

281

MEASURES AND MEASUREMENT

What is the **weight per gallon** of common **fuels**?

One Gallon of Fuel	Weight in Pounds/Kgs
Propane	4.23/1.9
Butane	4.86/2.2
Gasoline	6.00/2.7
Kerosene	6.75/3.1
Aviation gasoline	6.46–6.99/2.9–3.2

How much does a **barrel** of **oil weigh**?

A barrel of oil weighs about 306 pounds (139 kilograms).

How many **gallons** are in a **barrel of oil**?

The barrel, a common measure of crude oil, contains 42 U.S. gallons or 34.97 imperial gallons.

How many **gallons** of **gasoline** are in **one barrel** of oil?

U.S. refineries produce between 19 and 20 gallons of gasoline from one barrel (42 gallons) of crude oil. The remainder of the barrel yields distillate and residual fuel oils, jet fuel, and other petroleum products.

How do various **energy sources compare**?

Below are listed some comparisons (approximate equivalents) for the energy sources:

Energy Unit	Equivalent
1 BTU of energy	1 match tip
• 250 calories (International Steam Table)	
• 0.25 kilocalories (food calories)	
1,000 BTU of energy	2 5-ounce glasses of wine
• 250 kilocalories (food calories)	
• 0.8 peanut butter and jelly sandwiches	
1 million BTU of energy	90 pounds of coal
• 120 pounds of oven-dried hardwood	
• 8 gallons of motor gasoline	

Energy Unit	Equivalent
• 10 therms of dry natural gas	
• 11 gallons of propane	
• 2 months of the dietary intake of a laborer	
1 quadrillion BTU of energy	45 million short tons of coal
• 60 million short tons of oven-dried hardwood	
• 1 trillion cubic feet of dry natural gas	
• 170 million barrels of crude oil	
• 470 thousand barrels of crude oil per day for 1 year	
1 barrel of crude oil	5.6 thousand cubic feet of dry natural gas
• 0.26 short tons (520 pounds) of coal	
• 1,700 kilowatt-hours of electricity	
1 short ton of coal	3.8 barrels of crude oil
• 21,000 cubic feet of dry natural gas	
• 6,500 kilowatt-hours of electricity	
1,000 cubic feet of natural gas	0.18 barrels (7.4 gallons) of crude oil
• 0.05 short tons (93 pounds) of coal	
• 300 kilowatt-hours of electricity	
1,000 kilowatt-hours of electricity	0.59 barrels of crude oil
• 0.15 short tons (310 pounds) of coal	
• 3,300 cubic feet of dry natural gas	

Notes: One quadrillion equals 1,000,000,000,000,000. One therm equals 100,000 BTUs.

What are the approximate **heating values** of fuels?

Fuel	BTU	Unit of Measure
Oil	141,000	gallon
Coal	31,000	pound
Natural gas	1,000	cubic feet
Steam	1,000	cubic feet
Electricity	3,413	kilowatt hour
Gasoline	124,000	gallon

A BTU (British thermal unit), a common energy measurement, is defined as the amount of energy required to raise the temperature of one pound of water by 1°F.

What are the **fuel equivalents** to produce **one quad** of **energy**?

One quad (meaning one quadrillion) is equivalent to:

- 1×10^{15} BTU
- 252×10^{15} calories or 252×10^{12} K calories

In fossil fuels, one quad is equivalent to:
- 180 million gallons (681 million liters) of crude oil
- 0.98 trillion cubic feet (0.028 trillion cubic meters) of natural gas
- 37.88 million tons of anthracite coal
- 38.46 million tons of bituminous coal

In nuclear fuels, one quad is equivalent to 2,500 tons of triuranium octaoxide (U_3O_8) if only uranium-235 (U_{235}) is used.

In electrical output, one quad is equivalent to 2.93×10^{11} kilowatt-hours electric.

How much heat will **100 cubic feet of natural gas** provide?

One hundred cubic feet of natural gas can provide about 100,000 BTUs (British thermal units) of heat. A British thermal unit, a common energy measurement, is defined as the amount of energy required to raise one pound of water by 1°F.

How is a **heating degree day** defined?

Early in the twenty-first century engineers developed the concept of heating degree days as a useful index of heating fuel requirements. They found that when the daily mean temperature is lower than 65°F (18°C), most buildings require heat to maintain a 70°F (21°C) temperature. Each degree of mean temperature below 65°F (18°C) is counted as "one heating degree day." For every additional heating degree day, more fuel is needed to maintain a 70°F (21°C) indoor temperature. For example, a day with a mean temperature of 35°F (1.5°C) would be rated as 30 heating degree days and would require twice as much fuel as a day with a mean temperature of 50°F (10°C; 15 heating degree days). The heating degree concept has become a valuable tool for fuel companies for evaluation of fuel use rates and efficient scheduling of deliveries. Detailed daily, monthly, and seasonal totals are routinely computed for the stations of the National Weather Service.

What does the term **cooling degree day** mean?

It is a unit for estimating the energy needed for cooling a building. One unit is given for each degree Fahrenheit above the daily mean temperature when the mean temperature exceeds 75°F (24°C).

How many **BTUs** are **equivalent** to **one ton** of **cooling capacity**?

1 ton = 288,000 BTUs/24 hours or 12,000 BTUs/hour.

CONSUMPTION AND CONSERVATION

Does the **United States** currently produce **enough energy** to meet its **consumption needs**?

No. From 1958 forward, the United States has consumed more energy than it has produced, and the difference has been met by energy imports. In 2008, 73.71 quadrillion BTUs was the total United States energy production; in that same year 99.30 quadrillion BTUs were consumed.

Americans have become notorious for energy consumption, especially when it comes to transportation.

How much **energy consumption** in the **United States** is used by **transportation**?

Transportation accounts for only 25 percent of the total energy consumption in the United States; industries and buildings account for 42 percent and 33 percent, respectively.

Usage	Percent of Total U.S. Consumption
Industry	42
Metals	26
Chemicals	19
Fossil fuels	14
Paper	8
Stone, glass, clay	7
Food	5
Other industries	21

Usage	Percent of Total U.S. Consumption
Buildings	33
Heating, cooling, ventilation	64
Hot water heating	24
Lighting	12
Transportation	25
Motor vehicles	74
Air	14
Marine	7
Rail	5

Which **countries** consume the **most energy**?

Top Energy-Consuming Countries (2006, in quadrillion BTUs)

Country	Energy Consumption
United States	99.86
China	73.81
Russia	30.39
Japan	22.79
India	17.68
Germany	14.63
Canada	13.95
France	11.45
United Kingdom	9.80
Brazil	9.64

The United States consumes nearly 25 percent of the world's total energy consumption.

How much does it **cost** to **generate one kilowatt-hour** of electricity?

The cost to generate one kilowatt-hour of electricity varies by the energy source.

Energy Source	Cost (in cents)
Biomass	6–9
Coal	4–6
Geothermal	3–8
Hydropower	4–10
Natural Gas	5–7
Nuclear Power	2–12
Photovoltaics	20–25
Solar thermal	5–13
Wind	4–7

How long will fossil fuel reserves last?

Determining how long fossil fuel reserves, the non-renewable sources of energy, will last is dependent on production rates and the rate of consumption. Other factors complicating the determination of how long fossil fuel reserves will last is whether additional supplies will be discovered, the economics of recovery of fossil fuels, and the growth of renewable sources of energy. Different models have been proposed, but researchers are now predicting that oil and gas reserves will last another 35 and 37 years, respectively. Coal reserves are predicted to last 107 years. Based on this model, coal will be the only fossil fuel available after 2042 and would last until 2112.

How much **energy** does the average **person** in the **United States use** in a **year**?

In 2009, the United States, energy use per person was about 308 million BTUs (British Thermal Units). A BTU, a common energy measurement, is defined as the amount of energy required to raise one pound of water by 1°F. Below is energy use per person for selected years:

Year	Energy Consumption per Person (Million BTU)
1950	227
1960	250
1970	331
1980	344
1990	339
2000	351
2005	340
2008	327
2009	308

How much **electricity** does a **typical American home** use?

In 2008, the average monthly residential electricity consumption was 920 kilowatt-hours. More than half of the residential electricity consumption is for household appliances.

How does the **Energy Star program** promote energy efficiency?

Energy Star is a dynamic government/industry partnership that offers businesses and consumers energy-efficient solutions, making it easy to save money while protecting the environment for future generations. In 1992 the U.S. Environmental Protection Agency (EPA) introduced Energy Star as a voluntary labeling program designed to identify and promote energy-efficient products to reduce greenhouse gas emissions. Computers and monitors were the first labeled products. The Energy Star label is now on major appliances, office equipment, lighting, consumer electronics, and more. The EPA has also extended the label to cover new homes and commercial and industrial buildings.

Through its partnerships with more than 7,000 private and public sector organizations, Energy Star delivers the technical information and tools that organizations and consumers need to choose energy-efficient solutions and best management practices. Energy Star has successfully delivered energy and cost savings across the country, saving businesses, organizations, and consumers more than $5 billion a year. Energy Star has been instrumental in promoting the widespread use of such technological innovations as LED traffic lights, efficient fluorescent lighting, power management systems for office equipment, and low standby energy use.

How much **money** can be **saved** by **lowering** the setting on a home **furnace thermostat**?

Studies have shown that a 5°F reduction in the home thermostat setting for approximately eight hours will save up to ten percent in fuel costs.

How much **energy** is **saved** by **raising** the **setting** for a house **air conditioner**?

In general, for every 1°F the inside temperature is increased, the energy needed for air conditioning is reduced by three percent. If all consumers raised the settings on their air conditioners by 6°F, for example, 190,000 barrels of oil could be saved each day.

How much **fuel** can be **saved** when a home is properly **insulated**?

Insulation of a single-family house with EPS or XPS (extruded polystyrene) over a 50-year period has the potential to save 80 metric tons of heating oil. This in turn corresponds to the fuel consumption of a fully loaded jumbo jet during a flight from Frankfurt, Germany, to New York.

How much **energy** is required to use various **electrical appliances**?

The formula to estimate the amount of energy a specific appliance consumes is:

(wattage × hours used per day)/1,000 = daily kilowatt-hour (kWh) consumption.

To calculate the annual consumption, multiply this amount by the number of days you use the appliance during the year. The annual cost to run an appliance is calculated by multiplying the kWh per year by the local utility's rate per kWh.

Is it possible to compare the energy efficiency of different brands of appliances?

In 1980, the Federal Trade Commission's Appliance Labeling Rule became effective. It requires EnergyGuide labels be placed on all new refrigerators, freezers, water heaters, dishwashers, clothes washers, room air conditioners, heat pumps, furnaces, and boilers. The bright yellow labels with black lettering identify energy consumption and operating cost for each of the various household appliances. EnergyGuide labels show the estimated yearly electricity consumption to operate the product along with a scale of comparison among similar products. The comparison scale shows the least and most energy used by comparable models.

The table below indicates the wattage for various household electrical products.

Appliance	Wattage
Clock radio	10
Coffeemaker	900–1,200
Clothes washer	350–500
Clothes dryer	1,800–5,000
Dishwasher (using the drying feature greatly increases energy consumption)	1,200–2,400
Ceiling fan	65–175
Window fan	55–250
Furnace fan	750
Whole house fan	240–750
Hair dryer	1,200–1,875
Microwave oven	750–1100
PC CPU—awake/asleep	120/30 or less
PC Monitor—awake/asleep	150/30 or less
Laptop	50
Refrigerator (frost-free, 16 cubic feet)	725
19" color TV	65–100
27" color TV	113
36" color TV	133
53"–61" projection TV	170
Flat screen TV	120
Toaster oven	1,225
DVD player	20–25
Vacuum cleaner	1,000–1,440
Water heater (40 gal.)	4,500–5,500

Fluorescent bulbs are being touted as a smart and easy way to save a lot of electricity, both in the home and at work.

When was **gas lighting** invented?

In 1799, Philippe Lebon (1767–1804) patented a method of distilling gas from wood for use in a "Thermolamp," a type of lamp. By 1802, William Murdock (1754–1839) installed gas lighting in a factory in Birmingham, England. The introduction of widespread, reliable interior illumination enabled dramatic changes in commerce and manufacturing.

What are the advantages of **compact fluorescent light bulbs** over **incandescent** light bulbs?

Compact fluorescent light bulbs (CFLs) have the following advantages over incandescent light bulbs when used properly: they last up to ten times longer, use about a quarter of the energy, and produce 90 percent less heat, while producing more light per watt. For example, a 27-watt compact fluorescent lamp provides about 1,800 lumens, compared to 1,750 lumens from a 100-watt incandescent lamp. CFLs are most efficient in areas where lights are on for long periods of time.

What **changes** have been made to make **windows more efficient**?

Windows can account for 10 percent to 25 percent of a residential heating bill. During the winter months in cold climates, there may be significant heat loss through the windows. Conversely, during the summer and in warm climates, air conditioners must work harder to cool hot air from sunny windows. Until recently, clear glass was the primary glazing material used in windows. Although glass is durable and allows a high percentage of sunlight to enter buildings, it has very little resistance to heat flow.

Glazing technology has changed greatly during the past two decades. There are now several types of advanced glazing systems available to help control heat loss or gain. The advanced glazings include double- and triple-pane windows with such coatings as low-emissivity (low-e), spectrally selective, heat-absorbing (tinted), reflective, or a combination of these; windows can also be filled with a gas (typically xenon, argon, or krypton) that helps in insulation.

How does **driving speed** affect **gas mileage** for most automobiles?

Most automobiles get about 28 percent more miles per gallon of fuel at 50 miles (80.5 kilometers) per hour than at 70 miles (112 kilometers) per hour, and about 21 percent

---

When should a fluorescent light be turned off to save energy?

Fluorescent lights use a lot of electric current getting started, and frequently switching the light on and off will shorten the lamp's life and efficiency. It is energy-efficient to turn off a fluorescent light only if it will not be used again within 15 minutes.

---

more at 55 miles (88.5 kilometers) per hour than at 70 miles (112 kilometers) per hour.

Is it more **economical** to run an **automobile** with its **windows open** rather than using its **air conditioner**?

At speeds greater than 40 miles (64 kilometers) per hour, less fuel is used in driving an automobile with the air conditioner on and the windows up than with the windows rolled down. This is due to the air drag effect—the resistance that a vehicle encounters as it moves through a fluid medium, such as air. In automobiles, the amount of engine power required to overcome this drag force increases with the cube of the vehicle's speed—twice the speed requires eight times the power. For example, it takes 5 horsepower for the engine to overcome the air resistance at 40 miles (64 kilometers) per hour; but at 60 miles (97 kilometers) per hour, it takes 18 horsepower; at 80 miles (128 kilometers) per hour, it takes 42 horsepower. Improved aerodynamics, in which the drag coefficient (measure of air drag effect) is reduced, significantly increases fuel efficiency. The average automobile in 1990 had a drag coefficient of about 0.3 to 0.35. In the early 1960s it was 0.5 on average, and it was 0.47 in the 1970s. The lowest maximum level possible for wheeled vehicles is 0.15.

When is it **more economical** to **restart** an **automobile** rather than let it idle?

Tests by the Environmental Protection Agency have shown that it is more economical to turn an engine off rather than let it idle if the idle time exceeds 60 seconds.

How much gasoline is **wasted** by **underinflated tires**?

Underinflated tires waste as much as 1 gallon (4.5 liters) out of every 20 gallons (91 liters) of gasoline. To save fuel, follow the automaker's guidelines regarding recommended air pressure levels for the tires. However, greater fuel economy can be achieved by inflating tires to the maximum air pressure listed on the sidewall of the tire, resulting in less rolling resistance.

What is the **energy use** of various **modes** of **transportation**?

Mode of Transportation	Energy Use (BTUs)	Number of Passengers per Passenger Mile
Bicycle	1	80
Automobile—high economy	4	600
Motorcycle	1	2,100
Bus—inter-city	45	600
Subway train	1,000	900
747 jet plane	360	3,440

What is the **fuel consumption** of various **aircraft**?

Fuel consumption (gallons)	Aircraft per 1,000-mile flight	Number of passengers
737	128	1,600
737 Stretch	188	1,713
747–400	413	6,584
SST Concorde	126	6,400
Turboprop DHC-8	37	985
Fighter plane F-15	1	750
Military cargo plane C-17	126,000 pound of cargo	5,310

ENVIRONMENT

BIOMES, ECOLOGICAL CYCLES, AND ENVIRONMENTAL MILESTONES

What is the **ecological footprint**?

The ecological footprint is a measurement of how fast humans consume resources and generate waste compared to how fast nature can absorb the waste and generate new resources. Since the late 1970s, humanity has been in ecological overshoot; i.e., the annual demands on nature exceed what the earth can generate in a year. It currently takes the earth one year and five months to regenerate what is used in one year. Mathis Wackernagel (1962–) and William Rees (1943–) developed the concept of the ecological footprint in 1990.

What is **biogeography**?

Biogeography is the study of the distribution, both current and past, of individual species in specific environments. One of the first biogeographers was Carolus Linnaeus (1707–1778), a Swedish botanist who studied the distribution of plants. Biogeography specifically addresses the questions of evolution, extinction, and dispersal of organisms in specific ecosystems.

What is a **biome**?

A biome is a plant and animal ecosystem that covers a large geographical area. Complex interactions of climate, geology, soil types, water resources, and latitude all determine the kinds of plants and animals that thrive in different places. Fourteen major ecological zones, called "biomes," exist over five major climatic regions and eight zoogeographical regions. Important land biomes include tundra, coniferous forests, deciduous forests, grasslands, savannahs, deserts, chaparral, and tropical rain forests.

Biome	Temperature	Precipitation	Vegetation	Animals
Arctic tundra	-40°F to 64°F (-40°C to 18°C)	Dry season, wet season	Shrubs, grasses, lichens, mosses	Birds, insects, mammals
Deciduous forest	Warm summers, cold winters	Low, distributed throughout year	Trees, shrubs herbs, lichens, mosses	Mammals, birds, insects, reptiles
Desert	Hottest; great daily range	Driest <10 in (25 cm) of rain per year	Trees, shrubs, succulents, forbs	Birds, small mammals, reptiles
Taiga	Cold winters, cool summers	Moderate	Evergreens, tamarack	Birds, mammals
Tropical Rainforest	Hot	Wet season, short dry season	Trees, vines, epiphytes, fungi	Small mammals, birds, insects
Tropical Savannah	Hot	Wet season, dry season	Tall grasses shrubs, trees	Large mammals, birds, reptiles
Temperate grassland	Warm summers, cold winters	Seasonal drought, occasional fires	Tall grasses	Large mammals, birds, reptiles

What is **biodiversity**?

Biodiversity refers to genetic variability within a species, diversity of populations of a species, diversity of species within a natural community, or the wide array of natural communities and ecosystems throughout the world. Some scientists estimate that there may be between 15 and 100 million species throughout the world. Biodiversity is threatened at the present time more than at any other time in history. In the time since the North American continent was settled, as many as 500 plant and animal species have disappeared in North America. Some recent examples of threats to biodiversity in the United States include: 50 percent of the United States no longer supports its original vegetation; in the Great Plains, 99 percent of the original prairies are gone; and across the United States, we destroy 100,000 acres of wetlands each year.

Who owns the forests in the United States?

Private individuals own 57 percent of the forests in the United States. The remaining forest area land is owned by the federal government (20 percent), corporations (15 percent), and state and local governments (8 percent).

What **percentage** of Earth's surface is **forest area**?

The world's total forest area is just over 4 billion hectares or 31 percent of the total land area. More than half of the world's forest areas are in the five most forest-rich countries: the Russian Federation, Brazil, Canada, the United States, and China.

What **percentage** of Earth's surface is **tropical rain forest**?

Rain forests account for approximately seven percent of Earth's surface, or about 3 million square miles (7.7 million square kilometers).

What is the **largest rain forest**?

The Amazon Basin is the world's largest continuous tropical rain forest. It covers about 2.7 million square miles (6.9 million square kilometers).

What is the **importance** of the **rain forest**?

Half of all medicines prescribed worldwide are originally derived from wild products, and the U.S. National Cancer Institute has identified more than two thousand tropical rain forest plants with the potential to fight cancer. Rubber, timber, gums, resins and waxes, pesticides, lubricants, nuts and fruits, flavorings and dyestuffs, steroids, latexes, essential and edible oils, and bamboo are among the products that would be drastically affected by the depletion of the tropical forests. In addition, rain forests greatly influence patterns of rain deposition in tropical areas; smaller rain forests mean less rain. Large groups of plants, like those found in rain forests, also help control levels of carbon dioxide in the atmosphere.

What is the **rate** of **species extinction** in the tropical **rain forests**?

Biologists estimate that tropical rain forests contain approximately one-half of Earth's animal and plant species. These forests contain 155,000 of the 250,000 known plant species and innumerable insect and animal species. Nearly 100 species become extinct each day. This is equivalent to four species per hour. At the current rates, five to ten percent of the tropical rain forest species will become extinct every decade.

How **rapidly** is **deforestation** occurring?

Agriculture, excessive logging, and fires are major causes of deforestation. Afforestation and the natural expansion of forests help to decrease the rate of deforestation. The rate of deforestation has decreased from 16 million hectares (61,776 square miles) per year during the 1990s to 13 million hectares (50,193 square miles) per year in the decade 2000–2010 (according to preliminary data for the decade). The net change in forest area for 2000–2010 is estimated at –5.2 million hectares (20,007 square miles) per year (an area about the size of Costa Rica), down from –8.3 million hectares (32,046 square miles) per year in the 1990s.

What is a **wetland**?

A wetland is an area that is covered by water for at least part of the year and has characteristic soils and water-tolerant plants. Examples of wetlands include swamps, marshes, bogs, and estuaries.

Type of Wetland	Typical Features
Swamp	Tree species such as willow, cypress, and mangrove
Marsh	Grasses such as cattails, reeds, and wild rice
Bogs	Floating vegetation, including mosses and cranberries
Estuaries	Specially adapted flora and fauna, such as crustaceans, grasses, and certain types of fishes

How many **acres** of **wetlands** have been **lost** in the **United States**?

Wetlands are the lands between aquatic and terrestrial areas, such as bogs, marshes, swamps, and coastal waters. At one time considered wastelands, scientists now recognize the importance of wetlands to improve water quality, stabilize water levels, prevent flooding, regulate erosion, and sustain a variety of organisms. The United States has lost approximately 100 million acres of wetland areas between colonial times and the 1970s. The 1993 Wetlands Plan established a goal of reversing the trend of 100,000 acres of wetland loss to 100,000 acres of wetland recovery. The most recent Wetlands Status and Trends report (1998–2004), published in 2005 by the U.S. Fish and Wildlife Service, showed a gain of 191,750 (77,630 ha) wetland acres, or an average annual net gain of 32,000 acres of wetlands. The net gain in wetland area is attributed to wetlands created, enhanced, or restored.

What is **limnology**?

Limnology is the study of freshwater ecosystems—especially lakes, ponds, and streams. These ecosystems are more fragile than marine environments since they are

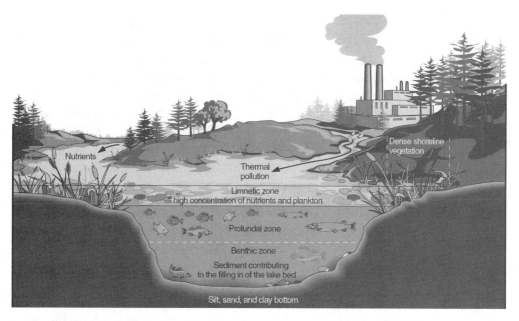

The structure of a eutrophic lake.

subject to great extremes in temperature. F.A. Forel (1848–1931), a Swiss professor, has been called the father of limnology.

What is an **estuary**?

Estuaries are places where freshwater streams and rivers meet the sea. The salinity of such areas is less than that of the open ocean but greater than that of a typical river, so organisms living in or near estuaries have special adaptations. Estuaries are rich sources of invertebrates, such as clams, shrimps, and crabs, as well as fishes such as striped bass, mullet, and menhaden. Unfortunately, estuaries are also popular locations for human habitation and businesses. Contamination from shipping, household pollutants, and power plants are carried to the sea by rivers and streams and threaten the ecological health of many estuaries.

What is **eutrophication**?

Eutrophication is a process in which the supply of plant nutrients in a lake or pond is increased. In time, the result of natural eutrophication may be dry land where water once flowed, caused by plant overgrowth. Natural fertilizers, washed from the soil, result in an accelerated growth of plants, producing overcrowding. As the plants die off, the dead and decaying vegetation depletes the lake's oxygen supply, causing fish to die. The accumulated dead plant and animal material eventually changes a deep lake to a shallow one, then to a swamp, and finally it becomes dry land. While the process of eutrophication is a natural one, it has been accelerated enormously by human activities. Fertilizers from farms, sewage, industrial wastes, and some detergents all contribute to the problem.

What does it mean when a **lake** is **brown** or **blue**?

When a lake is brown, it usually indicates that eutrophication is occurring. This process refers to the premature "aging" of a lake, when nutrients are added to the water, usually due to runoff, which may be either agricultural or industrial in origin. Due to this rich supply of nutrients, blue-green algae begin to take over the green algae in the lake, and food webs within the lake are disturbed, leading to an eventual loss of fish. When a lake is blue, this usually means that the lake has been damaged by acid precipitation. The gradual drop in pH caused by exposure to acid rain causes disruption of the food webs, eventually killing most organisms. The end result is clear water, which reflects the low productivity of the lake.

What is **red tide** and what causes it?

Red tide is a term used for a brownish or reddish discoloration occurring in ocean, river, or lake water. It is caused by the rapid reproduction of a variety of toxic organisms, especially the toxic red dinoflagellates, which are members of the genera *Gymnodidium* and *Gonyaulax*. Some red tides are harmless, but millions of fish may be killed during a "bloom," as the buildup is called. Other red tides can poison shellfish and the birds or humans who eat the contaminated food. Scientists do not fully understand why the "bloom" occurs.

What are **cyanobacterial blooms**?

Cyanobacterial (blue-green algae) blooms are occurring more frequently in freshwater lakes and along coastlines. They are often due to nutrient enrichment of the lakes, especially nitrates and phosphates, caused by human activities. Cyanobacteria grow faster at warmer temperatures. The thick masses of cyanobacteria absorb so much light that the water temperature increases. The algae at the base of the aquatic food webs prefer cooler temperatures and cannot compete successfully with rapidly growing cyanobacteria in the warmer temperatures. *Microcystis, Nostoc,* and *Anabaena* are examples of cyanobacteria that cause blooms as well as release neurotoxins. This category of blooms is referred to as Harmful Algal Blooms (HABs). They can occur in drinking water and survive typical water purification processes.

What is a **trophic level**?

A trophic level represents a step in the dynamics of energy flow through an ecosystem. The first trophic level is made up of the producers, those within the ecosystem that harvest energy from an outside source like the sun (or deep-sea thermal vents) and stabilize or "fix" it so that it remains in the system and is accessible to others within the system. The second level would comprise those who consume the producers, also known as the primary consumers. The next level would contain the secondary consumers (those who consume the primary consumers), and so on. Because of the limited amount of energy available to each level, these trophic pyramids rarely rise above a third or fourth level of structure. R. Lindeman (1915–1942) was one of the first ecologists to refer to the "trophic dynamics" of ecosystems, doing so in 1942.

How does a **food chain function**?

A food chain is the transfer of food energy from the source in plants through a series of organisms eating and being eaten by other organisms. Food chains overlap because many organisms eat more than one type of food, so that these chains can look more like food webs. In 1891 German zoologist Karl Semper (1832–1893) introduced the food chain concept.

What is a **food web**?

A food web consists of interconnecting food chains. Many animals feed on different foods rather than exclusively on one single species of prey or one type of plant. Animals that use a variety of food sources have a greater chance of survival than those with a single food source. Complex food webs provide greater stability to a living community.

Why do ecosystems **need decomposers**?

While energy flows through ecosystems in only one direction, entering at the producer level and exiting as heat and the transfer of energy (as biomass) to consumers, chemical compounds can be reused over and over again. In a well-functioning ecosystem, some organisms make their living by breaking down structures and recycling the compounds. These organisms are known as decomposers. Without these organisms, the chemicals used to build a tree would remain locked in the tree biomass for eternity instead of being returned to the soil after the tree's death. The chemicals in the soil cause new growth, beginning the cycle once again.

What is a **biogeochemical cycle**?

The elements that organisms need most (carbon, nitrogen, phosphorus, and sulfur) cycle through the physical environment, the organism, and then back to the environment. Each element has a distinctive cycle that depends on the physical and chemical properties of the element. Examples of biogeochemical cycles include the carbon and nitrogen cycles, both of which have a prominent gaseous phase. Examples of biogeochemical cycles with a prominent geologic phase include phosphorus and sulfur, where a large portion of the element may be stored in ocean sediments. Examples of cycles with a prominent atmospheric phase include carbon and nitrogen. These biogeochemical cycles involve biological, geologic, and chemical interactions.

What is the **hydrologic cycle**?

The hydrologic cycle takes place in the hydrosphere, which is the region containing all the water in the atmosphere and Earth's surface. It involves five phases: condensation, infiltration, runoff, evaporation, and precipitation. Rain, and other precipitation, is part of the hydrologic cycle.

What is the **carbon cycle**?

To survive, every organism must have access to carbon atoms. Carbon makes up about 49 percent of the dry weight of organisms. The carbon cycle includes movement of carbon from the gaseous phase (carbon dioxide in the atmosphere) to solid phase (carbon-containing compounds in living organisms) and then back to the atmosphere via decomposers. The atmosphere is the largest reservoir of carbon, containing 32 percent CO_2. Biological processes on land shuttle carbon between atmospheric and terrestrial compartments, with photosynthesis removing CO_2 from the atmosphere and cell respiration returning CO_2 to the atmosphere.

How do **plants obtain nitrogen**?

The primary way that plants obtain nitrogen compounds is via the nitrogen cycle, which is a series of reactions involving several different types of bacteria, including nitrogen-fixing bacteria and denitrifying bacteria. During nitrogen fixation, symbiotic bacteria, which live in association with the roots of legumes, are able through a series of enzymatic reactions to make nitrogen available for plants. Nitrogen is crucial to all organisms because it is an integral element of proteins and nucleic acids. Although Earth's atmosphere is 79 percent nitrogen, molecular nitrogen is very stable and does

> ### How is Henry David Thoreau associated with the environment?
>
> Henry David Thoreau (1817–1862) was a writer and naturalist from New England. His most familiar work, *Walden*, describes the time he spent in a cabin near Walden Pond in Massachusetts. He is also known for being one of the first to write and lecture on the topic of forest succession. His work, along with that of John Muir (1838–1914) and others, has served to inspire those others to understand the natural world and provide for its conservation.

not easily combine with other elements. Plants must use nitrogen in its fixed form, such as ammonia, urea, or the nitrate ion.

Who is considered the **founder** of **modern conservation**?

Scottish-born American naturalist John Muir (1838–1914) is the father of conservation and the founder of the Sierra Club. He fought for the preservation of the Sierra Nevada in California and the creation of Yosemite National Park. He directed most of the Sierra Club's conservation efforts and was a lobbyist for the Antiquities Act, which prohibited the removal or destruction of structures of historic significance from federal lands. Another prominent influence was George Perkins Marsh (1801–1882), a Vermont lawyer and scholar. His book *Man and Nature* emphasized the mistakes of past civilizations that resulted in the destruction of natural resources.

As the conservation movement swept through the country in the last three decades of the nineteenth century, a number of prominent citizens joined the efforts to conserve natural resources and to preserve wilderness areas. Writer John Burroughs (1837–1921), forester Gifford Pinchot (1865–1946), botanist Charles Sprague Sargent (1841–1927), and editor Robert Underwood Johnson (1857–1937) were early advocates of conservation.

What was the **first U.S. national park**?

On March 1, 1872, an Act of Congress signed by Ulysses S. Grant (1822–1885) established Yellowstone National Park as the first national park. The action inspired a worldwide national park movement.

When was the U.S. **National Park Service** established?

The National Park Service was created by an Act signed by President Woodrow Wilson (1856–1924) on August 25, 1916. Its responsibility was to administer national parks and monuments.

What was the environmental significance of Rachel Carson's *Silent Spring*?

In the book *Silent Spring*, published in 1962, Rachel Carson (1907–1964) exposed the dangers of pesticides, particularly DDT, to the reproduction of species that prey upon the insects for which the pesticide was intended. *Silent Spring* raised the public awareness and is considered a pivotal point at the beginning of the environmental movement.

Who started **Earth Day**?

The first Earth Day, April 22, 1970, was coordinated by Denis Hayes (1944–) at the request of Gaylord Nelson (1916–2005), U.S. senator from Wisconsin. Nelson is sometimes called the father of Earth Day. His main objective was to organize a nationwide public demonstration so large it would get the attention of politicians and force the environmental issue into the political dialogue of the nation. Important official actions that began soon after the celebration of the first Earth Day were the establishment of the Environmental Protection Agency (EPA); the creation of the President's Council on Environmental Quality; and the passage of the Clean Air Act, establishing national air quality standards. In 1995 Gaylord Nelson received the Presidential Medal of Freedom for his contributions to the environmental protection movement. Earth Day continues to be celebrated each spring.

When was the **Environmental Protection Agency** (EPA) **created** and what does it do?

In 1970 President Richard M. Nixon (1913–1994) signed an executive order that created the Environmental Protection Agency (EPA) as an independent agency of the U.S. government. The creation of a federal agency by executive order rather than by an act of the legislative branch is somewhat uncommon. The EPA was established in response to public concern about unhealthy air, polluted rivers and groundwater, unsafe drinking water, endangered species, and hazardous waste disposal. Responsibilities of the EPA include environmental research, monitoring, and enforcement of legislation regulating environmental activities. The EPA also manages the cleanup of toxic chemical sites as part of a program known as Superfund.

Why is **El Niño harmful**?

Along the west coast of South America, near the end of each calendar year, a warm current of nutrient-poor tropical water moves southward, replacing the cold, nutrient-rich surface water. Because this condition frequently occurs around Christmas, local residents call it El Niño (Spanish for child), referring to the Christ child. In most years the warming lasts for only a few weeks. However, when El Niño conditions last for many months, the economic results can be catastrophic. It is this extended episode of extremely warm water that scientists now refer to as El Niño.

During a severe El Niño, large numbers of fish and marine plants may die. Decomposition of the dead material depletes the water's oxygen supply, which leads to the bacterial production of huge amounts of smelly hydrogen sulfide. A greatly reduced fish (especially anchovy) harvest affects the world's fishmeal supply, leading to higher prices for poultry and other animals that normally are fed fishmeal. Anchovies and sardines are also major food sources for marine mammals, such as sea lions and seals. When the food source is in short supply, these animals travel farther from their homes in search of food. Not only do many sea lions and seals starve, but also a large proportion of the infant animals die. The major El Niño event of 1997–1998 indirectly caused 2,100 human deaths and $33 billion in damage globally.

POLLUTION AND WASTES

What is **air pollution**?

Air pollution is the contamination of Earth's atmosphere at levels high enough to harm humans, others organisms, or other materials. The major air pollutants are particulate matter, ozone, carbon monoxide, nitrogen oxides, sulfur dioxide, and lead. Primary air pollutants, including carbon monoxide, nitrogen oxides, sulfur dioxide, and particulate matter, enter the atmosphere directly. Secondary air pollutants are harmful chemicals that form from other substances released into the atmosphere. Ozone and sulfur trioxide are examples of secondary air pollutants.

What are the **sources** of **air pollution**?

Transportation is one of the major contributors to air pollution. Other significant sources of air pollution are fuel combustion (burning fossil fuels) and industrial processes, such as refineries, iron and steel mills, paper mills, and chemical plants.

What are the **health effects** of the major **air pollutants**?

Common health effects of exposure to even low levels of air pollutants are irritated eyes and inflammation of the respiratory tract. There is some evidence that exposure to air pollutants suppresses the immune system, increasing susceptibility to infections.

Pollutant	Source	Health Effects
Particulate matter	Industries, electric power plants, motor vehicles, construction, agriculture	Aggravates respiratory illnesses; long-term exposure may cause increased incidence of chronic conditions such as bronchitis; suppresses immune system; heavy metals and organic chemicals may cause cancer

Pollutant	Source	Health Effects
Nitrogen oxides	Motor vehicles, industries, heavily fertilized farmland	Irritate respiratory tract; aggravates asthma and chronic bronchitis
Sulfur oxides	Electric power plants and other industries	Irritate respiratory tract; long-term exposure may cause increased incidence of chronic conditions such as bronchitis; suppresses immune system; heavy metals and organic chemicals may cause cancer
Carbon monoxide	Motor vehicles, industries, fireplace	Reduces blood's ability to transport oxygen; low levels cause headaches and fatigue; higher levels cause mental impairment or death
Ozone	Formed in atmosphere	Irritates eyes and respiratory tract produces chest discomfort; aggravates asthma and chronic bronchitis

What is the **Pollutant Standard Index**?

The U.S. Environmental Protection Agency and the South Coast Air Quality Management District of El Monte, California, devised the Pollutant Standard Index to monitor concentrations of pollutants in the air and inform the public concerning related health effects. The scale measures the amount of pollution in parts per million, and has been in use nationwide since 1978.

PSI Index	Health Effects	Cautionary Status
0	Good	
50	Moderate	
100	Unhealthful	
200	Very unhealthful	Alert: elderly or ill should stay indoors and reduce physical activity.
300	Hazardous	Warning: General population should stay indoors and reduce physical activity.
400	Extremely hazardous	Emergency: all people remain indoors windows shut, no physical exertion.
500	Toxic	Significant harm; same as above.

What are the **greenhouse gases**?

Scientists recognize carbon dioxide (CO_2), methane (CH_4), chlorofluorocarbons (CFCs), nitrous oxide (N_2O), and water vapor as significant greenhouse gases. Green-

An atmosphere with natural levels of greenhouse gases (left) compared with an atmosphere of increased greenhouse effect (right).

house gases account for less than one percent of Earth's atmosphere. These gases trap heat in Earth's atmosphere, preventing the heat from escaping back into space. Human activities, such as using gasoline in automobiles for fuel, account for the release of carbon dioxide and nitrogen oxides.

Emissions of Greenhouse Gases in the United States, 1990–2008
(Million Metric Tons of Gas)

Gas	1990	1995	2000	2005	2008
Carbon Dioxide	5,090.0	5,417.3	5,965.3	6,089.0	5,905.3
Methane	613.4	613.2	586.1	553.2	567.3
Nitrous Oxide	322.7	343.2	346.5	329.4	319.1
HFCs*	36.9	62.2	103.2	119.3	130.2
PFCs*	20.8	15.6	13.5	6.2	7.5
SF_6*	32.8	28.1	19.2	17.9	16.5

*HFCs = hydrofluorocarbons, PFCs = perfluorocarbons, SF_6 = sulfur hexafluoride.

What is the **greenhouse effect**?

The greenhouse effect is a warming near Earth's surface that results when Earth's atmosphere traps the sun's heat. The atmosphere acts much like the glass walls and roof of a greenhouse. The effect was described by John Tyndall (1820–1893) in 1861. It was given the greenhouse analogy much later in 1896 by the Swedish chemist Svante Arrhenius (1859–1927). The greenhouse effect is what makes the earth habitable. Without the presence of water vapor, carbon dioxide, and other gases in the atmosphere, too much heat would escape and the earth would be too cold to sustain life. Carbon dioxide, methane, nitrous oxide, and other greenhouse gases absorb the infrared radiation rising from the earth and hold this heat in the atmosphere instead of reflecting it back into space.

In the twentieth century, the increased buildup of carbon dioxide, caused by the burning of fossil fuels, has been a matter of concern. There is some controversy concerning whether the increase noted in the earth's average temperature is due to the increased amount of carbon dioxide and other gases, or is due to other causes. Volcanic activity, destruction of the rain forests, use of aerosols, and increased agricultural activity may also be contributing factors.

Which **countries** emit the **most carbon dioxide** into the air?

Carbon dioxide is one of the greenhouse gases. In 2008 the ten countries that emitted the most carbon dioxide from the consumption and flaring of fossil fuels were:

Country	CO_2 emissions	Percentage of Global Total (in million metric tons)
China	6,550.5	22.3
United States	5,595.5	19.0
Russia	1,593.8	5.4
India	1,427.6	4.9
Japan	1,151.1	3.9
Germany	803.9	2.7
Canada	550.9	1.9
United Kingdom	510.6	1.7
Islamic Republic of Iran	505.0	1.7
Korea	501.3	1.7
World	29,381.4	

Is **ozone beneficial** or **harmful** to life on Earth?

Ozone, a form of oxygen with three atoms instead of the normal two, is highly toxic; less than one part per million of this blue-tinged gas is poisonous to humans. In Earth's upper atmosphere (stratosphere), it is a major factor in making life on Earth possible. About 90 percent of the planet's ozone is in the ozone layer. The ozone belt shields the Earth from, and filters the excessive ultraviolet (UV) radiation generated by, the sun. Scientists predict that a diminished or depleted ozone layer could lead to increased health problems for humans, such as skin cancer, cataracts, and weakened immune systems. Increased UV can also lead to reduced crop yields and disruption of aquatic ecosystems, including the marine food chain. While beneficial in the

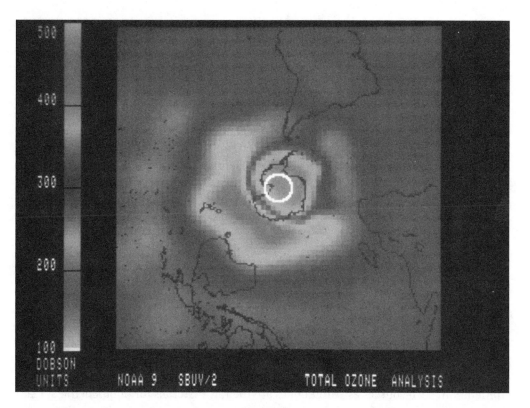

A 1987 satellite image of ozone depletion over the South Pole. Scientists have been concerned about the damage that chemicals such as CFCs have done to the ozone layer, which protects us from the sun's radiation.

stratosphere, near ground level it is a pollutant that helps form photochemical smog and acid rain.

How do **chlorofluorocarbons** affect Earth's **ozone layer**?

Chlorofluorocarbons (CFCs) are hydrocarbons, such as freon, in which part or all of the hydrogen atoms have been replaced by fluorine atoms. These can be liquids or gases, are non-flammable and heat-stable, and are used as refrigerants, aerosol propellants, and solvents. When released into the air, they slowly rise into Earth's upper atmosphere, where they are broken apart by ultraviolet rays from the sun. Some of the resultant molecular fragments react with the ozone in the atmosphere, reducing the amount of ozone. The CFC molecules' chlorine atoms act as catalysts in a complex set of reactions that convert two molecules of ozone into three molecules of ordinary oxygen.

This is depleting the beneficial ozone layer faster than it can be recharged by natural processes. The resultant "hole" lets through more ultraviolet light to Earth's surface and creates health problems for humans, such as cataracts and skin cancer, and disturbs del-

307

icate ecosystems (for example, making plants produce fewer seeds). In 1978 the U.S. government banned the use of fluorocarbon aerosols, and currently aerosol propellants have been changed from fluorocarbons to hydrocarbons, such as butane.

How large is the Antarctic ozone hole?

The term "hole" is widely used in popular media when reporting on ozone. However, the concept is more correctly described as a low concentration of ozone that occurs in August-October (springtime in the Southern Hemisphere). It was not observed until 1979. The first scientific article on ozone depletion in the Antarctic was published in *Nature* in 1985. The largest ozone hole ever observed, 11.4 million square miles (29.6 million square kilometers), occurred on September 24, 2006. The daily maximum ozone hole area for 2009 was 9.3 million square miles (24 million square kilometers).

What are the components of smog?

Smog, the most widespread pollutant in the United States, is a photochemical reaction resulting in ground-level ozone. Ozone, an odorless, tasteless gas in the presence of light can initiate a chain of chemical reactions. Ozone is a desirable gas in the stratospheric layer of the atmosphere, but it can be hazardous to health when found near Earth's surface in the troposphere. The hydrocarbons, hydrocarbon derivations, and nitric oxides emitted from such sources as automobiles are the raw materials for photochemical reactions. In the presence of oxygen and sunlight, the nitric oxides combine with organic compounds, such as the hydrocarbons from unburned gasoline, to produce a whitish haze, sometimes tinged with a yellow-brown color. In this process, a large number of new hydrocarbons and oxyhydrocarbons are produced. These secondary hydrocarbon products may comprise as much as 95 percent of the total organics in a severe smog episode.

What were the goals of the Montreal Protocol?

The Montreal Protocol was signed in 1987 by members of the United Nations to phase out and reduce production of CFCs and other ozone-damaging chemicals by 50 percent by 1998. Amendments and revisions to the Montreal Protocol were made in 1990, 1992, and 1997 to include other chemicals and accelerate the phase-out of certain chemicals. Production of CFCs, carbon tetrachloride, and methyl chlororform was phased out by 1996 in the United States and other highly developed countries.

What are some of the accomplishments achieved in reducing air pollution since the Clean Air Act was passed in 1970?

One of the goals of the EPA was to set national air quality standards for six common air pollutants—carbon monoxide, lead, nitrogen dioxide, ozone, particulate matter,

What is the Kyoto Protocol?

The Kyoto Protocol was an international summit held in Kyoto, Japan, in December 1997. Its goal was for governments around the world to reach an agreement regarding emissions of carbon dioxide and other greenhouse gases. The Kyoto Protocol called for the industrialized nations to reduce national emissions over the period 2008–2012 to five percent below the 1990 levels. The protocol covers these greenhouse gases: carbon dioxide, methane, and nitrous oxide. Other chemicals such as hydrofluorocarbons, perfluorocarbons, and sulfur hexafluoride were to be added in subsequent years.

and sulfur dioxide. Since passage of the Clean Air Act in 1970, the amount of these six pollutants in the air has decreased by more than 50 percent. Other accomplishments are:

- The reduction by nearly 70 percent of air toxics from large industrial sources (chemical plants, petroleum refineries, and paper mills)
- New cars are more than 90 percent cleaner
- Production of most ozone-depleting chemicals has ceased.

What is **acid rain**?

Acid deposition is the fallout of acidic compounds in solid, liquid, or gaseous forms. Wet deposition occurs as precipitation while dry deposition is the fallout of particulate matter. Acid rain is the best-known form of acid deposition. The term "acid rain" was coined by British chemist Robert Angus Smith (1817–1884) who, in 1872, published *Air & Rain: The Beginnings of a Chemical Climatology*. Since then, acid rain has become an increasingly used term for rain, snow, sleet, or other precipitation that has been polluted by acids such as sulfuric and nitric acids.

When gasoline, coal, or oil are burned, their waste products of sulfur dioxide and nitrogen dioxide combine in complex chemical reactions with water vapor in clouds to form acids. The United States alone discharges 40 million metric tons of sulfur and nitrogen oxides into the atmosphere. This, combined with natural emissions of sulfur and nitrogen compounds, has resulted in severe ecological damage. Hundreds of lakes in North America (especially northeastern Canada and United States) and in Scandinavia are so acidic that they cannot support fish life. Crops, forests, and building materials, such as marble, limestone, sandstone, and bronze, have been affected as well, but the extent is not as well documented. However, in Europe, where so many living trees are stunted or killed, the new word "Waldsterben" (forest death) has been coined to describe this new phenomenon.

In 1990, amendments to the U.S. Clean Air Act contained provisions to control emissions that cause acid rain. It included the reductions of sulfur dioxide emissions from 19 million tons to 9.1 million tons annually and the reduction of industrial nitrogen oxide emissions from 6 to 4 million tons annually.

Year	Sulfur Dioxide Emissions (million tons)	Nitrogen Oxide Emissions (million tons)
1990	23.1	25.5
1995	18.6	25.0
2000	16.3	22.6
2005	14.8	19.1
2008	11.4	16.3

How **acidic** is **acid rain**?

Acidity or alkalinity is measured by a scale known as the pH (potential for Hydrogen) scale. It runs from zero to 14. Since it is logarithmic, a change in one unit equals a tenfold increase or decrease. So a solution at pH 2 is 10 times more acidic than one at pH 3 and 100 times as acidic as a solution at pH 4. Zero is extremely acid, 7 is neutral, and 14 is very alkaline. Any rain below 5.0 is considered acid rain; some scientists use the value of 5.6 or less. Normal rain and snow containing dissolved carbon dioxide (a weak acid) measure about pH 5.6. Actual values vary according to geographical area. Eastern Europe and parts of Scandinavia have an average rain pH of 4.3 to 4.5; the rest of Europe is 4.5 to 5.1; eastern United States and Canada ranges from 4.2 to 4.6; and Mississippi Valley has a range of 4.6 to 4.8. The worst North American area, having 4.2, is centered around Lake Erie and Lake Ontario.

Which **pollutants** lead to **indoor air pollution**?

Indoor air pollution, also known as "sick building syndrome," results from conditions in modern, high energy efficiency buildings, that have reduced outside air exchange, or have inadequate ventilation, chemical contamination, and microbial contamination. Indoor air pollution can produce various symptoms, such as headache, nausea, and eye, nose, and throat irritation. In addition, houses are affected by indoor air pollution emanating from consumer and building products and from tobacco smoke. Below are listed some pollutants found in houses:

Pollutant	Sources	Effects
Asbestos	Old or damaged insulation, fireproofing, or acoustical tiles	Many years later, chest and abdominal cancers and lung diseases

Pollutant	Sources	Effects
Biological pollutants	Bacteria, mold, and mildew; viruses; animal dander and cat saliva; mites; cockroaches; pollen	Eye, nose, and throat irritation; shortness of breath; dizziness; lethargy; fever; digestive problems; asthma; influenza and other infectious diseases
Carbon monoxide	Unvented kerosene and gas heaters; leaking chimneys and furnaces; wood stoves and fireplaces; gas stoves; automobile exhaust from attached garages; tobacco smoke	At low levels, fatigue; at higher levels, impaired vision and coordination; headaches; dizziness; confusion; nausea; fatal at very high concentrations
Formaldehyde	Plywood, wall paneling, particle board, and fiber-board; foam insulation; fire and tobacco smoke; textiles and glues	Eye, nose, and throat irritations; wheezing and coughing; fatigue; skin rash; severe allergic reactions; may cause cancer
Lead	Automobile exhaust; sanding or burning of lead paint; soldering	Impaired mental and physical development in children; decreased coordination and mental abilities; kidneys, nervous system, and red blood cell damage
Mercury	Some latex paints	Vapors can cause kidney damage; long-term exposure can cause brain damage
Nitrogen dioxide	Kerosene heaters and unvented gas stoves and heaters; tobacco smoke	Eye, nose, and throat irritation; may impair lung function and increase respiratory infections in young children
Organic gases	Paints, paint strippers, solvents, and wood preservatives; aerosol sprays; cleansers and disinfectants; moth repellents; air fresheners; stored fuels; hobby; supplies dry-cleaned clothing	Eye, nose and throat irritation; headaches; loss of coordination; nausea; damage to liver, kidney, and nervous system; some organics cause cancer in animals and are suspected of causing cancer in humans
Pesticides	Products used to kill household pests and products used on lawns or gardens that drift or are tracked inside the house	Irritation to eye, nose, and throat; damage to nervous systems and kidneys; cancer

311

Pollutant	Sources	Effects
Radon	Earth and rock beneath the home; well water, building materials	No immediate symptoms; estimated to cause about ten percent of lung cancer deaths; smokers at higher risk

Why is exposure to **asbestos** a **health hazard**?

Asbestos fibers were used in building materials between 1900 and the early 1970s as insulation for walls and pipes, as fireproofing for walls and fireplaces, in soundproofing and acoustic ceiling tiles, as a strengthener for vinyl flooring and joint compounds, and as a paint texturizer. Asbestos poses a health hazard only if the tiny fibers are released into the air, but this can happen with any normal fraying or cracking. Asbestos removal aggravates this normal process and multiplies the danger level—it should only be handled by a contractor trained in handling asbestos. Once released, the particles can hang suspended in the air for more than 20 hours.

Exposure to asbestos has long been known to cause asbestosis. This is a chronic, restrictive lung disease caused by the inhalation of tiny mineral asbestos fibers that scar lung tissues. Asbestos has also been linked with cancers of the larynx, pharynx, oral cavity, pancreas, kidneys, ovaries, and gastrointestinal tract. The American Lung Association reports that prolonged exposure doubles the likelihood that a smoker will develop lung cancer. It takes cancer 15 to 30 years to develop from asbestos. Mesothelioma is a rare cancer affecting the surface lining of the pleura (lung) or peritoneum (abdomen), which generally spreads rapidly over large surfaces of either the thoracic or abdominal cavities. Current treatment methods include surgery, radiation, and chemotherapy although mesothelioma continues to be difficult to control.

What causes **formaldehyde contamination** in homes?

Formaldehyde contamination is related to the widespread construction use of wood products bonded with urea-formaldehyde resins and products containing formaldehyde. Major formaldehyde sources include subflooring of particle board; wall paneling made from hardwood plywood or particle board; and cabinets and furniture made from particle board, medium density fiberboard, hardwood plywood, or solid wood. Urea-formaldehyde foam insulation (UFFI) has received the most media notoriety and regulatory attention. Formaldehyde is also used in drapes, upholstery, carpeting, wallpaper adhesives, milk cartons, car bodies, household disinfectants, permanent-press clothing, and paper towels. In particular, mobile homes seem to have higher formaldehyde levels than other houses. The release of formaldehyde into the air by these products (called outgassing) can develop poisoning symptoms in humans. The EPA classifies formaldehyde as a potential human carcinogen (cancer-causing agent).

Why is **radon** a health hazard?

Radon is a colorless, odorless, tasteless, radioactive gaseous element produced by the decay of radium. It has three naturally occurring isotopes found in many natural materials, such as soil, rocks, well water, and building materials. Because the gas is continually released into the air, it makes up the largest source of radiation that humans receive. A National Academy of Sciences (NAS) report noted that radon was the second leading cause of lung cancer. It has been estimated that it may cause as much as 12 percent, or about 15,000 to 22,000 cases, of lung cancer deaths annually. Smokers seem to be at a higher risk than non-smokers.

The U.S. Environmental Protection Agency (EPA) recommends that in radon testing the level should not be more than four picocuries per liter. The estimated national average is 1.5 picocuries per liter. Because EPA's "safe level" is equivalent to 200 chest X rays per year, some experts believe that lower levels are appropriate. The American Society of Heating, Refrigeration, and Air-Conditioning Engineers (ASHRAE) recommends two picocuries/liter. The EPA estimates that nationally 1 in 15, or six percent, of all homes have radon levels that are above the four picocuries/liter limit. This is down from a 1987 survey that estimated 21 percent of homes were above this level.

How are **hazardous waste** materials **classified**?

There are four types of hazardous waste materials—corrosive, ignitable, reactive, and toxic.

- *Corrosive* materials can wear away or destroy a substance. Most acids are corrosive and can destroy metal, burn skin, and give off vapors that burn the eyes.
- *Ignitable* materials can burst into flames easily. These materials pose a fire hazard and can irritate the skin, eyes, and lungs. Gasoline, paint, and furniture polish are ignitable.
- *Reactive* materials can explode or create poisonous gas when combined with other chemicals. Combining chlorine bleach and ammonia, for example, creates a poisonous gas.
- *Toxic* materials or substances can poison humans and other life. They can cause illness or death if swallowed or absorbed through the skin. Pesticides and household cleaning agents are toxic.

What is the **Toxic Release Inventory** (TRI)?

TRI is a government mandated, publicly available compilation of information on the release of nearly 650 individual toxic chemicals and toxic chemical categories by manufacturing facilities in the United States. The law requires manufacturers to state the amounts of chemicals they release directly to air, land, or water, or that they transfer to off-site facilities that treat or dispose of wastes. The U.S. Environmental Protection

313

Agency compiles these reports into an annual inventory and makes the information available in a computerized database. In 2008, 21,695 facilities released 3.9 billion pounds (1.8 billion kilograms) of toxic chemicals into the environment. The total surface water discharges amounted to 246,321,372 pounds (111,729,495 kilograms); while 1,144,615,081 pounds (519,188,667 kilograms) were emitted into the air; 1,786,070,607 pounds (810,148,000 kilograms) were released on-site to land; and 196,339,106 pounds (89,057,920 kilograms) were injected into underground wells. The total amount of toxic chemicals released in 2008 was six percent (257 million pounds [116.5 million kilograms]) lower than the amount released in 2007.

Which **industries** release the **most toxic** chemicals?

The metal mining industry released the most toxic chemicals for the year 2008, accounting for 30 percent of total chemical releases.

Industry	Total Releases (pounds)	Percent of Total
Metal mining	1,157,710,105	30
Electric utilities	905,697,596	23
Chemicals	452,245,439	12
Primary metals	440,462,686	11
Paper	186,356,909	5
Hazardous waste/solvent	168,601,834	4
Food	167,223,768	4
All others	538,718,584	11

DDT in the environment caused the decline of bald eagles in North America. The chemical got into the food chain, and though it did not kill adult birds it softened egg shells and prevented chicks from hatching.

What **chemicals** were initially **banned** by the **Stockholm Convention** on Persistent Organic Pollutants?

A group of chemicals known as the "dirty dozen" were the original group of chemicals banned by the Stockholm Convention on Persistent Organic Pollutants (POPs). All of these chemicals possess toxic properties, resist degradation, and are transported across international boundaries via air, water, and migratory species. The Convention went into effect in 2004. Since then, other chemical compounds have been added or are being considered for inclusion on the list since the original 12. Health effects of these compounds include disruption of the endocrine system, cancer, and adverse effects in the developmental processes of organisms.

The "Dirty Dozen"

Persistent Organic Pollutant (POP)	Use
Aldrin	Insecticide
Chlordane	Insecticide
DDT (dichlorodiphenyl-trichloroethane)	Insecticide
Dieldrin	Insecticide
Endrin	Rodentcide and insecticide
Heptachlor	Fungicide
Hexachlorobenzene	Insecticide and fire retardant
Mirex™	Insecticide
Toxaphene™	Insecticide
PCBs (Polychlorinated biphenyls)	Industrial chemical
Dioxins	By-product of certain manufacturing processes
Furans (dibenzofurans)	By-product of certain manufacturing processes

How did **DDT affect** the **environment**?

Although DDT was synthesized as early as 1874 by Othmar Zeidler (1859–1911), it was the Swiss chemist Paul Müller (1899–1965) who recognized its insecticidal properties in 1939. He was awarded the 1948 Nobel Prize in Physiology or Medicine for his development of dichloro-diphenyl-trichloro-ethene, or DDT. Unlike the arsenic-based compounds then in use, DDT was effective in killing insects and seemed not to harm plants and animals. In the following 20 years it proved to be effective in controlling disease-carrying insects (mosquitoes that carry malaria and yellow fever, and lice that carry typhus) and in killing many plant crop destroyers. Publication of Rachel Carson's *Silent Spring* in 1962 alerted scientists to the detrimental effects of DDT. Increasingly, DDT-resistant insect species and the accumulative hazardous effects of DDT on plant and animal life cycles led to its disuse in many countries during the 1970s.

What are **PCBs**?

Polychlorinated biphenyls (PCBs) are a group of chemicals that were widely used before 1970 in the electrical industry as coolants for transformers and in capacitors and other electrical devices. They caused environmental problems because they do not break down and can spread through the water, soil, and air. They have been linked by some scientists to cancer and reproductive disorders and have been shown to cause

315

liver function abnormalities. Government action has resulted in the control of the use, disposal, and production of PCBs in nearly all areas of the world, including the United States.

What problems may be encountered when **polyvinyl chloride** (PVC) plastics are **burned**?

Chlorinated plastics, such as PVC, contribute to the formation of hydrochloric acid gases. They also may be a part of a mix of substances containing chlorine that form a precursor to dioxin in the burning process. Polystyrene, polyethylene, and polyethylene terephthalate (PET) do not produce these pollutants.

What are **Operation Ranch Hand** and **Agent Orange**?

Operation Ranch Hand was the tactical military project for the aerial spraying of herbicides in South Vietnam during the Vietnam Conflict (1961–1975). In these operations Agent Orange, the collective name for the herbicides 2,4-D and 2,4,5-T, was used for defoliation. The name derives from the color-coded drums in which the herbicides were stored. In all, U.S. troops sprayed approximately 19 million gallons (72 million liters) of herbicides over 4 million acres (1.6 million hectare).

Concerns about the health effects of Agent Orange were initially voiced in 1970, and since then the issue has been complicated by scientific and political debate. In 1993, a 16-member panel of experts reviewed the existing scientific evidence and found strong evidence of a statistical association between herbicides and soft-tissue sarcoma, non-Hodgkin's lymphoma, Hodgkin's disease, and chloracne. On the other hand, they concluded that no connection appeared to exist between exposure to Agent Orange and skin cancer, bladder cancer, brain tumors, or stomach cancer.

What is the **Toxic Substances Control Act** (TOSCA)?

In 1976, the U.S. Congress passed the Toxic Substances Control Act (TOSCA). This act requires the premarket testing of toxic substances. When a chemical substance is planned to be manufactured, the producer must notify the Environmental Protection Agency (EPA), and, if the data presented is determined to be inadequate to approve its use, the EPA will require the manufacturer to conduct further tests. Or, if it is later determined that a chemical is present at a level that presents an unreasonable public or environmental risk, or if there is insufficient data to know the chemical's effects, manufacturers have the burden of evaluating the chemical's characteristics and risks. If testing does not convince the EPA of the chemical's safety, the chemical's manufacturing, sale, or use can be limited or prohibited.

What is the Superfund Act?

In 1980, the U.S. Congress passed the Comprehensive Environmental Response, Compensation, and Liability Act, commonly known as the Superfund program. This law (along with amendments in 1986 and 1990) established a $16.3-billion Superfund financed jointly by federal and state governments and by special taxes on chemical and petrochemical industries (which provide 86 percent of the funding). The purpose of the Superfund is to identify and clean up abandoned hazardous-waste dump sites and leaking underground tanks that threaten human health and the environment.

To keep taxpayers from footing most of the bill, cleanups are based on the polluter-pays principle. The EPA is charged with locating dangerous dump sites, finding the potentially liable culprits, ordering them to pay for the entire cleanup and suing them if they don't. When the EPA can find no responsible party, it draws money out of the Superfund for cleanup.

Which **states** have the **most hazardous waste** (Superfund) sites?

As of 2010, there were 1,279 hazardous waste sites and 61 proposed sites. There are hazardous waste sites in all of the 50 states except North Dakota. The states with the most hazardous waste sites are:

- New Jersey: 112
- Pennsylvania: 95
- California: 94
- New York: 86
- Michigan: 66
- Florida: 52
- Texas: 48
- Washington: 48
- Illinois: 44
- Wisconsin: 38

The average cost to clean up a site is $20 million. Over 300 sites have been cleaned up enough to be taken off the National Priorities List.

What are **Brownfields**?

The Environmental Protection Agency defines Brownfields as abandoned, idled, or underused industrial or commercial sites where expansion or redevelopment is complicated by real or perceived environmental contamination. Real estate developers perceive Brownfields as inappropriate sites for redevelopment. There are approximately 450,000 Brownfields in the United States, with the heaviest concentrations being in the Northeast and Midwest.

What are the **sources** of **oil pollution** in the **oceans**?

Most oil pollution results from minor spillage from tankers and accidental discharges of oil when oil tankers are loaded and unloaded. Other sources of oil pollution include improper disposal of used motor oil, oil leaks from motor vehicles, routine ship maintenance, leaks in pipelines that transport oil, and accidents at storage facilities and refineries.

Source	Percent of Total Oil Spillage
Natural seepage	46
Discharges from consumption of oils (operational discharges from ships and discharges from land-based sources)	37
Accidental spills from ships	12
Extraction of oil	3

Where did the **first major oil spill** occur?

The first major commercial oil spill occurred on March 18, 1967, when the tanker *Torrey Canyon* grounded on the Seven Stones Shoal off the coast of Cornwall, England, spilling 830,000 barrels (119,000 tons) of Kuwaiti oil into the sea. This was the first major tanker accident. However, during World War II, German U-boat attacks on tankers, between January and June of 1942, off the United States East Coast, spilled 590,000 tons of oil.

Although the Exxon *Valdez* was widely publicized as a major spill of 35,000 tons in 1989, it is dwarfed by the deliberate dumping of oil from Sea Island into the Persian Gulf on January 25, 1991. It is estimated that the spill equaled almost 1.5 million tons of oil. A major spill also occurred in Russia in October 1994 in the Komi region of the Arctic. The size of the spill was reported to be as much as 2 million barrels (286,000 tons).

Miles and miles of oil spill booms were placed by delicate Gulf Coast wetlands to protect the fragile wildlife habitat from the 2010 toxic British Petroleum oil spill.

In addition to the large disasters, day-to-day pollution occurs from drilling platforms where waste generated from platform life, including human waste, and oils, chemicals, mud, and rock from drilling are discharged into the water.

Date	Cause	Tons Spilled
1/42–6/42	German U-boat attacks on tankers off the East Coast of the United States during World War II	590,000
3/18/67	Tanker *Torrey Canyon* grounds off Land's End in the English Channel	119,000
3/20/70	Tanker *Othello* collides with another ship in Tralhavet Bay, Sweden	60–100,000
12/19/72	Tanker *Sea Star* collides with another ship in Gulf of Oman	115,000
5/12/76	*Urquiola* grounds at La Coruna, Spain	100,000
3/16/78	Tanker *Amoco Cadiz* grounds off northwest France	223,000
6/3/79	Itox I oil well blows in southern Gulf of Mexico	600,000
7/79	Tankers *Atlantic Express* and *Aegean Captain* collide off Trinidad and Tobago	300,000
2/19/83	Blowout in Norwuz oil field in the Persian Gulf	600,000
8/6/83	Fire aboard *Castillo de Beliver* off Cape Town, South Africa	250,000
3/24/89	Exxon *Valdez* accident in Alaska	35,000
1/25/91	Iraq begins deliberately dumping oil into Persian Gulf from Sea Island, Kuwait	1,450,000
2/94	A structure to prevent pipeline leaks fails, spilling oil in the Komi Republic in northern Russia	almost 102,000,000
2/4/99	Tanker *New Carissa* spills some of its oil in Coos Bay, Oregon	238
11/13/02	Tanker *Prestige* splits in half off Galicia, Spain	67,000
4/20/10	BP offshore oil rig spill after an explosion in Gulf of Mexico	682,000*

* Estimated. Some experts believe a large amount of oil is still at the bottom of the ocean, but it is so far down under the ocean surface that it is difficult to measure quanities.

What are the **most commonly collected debris** items found along **ocean coasts**?

Debris found along ocean coasts is categorized as land-based, ocean-based, or general source debris. Land-based debris blows, washes, or is discharged into the water from land areas. Land-based debris originates with beachgoers, fishermen and shore-based manufacturers, processing facilities and waste-management facilities. Boating and

How much oil was spilled into the Caribbean Sea when the British Petroleum offshore oil rig exploded in 2010?

The largest accidental oil spill in the world began April 20, 2010 when the BP Deepwater Horizon rig exploded in the Gulf of Mexico. Initial estimates reported that only 1,000 barrels of oil per day were being released into the Gulf. These estimates quickly changed to 5,000 barrels per day and then 12,000–19,000 barrels per day by late May. As researchers continued to evaluate the situation, the estimate continued to rise until they determined that an average of 53,000 barrels of oil were released per day between April 20 and July 15, 2010. The final estimate was 5 million barrels of oil flowed into the Gulf of Mexico making it the largest oil spill ever.

fishing activities are often sources of ocean-based debris. During a five-year study from 2001 to 2006, a total of 238,103 items of debris were collected from various sites along the nation's coasts. Land-based debris accounted for 52 percent of the total, general-source debris accounted for 34 percent of the total, and ocean-based debris accounted for 14 percent of the total. The most frequently collected items were:

Debris along Ocean Coasts during 2001–2006 Study

Source	Number of Items Collected	Percent of Total
Land-based		
Straws	65,384	27.5
Balloons	18,509	7.8
Metal beverage cans	17,705	7.4
General sources		
Plastic beverage bottles	30,858	13
Plastic bags with seam < 1 meter	21,477	9
Plastic food bottles	8,355	3.5
Ocean-based		
Rope >1 meter	13,023	5.5
Fishing line	8,032	3.4
Floats/buoys	3,488	1.5

How much **garbage** does the **average American** generate?

According to the Environmental Protection Agency, about 250 million tons of municipal waste was generated in 2008. This is equivalent to 4.5 pounds (2.0 kilograms) per person per day, or approximately 1,650 pounds (730 kilograms) per year.

How does the **per capita solid waste** generation in the **United States** compare to **other countries**?

The reliance on nondurable goods and excess packaging suggests that the United States is a throw-away society. One of the differences in the per capita amount of garbage generated throughout the world is due to the cost of waste disposal. Among the developed nations, people have the incentive to waste less when disposal is more expensive. Developing nations generally consume less and consequently generate less waste.

Country	Per capita waste generation (pounds per day)	Per capita waste generation (kilograms per day)
United States	2.0	4.5
Canada	1.7	3.75
Netherlands	1.4	3.0
Japan	1.1	2.4
Germany	0.9	2.0
Sweden	0.9	2.0

What are the **components** of **municipal solid waste**?

Municipal solid waste consists of things we commonly use and then throw away; such as paper and packaging, food scraps, yard waste and large household items, including old sofas, appliances, and computers. It does not include industrial, hazardous, or construction waste. The distribution of the municipal solid waste generated in 2008 is illustrated below:

Waste Product	Weight (millions of tons)	Percent of Total
Paper and paperboard	77.4	31.0
Yard wastes	32.9	13.2
Food scraps	31.8	12.7
Plastics	30.0	12.0
Metals	20.9	8.4
Rubber, leather, and textiles	19.8	7.9
Wood	16.4	6.6
Glass	12.2	4.9
Other wastes	8.0	3.3

When was the **first garbage incinerator** built?

The first garbage incinerator in the United States was built in 1885 on Governor's Island in New York Harbor.

How much **methane fuel** does a **ton of garbage** make?

Over a period of 10 to 15 years, a ton of garbage will make 14,126 cubic feet (400 cubic meters) of fuel, although a landfill site will generate smaller amounts for 50 to 100 years. One ton of garbage can produce more than 100 times its volume in methane over a decade. Landfill operators tend not to maximize methane production.

How is **municipal waste managed**?

More than half of the municipal waste generated in the United States is discarded in landfills. The balance is either recovered through recycling programs or combusted with energy recovery.

Management of Municipal Waste (2008)

Method of Disposal	Amount (millions of tons)	Percent of Total
Discarded in landfills	135	54
Recovered through recycling and composting	83	33
Combustion with energy recovery	32	13
Total	250	100

How has **disposal** of **solid waste** to landfill facilities **changed**?

Landfilling has been an essential component of waste management for several decades. In 1960, 94 percent of all garbage was sent to landfills. During the following decades, although the amount of municipal solid waste increased, the amount going to landfills decreased. During the years 1990 to 2008, the total amount of waste going to landfills decreased from 142.3 million tons to 135.1 million tons.

Year	Percent of Municipal Solid Waste in Landfills
1960	93.6
1970	93.1
1980	88.6
1990	69.3
2000	56.9
2005	55.6
2008	54.2

Much of this decrease was due to an increased amount of waste being recycled. As of 2008, there were 1,812 landfills in the continental United States. Although the number of landfills in the United States has decreased, the average landfill size has increased. At the national level, capacity of existing landfills is sufficient.

What is **bioremediation**?

Bioremediation is the degradation, decomposition, or stabilization of pollutants by microorganisms such as bacteria, fungi, and cyanobacteria. Oxygen and organisms are injected into contaminated soil and/or water (e.g., oil spills). The microorganisms feed on and eliminate the pollutants. When the pollutants are gone, the organisms die.

What is **phytoremediation**?

Phytoremediation, a newer green technology, is the use of green plants to remove pollutants from the soil or render them harmless. Certain plants, such as alpine penny-cress (*Thlaspi caerulescens*), have been identified as metal hyperaccumulators. Researchers have found that these plants can grow in soils contaminated with toxic heavy metals, including cadmium, zinc, and nickel. The plants extract the toxic heavy metals from the soil and concentrate them in the stems, shoots, and leaves. These plant tissues may then be collected and disposed of in a hazardous waste landfill. Researchers are investigating ways to recover the metals by extracting them from the plants. Although phytoremediation is limited by how deep the roots of the plants grow in the soil, it is a less costly alternative way to clean up hazardous waste sites.

RECYCLING AND CONSERVATION

What is the significance of the **recycling symbol**?

The familiar recycling symbol consists of three arrows representing the three components of recycling. The three steps are:

1. Collection and processing of recyclable materials
2. Use of recyclables by industry to manufacture new products
3. Consumer purchase of products made from recycled materials

What is the **Resource Conservation and Recovery Act**?

In 1976, the U.S. Congress passed the Resource Conservation and Recovery Act (RCRA), which was amended in 1984 and 1986. This law requires the EPA to identify hazardous wastes and set standards for their management including generation, transportation, treatment, storage, and disposal of hazardous waste. The law requires all firms that store, treat, or dispose of more than 220 pounds (100 kilograms) of hazardous wastes per month to have a permit stating how such wastes are managed.

How much has **recycling** of **municipal solid waste grown** since 1960?

In 1960 the recycling rate was 6.4 percent. It has grown to 33.2 percent in 2008. On average, Americans recycled and composted 1.5 pounds (0.7 kilograms) of our individual waste generation of 4.5 pounds (2 kilograms) per person per day.

323

Recycling in the United States, 1960–2008

Year	Weight (thousands of tons)	Percent of Total Solid Waste
1960	5,610	6.4
1970	8,020	6.6
1980	14,520	9.6
1990	33,240	16.2
2000	69,330	29.0
2005	79,180	31.7
2008	82,870	33.2

Which **U.S. states recycle** the **most** per year?

There are differences in the amount each state recycles. California recycles the most tonnage per year.

State	Municipal Waste Recycled (Tons)	Percent of Total
California	19,409,000	39
Illinois	9,773,194	37
New York	7,694,160	35
Florida	6,775,329	29
Texas	5,900,000	19

What **state** was **first** to create **mandatory deposits** on beverage containers?

In 1971, the state of Oregon was the first to create legislation mandating deposits on beverage containers. The deposit was five cents per container.

Which **products** have the **highest recycling** rates?

Auto batteries, office-type papers, and yard trimmings have the highest recycling rates.

Product	Percent Recycled (2008)
Auto batteries	99.2
Office paper	70.9
Yard trimmings	64.7
Steel cans	62.8
Aluminum cans	48.2
Tires	35.4
Plastic bottles (HDPE transluscent)	29.3
Glass containers	28.0
Polyethylene terephthalate (PET) bottles & jars	27.7

When was **metal recycling started**?

The first metal recycling in the United States occurred in 1776 when patriots in New York City toppled a statue of King George III (1738–1820), which was melted down and made into 42,088 bullets.

When did **aluminum recycling begin**?

Aluminum recycling dates back to 1888 when the smelting process was invented. Since aluminum is so valuable, companies involved in manufacturing aluminum were motivated to discover ways to make aluminum from aluminum. Recycling aluminum saves 95 percent of the energy required to make new aluminum since it eliminates the mining of new ore. It is estimated that 73 percent of all aluminum ever produced is still in use.

How long does it take for a discarded aluminum **beverage can** to be recycled and returned as a **new beverage can**?

A recycled aluminum beverage can return to the shelf as a new beverage can in as little as 60 days.

How can **plastics** be made **biodegradable**?

Plastic does not rust or rot. This is an advantage in its usage, but when it comes to disposal of plastic the advantage turns into a liability. Degradable plastic has starch in it so that it can be attacked by starch-eating bacteria to eventually disintegrate the plastic into bits. Chemically degradable plastic can be broken up with a chemical solution that dissolves it. Used in surgery, biodegradable plastic stitches slowly dissolve in the body fluids. Photodegradable plastic contains chemicals that disintegrate over a period of one to three years when exposed to light. One-quarter of the plastic yokes used to package beverages are made from a plastic called Ecolyte, which is photodegradable.

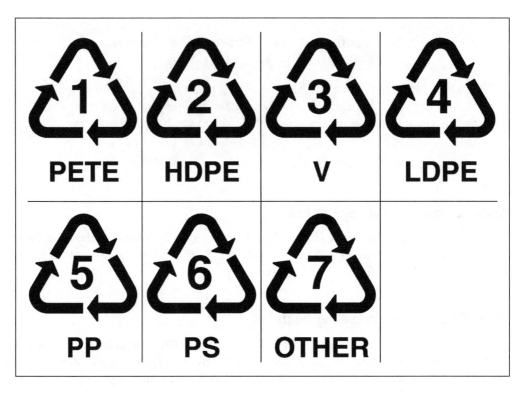

The logos used to indicate recyclable materials. Most curbside recycling programs accept 1s and 2s, but other types of materials are more difficult to recycle efficiently.

What do the **numbers** inside the **recycling symbol** on plastic containers **mean**?

The Society of the Plastics Industry developed a voluntary coding system for plastic containers to assist recyclers in sorting plastic containers. The symbol is designed to be imprinted on the bottom of the plastic containers. The numerical code appears inside three arrows that form a triangle. A guide to what the numbers mean is listed below. The most commonly recycled plastics are polyethylene terephthalate (PET) and high-density polyethylene (HDPE).

Code	Material	Examples
1	Polyethylene terephthalate (PET/PETE)	2-liter soft drink bottle
2	High-density polyethylene (HDPE)	Milk and water jugs
3	Vinyl (PVC)	Plastic pipes, shampoo bottles
4	Low-density polyethylene (LDPE)	Produce bags, food storage containers
5	Polypropylene (PP)	Squeeze bottles, drinking straws
6	Polystyrene (PS)	Fast-food packaging, other packaging
7	Other	Food containers

What **products** are made from **recycled plastic**?

Resin	Common Uses	Products Made from Recycled Resin
HDPE	Beverage bottles, milk jugs, detergent bottles, milk and soft drink crates, pipe, cable, film	Motor oil bottles, pipes, and pails
LDPE	Film bags such as trash bags, coatings, plastic bottles	New trash bags, pallets, carpets, fiberfill, non-food
PET	Soft drink, detergent, and juice bottles	Bottles/containers
PP	Auto battery cases, screw-on caps and lids; some yogurt and margarine tubs, plastic film	Auto parts, batteries, carpets
PS	Housewares, electronics, fast food carry-out packaging, plastic utensils	Insulation board, office equipment, reusable cafeteria trays
PVC	Sporting goods, luggage, pipes, auto parts; in packaging for shampoo bottles, blister packaging, and films	Drainage pipes, fencing, house siding

A new clothing fiber called Fortrel EcoSpun is made from recycled plastic soda bottles. The fiber is knit or woven into garments such as fleece for outerwear or long underwear. The processor estimates that every pound of Fortrel EcoSpun fiber results in ten plastic bottles being kept out of landfills.

What **natural resources** are **saved** by recycling paper?

One ton (907 kilograms) of recycled waste paper would save an average of 7,000 gallons (26,460 liters) of water, 3.3 cubic yards (2.5 cubic meters) of landfill space, 3 barrels of oil, 17 trees, and 4,000 kilowatt-hours of electricity—enough energy to power the average home for six months. It would also reduce air pollution by 74 percent.

When offered a choice between **plastic** or **paper bags** for your groceries, which should you choose?

The answer is neither. Both are environmentally harmful and the question of which is the more damaging has no clear-cut answer. Twelve million barrels of oil (a nonrenewable resource) are required to produce 100 billion plastic bags. Plastic bags degrade slowly in landfills and can harm wildlife if swallowed, and producing them pollutes the environment. In contrast, 35 million trees are cut down to produce 25 million brown paper bags accompanied by air and water pollution during the manufacturing process. Although each can be recycled, the EPA estimates that only one percent of plastic bags and 20 percent of paper bags are recycled. Instead of choosing between paper and plastic bags, bring your own reusable canvas or string containers to the store, and save and reuse any paper or plastic bags you get.

When was **paper recycling started**?

Paper recycling was actually born in 1690 in the United States when the first paper mill was established by the Rittenhouse family on the banks of Wissahickon Creek, near Philadelphia. The paper at this mill was made from recycled rags.

How much newspaper must be recycled to save one tree?

One 35 to 40 foot (10.6 to 12 meter) tree produces a stack of newspapers 4 feet (1.2 meters) thick; this much newspaper must be recycled to save a tree.

How much **paper** is **recycled** in the **United States**?

In 2008, 43 million tons, or 56 percent, of all paper used was recovered for recycling. This amounts to 340 pounds of paper for every man, woman, and child in the United States.

How much **water** does the **average family** in the United States use **per day**?

The average family in the United States uses 69.3 gallons (262 liters) of water per day. This includes showers, toilets, faucets, dish washing, clothes washing, faucets, and other uses.

Average Daily per Capita Water Usage (U.S.)

Use/Activity	Gallons per Capita	Liters per Capita	Percent Total Use
Toilets	18.5	70	26.7
Clothes washers	15.0	57	21.7

Use/Activity	Gallons per Capita	Liters per Capita	Percent Total Use
Showers	11.6	44	16.8
Faucets	10.9	41	15.7
Leaks	9.5	36	13.7
Baths	1.2	4.5	1.7
Dishwashers	1.0	3.8	1.4
Other household uses	1.6	6	2.2

How much water can the average family in the United States save by installing efficient water fixtures and checking regularly for leaks?

The average family in the United States would use only 45.2 gallons (171 liters) of water per day, saving 24.1 gallons (91 liters) per day, by installing more efficient water fixtures and checking for leaks. This reduces daily per capita water use by 35 percent.

Average Daily per Capita Water Usage (U.S.)
(using more efficient fixtures)

Use/Activity	Gallons per Capita	Liters per Capita	Percent Total Use
Faucets	10.8	41	23.9
Clothes washers	10.0	38	22.1
Showers	8.8	33	19.5
Toilets	8.2	31	18
Leaks	4.0	15	8.8
Baths	1.2	4.5	2.7
Dishwashers	0.7	2.7	1.5
Other household uses	1.6	6	2.2

Is washing dishes by hand better for the environment than using an automatic dishwasher?

Dishwashers often save energy and water compared to hand washing. Depending on the brand, dishwashers typically consume 7.5 to 12 gallons (28 to 45 liters) of water per normal wash. Hand-washing a day's worth of dishes may use up to 15 gallons (57 liters) of water. One university study found that dishwashers consume about 37 percent less water than washing by hand.

Several steps can be taken for additional energy savings when using a dishwasher. The setting on a home's water heater can be turned down to 120°F (49°C) if the dishwasher has a booster heater. While some machines feature a no-heat, air-dry setting, simply opening the door after the final rinse to let the dishes air dry will save energy.

What is eCycling?

ECycling is the reuse and recycling of electronic equipment. Donating used electronic equipment for use by others is the environmentally preferred alternative to discarding used electronics. If an electronic device cannot be reused, it should be recycled. The recycling rate for electronic devices is only 18 percent, with the remaining amount of disposed electronics being discarded in landfills. One of the goals of The Plug-In to eCycling Campaign, a partnership between the EPA and leading consumer electronics manufacturers and retailers, is to increase the recycling rate of electronics to 35 percent. In 2008, the Plug-In to eCycling partners recycled 34,000 tons (30,844 metric tons) of used consumer electronics.

Prewashing the dishes before loading generally wastes water since most machines can handle even heavily soiled plates.

What are the **uses** of **discarded tires**?

During 2007, 4,596 thousand tons of tires were scrapped in the United States. Nearly 90 percent, or 4,106 thousand tons, of tires were sent to one of the three major markets for scrap tires—tire-derived fuel, ground rubber applications, and civil engineering. Tire-derived fuel (TDF) accounted for 54 percent or 2,484 thousand tons of the scrap tires generated. TDF is used in a variety of combustion technologies, including cement kilns, pulp and paper mill boilers, utility and industrial boilers. The TDF market is expected to continue to grow. Ground rubber applications, including new rubber products, playground and other sports surfacing, and rubber-modified asphalt, consumed 789 thousand tons (17 percent of the total) of scrapped tires. Another 562 tons (12 percent of the total) of tires were used in civil engineering applications. These include tire shreds used in road and landfill construction, septic tank leach fields, and other construction projects. An additional 594 thousand tons of tires were sent to landfills in 2007.

Tires may be disposed in either a landfill or a monofill, a separate landfill only for tires. In certain areas of the country, especially in the Western states, landfills are a more efficient option than other scrap tire markets. Landfills are also the only option for tires that are in such poor condition they are not candidates for the scrap tire market. Furthermore, landfills are an important disposal option for the residue and by-products from tire shredders.

How much **electronic waste** is generated in the **United States**?

Electronic waste (e-waste), consisting of TVs and other video equipment, computers and assorted peripheral equipment, audio equipment and cell phones, accounts for

Electronics such as computers and iPhones contain heavy metals and other harmful pollutants. Efforts are growing to recycle these products and keep them from being tossed into landfills.

less than two percent of the total municipal solid waste. However, the amount of electronic equipment that is generated is increasing steadily. The National Safety Council Study of 1998 estimated that 20 million computers became obsolete. Just seven years later, in 2005, the EPA estimated 26–37 million computers became obsolete. More recently, in 2007, the EPA estimated the number of computers that became obsolete had doubled since 1998. Furthermore, the Consumer Electronics Association estimated 304 million electronics were removed from U.S. households in 2005. In 2007, approximately 414,000 tons (375,574 metric tons) of electronics were collected in the United States for recycling.

What is the **environmental impact** of **disposing electronics** in landfills?

The EPA believes that the disposal of electronics in properly managed municipal solid waste landfills does not threaten human health and the environment. However, recycling electronics will decrease the demand for additional mining of valuable resources and manufacturing new products. Recycling electronics recovers valuable materials, such as copper and engineered plastics, and as a result reduces greenhouse gas emissions, pollution, saves energy, and saves resources by extracting fewer raw materials from the earth.

What are some **environmental advantages** to **recycling electronics**?

- Recycling one million desktop computers prevents the release of greenhouse gases equivalent to the annual emissions of 16,000 passenger cars.

331

- Recycling one million laptop computers saves the energy equivalent to the electricity used by 3,657 U.S. homes in one year.
- Recycling the 414,000 tons of electronics that were collected in the United States in 2007 prevented the release of greenhouse gases equivalent to the annual emissions of more than 178,000 cars.
- One metric ton of circuit boards can contain 40 to 800 times the concentration of gold ore mined in the United States.
- One metric tron of circuit boards can contain 30 to 40 times the concentration of copper ore mined in the United States.
- Recycling one million cell phones could recover 7,500 pounds (3,402 kilograms) of gold which could be used in new products rather than having to mine more gold.

What is a **"green product"**?

Green products are environmentally safe products that contain no chlorofluorocarbons, are degradable (can decompose), and are made from recycled materials. "Deep-green" products are those from small suppliers who build their identities around their claimed environmental virtues. "Greened-up" products come from the industry giants and are environmentally improved versions of established brands.

Is it possible to buy **green electronics**?

Electronic manufacturers are designing products that are more environmentally friendly. Consumers should look for products that:

- Contain fewer toxic constituents
- Use recycled materials in the new product
- Are designed for easy upgrading and disassembly
- Are energy efficient
- Use minimal packaging
- Have leasing or takeback options for reuse or recycling
- Meet performance criteria that show they are environmentally preferable.

Which parts of the world have the **largest protected areas**?

Protected areas include national parks, nature reserves, national monuments, and other sites. There are at least 114,296 protected areas around the world, covering nearly 13 percent of the total land area. The regions with the largest amount of protected areas are:

Regions of the World with Largest Amount of Protected Areas (by actual area)

Region	Square Miles/Square Kilometers (in thousands)
North America	1,586 / 4,109
South America (excluding Brazil)	810 / 2,098
East Asia	681 / 1,765
North Eurasia	678 / 1,755
Eastern and Southern Africa	652 / 1,689
Brazil	622 / 1,612
Australia and New Zealand	594 / 1,538
North Africa and Middle East	497 / 1,286
Western and Central Africa	433 / 1,121
Europe	337 / 874

Regions of the World with Largest Amount of Protected Areas (by percent of total land area)

Region	Percent
Central America	30.28
Caribbean	29.05
South America (excluding Brazil)	22.55
Australia and New Zealand	19.19
Brazil	18.85
Southeastern Asia	18.60
North America	17.31
Europe	16.72
East Asia	15.00
Eastern and Southern Africa	14.70

What are some of the **largest national parks** in the United States?

The largest national parks are in Alaska:

Park	Area (acres)
Wrangell-St. Elias	8,323,148
Gates of the Arctic	7,523,898
Denali	4,740,912
Katmai	3,674,540
Glacier Bay	3,224,840
Lake Clark	2,619,733
Kobuk Valley	1,750,717

When was the symbol of Smokey Bear first used to encourage forest fire prevention?

The origin of Smokey Bear can be traced to World War II when the U.S. Forest Service, concerned about maintaining a steady lumber supply for the war effort, wished to educate the public about the dangers of forest fires. They sought volunteer advertising support from the War Advertising Council, and on August 9, 1944, Albert Staehle (1899–1974), a noted illustrator of animals, created Smokey Bear.

Since 1944, Smokey Bear has been a national symbol of forest fire prevention not only in America, but also in Canada and Mexico, where he is known as Simon. This public service advertising (PSA) campaign is the longest running PSA campaign in U.S. history. In 1947, a Los Angeles advertising agency coined the slogan "Only you can prevent forest fires." On April 23, 2001, after more than 50 years, the famous ad slogan was revised to "Only you can prevent wildfires" in response to the wildfire outbreaks during 2000. The campaign gained a living mascot in 1950 when a firefighting crew rescued a male bear cub from a forest fire in the Capital Mountains of New Mexico. Sent to the National Zoo in Washington, D.C., to become Smokey Bear, the animal was a living symbol of forest fire protection until his death in 1976. His remains are buried at the Smokey Bear State Historical Park in Capitan, New Mexico.

The five largest parks in the 48 contiguous states are:

Park	Location	Area (acres)
Death Valley	California	3,373,042
Yellowstone	Idaho, Montana, Wyoming	2,219,791
Everglades	Florida	1,398,893
Grand Canyon	Arizona	1,217,403
Glacier	Montana	1,013,322

Which **national parks** in the United States are the **most popular**?

Park/Location	Visitors (2009)
Great Smoky Mountains National Park (N. Carolina & Tennessee)	9,491,437
Grand Canyon National Park (Arizona)	4,348,068
Yosemite National Park (California)	3,737,472
Yellowstone National Park (Wyoming)	3,295,187

Park/Location	Visitors (2009)
Olympic National Park (Washington)	3,276,459
Rocky Mountain National Park (Colorado)	2,822,325
Zion National Park (Utah)	2,735,402
Cuyahoga Valley (Ohio)	2,589,288
Grand Teton National Park (Wyoming)	2,580,081
Acadia National Park (Maine)	2,227,698

What **causes** the **most forest fires** in the **western United States**?

Lightning is the single largest cause of forest fires in the western states.

EXTINCT AND ENDANGERED PLANTS AND ANIMALS

When was the **term "dinosaur"** first used?

The term "dinosaur" was first used by Richard Owen (1804–1892) in 1841 in his report on British fossil reptiles. The term, meaning "fearful lizard" was used to describe the group of large extinct reptiles whose fossil remains had been found by many collectors.

What is the name of the **early Jurassic mammal** that is now extinct?

The fossil site of the mammal *Hadrocodium wui* was in Yunnan Province, China. This newly described mammal is at least 195 million years old. The estimated weight of the whole mammal is about 0.07 ounces (2 grams). Its tiny skull was smaller than a human thumbnail.

Did **dinosaurs** and **humans** ever **coexist**?

No. Dinosaurs first appeared in the Triassic Period (about 220 million years ago) and disappeared at the end of the Cretaceous Period (about 65 million years ago). Modern humans (*Homo sapiens*) appeared only about 25,000 years ago. Movies that show humans and dinosaurs existing together are only Hollywood fantasies.

What were the **smallest** and **largest dinosaurs**?

Compsognathus, a carnivore from the late Jurassic period (131 million years ago), was about the size of a chicken and measured, at most, 35 inches (89 centimeters) from

335

the tip of its snout to the tip of its tail. The average weight was about 6 pounds 8 ounces (3 kilograms), but they could be as much as 15 pounds (6.8 kilograms).

The largest species for which a whole skeleton is known is *Brachiosaurus*. A specimen in the Humboldt Museum in Berlin measures 72.75 feet (22.2 meters) long and 46 feet (14 meters) high. It weighed an estimated 34.7 tons (31,480 kilograms). *Brachiosaurus* was a four-footed, plant-eating dinosaur with a long neck and a long tail and lived from about 121 to 155 million years ago.

How does a **mastodon differ** from a **mammoth**?

Although the words are sometimes used interchangeably, the mammoth and the mastodon were two different animals. The mastodon seems to have appeared first, while a side branch may have led to the mammoth.

The *mastodon* lived in Africa, Europe, Asia, and North and South America. It appeared in the Oligocene epoch (25 to 38 million years ago) and survived until less than one million years ago. It stood a maximum of 10 feet (3 meters) tall and was covered with dense woolly hair. Its tusks were aligned straight forward and were nearly parallel to each other.

The *mammoth* evolved less than two million years ago and died out about ten thousand years ago. It lived in North America, Europe, and Asia. Like the mastodon, the mammoth was covered with dense, woolly hair, with a long, coarse layer of outer hair to protect it from the cold. It was somewhat larger than the mastodon, standing 9 to 15 feet (2.7 to 4.5 meters). The mammoth's tusks tended to spiral outward, then up.

The gradual warming of Earth's climate and the change in environment were probably primary factors in the animals' extinction. Early man killed many of them as well, perhaps hastening the process.

Why did **dinosaurs** become **extinct**?

There are many theories as to why dinosaurs disappeared from Earth about 65 million years ago. Scientists debate whether dinosaurs became extinct gradually or all at once. The gradualists believe that the dinosaur population steadily declined at the end of

Cretaceous Period. Numerous reasons have been proposed for this. Some claim the dinosaurs' extinction was caused by biological changes that made them less competitive with other organisms, especially the mammals that were just beginning to appear. Overpopulation has been argued, as has the theory that mammals ate too many dinosaur eggs for the animals to reproduce themselves. Others believe that disease—everything from rickets to constipation—wiped them out. Changes in climate, continental drift, volcanic eruptions, and shifts in Earth's axis, orbit, and/or magnetic field have also been held responsible.

The catastrophists argue that a single disastrous event caused the extinction not only of the dinosaurs but also of a large number of other species that coexisted with them. In 1980, American physicist Luis Alvarez (1911–1988) and his geologist son, Walter Alvarez (1940–), proposed that a large comet or meteoroid struck Earth 65 million years ago. They pointed out that there is a high concentration of the element iridium in the sediments at the boundary between the Cretaceous and Tertiary Periods. Iridium is rare on Earth, so the only source of such a large amount of it had to be outer space. This iridium anomaly has since been discovered at over 50 sites around the world. In 1990, tiny glass fragments, which could have been caused by the extreme heat of an impact, were identified in Haiti. A 110-mile (177-kilometer) wide crater in the Yucatan Peninsula, long covered by sediments, has been dated to 64.98 million years ago, making it a leading candidate for the site of this impact.

A hit by a large extraterrestrial object, perhaps as much as 6 miles (9.3 kilometers) wide, would have had a catastrophic effect upon the world's climate. Huge amounts of dust and debris would have been thrown into the atmosphere, reducing the amount of sunlight reaching the surface. Heat from the blast may also have caused large forest fires, which would have added smoke and ash to the air. Lack of sunlight would kill off plants and have a domino-like effect on other organisms in the food chain, including the dinosaurs.

It is possible that the reason for the dinosaurs' extinction may have been a combination of both theories. The dinosaurs may have been gradually declining, for whatever reason. The impact of a large object from space merely delivered the final devastating blow.

The fact that dinosaurs became extinct has been cited as proof of their inferiority and that they were evolutionary failures. However, these animals flourished for 150 million years. By comparison, the earliest ancestors of humanity appeared only about three million years ago. Humans have a long way to go before they can claim the same sort of success as the dinosaurs.

How did the **dodo** become **extinct**?

The dodo became extinct around 1800. Thousands were slaughtered for meat, but pigs and monkeys, which destroyed dodo eggs, were probably most responsible for the

A nineteenth-century engraving of the extinct Dodo bird. The bird has become a poster child, of sorts, for extinct species that died out as a result of hunting and other human activities.

dodo's extinction. Dodos were native to the Mascarene Islands in the Central Indian Ocean. They became extinct on Mauritius soon after 1680 and on Réunion about 1750. They remained on Rodriguez until 1800.

When did the **last passenger pigeon** die?

At one time, 200 years ago, the passenger pigeon (*Ectopistes migratorius*) was the world's most abundant bird. Although the species was found only in eastern North America, it had a population of three to five billion birds (25 percent of the North American land bird population). Over-hunting caused a chain of events that reduced their numbers below minimum threshold for viability. In the 1890s several states passed laws to protect the pigeon, but it was too late. The last known wild bird was shot in 1900. The last passenger pigeon, named Martha, died on September 1, 1914, in the Cincinnati Zoo. In a span of just 200 years the passenger pigeon passed from the world's most abundant bird species into extinction.

What is the difference between an **"endangered" species** and a **"threatened" species**?

An "endangered" species is one that is in danger of extinction throughout all or a significant portion of its range. A "threatened" species is one that is likely to become endangered in the foreseeable future due to declining numbers.

How is it **determined** that a species is **"endangered"**?

This determination is a complex process that has no set of fixed criteria that can be applied consistently to all species. The known number of living members in a species is not the sole factor. A species with a million members known to be alive but living in only one small area could be considered endangered, whereas another species having a smaller number of members, but spread out in a broad area, would not be considered so threatened. Reproduction data—the frequency of reproduction, the average number of offspring born, the survival rate, etc.—enter into such determinations. In the United States, the director of the U.S. Fish and Wildlife Service (within the Department of the Interior) determines which species are to be considered endangered, based on research and field data from specialists, biologists, botanists, and naturalists.

According to the Endangered Species Act of 1973, a species can be listed if it is threatened by any of the following:

1. The present or threatened destruction, modification, or curtailment of its habitat or range;

2. Utilization for commercial, sporting, scientific, or educational purposes at levels that detrimentally affect it;

3. Disease or predation;

4. Absence of regulatory mechanisms adequate to prevent the decline of a species or degradation of its habitat;

5. Other natural or man-made factors affecting its continued existence.

If the species is so threatened, the director then determines the "critical habitat," which is the species' inhabitation areas that contain the essential physical or biological features necessary for the species' preservation. The critical habitat can include non-habitation areas, which are deemed necessary for the protection of the species.

Which species have become **extinct** since the **Endangered Species Act** was **passed** in 1973?

Nine domestic species have become extinct.

First Listed	Date Delisted	Species Name
8/27/1984	2/23/2004	Broadbill, Guam (*Myiagra freycineti*)
3/11/1967	9/2/1983	Cisco, longjaw (*Coregonus alpenae*)
4/30/1980	12/4/1987	Gambusia, Amistad (*Gambusia amistadensis*)
12/8/1977	2/23/2004	Mallard, Mariana (*Anas oustaleti*)
6/14/1976	1/9/1984	Pearlymussel, Sampson's (*Epioblasma sampsonii*)
3/11/1967	9/2/1983	Pike, blue (*Stizostedion vitreum glaucum*)
10/13/1970	1/15/1982	Pupfish, Tecopa (*Cyprinodon nevadensis calidae*)
3/11/1967	12/12/1990	Sparrow, dusky seaside (*Ammodramus maritimus nigrescens*)
6/4/1973	10/12/1983	Sparrow, Santa Barbara song (*Melospiza melodia graminea*)

Which **species** have been **removed** from the **Endangered Species List** because they have **recovered**?

Twenty-two species have been removed from the Endangered Species List because they have recovered.

First Listed	Date Delisted	Species Name
3/11/1967	6/04/1987	Alligator, American (*Alligator mississippiensis*)
9/17/1980	8/27/2002	Cinquefoil, Robbins' (*Potentilla robbinsiana*)
3/11/1967	7/24/2003	Deer, Columbian white-tailed (Douglas County, Oregon) (*Odocoileus virginianus leucurus*)
6/2/1970	9/12/1985	Dove, Palau ground (*Gallicolumba canifrons*)
3/11/1967	8/8/2007	Eagle, bald (lower 48 states) (*Haliaeetus leucocephalus*)
6/2/1970	8/25/1999	Falcon, American peregrine (*Falco peregrinus anatum*)
6/2/1970	10/5/1994	Falcon, Arctic peregrine (*Falco peregrinus tundrius*)
6/2/1970	9/12/1985	Flycatcher, Palau fantail (*Rhipidura lepida*)
3/11/1967	3/20/2001	Goose, Aleutian Canada (*Branta canadensis leucopareia*)
12/30/1974	3/9/1995	Kangaroo, eastern gray (*Macropus giganteus*)
12/30/1974	3/9/1995	Kangaroo, red (*Macropus rufus*)
12/30/1974	3/9/1995	Kangaroo, western gray (*Macropus fuliginosus*)
4/26/1978	9/14/1989	Milk-vetch, Rydberg (*Astragalus perianus*)
6/2/1970	9/21/2004	Monarch, Tinian (old world flycatcher) (*Monarcha takatsukasae*)
6/2/1970	9/12/1985	Owl, Palau (*Pyroglaux podargina*)
6/2/1970	2/4/1985	Pelican, brown (U.S. Atlantic coast, FL, AL) (*Pelecanus occidentalis*)
6/2/1970	12/17/2009	Pelican, brown (except U.S. Atlantic coast) (*Pelecanus occidentalis*)
7/1/1985	9/25/2008	Squirrel, Virginia northern flying (*Glaucomys sabrinus fuscus*)
5/22/1997	8/18/2005	Sunflower, Eggert's (*Helianthus eggertii*)
6/2/1970	6/16/1994	Whale, gray (except where listed) (*Eschrichtius robustus*)
3/28/2008	4/2/2009	Wolf, gray (Northern Rocky Mountain DPS) (*Canis lupus*)
7/19/1990	10/7/2008	Woolly-star, Hoover's (*Eriastrum hooveri*)

The status of seven species has been changed due to taxonomic revision. New information has been discovered for four other species.

How **many species** of **plants** and **animals** are **threatened** or **endangered** in the world?

Summary of Listed Species
Species and Recovery Plans as of 4/26/2010

	Endangered U.S./Foreign		Threatened U.S./Foreign		Total Species	U.S. Species in Recovery Plans
Mammals	70	255	15	20	360	60
Birds	76	182	16	7	281	85
Reptiles	13	66	24	16	119	38
Amphibians	14	8	11	1	34	17
Fishes	74	11	66	1	152	102
Clams	62	2	8	0	72	70
Snails	24	1	11	0	36	30
Insects	48	4	10	0	62	40
Arachnids	12	0	0	0	12	12
Crustaceans	18	0	3	0	22	18
Corals	0	0	2	0	2	0
Animal Subtotal	412	529	166	45	1,152	472
Flowering Plants	616	1	146	0	763	636
Conifers and Cycads	2	0	1	2	5	3
Ferns and Allies	27	0	2	0	29	26
Lichens	2	0	0	0	2	2
Plant Subtotal	647	1	149	2	799	667
Grand Total	1,059	530	315	47	1,951	1,139

What is the **status** of the **African elephant**?

From 1979 to 1989, Africa lost half of its elephants from poaching and illegal ivory trade, with the population decreasing from an estimated 1.3 million to 600,000. This led to the transfer of the African elephant from threatened to endangered status in October 1989 by the Convention on International Trade in Endangered Species (CITES). An ivory ban took effect on January 18, 1990. Botswana, Namibia, and Zimbabwe have agreed to restrict the sale of ivory to a single, government-controlled center in each country. All countries have further pledged to allow independent monitoring of the sale, packing, and shipping processes to ensure compliance with all conditions. Finally, all three countries have promised that all net revenues from the sale of ivory will be directed back into elephant conservation for use in monitoring,

341

Are tigers in danger of becoming extinct?

Tigers are listed as endangered by the U.S. Fish and Wildlife Service and the World Conservation Union (IUCN) and are included in the Convention on International Trade in Endangered Species of Wild Fauna and Flora. They are found in isolated regions of India, Bangladesh, Nepal, Bhutan, Southeast Asia, Manchuria, China, Korea, Russia, and Indonesia. Four subspecies of tiger—Balinese tiger (*Panthera tigris balica*), South China tiger (*Panthera tigris amoyensis*), Caspian tiger (*Panthera tigris virgata*), and Javan tiger (*Panthera tigris sondiaca*) have become extinct due to habitat loss, poaching, and overhunting.

The Siberian tiger, also known as the Amur tiger (*Panthera tigris altaica*) is one subspecies that has made a comeback in recent years. Its total worldwide population had dropped to 20 or 30 individuals in 1940, but there are over 500 in the wild today. Since mid-1990s there has been an increase in the poaching of the Bengal tiger (*Panthera tigris tigris*) since the bones of this subspecies are a valuable commodity on the black market. Tiger bones are used in Chinese medicines.

research, law enforcement, other management expenses, or community-based conservation programs within elephant range.

Current population estimates are 470,000 to 690,000 elephants throughout Africa. A newer concern is the reduction of their natural habitat. The human populations are expanding to areas of elephant habitat. New land areas, that were once elephant habitat areas, are now being used for agriculture.

Are **turtles endangered**?

Worldwide turtle populations have declined due to several reasons, including habitat destruction; exploitation of species by humans for their eggs, leather, and meat; and their becoming accidentally caught in the nets of fishermen. In particular danger are sea turtles, such as Kemp's ridley sea turtle (*Lepidochelys kempii*), which is believed to have a population of only a few hundred. Other species include the Central American river turtle (*Dermatemys mawii*), the green sea turtle (*Chelonia mydas*), and the leatherback sea turtle (*Dermochelys coriacea*). Endangered tortoises include the angulated tortoise (*Geochelone yniphora*), the desert tortoise (*Gopherus agassizii*), and the Galapagos tortoise (*Geochelone elephantopus*).

What is a **dolphin-safe tuna**?

The order *Cetacea*, composed of whales, dolphins, and porpoises, were spared from the extinction of large mammals at the end of the Pleistocene about 10,000 years ago.

What is the **current population** and status of the **great whales**?

Species	Original Population	Current Population	Status
Sperm (*Physeter macro cephalas*)	2,400,000	200,000 (est.)	Endangered
Blue (*Balenoptera musculus*)	226,000	2,300	Endangered
Finback (*Balenoptera physalus*)	543,000	30,000	Endangered
Humpback (*Megaptera novaeangliae*)	146,000	28,000	Least concern (lower risk)
Northern Atlantic Right (*Eubalaena glacialis*)	120,000	300–350	Endangered
Southern Right (*Eubalaena australis*)	70,000	7,000	Conservation dependent
Sei (*Balaenoptera borealis*)	254,000	24,000 in Southern Hemisphere 4,600 in the North Atlantic 22,000–37,000 in the Pacific	Endangered
Gray (*Eschrichtius robustus*)	20,000	26,000 but only 100 in the Western North Pacific	Endangered in western North Pacific; recovered in eastern North Paciric
Bowhead (*Balaena mysticetus*)	20,000	10,000	Endangered but dependent on location
Bryde's (*Balaenoptera edeni*)	92,000	40,000–80,000	Insufficient data
Minke (*Balaenoptera acutorostrata*)	295,000	610,000–1,284,000	Lower risk

The twentieth century, with its many technological improvements, was the most destructive period for the *Cetacea*. In 1972, the U.S. Congress passed the Marine Mammal Protection Act; one of its goals was to reduce the number of small cetaceans (notably Stenalla and Delphinus) killed and injured during commercial fishing operations, such as the incidental catch of dolphins in tuna purse-seines (nets that close up to form a huge ball to be hoisted aboard a ship).

Dolphins are often found swimming with schools of yellowfin tuna and are caught along with the tuna by fisherman who use purse-seine nets. The dolphins are drowned by this fishing method because they must be able to breathe air to survive. The Earth Island Institute established five standards in order for tuna to be considered "dolphin safe". They are:

- No intentional chasing, netting, or encirclement of dolphins during an entire tuna fishing trip
- No use of drift gill nets to catch tuna
- No accidental killing or serious injury to any dolphins during net sets
- No mixing of dolphin-safe and dolphin-deadly tuna in individual boat wells (for accidental kill of dolphins), or in processing or storage facilities
- Each trip in the Eastern Tropical Pacific Ocean (ETP) by vessels 400 gross tons and above must have an independent observer onboard attesting to the compliance with points (1) through (4) above

More than 90 percent of the world's tuna companies adhere to these standards.

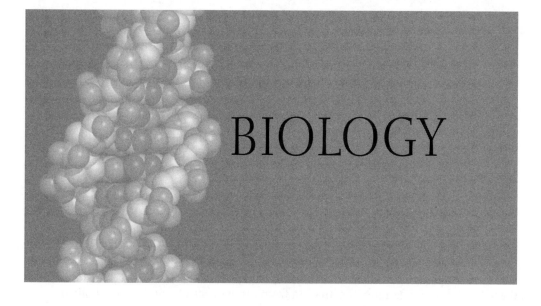

BIOLOGY

CELLS

What is the **cell theory**?

The cell theory is the concept that all living things are made up of essential units called "cells." The cell is the simplest collection of matter that can live. There are diverse forms of life existing as single-celled organisms. More complex organisms, including plants and animals, are multicellular—cooperatives of many kinds of specialized cells that could not survive for long on their own. All cells come from preexisting cells and are related by division to earlier cells that have been modified in various ways during the long evolutionary history of life on Earth. Everything an organism does occurs fundamentally at the cellular level.

Which **scientists** made **important discoveries** associated with the **cell**?

In the late 1600s, Robert Hooke (1635–1703) was the first to see a cell, initially in a section of cork, and then in bones and plants. In 1824, Henri Dutrochet (1776–1847) proposed that animals and plants had similar cell structures. Robert Brown (1773–1858) discovered the cell nucleus in 1831, and Matthias Schleiden (1804–1881) named the nucleolus (the structure within the nucleus now known to be involved in the production of ribosomes) around that time. Schleiden and Theodor Schwann (1810–1882) described a general cell theory in 1839, the former stating that cells were the basic unit of plants and Schwann extending the idea to animals. Robert Remak (1815–1865) was the first to describe cell division in 1855. Chromosomes were named and observed in the nucleus of a cell in 1888 by Wilhelm von Waldeyen-Hartz (1836–1921). Walther Flemming (1843–1905) was the first individual to follow chromosomes through the entire process of cell division.

345

What is the **origin** of the **term "cell"**?

The term "cell" was first used by Robert Hooke, an English scientist who described cells he observed in a slice of cork in 1665. Using a microscope that magnified 30 times, Hooke identified little chambers or compartments in the cork that he called *cellulae*, a Latin term meaning "little rooms" because they reminded him of the cells inhabited by monks. It is from this word that we got the modern term "cell." He calculated that one square inch of cork would contain 1,259,712,000 of these tiny chambers or cells!

What are the **differences** between **prokaryotic** and **eukaryotic** cells?

Eukaryotic cells are much more complex than prokaryotic cells, having compartmentalized interiors and membrane-bound organelles within their cytoplasm. The major feature of a eukaryotic cell is its membrane-bound nucleus, which compartmentalizes the activities of the genetic information from other types of cellular metabolism.

Comparison of Prokaryotic and Eukaryotic Cells

Characteristic	Prokaryotic Cells	Eukaryotic Cells
Organisms	Eubacteria and Archaebacteria	Protista, Fungi, Plants, Animals
Cell size	Usually 1–10μ across	Usually 10–100μ across
Membrane-bound organelles	No	Yes
Ribosomes	Yes	Yes
Mode of cell division	Cell fission	Mitosis and meiosis
DNA location	Nucleoid	Nucleus
Membranes	Some	Many
Cytoskeleton	No	Yes

What groups of **organisms** have **prokaryotic cells** and which ones have **eukaryotic cells**?

All living organisms are grouped into three large groups called domains. They are: Bacteria, Archae, and Eukarya. The domains Bacteria (eubacteria or "true" bacteria) and Archae (archaebacteria or "ancient" bacteria) consist of unicellular organisms with prokaryotic cells. The domain Eukarya consists of four kingdoms: Protista, Fungi, Plantae, and Animalia. Organisms in these groups have eukaryotic cells. Eukaryon means "true nucleus."

What are **organelles**?

Organelles—frequently called "little organs"— are found in all eukaryotic cells; they are specialized, membrane bound, cellular structures that perform a specific function.

> ## What are the largest and smallest organelles in a cell?
>
> The largest organelle in a cell is the nucleus. The next largest organelle would be the chloroplast, which is substantially bigger than a mitochondrion. The smallest organelle in a cell is the ribosome.

Eukaryotic cells contain several kinds of organelles, including the nucleus, mitochondria, chloroplasts, endoplasmic reticulum, and Golgi apparatus.

What **cell structures** are **unique** to **plant** cells and which ones are unique to **animal** cells?

The chloroplast, central vacuole, tonoplast, cell wall, and plasmodesmata are only found to occur in plant cells. Lysosomes and centrioles are found only in animal cells.

What are the major **components** of the **eukaryotic cell**?

Structure	Description
Cell Nucleus	
Nucleus	Large structure surrounded by double membrane
Nucleolus	Special body within nucleus; consists of RNA and protein
Chromosomes	Composed of a complex of DNA and protein known as chromatin; resemble rod-like structures after cell division
Cytoplasmic Organelles	
Plasma membrane	Membrane boundary of living cell
Endoplasmic reticulum (ER)	Network of internal membranes extending through cytoplasm
Smooth endoplasmic reticulum	Lacks ribosomes on the outer surface
Rough endoplasmic reticulum	Ribosomes stud outer surface
Ribosomes	Granules composed of RNA and protein; some attached to ER and some are free in cytosol
Golgi complex	Stacks of flattened membrane sacs
Lysosomes	Membranous sacs (in animals)
Vacuoles	Membranous sacs (mostly in plants, fungi, and algae)
Microbodies (e.g., peroxisomes)	Membranous sacs containing a variety of enzymes

347

Structure	Description
Mitochondria	Sacs consisting of two membranes; inner membrane is folded to form cristae and encloses matrix
Plastids (e.g., chloroplasts)	Double membrane structure enclosing internal thylakoid membranes; chloroplasts contain chlorophyll in thylakoid membranes
The Cytoskeleton	
Microtubules	Hollow tubes made of subunits of tubulin protein
Microfilaments	Solid, rod-like structures consisting of actin protein
Centrioles	Pair of hollow cylinders located near center of cell; each centriole consists of nine microtubule triplets (9 × 3 structure)
Cilia	Relatively short projections extending from surface of cell; covered by plasma membrane; made of two central and nine peripheral microtubules (9 + 2 structure)
Flagella	Long projections made of two central and nine peripheral microtubules (9 + 2 structure); extend from surface of cell; covered by plasma membrane

How do the **cells** of bacteria, plants, and animals **compare** to each other?

	Bacterium	Plant (Eukaryote)	Animal (Eukaryote)
Cell wall	Present (protein polysaccharide)	Present (cellulose)	Absent
Plasma membrane	Present	Present	Present
Flagella and cilia	May be present	Absent except in sperm of a few species	Frequently present
Endoplasmic reticulum	Absent	Usually present	Usually present
Ribosome	Present	Present	Present
Microtubule	Absent	Present	Present
Centriole	Absent	Absent	Present
Golgi apparatus	Absent	Absent	Present
Cytoskeleton	Absent	Present	Present
Nucleus	Absent	Present	Present

	Bacterium	Plant (Eukaryote)	Animal (Eukaryote)
Mitochondrion	Absent	Present	Present
Chloroplast	Absent	Present	Absent
Nucleolus	Absent	Present	Present
Chromosome	A single circle of naked DNA	Multiple; DNA-protein complex	Multiple; DNA-protein complex
Microbody	Absent	Present	Present
Lysosome	Absent	Absent	Present
Vacuole	Absent	Usually a large single vacuole	Absent

What is the **major function** of the **nucleus**?

The nucleus is the information center for the cell and the storehouse of the genetic information (DNA) that directs all of the activities of a living eukaryotic cell. It is usually the largest organelle in a eukaryotic cell and contains the chromosomes.

When was the **nucleus first described**?

The Scottish botanist Robert Brown (1773–1858) first named and described the nucleus in 1831, while studying orchids. Brown called this structure the nucleus, from the Latin nucula, meaning "little nut" or "kernel."

Do **all cells** have a **nucleus**?

Prokaryotic cells do not have an organized nucleus. Most eukaryotic cells have a single organized nucleus. Red blood cells are the only mammalian cells that do not have a nucleus.

How thick is the **plasma membrane**?

The plasma membrane is only about 8 nanometers thick. It would take over 8,000 plasma membranes to equal the thickness of an average piece of paper.

What are the main **components** of the **plasma membrane**?

Component	Function
Cell surface markers	Self-recognition; tissue recognition
Interior protein network	Determines shape of cell
Phospholipid molecules	Provide permeability barrier, matrix for proteins
Transmembrane proteins	Transport molecules across membrane and against gradient

What are the main **functions** of the **plasma membrane**?

The main purpose of the plasma membrane is to provide a barrier that keeps cellular components inside the cell while simultaneously keeping unwanted substances from entering the cell. The membrane allows essential nutrients to be transported into the cell and aids in the removal of waste products from the cell. The specific functions of a membrane depend on the kinds of phospholipids and proteins present in the plasma membrane.

What are **stem cells** and what are some potential uses of such cells?

Stem cells are undifferentiated cells—meaning that they do not have a specific function—that are capable, under certain conditions, of producing cells that can become a specific type of tissue. Stem cells could be used to grow new hearts that could be transplanted without fear of rejection. They could be used to renew the function of injured structures like the spinal cord. They could be used as cell models for drug testing, thereby increasing the speed for finding cures. The potential benefits of stem cells have made the laboratory studies and research investigating these cells very exciting, yet controversial.

What is **mitosis** and what are the **stages** of this process?

Mitosis involves the replication of DNA and its separation into two new daughter cells which are genetically identical to the parent cell. While only four phases of mitosis are often listed, the entire process is actually composed of six phases:

- Interphase: Involves extensive preparation for the division process
- Prophase: The condensation of chromosomes; the nuclear membrane disappears; formation of the spindle apparatus; chromosomes attach to spindle fibers
- Metaphase: Chromosomes, attached by spindle fibers, align along the midline of a cell
- Anaphase: The entromere splits and chromatids move apart
- Telophase: The nuclear membrane reforms around newly divided chromosomes
- Cytokinesis: The division of cytoplasm, cell membranes, and organelles occur. In plants, a new cell wall forms

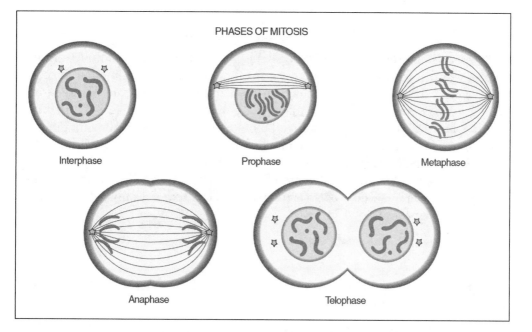

PHASES OF MITOSIS

Interphase

Prophase

Metaphase

Anaphase

Telophase

The stages of mitosis.

What is **photosynthesis** and why is it important?

Photosynthesis (from the Greek word *photo*, meaning "light," and *synthesis*, from the Greek work *syntithenai*, which means "to put together") is the process by which plants use energy derived from light in order to make food molecules from carbon dioxide and water. Oxygen (O_2) is produced as a waste product of this process. Photosynthesis is a dual-staged process with multiple components. Ultimately, photosynthesis is the process that provides food for the entire world. Each year more than 250 billion metric tons of sugar are created through photosynthesis. Photosynthesis is a source of food not only for plants, but also all organisms that are not capable of internally producing their own food, including humans.

Which **scientists** made significant **discoveries** toward the understanding of **photosynthesis**?

The ancient Greeks and Romans believed that plants derived their food from the soil. The earliest experiment to test this hypothesis was performed by the Belgian scientist Jan Baptista van Helmont (1577–1644), who grew a willow tree in a container of soil and fed it only water. At the end of five years, the weight of the willow tree had increased by 164 pounds (74.4 kilograms), while the weight of the soil had decreased by 2 ounces (57 grams). Van Helmont concluded the plant had received all its nourish-

Can artificial cells be made?

Research in progress at the National Aeronautical and Space Administration (NASA) is focused on cells as a means to deliver medicine in outer space; these cells are able to withstand dehydration and thus can be safely stored for long periods. Artificial cells are made of a polymer that acts like a cell membrane, but the polymer is stronger and more manageable than real membranes. These polymers are called polymersomes and can be made to cross-link with other polymers. Researchers think that many different kinds of molecules can be encapsulated within these polymersomes and then delivered to specific target organs. An example would be an artificial blood cell that not only delivers oxygen but also medication.

ment from the water and none from the soil. Joseph Priestley (1733–1804) demonstrated that air was "restored" by plants.

In 1771, Priestly conducted an experiment in which he placed a lighted candle in a glass container and allowed it to burn until extinguished by lack of oxygen. He then put a plant into the same chamber and allowed it to grow for a month. Repeating the candle experiment a month later, he found that the candle would now burn. Priestley's experiments showed that plants release oxygen (O_2) and take in carbon dioxide (CO_2) produced by combustion. The Dutch physician Jan Ingenhousz (1730–1799) confirmed Priestley's ideas, emphasizing that air is "restored" only by green plants in the presence of sunlight.

Evidence of photosynthesis's two-stage process was first presented by F.F. Blackman (1866–1947) in 1905. Blackman had identified that both a light-dependent stage and a light-independent stage occur during photosynthesis. In 1930 C.B. van Niel (1923–1977) became the first person to propose that water, rather than carbon dioxide, was the source of the oxygen that resulted from photosynthesis. In 1937 Robert Hill (1899–1991) discovered that chloroplasts are capable of producing oxygen in the absence of carbon dioxide only when the chloroplasts are illuminated and provided with an artificial electron acceptor.

How do **cells communicate** with each other?

Cells communicate with each other via small, signaling molecules that are produced by specific cells and received by target cells. This communication system operates on both a local and long-distance level. The signaling molecules can be proteins, fatty acid derivatives, or gases. Nitric oxide is an example of a gas that is part of a locally based signaling system and is able to signal for a human being's blood pressure to be lowered. Hormones are long-distance signaling molecules that must be transported via the circulatory system from their production site to their target cells. Plant cells, because of their rigid cell walls, have cytoplasmic bridges called plasmodesmata that allow cell-to-cell communication. Animals use gap junctions to transfer material between adjacent cells.

What is **cell cloning** and how is it used in scientific research?

Cell cloning is the process by which an exact copy is made of a cell. This cellular process is known as mitosis and is required for the growth and repair of multicellular organisms. Different types of cells in the body differ in their ability to perform mitosis. Some cells, like skin cells, produce clones quite often. Others, like those of the nervous system, will not reproduce after they have reached maturity and have differentiated. The scientific purpose of cloning is to produce many copies of certain types of cells, which can then be used for a variety of purposes, like basic research or the growth of replacement organs.

Can **cells** ever **change functions**?

The more specialized function that a cell performs, the less likely it is for the cell to change jobs within an organism. However, there are some cells that have unspecialized functions and are able to adapt to the changing needs of the body. In mammals, a good example of cells with adapting functions would be bone marrow cells, which are responsible for producing different types of cells in the blood. Bone marrow cells produce red blood cells and five types of white blood cells. Slime molds of the kingdom Protista have cells that are capable of drastically changing their function. The cellular adaptations that can occur in slime molds allow them to change from single-celled amoebas to multicellular, reproductive spore producers.

What are the **oldest** living **cultured human cells**?

The oldest, living, cultured human cells are the HeLa cell line. All HeLa cells are derived from Henrietta Lacks (1920–1951), a 31-year-old woman from Baltimore, Maryland, who died of cervical cancer in 1951. Epithelial tissue obtained by biopsy became the first continuously cultured human malignant cells. From this culture, scientists were able to discover that 80 to 90 percent of cervical carcinomas contain human papillomavirus DNA. HeLa cells are used in many biomedical experiments.

Why do **cells die**?

Cells die for a variety of reasons, many of which are not deliberate. For example, cells can starve to death, asphyxiate, or die from trauma. Cells that sustain some sort of damage, such as DNA alteration or viral infection, frequently undergo programmed

353

cell death. This process eliminates cells with a potentially lethal mutation or limits the spread of the virus. Programmed cell death can also be a normal part of embryonic development. Frogs undergo cell death that results in the elimination of tissues, allowing a tadpole to morph into an adult frog.

VIRUSES, BACTERIA, PROTISTS, AND FUNGI

What is a **virus**?

A virus is an infectious, protein-coated fragment of DNA or RNA. Viruses replicate by invading host cells and taking over the cell's "machinery" for DNA replication. Viral particles can then break out of the cells, spreading disease.

What is the **average size** of a **virus**?

The smallest viruses are about 17 nanometers in diameter, and the largest viruses are up to 1,000 nanometers (1 micrometer) in length. By comparison, the bacterium *Escherichia coli* is 2,000 nanometers in length, a cell nucleus is 2,800 nanometers in diameter, and an average eukaryotic cell is 10,000 nanometers in length.

Average Size of Common Viruses

Virus	Size (in nanometers)
Smallpox	250
Tobacco mosaic	240
Rabies	150
Influenza	100
Bacteriophage	95
Common cold	70
Polio	27
Parvovirus	20

Where are **viruses found**?

Viruses lie dormant in any environment (land, soil, air) and on any material. They infect every type of cell—plant, animal, bacterial, and fungal.

What was the **first virus** to be **isolated** in a **laboratory**?

In 1935, Wendell Stanley (1904–1971) of the Rockefeller Institute (known today as Rockefeller University) prepared an extract of the tobacco mosaic virus and purified it.

> ## Are viruses living organisms?
>
> **V**iruses cannot grow or replicate on their own and are inert outside their living host cell. Once they enter a host cell they become active. As such, they are between life and nonlife and are not considered living organisms.

The purified virus precipitated in the form of crystals. During this investigation Stanley was able to demonstrate that viruses can be regarded as chemical matter rather than as living organisms. The purified crystals retained the ability to infect healthy tobacco plants, thus characterizing them as viruses, not merely chemical compounds derived from a virus. Subsequent studies showed that the tobacco mosaic virus consisted of a protein and a nucleic acid. Further studies showed that this virus consisted of RNA (ribonucleic acid) surrounded by a protein coat. Stanley was awarded the Nobel Prize in Chemistry in 1946 for his discovery.

Do **viruses** contain both **DNA** and **RNA**?

Viruses have either DNA or RNA as their genomic material, whereas cells—including bacteria—have both.

What is the **difference** between a **virus** and a **retrovirus**?

A virus is a rudimentary biosystem that has some of the aspects of a living system such as having a genome (genetic code) and the ability to adapt to its environment. A virus, however, cannot acquire and store energy and is therefore not functional outside of its hosts. Viruses and retroviruses infect cells by attaching themselves to the host cell and either entering themselves or injecting their genetic material into the cell and then reproducing its genetic material within the host cell. The reproduced virus then is released to find and attack more host cells. The difference between a virus and retrovirus is a function of how each replicates its genetic material. A virus has a single strand of genetic material—either DNA or RNA. A retrovirus consists of a single strand of RNA. Once a retrovirus enters a cell, it collects nucleotides and assembles itself as a double strand of DNA that splices itself into the host's genetic material. Retroviruses were first identified by David Baltimore (1938–) and Howard Temin (1934–). They were awarded the Nobel Prize in Physiology or Medicine in 1975 for their discovery.

What was the **first retrovirus discovered**?

Dr. Robert Gallo (1937–) discovered the first retrovirus, human T cell lymphoma virus (HTLV), in 1979. The second human retrovirus to be discovered was human immunodeficiency virus, HIV.

355

What is a **bacteriophage**?

A bacteriophage, also called a phage, is a virus that infects bacteria. The term "bacteriophage" means "bacteria eater" (from the Greek word *phagein*, which means "to devour"). Phages consist of a long nucleic acid molecule (usually DNA) coiled within a polyhedral-shaped protein head. Many phages have a tail attached to the head. Fibers extending from the tail may be used to attach the virus to the bacterium.

What is the **difference** between a **virus** and a **viroid**?

Viroids are small fragments of nucleic acid (RNA) without a protein coat. They are usually associated with plant diseases and are several thousand times smaller than a virus.

Who first used the **term "prion"**?

Prions are abnormal forms of natural proteins. Stanley Prusiner (1942–) first used the term "prion" in 1982 in place of the expression "proteinaceous infectious particle" when describing an infectious agent. Prusiner was awarded the Nobel Prize in Physiology or Medicine in 1997. Current research indicates that a prion is composed of about 250 amino acids. Despite extensive and continuing investigations, no nucleic acid component has been found. Like viruses, prions are infectious agents.

How do **prions work** and what diseases have been linked to prions?

Scientists have not discovered exactly how prions work. Current research shows that prions accumulate in lysosomes. In the brain, it is possible that the filled lysosomes burst and damage cells. As diseased cells die, the prions contained in the cells are released and are able to attack other cells. It is thought that prions are responsible for the group of brain diseases known as transmissible spongiform encephalopathies (TSEs). This group includes the disease that is referred to as bovine spongiform encephalopathy (mad cow disease) when it occurs in cattle and Creutzfeldt-Jakob disease when it occurs in humans.

What is the **difference** between a **plasmid** and a **prion**?

A plasmid is a small, circular, self-replicating DNA molecule separate from the bacterial chromosome. Plasmids do not normally exist outside the cell and are generally beneficial to the bacterial cell. Plasmids are often used to pick up foreign DNA for use in genetic engineering. A prion is an infectious form of a protein or malformed protein that may increase in number by converting related proteins to more prions. Prions may cause a number of degenerative brain diseases such as "mad cow disease" or Creutzfeldt-Jakob disease in humans.

When were **bacteria discovered**?

Anton von Leeuwenhoek (1632–1723), a Dutch fabric merchant and civil servant, discovered bacteria and other microorganisms in 1674 when he looked at a drop of pond water through a glass lens. Early, single-lens instruments produced magnifications of 50–300 times real size (approximately one-third of the magnification produced by modern light microscopes). Primitive microscopes provided a perspective into the previously unknown world of small organisms, which von Leeuwenhoek called "animalcules" in a letter he wrote to the Royal Society of London. Because of these early investigations, von Leeuwenhoek is considered to be the "father of microbiology."

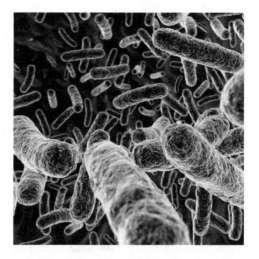

Bacteria were first discovered in 1674 by Anton von Leeuwenhoek. Since then, many species of bacteria have been found that cause a variety of diseases.

How many **groups** are **identified** in the domain **Bacteria**?

Biologists recognize at least a dozen different groups of bacteria.

Major Group	Gram Reaction	Characteristics	Examples
Actinomycetes	Positive	Produce spores and antibiotics; live in soil environment	*Streptomyces*
Chemoautotrophs	Negative	Live in soil environment; important in the nitrogen cycle	*Nitrosomonas*
Cyanobacteria	Negative	Contain chlorophyll and are capable of photosynthesis; live in aquatic environment	*Anabaena*
Enterobacteria	Negative	Live in intestinal and respiratory tracts; ability to decompose materials; do not form spores; pathogenic	*Escherichia, Salmonella, Vibrio*
Gram-positive cocci	Positive	Live in soil environment; inhabit the skin and mucous membranes of animals; pathogenic to humans	*Streptococcus, Staphylococcus*

357

Major Group	Gram Reaction	Characteristics	Examples
Gram-positive rods	Positive	Live in soil environment, or animal intestinal tracts; anaerobic; disease-causing	*Clostridia, Bacillus*
Lactic acid bacteria	Positive	Important in food production, especially dairy products; pathogenic to animals	*Lactobacillus, Listeria*
Myxobacteria	Negative	Move by secreting slime and gliding; ability to decompose materials	*Chondromyces*
Pseudomonads	Negative	Aerobic rods and cocci; live in soil environment	*Pseudomonas*
Rickettsias and chlamydias	Negative	Very small, intracellular parasites; pathogenic to humans	*Rickettsia, Chlamydia*
Spirochetes	Negative	Spiral-shaped; live in aquatic environment	*Treponema, Borrelia*

What are **Archaebacteria**?

Archaebacteria (domain Archaea) are primitive bacteria that often live in extreme environments. This domain includes the following: 1) Thermophiles ("heat lovers"), which live in very hot environments, including the hot sulfur springs of Yellowstone National Park, which reach temperatures ranging from 140–176°F (60–80°C). 2) Halophiles ("salt lovers"), which live in locations with high concentrations of salinity, such as the Great Salt Lake in Utah, which has salinity levels that range from 15–20 percent. Seawater normally has a level of salinity of 3 percent. 3) Methanogens, which obtain their energy by using hydrogen gas (H_2) to reduce carbon dioxide (CO_2) to methane gas (CH_4).

What is the **most abundant group** of **organisms**?

The eubacteria are the most abundant group of organisms on Earth. More living eubacteria inhabit the human mouth than the total number of mammals living on Earth.

How did the **discovery** of **bacteria impact** the **theory of spontaneous generation**?

The theory of spontaneous generation proposes that life can arise spontaneously from nonliving matter. One of the first scientists to challenge the theory of spontaneous generation was the Italian physician Francesco Redi (1626–1698). Redi performed an experiment to show that meat placed in covered containers (either glass-covered or

Which criteria were used to form what is referred to as the classical approach of bacterial taxonomy?

In this approach bacteria are grouped into genera and species on the basis of the following characteristics: 1) Structural and morphological characteristics including shape, size, arrangement, capsules, flagella, endospores, and Gram stain; 2) Biochemical and physiological traits such as optimum temperature and pH ranges for growth, oxygen requirements, growth factor requirements, respiration and the fermentation of end products, antibiotic sensitivities, and types of carbohydrates used as an energy sources.

gauze-covered) remained free of maggots, while meat left in an uncovered container eventually became infested with maggots from flies laying their eggs on the meat. After the discovery of microorganisms by Anton von Leeuwenhoek, the controversy surrounding spontaneous generation was renewed, as it had been assumed that food became spoiled by organisms arising spontaneously within food. In 1776 Lazzaro Spallanzani (1729–1799) showed that no growth occurred in flasks that were boiled after sealing. The controversy over the theory of spontaneous generation was finally resolved in 1861 by Louis Pasteur (1822–1895). He showed that the microorganisms found in spoiled food were similar to those found in the air. He concluded that the microorganisms that caused food to spoil were from the air and did not spontaneously arise.

Who were the **founders** of **modern bacteriology**?

German bacteriologist Robert Koch (1843–1910) and French chemist Louis Pasteur are considered the founders of bacteriology. In 1864 Pasteur devised a way to slowly heat foods and beverages to a temperature that was high enough to kill most of the microorganisms that would cause spoilage and disease, but would not ruin or curdle the food. This process is called pasteurization.

By demonstrating that tuberculosis was an infectious disease caused by a specific species of *Bacillus*, Koch in 1882 set the groundwork for public-health measures that would go on to significantly reduce the occurrences of many diseases. His laboratory procedures, methodologies for isolating microorganisms, and four postulates for determining agents of disease gave medical investigators valuable insights into the control of bacterial infections.

What period of time has come to be known as the **golden age of microbiology**?

The era known as the "golden age" of microbiology began in 1857, with the work of Louis Pasteur and Robert Koch, and lasted about 60 years. During this period of time, there were many important scientific discoveries. Joseph Lister's (1827–1912) practice of treating surgical wounds with a phenol solution led to the advent of aseptic surgery. The advancements Paul Ehrlich (1854–1915) made to the theory of immunity synthesized the "magic bullet", an arsenic compound that proved effective in treating syphilis in humans.

In 1884 Elie Metchnikoff (1845–1916), an associate of Pasteur, published a report on phagocytosis. The report explained the defensive process in which the body's white blood cells engulf and destroy microorganisms. In 1897 Masaki Ogata reported that rat fleas transmitted bubonic plague, ending the centuries-old mystery of how plague was transmitted. The following year, Kiyoshi Shiga (1871–1957) isolated the bacterium responsible for bacterial dysentery. This organism was eventually named *Shigella dysenteriae*.

During the "golden age" of microbiology, researchers identified the specific microorganism responsible for numerous infectious diseases. The following chart identifies many of these diseases, their infectious agent, who discovered them, and the year they were discovered.

Diseases and Their Discoverers

Disease	Infectious Agent	Discoverer	Year Discovered
Anthrax	*Bacillus anthracis*	Robert Koch	1876
Gonorrhea	*Neisseria gonorrhoeae*	Albert L. S. Neisser	1879
Malaria	*Plasmodium malariae*	Charles-Louis Alphonse Laveran	1880
Wound infections	*Staphylococcus aureus*	Sir Alexander Ogston	1881
Erysipelas	*Streptococcus pyagenes*	Friedrich Fehleisen	1882
Tuberculosis	*Mycabacterium tuberculosis*	Robert Koch	1882
Cholera	*Vibrio cholerae*	Robert Koch	1883
Diphtheria	*Carynebacterium diphtheriae*	Edwin Klebs and Friederich Löeffler	1883–1884
Typhoid fever	*Salmonella typhi*	Karl Eberth and Georg Gaffky	1884
Bladder infections	*Escherichia coli*	Theodor Escherich	1885
Salmonellosis	*Salmonella enteritidis*	August Gaertner	1888
Tetanus	*Clostridium tetani*	Shibasaburo Kitasato	1889
Gas gangrene	*Clostridium perfringens*	William Henry Welch and George Henry Falkiner Nuttall	1892
Plague	*Yersinia pestis*	Alexandre Yersin and Shibasaburo Kitasato	1894

Disease	Infectious Agent	Discoverer	Year Discovered
Botulism	*Clostridium botulinum*	Emit Van Ermengem	1897
Shigellosis	*Shigella dysenteriae*	Kiyoshi Shiga	1898
Syphilis	*Treponema pallidum*	Fritz R. Schaudinn and P. Erich Hoffman	1905
Whooping cough	*Bordetella pertussis*	Jules Bordet and Octave Gengou	1906

What are the main **components** of a **bacterial cell**?

The major components of a bacterial cell are the plasma membrane, cell wall, and a nuclear region containing a single, circular DNA molecule. Plasmids—small circular pieces of DNA that exist independently of the bacterial chromosome—are also present in a bacterial cell. In addition, some bacteria may have flagella, which aids in movement; pili or fimbriae, which are short, hairlike appendages that help bacteria adhere to various surfaces, including the cells that they infect; or a capsule of slime around the cell wall that protects it from other microorganisms.

How do **viruses compare** to **bacteria**?

Characteristic	Bacteria	Viruses
Able to pass through bacteriological filters	No	Yes
Contains a plasma membrane	Yes	No
Contains ribosomes	Yes	No
Possesses genetic material	No	Yes
Requires a living host to multiply	No	Yes
Sensitive to antibiotics	Yes	No
Sensitive to interferon	No	Yes

How many **genes** are in a **typical bacterial cell**?

The bacterium *Escherichia coli* has about 5,000 genes.

Do **bacteria** all have the **same shape**?

Bacteria have three main shapes—spherical, rod-shaped, and spiral. Spherical bacteria, known as cocci, occur singularly in some species and as groups in other species. Cocci have the ability to stick together and form a pair (diplococci); when they stick together in long chains, they are called streptococci. Irregularly shaped clumps or clusters of bacteria are called staphylococci. Rod-shaped bacteria, called bacilli, occur

361

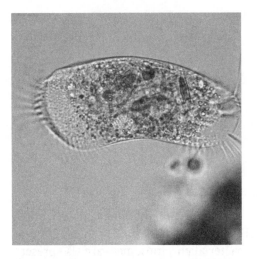

The flagella on this *Stylonychia* specimen can clearly be seen at the left and lower right. Flagella aid in locomotion of single-celled organisms like this one.

as single rods or as long chains of rods. Spiral- or helical-shaped bacteria are called spirilla.

How do **bacteria reproduce**?

Bacteria reproduce asexually, by binary fission—a process in which one cell divides into two similar cells. First the circular, bacterial DNA replicates, and then a transverse wall is formed by an ingrowth of both plasma membrane and the cell wall.

How **quickly** do **bacteria reproduce**?

Bacteria can reproduce very rapidly in favorable environments in both laboratory cultures and natural habitats. The time required for a bacterial population to double is called generation time. For example, under optimal conditions, *Escherichia coli* can divide every 17 minutes. A laboratory culture started with a single cell can produce a colony of 10^7 to 10^8 bacteria in about 12 hours.

Generation Time for Selected Bacteria

Bacterium	Generation time (in minutes)
Escherichia coli	17
Salmonella typhimurium	24
Staphylococcus aureus	32
Clostridium botulinum	35
Streptococcus lactis	48
Lactobacillus acidophilus	66

Where are **bacteria found**?

Bacteria inhabit every place on Earth—including places where no other organism can survive. Bacteria have been detected as high as 20 miles (32 kilometers) above the earth and 7 miles (11 kilometers) deep in the waters of the Pacific Ocean. They are found in extreme environments, such as the Arctic tundra, boiling hot springs, and our bodies. Heat-tolerant bacteria have been found at a gold mine in South Africa at a level of 2.17 miles (3.5 kilometers) below Earth's surface where the temperature in the mine was 149°F (65°C).

> ## Are any bacteria visible to the naked eye?
>
> *E* *pulopiscium fishelsoni*, which lives in the gut of the brown surgeonfish (*Acanthurus nigrofuscus*), is visible to naked eye. It was first identified in 1985 and mistakenly classified as a protozoan. Later studies analyzed the organism's genetic material and proved it to be a bacterium of unprecedented size: 0.015 inches (0.38 millimeters) in diameter, or about the size of a period in a small-print book.

What are **Koch's postulates**?

Robert Koch was the first to identify that various microorganisms are the cause of disease. His four basic criteria of bacteriology, known as Koch's postulates, are still considered fundamental principles of bacteriology. The characteristics are as follows: 1) The organism must be found in tissues of animals that have been infected with the disease, rather than in disease-free animals. 2) The organism must be isolated from the diseased animal and grown in a pure culture or *in vitro*. 3) The cultured organism must be able to be transferred to a healthy animal, which will show signs of the disease after having been exposed to the organism. 4) The organism must be able to be isolated from the infected animal. Koch was awarded the Nobel Prize in Physiology or Medicine in 1905 for his research on tuberculosis.

How **dangerous** is *Clostridium botulinum*?

The bacterium *Clostridium botulinum* can grow in food products and produce a toxin called botulinum, the most toxic substance known. Microbiologists estimate that one gram of this toxin can kill 14 million adults! This bacterium can withstand boiling water (212°F or [100°C]) but is killed in five minutes in 248°F (120°C). This tolerance makes *Clostridium botulinum* a serious concern for people who can vegetables at home. If home canning is not done properly, this bacterium will grow in the anaerobic conditions of the sealed container and create extremely poisonous food. The endospores of *Clostridium botulinum* can germinate in poorly prepared canned goods, so individuals should never eat food from a can that appears swollen, as it is a sign that the can has become filled with gas released during germination. Consuming food from a can containing endospores that have undergone germination can lead to nerve paralysis, severe vomiting, and even death.

How are **bacteria cultured**?

Bacteria are usually cultured in Petri dishes that contain a culture medium, usually nutrient agar. Petri dishes were developed in 1887 by Julius Richard Petri (1852–1921), a member of Robert Koch's laboratory. The top of the dish is larger than the bottom so that when the dish is closed a strong seal is created, preventing contamination of the culture. Agar was developed as a culture media for bacteria by Robert Koch. Koch was interested in the isolation of bacteria in pure culture. Because isolation was difficult in liquid media, he began to study ways in which bacteria could be grown on solid media. After sterile, boiled potatoes proved unsatisfactory, a better alternative was suggested by Fannie E. Hesse (1850–1934), the wife of Walther Hesse (1846–1911), who was one of Koch's assistants. She suggested that agar, which she had used to thicken sauces, jams, and jellies, be used to solidify liquid nutrient broth. Agar is generally inexpensive and, once jelled, does not melt until reaching a temperature of 212°F (100°C). If 1–2 grams of agar are added to 100 milliliter of nutrient broth, it produces a solid medium that is not degraded by most bacteria. Stacks of Petri dishes culturing bacteria are one of the most common items in a microbiology laboratory.

Who first proposed the kingdom Protista?

The German zoologist Ernst Haeckel (1834–1919) first proposed the kingdom Protista in 1866, for the newly discovered organisms that were neither plant nor animal. The term "protest" is derived from the Greek term *protistos*, meaning "the very first."

What are the characteristics of the protists?

Protists are a diverse group of organisms. All protists are eukaryotic. Many are unicellular, but they may be multicellular, multinucleate, or exhibit a colonial organization. Although most are microscopic, some are much larger, reaching lengths of nearly 200 feet (60 meters). In early, traditional taxonomic schemes, they were united on the basis of being neither plant nor animal nor fungus. Current evidence suggests that protists exhibit characteristics of the plant, animal, and fungi kingdoms.

What are the major groups of organisms in the kingdom Protista?

There is little agreement among taxonomists on how to classify the protists, but they may be conveniently divided into seven general groups that share certain characteristics of locomotion, nutrition, and reproduction. The following list exhibits the general groupings:

- Sarcodinas: Amoebas and related organisms that have no permanent locomotive structure
- Algae: Single-celled and multicellular organisms that are photosynthetic
- Diatoms: Photosynthetic organisms with hard shells formed of silica
- Flagellates: Organisms that propel themselves through water with flagella
- Sporozoans: Nonmotile parasites that spread by forming spores

> ## How did the protist *Phytophthora infestans* influence Irish history?
>
> *P*hytophthora infestans, one of the potato's most lethal pathogens, causes the late blight of potato disease. This pathogen was responsible for the Irish potato famine of 1845–1849. *P. infestans* causes the leaves and stem of the potato plant to decay, eventually causing the tuber to stop growing. In addition, the tubers are attacked by the pathogen and rot. It has been estimated that during the potato famine 1.5 million Irish people emigrated from their country and moved to various parts of the world, but most immigrated to the United States. An estimated 400,000 people perished during the famine due to malnutrition.

- Ciliates: Organisms that have many short, hairlike structures on their cell surface associated with locomotion
- Molds: Heterotrophs with restricted mobility that have cell walls composed of carbohydrates

Which **protist** is an **indicator** of **polluted water**?

Euglenoids are unicellular flagellates; many euglenoids are capable of photosynthesis and are autotrophic. They are commonly found in freshwater ponds and puddles. Others do not carry on photosynthesis and are heterotrophic, often found in water with large amounts of organic material. Euglenoids frequently serve as bioindicators and are found in large numbers in polluted waters.

What **slime mold** serves as a **model organism** in **developmental biology**?

Dictyostelium discoideum has been studied as a model for the developmental biology of complex organisms. Under optimum conditions, this organism lives as individual, amoeboid cells. When food is scarce, the cells stream together into a moving mass, resembling a slug that differentiates into a stalk with a spore-bearing body at its top. This structure releases spores that can grow into a new amoeboid cell. The development from identical, free-living cells to a multicellular organism simulates many of the properties of more complex and complicated organisms.

What **evidence** has led scientists to believe **land plants evolved** from **green algae**?

Many scientists believe that ancient green algae evolved into land plants. The chloroplasts present in green algae are the same as those of land plants. In addition, green algae have cell walls of similar composition to land plants; both store food, such as

365

People usually think of mushrooms like these when the word "fungi" is mentioned, but there are many types of fungi, including yeasts and molds.

starch, in the same manner. Most green algae live in freshwater habitats with highly variable conditions. The ongoing changes in their environment have made them highly adaptable.

What **characteristics** do all **fungi share?**

In the earliest classification systems, fungi were classified as plants. The first classification system to recognize fungi as a separate kingdom was proposed in 1784. Researchers identified four characteristics shared by all fungi: fungi lack chlorophyll; the cell walls of fungi contain the carbohydrate chitin (the same tough material a crab shell is made of); fungi are not truly multicellular since the cytoplasm of one fungal cell mingles with the cytoplasm of adjacent cells; and fungi are heterotrophic eukaryotes (unable to produce their own food from inorganic matter) while plants are autotrophic eukaryotes.

What **organisms** are included in the **kingdom Fungi?**

Members of the kingdom Fungi range from single-celled yeasts to *Armillaria ostoyea*, a species that covers 2,220 acres (890 hectares)! Also included are mushrooms that are commonly consumed, the black mold that forms on stale bread, the mildew that grows on damp shower curtains, rusts, smuts, puffballs, toadstools, shelf fungi, and the death cap mushroom, *Amanita phalloides*. Of the bewildering variety of organisms that live on the planet Earth, perhaps the most unusual and peculiarly different from human beings are fungi. Fungi are able to rot timber, attack living plants, spoil food, and afflict humans with athlete's foot and even worse maladies. Fungi also decompose dead organisms, fallen leaves, and other organic materials. In addition, they produce antibiotics and other drugs, make bread rise, and ferment beer and wine.

Where are **fungi found?**

Fungi grow best in dark, moist habitats, but they can be found wherever organic material is available. Moisture is necessary for their growth, and they can obtain water from the atmosphere as well as from the medium upon which they live. When the environment becomes very dry, fungi survive by going into a resting stage or by producing spores that are resistant to drying. The optimum pH for most species is 5.6, but some fungi can tolerate and grow in pH ranging from 2 to 9. Certain fungi can

Which children's author studied and drew illustrations of fungi?

Beatrix Potter (1866–1943), perhaps best known for having written *The Tale of Peter Rabbit* in 1902, began drawing and painting fungi in 1888. She eventually completed a collection of almost 300 detailed watercolors, which are now in the Armitt Library in Ambleside, England. In 1897 she prepared a scientific paper on the germination of *Aaricineae* spores for a meeting of the Linnean Society of London. Although her findings were originally rejected, experts now consider her ideas correct.

grow in concentrated salt solutions or sugar solutions, such as jelly or jam, which prevents bacterial growth. Fungi also thrive over a wide temperature range. Even refrigerated food may be susceptible to fungal invasion.

Since **fungi lack chlorophyll** necessary to produce their own food how do they **obtain food**?

Fungi are saprobes that absorb nutrients from wastes and dead organisms. Instead of taking food inside its body and then digesting it as an animal would, a fungus digests food outside its body by secreting strong hydrolytic enzymes onto the food. In this way, complex organic compounds are broken down into simpler compounds that the fungus can absorb through the cell wall and cell membrane.

How many **kinds of mushrooms** are **edible**?

Among the basidiomycetes, there are approximately 200 varieties of edible mushrooms and about 70 species of poisonous ones. Some edible mushrooms are cultivated commercially; more than 844 million pounds (382,832 metric tons) are produced in the United States each year.

What is unusual about *Amanita* mushrooms?

Some of the most poisonous mushrooms belong to the genus *Amanita*. Toxic species of this genus have been called such names as "death angel" (*Amanita phalloides*) and "destroying angel" (*Amanita virosa*). Ingestion of a single cap can kill a healthy, adult human! Even ingesting a tiny bit of the amatoxin—the toxin present in species of this genus—may result in liver ailments that will last the rest of a person's life.

What **antidote** is available for **mushroom poisoning**?

No effective antidote for human poisoning by mushrooms has been discovered. The toxins produced by mushrooms accumulate in the liver and lead to irreversible liver dam-

Which mushrooms were considered sacred by the Aztecs?

Mushrooms of the genera, *Conocybe* and *Psilocybe*, both of which have hallucinogenic properties, were considered sacred by the Aztecs. These mushrooms are still used in religious ceremonies by the descendents of the Aztecs. Psilocybin, which is chemically related to lysengic acid diethylamide (LSD), is a component to both genera and is responsible for the trance-like state and colorful visions experienced by those who eat these mushrooms.

age. Unfortunately, there may be no indication of poisoning for several hours after ingesting a toxic mushroom. When the symptoms do present, they often resemble typical food poisoning. Liver failure becomes apparent three to six days after ingesting the poisonous mushroom. Oftentimes a liver transplant may be the only possible treatment.

What are **truffles**?

Truffles, a delight of gourmets, are arguably the most-prized edible fungi. Found mainly in Western Europe, they grow near the roots of trees (particularly oak, but also chestnut, hazel, and beech) in open woodlands. Unlike typical mushrooms, truffles develop 3 to 12 inches (7.6 to 30.5 centimeters) underground making them difficult to find. Truffle hunters use specially trained dogs and pigs to find the flavorful morsels. Both animals have a keen sense of smell and are attracted by the strong, nut-like aroma of truffles. In fact, trained pigs are able to pick up the scent a truffle from 20 feet (6.1 meters) away. After catching a whiff of a truffle's scent, the animals rush to the origin of the aroma and quickly root out the precious prize. Once the truffle is found, the truffle hunter (referred to in French as *trufficulteur*) carefully scrapes back the earth to reveal the fungus. Truffles should not be touched by human skin, as doing so can cause the fungus to rot.

What do **truffles look like**?

A truffle has a rather unappealing appearance—round and irregularly shaped with a thick, rough, wrinkled skin that varies in color from almost black to off-white. The fruiting bodies present on truffles are fragrant, fleshy structures, that usually grow to about the size of a golf ball; they range from white to gray or brown to nearly black in color. There are nearly 70 known varieties of truffles, but the most desirable is the black truffle—known as black diamond—that grows in France's Perigord and Quercy regions as well as Italy's Umbria region. The flesh of the black diamond appears to be black, but it is actually dark brown, and contains white striations. The flesh has an aroma that is extremely pungent. The next most popular is the white truffle (actually off-white or beige) of Italy's Piedmont region. Both the aroma and flavor of this truffle are earthy and garlicky. Fresh truffles are available from late fall to midwinter and can be stored in the refrigerator for up to three days. Dark truffles are generally used to flavor foods such as omelets, polentas, risottos, and sauces. White truffles are usually served raw; they are often grated over foods such as pasta or dishes containing cheese, as their flavors are complementary. They are also added at the last minute to cooked dishes.

What are **lichens**?

Lichens are organisms that grow on rocks, tree branches, and bare ground. They are composed of two different entities living together in a symbiotic relationship: 1) a population of either algal or cyanobacterial cells that are single or filamentous; and 2) fungi. Lichens do not have roots, stems, flowers, or leaves. The fungal component of the lichen is called the mycobiont (Greek *mykes*, "fungus" and *bios*, "life") and the photosynthetic component is called the photobiont (Greek *photo*, light and *bios*, life). The scientific name given to the lichen is the name of the fungus and is most often an ascomycete. As the fungus has no chlorophyll, it cannot manufacture its own food, but can absorb food from algae. Lichens and algae enjoy a symbiotic relationship. Lichens can often be found growing around and on top of algae, providing the algae protection from the sun thus decreasing the loss of moisture. Fungi and algae were the first organisms recognized as having a symbiotic relationship. A unique feature of this relationship is that it is so perfectly developed and balanced that the two organisms behave as a single organism.

What is the **relationship** between **lichens** and **pollution**?

Lichens are extremely sensitive to pollutants in the atmosphere and can be used as bioindicators of air quality. They absorb minerals from the air, from rainwater, and directly from their substrate. Lichen growth has been used as an indicator of air pollution, especially sulfur dioxide. Pollutants are absorbed by lichens, causing the destruction of their chlorophyll, which leads to a decrease in the occurrence of photosynthesis and changes in membrane permeability. Lichens are generally absent in and around cities, even though suitable substrates exist; the reason for this is the polluted exhaust from automobiles and industrial activity. They are beginning to disappear from national parks and other relatively remote areas that are becoming increasingly contaminated by industrial pollution. The return of lichens to an area frequently indicates a reduction in air pollution.

Lichens are also used to assess radioactive pollution levels in the vicinity of uranium mines, environments where nuclear-powered satellites have crashed, former nuclear bomb testing sites, and power stations that have incurred accidents. Following the Chernobyl nuclear power station disaster in 1986, arctic lichens as far away as Lapland were tested and showed levels of radioactive dust that were as much as165 times higher than had been previously recorded.

369

What is the role of **yeast** in **beer production**?

Beer is made by fermenting water, malt, sugar, hops, yeast (species *Saccaromyces*), salt, and citric acid. Each ingredient has a specific role in the creation of beer. Malt is produced from a grain—usually barley—that has sprouted, been dried in a kiln, and ground into a powder. Malt gives beer its characteristic body and flavor. Hops is made from the fruit that grows on the herb *Humulus lupulus* (a member of the mulberry family) The fruit is picked when ripe and is then dried; the ingredient gives beer a slightly bitter flavor. Yeast is used for the fermentation process.

Making beer is a complex process. One method begins by mixing and mashing malted barley with a cooked cereal grain such as corn. This mixture called "wort," is filtered before hops is added to it. The wort is then heated until it is completely soluble. The hops is removed, and after the mixture is cooled, yeast is added. The beer ferments for 8 to 11 days at temperatures that range between 50°–70°F (10°–21°C). The beer is then stored and kept at a state that is close to freezing. During the next few months the liquid takes on its final character before carbon dioxide is added for effervescence. The beer is then refrigerated, filtered, and pasteurized in preparation for bottling or canning.

Is the **same strain** of **yeast** used to make **lager beers** and **ales**?

Two common strains of yeast are used to ferment beer: *Saccharomyces carlsbergensis* and *Saccharomyces cerevisiae*. *Saccharomyces carlsbergensis*, also known as bottom yeast, sinks to the bottom of the fermentation vat. Strains of bottom yeast ferment best at 42.8°F–53.6°F (6°–12°C) and take 8 to 14 days to produce lager beers. *Saccha-*

romyces cerevisiae, also known as top-fermenting yeast, is distributed throughout the wort and is carried to the top of the fermenting vat by the carbon dioxide (CO_2). Top-fermenting yeast ferment at a higher temperature (57.2°F–73.4°F/14°–23°C) over only five to seven days. Top-fermenting yeasts produce ales, porter, and stout beers.

Which **tree** that is **native** to the **United States** almost became **extinct due** to **fungus**?

The American chestnut (*Castenea dentata*) was widespread across eastern North America until the early 1900s. It was one of the most important trees of the eastern hardwood forest and made up almost half of the population of trees in central and southern Pennsylvania, New Jersey, and southern New England. In its entire range, the species dominated the deciduous forests, making up almost one quarter of the trees. The fungus *Cryphonectria parasitica*, commonly known as the chestnut blight, caused the near extinction of the American chestnut tree in North America. Through ongoing genetic studies and improvements in plant breeding techniques, the American Chestnut Foundation is working to restore this tree to its native range within the woodlands of the eastern United States.

What **cheeses** are associated with **fungi**?

The unique flavor of cheeses such as Roquefort, Camembert, and Brie is produced by the action of members of the genus *Penicillium*. Roquefort is often referred to as "the king of cheeses"; it is one of the oldest and best-known in the world. This "blue cheese" has been enjoyed since Roman times and was a favorite of Charlemagne, king of the Franks and emperor of the Holy Roman Empire (742–814). Roquefort is made from sheep's milk that has been exposed to the mold *Penicillium roqueforti* and aged for three months or more in the limestone caverns of Mount Combalou, near the village of Roquefort in southwestern France. This is the only place true Roquefort can be aged. It has a

Certain species of fungi give some of our favorite cheeses their distinctive flavor. Sheep's milk is exposed to *Penicillium roqueforti* to give us Roquefort cheese (top), and *Penicillium camemberti* is used in making brie cheese (bottom) and camembert.

371

How were fungi involved in World War I?

During World War I, the Germans needed glycerol to make nitroglycerin, which is used in the production of explosives such as dynamite. Before the war, the Germans had imported their glycerol, but this impact was prevented by the British naval blockade during the war. The German scientist Carl Neuberg (1877–1956) knew that trace levels of glycerol are produced when *Saccharomyces cerevisiae* is used during the alcoholic fermentation of sugar. He sought and developed a modified fermentation process in which the yeast would produce significant quantities of glycerol and less ethanol. The production of glycerol was improved by adding 3.5 percent sodium sulfite at pH 7 to the fermentation process, which blocked one chemical reaction in the metabolic pathway. Neuberg's procedure was implemented with the conversion of German beer breweries to glycerol plants. The plants produced 1,000 tons of glycerol per month. After the war ended, the production of glycerol was not in demand, so it was suspended.

creamy, rich texture and is pungent, piquant, and salty. It has a creamy white interior with blue veins; the cheese is held together with a snowy white rind. True Roquefort is authenticated by the presence of a red sheep on the emblem of the cheese's wrapper.

Penicillium camemberti give Camembert and Brie cheeses their special qualities. Napoleon is said to have christened Camembert cheese with its name; supposedly the name comes from the Norman village where a farmer's wife first served it to Napoleon. This cheese is formed of cow's milk cheese and has a white, downy rind and a smooth, creamy interior. When perfectly ripe and served at room temperature, the cheese should ooze thickly. Although Brie is made in many places, Brie from the region of the same name east of Paris is considered one of the world's finest cheeses by connoisseurs. Similar to Camembert, it has a white, surface-ripened rind and smooth, buttery interior.

How was it **discovered** that **microorganisms** are **effective against bacterial infections**?

British microbiologist Alexander Fleming (1881–1955) happened upon the discovery of penicillin's use as an antibacterial agent. In 1928 Fleming was researching staphylococci at St. Mary's Hospital in London. As part of his investigation, he had spread staphylococci on several Petri dishes before going on vacation. Upon his return he noticed a green-yellow mold contaminating one of the Petri dishes. The staphylococci had failed to grow near the mold. He identified the mold as being of the species of *Penicillium notatum*. Further investigation proved that the staphylococci and other

Gram-positive organisms are killed by *P. notatum*. It was not until the 1940s that Howard Florey (1898–1968) and Ernst Boris Chain (1906–1979) rediscovered penicillin and were able to isolate it for medical use. In 1945 Fleming, Florey, and Chain shared the Nobel Prize in Physiology or Medicine for their work on penicillin.

DNA, RNA, AND CHROMOSOMES

What is **DNA**?

Deoxyribonucleic acid (DNA) is the genetic material for all cellular organisms. The discovery of the structure of DNA is considered the most important molecular discovery of the twentieth century.

What are the **component molecules** of **DNA**?

The full name of DNA is deoxyribonucleic acid, with the "nucleic" part coming from the location of DNA in the nuclei of eukaryotic cells. DNA is actually a polymer (long strand) of nucleotides. A nucleotide has three component parts: a phosphate group, a five-carbon sugar (deoxyribose), and a nitrogen base. If you visualize DNA as a ladder, the sides of the ladder are made of the phosphate and deoxyribose molecules, and the rungs are made of two different nitrogen bases. The nitrogen bases are the crucial part of the molecule with regard to genes. Specific sequences of nitrogen bases make up a gene.

How is a **DNA** molecule **held together**?

Although DNA is held together by several different kinds of chemical interactions, it is still a rather fragile molecule. The nitrogen bases that constitute the "rungs" of the ladder are held together by hydrogen bonds. The "sides" of the ladder (the phosphate and deoxyribose molecules) are held together by a type of covalent bond called a phosphodiester bond. Because part of the DNA molecule is polar (the outside of the ladder), and the rungs (nitrogen bases) are nonpolar, there are other interactions—called hydrostatic interactions—that occur between the hydrogen and oxygen atoms of DNA and water. The internal part of the DNA tends to repel water, while the external sugar-phosphate molecules tend to attract water. This creates a kind of molecular pressure that glues the helix together.

What are the **nitrogenous bases** of **DNA**?

The nitrogenous bases have a ring of nitrogen and carbon atoms with various functional groups attached. There are two types of nitrogenous bases. They differ in their structure: thymine and cytosine have a single-ring structure, while adenine and gua-

373

What term was originally used for DNA?

DNA was originally called nuclein because it was first isolated in 1869 from the nuclei of cells. In the 1860s Johann Frederick Miescher (1844–1895), a Swiss biochemist working in Germany at the University of Tubingen lab of Felix Hoppe-Seyler (1825–1895), was given the task of researching the composition of white blood cells. He found a good source of white blood cells from the used bandages that he obtained at a nearby hospital. He washed off the pus and isolated a new molecule from the cell nucleus; white blood cells have very large nuclei. He called the substance nuclein.

nine have double-ring structures. When James Watson (1928–) and Francis Crick (1916–2004) were imagining how the bases would join together, they knew that the pairing had to be such that there was always a uniform diameter for the molecule. It therefore became apparent that a double-ring base must always be paired with a single-ring base on the opposite strand.

What is the law of **complementary base pairing**?

The law of complementary base pairing refers to the pairing of nitrogenous bases in a specific manner: purines pair with pyrimidines. More specifically, adenine must always pair with thymine, and guanine must always pair with cytosine. The basis of this law came from the data obtained by Edwin Chargaff (1905–2002) and is known as Chargaff's law, or rule.

How is **DNA unzipped**?

DNA is unzipped during its replication process; the two strands of the double helix are separated and a new complementary DNA strand is synthesized from the parent strands. Also, during DNA transcription, one DNA strand, known as the template strand, is transcribed (copied) into an RNA strand. In order for the two strands of DNA to separate, the hydrogen bonds between the nitrogen bases must be broken. DNA helicase breaks the bonds. However, the enzyme does not actually unwind the DNA; there are special proteins that first separate the DNA strands at a specific site on the chromosome. These are called initiator proteins.

What is **needed** for **DNA replication**?

DNA replication is a complex process requiring more than a dozen enzymes, nucleotides, and energy. In eukaryotic cells there are multiple sites called origins of replication; at these sites, enzymes unwind the helix by breaking the double bonds

> ## How is mutation rate calculated?
>
> The mutation rate is generally expressed as the number of mutations per cell divisions, per gamete, or per round of replication.

between the nitrogen bases. Once the molecule is opened, separate strands are kept from rejoining by DNA stabilizing proteins. DNA polymerase molecules read the sequences in the strands being copied and catalyze the addition of complementary bases to form new strands.

Is **DNA** always **copied exactly**?

Considering how many cells there are in the human body and how often it occurs, DNA replication is fairly accurate. Spontaneous damage to DNA is low, occurring at the rate of 1–100 mutations per 10 billion cells in bacteria. The rate for eukaryotic genes is higher, about 1–10 mutations per million gametes. The rate of mutation can vary according to different genes in different organisms.

What is a **mutation**?

A mutation is an alteration in the DNA sequence of a gene. Mutations are a source of variation to a population, but they can have detrimental effects in that they may cause diseases and disorders. One example of a disease caused by a mutation is sickle cell disease, in which there is a change in the amino acid sequence (valine is substituted for glutamic acid) of two of the four polypeptide chains that make up the oxygen-carrying protein known as hemoglobin.

How **fast** is **DNA copied**?

In prokaryotes, about 1,000 nucleotides can be copied per second, so all of the 4.7 megabytes of *Escherichia coli* can be copied in about 40 minutes. Since the eukaryotic genome is immense compared to the prokaryotic genome, one might think that the eukaryotic DNA replication would take a very long time. However, actual measurements show that the chromosomes in eukaryotes have multiple replication sites per chromosome. Eukaryotic cells can replicate about 500–5,000 bases per minute; the actual time to copy the entire genome would depend on the size of their genome.

What is **polymerase chain reaction**?

Polymerase chain reaction, or PCR, is a laboratory technique that amplifies or copies any piece of DNA very quickly without using cells. The DNA is incubated in a test tube   375 with a special kind of DNA polymerase, a supply of nucleotides, and short pieces of synthetic single-strand DNA that serve a primers for DNA synthesis. With automation, PCR can make billions of copies of a particular segment of DNA in a few hours. Each cycle of the PCR procedure takes only about five minutes. At the end of the cycle, the DNA segment—even one with hundreds of base pairs—has been doubled. A PCR machine repeats the cycle over and over. PCR is much faster than the days it takes to clone a piece of DNA by making a recombinant plasmid and letting it replicate within bacteria.

PCR was developed by the biochemist Kary Mullis (1944–) in 1983 while working for Cetus Corporation, a California biotechnology firm. In 1993, Mullis won the Nobel Prize in Chemistry for developing PCR.

How is **DNA organized** in the **nucleus**?

Within the nucleus DNA is organized with proteins, called histones, into a fibrous material called chromatin. As a cell prepares to divide or reproduce, the thin chromatin fibers condense, becoming thick enough to be seen as separate structures, which are called chromosomes.

Is **all DNA** found in the **nucleus**?

In addition to the nuclear DNA of eukaryotic cells, mitochondria (an organelle found in both plant and animal cells) and chloroplasts (found in plant and algal cells) both contain DNA. Mitochondrial DNA contains genes essential to cellular metabolism. Chloroplast DNA contains genetic information essential to photosynthesis.

How **much DNA** is in a **typical human cell**?

If the DNA in a single human cell were stretched out and laid end-to-end, it would measure approximately 6.5 feet (2 meter). The average human body contains 10 to 20 billion miles (16 to 32 billion kilometers) of DNA distributed among trillions of cells. If the total DNA in all the cells from one human were unraveled, it would stretch to the sun and back more than five hundred times.

What is the **difference** between **DNA** and **RNA**?

DNA (deoxyribonucleic acid) is a nucleic acid formed from a repetition of simple building blocks called nucleotides. The nucleotides consist of phosphate (PO_4),

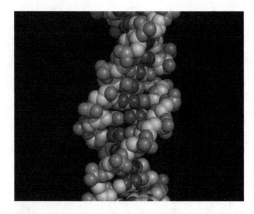
A model showing the double-helix structure of a DNA molecule.

How did scientists decide that DNA was the genetic material for all cellular organisms?

The proof that the material basis for a gene is DNA came from the work of Oswald T. Avery (1877–1955), Colin M. MacLeod (1909–1972), and Maclyn McCarty (1911–2005) in a paper published in 1944. This group of scientists followed the work of Frederick Griffith (1879–1941) in order to discover what causes nonlethal bacteria to transform to a lethal strain. Using specific enzymes, all parts of the S (lethal) bacteria were degraded, including the sugarlike coat, the proteins, and the RNA. The degradation of these substances by enzymes did not affect the transformation process. Finally, when the lethal bacteria were exposed to DNase, an enzyme that destroys DNA, all transformation activity ceased. The transforming factor was DNA.

sugar (deoxyribose), and a base that is either adenine (A), thymine (T), guanine (G), or cytosine (C). In a DNA molecule, this basic unit is repeated in a double helix structure made from two chains of nucleotides linked between the bases. The links are either between A and T or between G and C. The structure of the bases does not allow other kinds of links. The famous double helix structure resembles a twisted ladder. The 1962 Nobel Prize in Physiology or Medicine was awarded to James Watson, Francis Crick, and Maurice Wilkins (1916–2004) for determining the molecular structure of DNA.

RNA (ribonucleic acid) is also a nucleic acid, but it consists of a single chain and the sugar is ribose rather than deoxyribose. The bases are the same except that the thymine (T), which appears in DNA, is replaced by another base called uracil (U), which links only to adenine (A).

When was **RNA discovered**?

By the 1940s it was known that there was another kind of nucleic acid other than DNA, this one called RNA. Phoebus Levene (1869–1940), a Russian-born chemist, further refined the work of Albrecht Kossel (1853–1927). Kossel was awarded the Nobel Prize in Physiology or Medicine in 1910 for determining the composition of nuclein. At the time of Kossel's work, it was not clear that DNA and RNA were different substances. In 1909, Levene isolated the carbohydrate portion of nucleic acid from yeast and identified it as the pentose sugar ribose. In 1929, he succeeded in identifying the carbohydrate portion of the nucleic acid isolated from the thymus of an animal. It was also a pentose sugar, but it differed from ribose in that it lacked one oxygen atom. Levene called the new substance deoxyribose. These studies defined the chemical differences between DNA and RNA by their sugar molecules.

Where is **RNA formed**?

All RNA is formed in the nucleus (eukaryotic cells) or in the nucleoid region (prokaryotic cells). The principal enzyme responsible for RNA synthesis is RNA polymerase.

How many **types** of **RNA** are found in **eukaryotic cells**?

There are five major types of RNA found in eukaryotic cells: 1) heterogeneous nuclear RNA (hnRNA), 2) messenger RNA (mRNA), 3) transfer RNA (tRNA), 4) ribosomal RNA (rRNA), and 5) small nuclear RNA. The primary types of RNA are mRNA, tRNA, and rRNA. Messenger RNA, a single strand copied from a DNA strand, carries the genetic code from the DNA to the site of protein synthesis on the ribosomes. The most abundant type of RNA, rRNA, participates in protein synthesis in the ribosomes. Transfer (tRNA) is the translation molecule. Each tRNA molecule carries a specific anticodon, picks up a specific amino acid, and conveys the amino acid to the appropriate codon on mRNA.

How are **chromosomes assembled**?

Chromosomes are assembled on a scaffold of proteins, called histones, that allow DNA to be tightly packed. There are five major types of histones, all of which have a positive charge; the positive charges of the histones attract the negative charges on the phosphates of DNA, thus holding the DNA in contact with the histones. These thicker strands of DNA and proteins are called chromatin. Chromatin is then packed to form the familiar structure of a chromosome. During mitosis, chromosomes acquire characteristic shapes that allow them to be counted and identified.

When were **chromosomes first observed**?

Chromosomes were observed as early as 1872, when Edmund Russow (1841–1897) described seeing items that resembled small rods during cell division; he named the rods "Stäbchen." Edouard van Beneden (1846–1910) used the term *bâtonnet* in 1875 to describe nuclear duplication. The following year, 1876, Edouard Balbiani (1825–1899) described that at the time of cell division the nucleus dissolved into a collection of *bâtonnets étroits* ("narrow little rods"). Walther Flemming (1843–1905) discovered that the chromosomal "threads" or *Fäden* split longitudinally during mitosis.

Which **organism** has the **largest number** of **chromosomes**?

Ophioglossum reticulatum, a species of fern, has the largest number of chromosomes with more than 1,260 (630 pairs).

How many **chromosomes** are in a **human body cell**?

A human being normally has 46 chromosomes (23 pairs) in all but the sex cells. Half of each pair is inherited from the mother's egg; the other, from the father's sperm. When the sperm and egg unite in fertilization, they create a single cell, or zygote, with 46 chromosomes. When cell division occurs, the 46 chromosomes are duplicated; this process is repeated billions of times over, with each of the cells containing the identical set of chromosomes. Only the gametes, or sex cells, are different. In their cell division, the members of each pair of chromosomes are separated and distributed to different cells. Each gamete has only 23 chromosomes.

What are the **components** of a **gene**?

The term "gene" describes a section of DNA that will be used as a template to build a strand of RNA or protein. In addition to this information, each gene also contains a promoter region, which indicates where the coding information actually begins, and a terminator, which delineates the end of the gene.

What is the average **size** of a **gene**?

The average size of a vertebrate gene is about 30,000 base pairs. Bacteria, because their sequences contain only coding material, have smaller genes of about 1,000 base pairs each. Human genes are in the 20,000–50,000 base pairs range, although sizes greater than 100,000 have been suggested as well.

What is the **difference** between a **gene** and a **chromosome**?

The human genome contains 24 distinct, physically separate units called chromosomes. Arranged linearly along the chromosomes are tens of thousands of genes. The term "gene" refers to a particular part of a DNA molecule defined by a specific sequence of nucleotides. It is the specific sequence of the nitrogen bases that encodes a gene. The human genome contains about three billion base pairs, and the length of genes varies widely.

How are **genes controlled**?

Genes are controlled by regulatory mechanisms that vary by whether the organism is a prokaryote or a eukaryote. Bacteria (prokaryote) genes can be regulated by DNA binding proteins that influence the rate of transcription, or by global regulatory mechanisms that refer to an organism's response to specific environmental stimuli such as heat shock. This is particularly important in bacteria. Gene control in eukaryotes depends on a complex set of regulatory elements that turn genes off and on at

Do both sides of DNA contain genes?

One strand of DNA contains the information that codes for genes, and it is called the "antisense strand" or "noncoding strand." It is this strand that is transcribed into mRNA and is designated as the template strand. The other, complementary strand is called the "coding strand" (because it contains codons) or "sense strand." Its sequence is identical to the mRNA strand, except for the substitution of U (uracil) for T (thymine).

specific times. Among these regulatory elements are DNA binding proteins as well as proteins that, in turn, control the activity of the DNA binding proteins.

What is the **one gene–one enzyme hypothesis**?

In the 1930s, George Beadle (1903–1989) and Boris Ephrussi (1901–1979) theorized that the variety of fruit fly mutations might be due to mutations of individual genes that code for each of the enzymes involved in a given pathway. Subsequently, Beadle and Edward Tatum (1909–1975) performed a series of experiments with the orange bread mold Neurospora that elucidated the enzymatic pathway required by the fungus to produce a specific nutritional requirement, arginine. The researchers were able to create a series of mutants, each lacking in a different enzyme in the pathway. In this way they were able to piece together the sequence of events required for the production of arginine and thereby show where each mutant fit. The work of Beadle and Tatum provided important support for the one gene one enzyme hypothesis, which holds that the function of a gene is to produce a specific enzyme. Their work garnered a Nobel Prize in Physiology or Medicine in 1958.

Who discovered **jumping genes**?

Barbara McClintock (1902–1992), who worked on the cytogenetics of maize during the 1950s at Cold Spring Harbor Laboratory in New York, discovered that certain mutable genes were transferred from cell to cell during development of the corn kernel. McClintock made this inference based on observations of changing patterns of coloration in maize kernels over many generations of controlled crosses. She was awarded the Nobel Prize in Physiology or Medicine in 1983 for her work.

What is the biological basis for **DNA fingerprinting**?

British geneticist, Alec Jeffreys (1950–) formulated the method of DNA fingerprinting, also known as DNA typing or DNA profiling, based on the fact that unique genetic

What is the p53 gene?

Discovered in 1979, p53—sometimes referred to as "The Guardian Angel of the Genome"—is a gene that, when a cell's DNA is damaged, acts as an "emergency brake" to halt the resulting cycle of cell division that can lead to tumor growth and cancer. It also acts as an executioner, programming damaged cells to self-destruct before their altered DNA can be replicated. However, when it mutates, p53 can lose its suppressive powers or have the devastating effect of actually promoting abnormal cell growth. Indeed, p53 is the most commonly mutated gene found in human tumors. Scientists have discovered a compound that could restore function to a mutant p53. Such a discovery may lead to the development of anti-cancer drugs targeting the mutant p53 gene.

differences exist between individuals. Most DNA sequences are identical, but out of 100 base pairs (of DNA), two people will generally differ by one base pair. Since there are three billion base pairs in human DNA, one individual's DNA will differ from another's by three million base pairs. To examine an individual's DNA fingerprint, the DNA sample is cut with a restriction endonuclease, and the fragments are separated by gel electrophoresis. The fragments are then transferred to a nylon membrane, where they are incubated in a solution containing a radioactive DNA probe that is complementary to specific polymorphic sequences.

How can the **entire genome fit** into the **nucleus** of a cell?

The average nucleus has a diameter of less than 5 micrometers, and the DNA (eukaryotic) has a length of 1–2 micrometers. In order to fit DNA into a nucleus, DNA and proteins are tightly packed to form threads called chromatin. These threads are so thick that they actually become visible with a light microscope.

GENETICS AND EVOLUTION

Who is generally known as the **founder of genetics**?

Gregor Mendel (1822–1884), an Austrian monk and biologist, is considered the founder of genetics. Mendel was the first to demonstrate transmission of distinct physical characteristics from generation to generation through his work with the garden pea. It was the English biologist William Bateson (1861–1926), however, who brought Mendel's work to the attention of the scientific world and who coined the term "genetics."

What is meant by **Mendelian inheritance**?

Mendelian inheritance refers to genetic traits carried through heredity; the process was studied and described by Austrian monk Gregor Mendel. Mendel was the first to deduce correctly the basic principles of heredity. Mendelian traits are also called single gene or monogenic traits, because they are controlled by the action of a single gene or gene pair. More than 4,300 human disorders are known or suspected to be inherited as Mendelian traits, encompassing autosomal dominant (e.g., neurofibromatosis), autosomal recessive (e.g., cystic fibrosis), and sex-linked dominant and recessive conditions (e.g., color-blindness and hemophilia).

Overall, incidence of Mendelian disorders in the human population is about one percent. Many non-anomalous characteristics that make up human variation are also inherited in Mendelian fashion.

Did **Darwin** and **Mendel** know each other?

Although both men lived during the same time period in the nineteenth century, they did not know each other. Charles Darwin's (1809–1882) publication *On the Origin of Species* (1859) popularized his theory of natural selection but raised many questions as to how organisms could display modified or new traits. In 1865 Mendel published his landmark paper "Experiments in Plant Hybridization." Researchers have been unable to document whether either man used the other's work in the development of their respective theories.

What is a **pedigree**?

A pedigree is a genetic history of a family, which shows the inheritance of traits through several generations. Information that can be obtained from a pedigree includes birth order, sex of children, twins, marriages, deaths, stillbirths, and pattern of inheritance of specific genetic traits.

What is the **most famous human pedigree**?

The sex-linked pattern of the inheritance of hemophilia within the royal family of Queen Victoria (1819–1901) and Prince Albert (1819–1861) is perhaps the most famous human pedigree.

What is meant by the **modern era** of **genetics**?

Mendel's work was really not appreciated until advances in cytology enabled scientists to better study cells. In 1900, Hugo deVries (1848–1935) of Holland, Carl Correns (1864–1933) of Germany, and Erich von Tschermak (1871–1962) of Austria examined Mendel's original 1865 paper and repeated the experiments. In the following years chromosomes were discovered as discrete structures within the nucleus of a cell. In 1917, Thomas Hunt Morgan (1866–1945), a fruit fly geneticist at Columbia University, extended Mendel's findings to the structure and function of chromosomes. This and subsequent findings in the 1950s were the beginning of the modern era of genetics.

Why are some **species** more commonly **used** for **genetic studies** than others?

Species with a relatively small genome, with a short generation time from seed to seed, and that are adaptable to living in captivity are appealing as experimental organisms. Even though many of these species bear little physical resemblance to humans, they do share part of our genome and so can answer some of the questions we have about genetic inheritance and gene expression.

Species Commonly Used in Genetics Research

Species	Kingdom	Genome Size (in millions of base pairs)
Arabidopsis thaliana (flowering plant)	Plant	120
Neurospora (orange bread mold)	Fungi	40
Escherichia coli (bacteria)	Monera	4.64
Drosophila melanogaster (fruit fly)	Animal	170
Caenorhabditis elegans (roundworm)	Animal	97

What was the **Human Genome Project** (HGP), and what were its goals?

The HGP was begun in 1990 as a 13-year effort and completed in 2003. According to the official HGP web site (http://www.doegenomes.org/), the goals were as follows:

- Identify all of the approximately 30,000–40,000 genes in human DNA
- Determine the sequence of the 3 billion chemical pairs of human DNA
- Store this information in public databases

- Improve tools for data analysis
- Transfer related technologies to the private sector
- Address the ethical, legal, and social issues that may stem from the project

What is a **genome**?

A genome is the complete set of genes inherited from one's parents. Genome sizes vary from one species to another. The final number for humans is yet to be determined.

Genome Sizes by Species

Type	Species	Genome Size (Bases)	Estimated Genes
Human immuno-deficiency virus	*HIV*	9,700	9
Bacterium	*Escherichia coli*	4.6 million	3,200
Yeast	*Saccharomyces cerevisiae*	12.1 million	6,000
Worm	*Caenorhabiditis elegans*	97 million	19,099
Mustard weed	*Arabidopsis thaliana*	100 million	25,000
Fruit fly	*Drosophila melanogaster*	137 million	13,000
Pufferfish	*Fugu rubripes*	400 million	38,000
Mouse	*Mus musculus*	2.6 billion	30,000
Human	*Homo sapiens*	3 billion	30,000

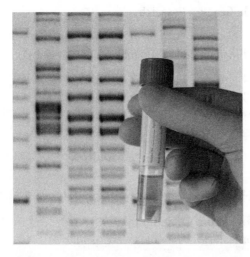

Because scientists have mapped the entire human genome, immense progress is being made in discovering the causes of some genetic diseases. Genetic analysis is also highly useful in crime labs.

How are **genes physically found** in a specific genome?

Finding one gene out of a possible 30,000 to 40,000 genes in the human genome is a difficult task. However, the process is made easier if the protein product of the gene is known. As an example, if a researcher is looking to find the gene for mouse hemoglobin, he or she would isolate the hemoglobin from mouse blood and determine the amino acid sequence. The amino acid sequence could then be used as a template to generate the nucleotide sequence. Working backward again, a complementary DNA probe to the sequence would be used to identify DNA molecules with the same sequence from the entire mouse genomic library.

How have genetics been linked to the Salem witch trials held in 1692 in Salem, Massachusetts?

It is believed that some of the early English colonists that settled in New England may have had Huntington's disease. Huntington's disease is an autosomal dominant disorder characterized by late onset symptoms (age 40 to 50) such as mild behavioral and neurological changes; as the disease progresses, psychiatric problems develop that frequently lead to insanity. Early descriptions of the odd behavior included names such as "that disorder" and "Saint Vitus's dance" to describe involuntary muscle jerks and twitches. Many of the witches who were on trial for possession may have had Huntington's disease, which causes uncontrollable movements and odd behavior.

However, if the protein product is not known, the task is more difficult. An example of this would be that of finding the susceptibility gene for late-onset Alzheimer's disease. DNA samples would be collected from family members of a patient with late-onset Alzheimer's disease. The DNA would be cut with restriction endonucleases, and restriction fragment length polymorphisms (RFLPs) would be compared among the family. If certain RFLPs are only found when the disease gene is present, then it is assumed that the distinctive fragments are markers for the gene. Geneticists then sequence the DNA in the same area of the chromosome where the marker was found, looking for potential gene candidates.

What is **genetic engineering**?

Genetic engineering, also known popularly as molecular cloning or gene cloning, is the artificial recombination of nucleic acid molecules in the test tube, their insertion into a virus, bacterial plasmid, or other vector system, and the subsequent incorporation of the chimeric molecules into a host organism in which they are capable of continued propagation. The construction of such molecules has also been termed gene manipulation because it usually involves the production of novel genetic combinations by biochemical means.

Genetic engineering techniques include cell fusion and the use of recombinant DNA (RNA) or gene-splicing. In cell fusion, the tough outer membranes of sperm and egg cells are stripped off by enzymes, and then the fragile cells are mixed and combined with the aid of chemicals or viruses. The result may be the creation of a new life form from two species. Recombinant DNA techniques transfer a specific genetic activity from one organism to the next through the use of bacterial plasmids (small circular pieces of DNA lying outside the main bacterial chromosome) and enzymes, such as restriction endonucleases (which cut the DNA strands); reverse transcriptase (which makes a DNA strand from an RNA strand); DNA ligase (which joins DNA strands together); and tag polymerase (which can make a double-stranded DNA molecule from a single stranded "primer" molecule).

The process begins with the isolation of suitable DNA strands and fragmenting them. After these fragments are combined with vectors, they are carried into bacterial cells where the DNA fragments are "spliced" on to plasmid DNA that has been opened up. These hybrid plasmids are now mixed with host cells to form transformed cells. Since only some of the transformed cells will exhibit the desired characteristic or gene activity, the transformed cells are separated and grown individually in cultures. This methodology has been successful in producing large quantities of hormones (such as insulin) for the biotechnology industry. However, it is more difficult to transform animal and plant cells. Yet the technique exists to make plants resistant to diseases and to make animals grow larger. Because genetic engineering interferes with the processes of heredity and can alter the genetic structure of our own species, there is much concern over the ethical ramifications of such power, as well as the possible health and ecological consequences of the creation of these bacterial forms. Some applications of genetic engineering in the various fields are listed below:

Agriculture—Crops having larger yields, disease- and drought-resistancy; bacterial sprays to prevent crop damage from freezing temperatures; and livestock improvement through changes in animal traits.

Industry—Use of bacteria to convert old newspaper and wood chips into sugar; oil- and toxin-absorbing bacteria for oil spill or toxic waste cleanups; and yeasts to accelerate wine fermentation.

Medicine—Alteration of human genes to eliminate disease (experimental stage); faster and more economical production of vital human substances to alleviate deficiency and disease symptoms (but not to cure them) such as insulin, interferon (cancer therapy), vitamins, human growth hormone ADA, antibodies, vaccines, and antibiotics.

Research—Modification of gene structure in medical research, especially cancer research.

Food Processing—Rennin (enzyme) in cheese aging.

What was the first **commercial use** of **genetic engineering**?

Recombinant DNA technology was first used commercially to produce human insulin from bacteria. In 1982, genetically engineered insulin was approved for use by diabetics. Insulin is normally produced by the pancreas, and the pancreas of slaughtered animals such as swine or sheep was used as a source of insulin. To provide a reliable source of human insulin, researchers obtained DNA from human cells carrying the gene with the information for making human insulin. Researchers made a copy of DNA carrying this insulin gene and moved it into a bacterium. When the bacterium was grown in the lab, the microbe split from one cell into two cells, and both cells got a copy of the insulin gene. Those two microbes grew, then divided into four, those four into eight, the eight into sixteen, and so forth. With each cell division, the two new cells each had a copy of the gene for human insulin. And because the cells had a copy of the genetic "recipe card" for insulin, they could make the insulin protein.

What are some **examples** of **genetic engineering** in animals and microbes?

One of the earliest applications of biotechnology was the genetic engineering of a growth hormone (bovine GH) produced naturally in the bovine pituitary. Bovine GH can increase milk production in lactating cows. Using biotechnology, scientists bioengineered the gene that controls bovine GH production into *E. coli*, grew the bacteria in fermentation chambers, and thus produced large quantities of bovine GH. The bioengineered bovine GH,when injected into lactating cows, resulted in an increase of up to 20 percent in national milk production. Using bovine GH, farmers are able to stabilize milk production in their herds, avoiding fluctuations in production levels. The U.S. Food and Drug Administration (FDA) approved use of recombinant bovine growth hormone in 1993 after ten years of study. A similar regimen was adapted using the pig equivalent of growth hormone (porcine GH). Injected in pigs, porcine GH reduced back fat and increased muscle (meat) gain. Pig growth hormone has been test marketed in a select population with no ill effects; however, it has not yet been approved by the FDA.

The first transgenic animal available as a food source on a large scale was the salmon, which reached U.S. food markets in 2001, following rigid evaluations of consumer and environmental safety. These salmon have the capability of growing from egg to market size (6 to 10 pounds) in 18 months, as compared to conventional fish breeding, which takes up to 36 months to bring a fish to market size. The use of transgenic salmon can help reduce overfishing of wild salmon stocks.

What is **cloning**?

A clone is a group of cells derived from the original cell by fission (one cell dividing into two cells) or by mitosis (cell nucleus division with each chromosome splitting into two). Cloning perpetuates an existing organism's genetic make-up. Gardeners have been making clones of plants for centuries by taking cuttings of plants to make genetically identical copies. For plants that refuse to grow from cuttings, or for the animal world, modern scientific techniques have greatly extended the range of cloning. The technique for plants starts with taking a cutting of a plant that best satisfies the criteria for reproductive success, beauty, or some other standard. Since all of the plant's cells contain the genetic information from which the entire plant can be reconstructed, the cutting can be taken from any part of the plant. Placed in a culture medium having nutritious chemicals and a growth hormone, the cells in the cutting divide, doubling in size every six weeks until the mass of cells produces small white globular points called embryoids. These embryoids develop roots, or shoots, and begin to look like tiny plants. Transplanted into compost, these plants grow into exact copies of the parent plant. The whole process takes 18 months. This process, called tissue culture, has been used to make clones of oil palm, asparagus, pineapples, strawberries, brussels sprouts, cauliflower, bananas, carnations, ferns, and others. Besides making highly productive copies of the best plant available, this method controls viral diseases that are passed through normal seed generations.

What was the **first mammal** to be successfully **cloned**?

The first mammal cloned from adult cells was Dolly, an ewe born 0n July 5, 1996. Dolly was born in a research facility in Scotland. Ian Wilmut (1944–) led the team of biologists that removed a nucleus from a mammary cell of an adult ewe and transplanted it into an enucleated egg extracted from a second ewe. Electrical pulses were administered to fuse the nucleus with its new host. When the egg began to divide and develop into an embryo, it was transplanted into a surrogate mother ewe. Dolly was the genetic twin of the ewe that donated the mammary cell nucleus. On April 13, 1998, Dolly gave birth to Bonnie—the product of a normal mating between Dolly and a Welsh mountain ram. This event demonstrates that Dolly was a healthy, fertile sheep, able to produce healthy offspring. Dolly was euthanized on February 14, 2003, because she had severe arthritis and a progressive lung disease that was caused by the retrovirus JSRV.

What is **evolution**?

Although it was originally defined in the nineteenth century as "descent with modification," evolution is currently described as the change in frequency of genetic traits (also known as the allelic frequency) within populations over time.

What were **early ideas** on **evolution**?

While some Greek philosophers had theories about the gradual evolution of life, Plato (427–347 B.C.E.) and Aristotle (384–322 B.C.E.) were not among them. In the 1700s, "natural theology" (the explanation of life as the manifestation of the creator's plan) held sway in Europe. This idea was the motive force behind the work of Carl Linnaeus

(1707–1778), who was the first to classify all known living things by kingdom. Also popular prior to the work of Charles Darwin (1809–1882) were the theories of "special creation" (creationism), "blending inheritance" (that offspring were always the mixture of the traits of their two parents), and "acquired characteristics."

What is **Lamarckian evolution**?

The French biologist Jean-Baptiste de Lamarck (1744–1829) is credited as the first person to propose a theory that attempts to explain how and why evolutionary change occurs in living organisms. The mechanism Lamarck proposed is known as "the inheritance of acquired characteristics," meaning that what individuals experience during their lifetime will be passed along to their offspring as genetic traits. This is sometimes referred to as the theory of "use and disuse."A classic example of this would be the giraffe's neck. Lamarckian evolution would predict that as giraffes stretch their necks to reach higher branches on trees, their necks grow longer. As a result, this increase in neck length will be transmitted to egg and sperm such that the offspring of giraffes whose necks have grown will also have long necks. While Lamarck's idea was analytically based on available data (giraffes have long necks and give birth to offspring with long necks as well), he did not know that, in general, environmental factors do not change genetic sequences in such a direct fashion.

Who was **Charles Darwin**?

The theory of natural selection proposed by Charles Darwin (1809–1882) revolutionized all aspects of natural science. Darwin was born into a family of physicians and planned to follow his father and grandfather in that profession. Unable to stand the sight of blood, he studied divinity at Cambridge and received a degree from the university in 1830.

What were the ***Beagle*** voyages?

The HMS *Beagle* was a naval survey ship that left England in December 1831 to chart the coastal waters of Patagonia, Peru, and Chile. On a voyage that would last five years, Darwin's job as unpaid companion to the captain onboard the *Beagle* allowed him to satisfy his interests in natural history. On its way to Asia, the ship spent time in the Galapagos Islands off the coast of Ecuador; Darwin's observations there caused him to generate his theory of natural selection.

What is the significance of **Darwin's finches**?

In his studies on the Galapagos Islands, Charles Darwin observed patterns in animals and plants that suggested to him that species changed over time to produce new species. Darwin collected several species of finches. The species were all similar, but

Three years after his death in 1882, a statue of Charles Darwin was erected at the London, England, Natural History Museum.

each had developed beaks and bills specialized to catch food in a different way. Some species had heavy bills for cracking open tough seeds. Others had slender bills for catching insects. One species used twigs to probe for insects in tree cavities. All the species resembled one species of South American finch. In fact, all the plants and animals of the Galapagos Islands were similar to those of the nearby (600 miles/1,000 kilometers away) coast of South America. Darwin felt that the simplest explanation for this similarity was that a few species of plants and animals from South America must have migrated to the Galapagos Islands. These few plants and animals then changed during the years they lived in their new home, giving rise to many new species. Evolutionary theory proposes that species change over time in response to environmental challenges.

How did **geology influence Darwin**?

While traveling aboard the HMS *Beagle*, Charles Darwin read the *Principles of Geology* by Charles Lyell (1797–1875). Catastrophism was the popular theory of the time about the forces driving geological change. Lyell's theory suggested that geologic change was not solely the result of random catastrophes. Rather, he proposed that geologic formations were most often the result of everyday occurrences like storms, waves, volcanic eruptions, and earthquakes that could be observed within an individual lifetime. This idea, that the same geologic processes at work today were also present during our evolutionary past, is known as Uniformitarianism. This conclusion also led Lyell and, before him, James Hutton (1726–1797), to suggest that Earth must be much older than the previously accepted age of 6,000 years, because these uniform processes would have required many millions of years to generate the structures he observed. Reading Lyell's work gave Darwin a new perspective as he traveled through South America and sought a mechanism by which he could explain his thoughts on evolution.

Who was **Alfred Russel Wallace**?

Alfred Russel Wallace (1823–1913) was a naturalist whose work was presented with Charles Darwin's at the Linnaean Society of London in 1858. After extensive travels in the Amazon basin, Wallace independently came to the same conclusions as Darwin on the significance of natural selection in driving the diversification of species. Wallace also worked as a natural history specimen collector in Indonesia. Wallace, like Darwin, also read the work of Thomas Malthus (1766–1834). During an attack of malaria in Indonesia, Wallace made the connection between the Malthusian concept of the struggle for existence and a mechanism for change within populations. From this, Wallace wrote the essay that was eventually presented with Darwin's work in 1858.

Who was **"Darwin's bulldog"**?

Thomas Huxley (1825–1895) was a staunch supporter of Darwin's work; in fact, Huxley wrote a favorable review of Darwin's *On the Origin of Species* that appeared soon after its publication. When the firestorm of controversy began after the appearance of Darwin's book, Huxley was ready and able to defend Darwin, whose chronic public reticence about his theories was at that time exacerbated by illness. Huxley's defense of Darwin was so vigorous during a debate with Bishop Samuel Wilberforce (1805–1873) at the British Association for the Advancement of Science in 1860 that he earned the title "Darwin's bulldog."

What is the significance of *On the Origin of Species*?

Charles Darwin (1809–1882) first proposed a theory of evolution based on natural selection in his treatise *On the Origin of Species*. The publication of *On the Origin of Species* ushered in a new era in our thinking about the nature of man. The intellectual revolution it caused and the impact it had on man's concept of himself and the world were considered by many to be greater than those caused by the works of Isaac Newton (1642–1727) and other individuals. The effect was immediate—the first edition sold out on the day of publication (November 24, 1859). *Origin* has been referred to as "the book that shook the world." Every modern discussion of man's future, the population explosion, the struggle for existence, the purpose of man and the universe, and man's place in nature rests on Darwin.

The work was a product of his analyses and interpretations of his findings from his voyages on the HMS *Beagle*. In Darwin's day the prevailing explanation for organic diversity was the story of creation in the book of Genesis in the Bible. *Origin* was the first publication to present scientifically sound, well-organized evidence for the theory of evolution. Darwin's theory was based on natural selection in which the best, or fittest individuals, survive more often than those who are less fit. If there is a difference in the genetic endowment among these individuals that correlates with fitness, the species will change over time, and will eventually resemble more closely (as a group) the fittest individuals. It is a two-step process: the first consists of the production of variation, and the second, of the sorting of this variability by natural selection in which the favorable variations tend to be preserved.

391

Did Charles Darwin have any nicknames?

Darwin had several nicknames. As a young naturalist on board the HMS *Beagle*, he was called "Philos" because of his intellectual pursuits and "Flycatcher" when his shipmates tired of him filling the ship with his collections. Later in his life, when he became a leader in the scientific community, journalists referred to him as the "Sage of Down" or the "Saint of Science," but his friend Thomas Henry Huxley privately called him the "Czar of Down" and the "Pope of Science." His own favorite nickname was "Stultis the Fool," and he often signed letters to scientific friends with "Stultis." This name referred to his habit of trying experiments most people would prejudge to be fruitless or fool's experiments.

Who coined the phrase "survival of the fittest"?

Although frequently associated with Darwinism, this phrase was coined by Herbert Spencer (1820–1903), an English sociologist. It is the process by which organisms that are less well adapted to their environment tend to perish and better-adapted organisms tend to survive.

What is the Darwin-Wallace theory?

The Darwin-Wallace theory can be summarized as the following: Species as a whole demonstrate descent with modification from common ancestors, and natural selection is the sum of the environmental forces that drive those modifications. The modifications or adaptations make the individuals in the population better suited to survival in their environment, more "fit" as it were.

The four postulates presented by Darwin in *On the Origin of Species by Natural Selection* are as follows: 1) Individuals within species are variable. 2) Some of these variations are passed on to offspring. 3) In every generation more offspring are produced than can survive. 4) The survival and reproduction of individuals are not random; the individuals who survive and go on to reproduce the most are those with the most favorable variation. They are naturally selected. It follows logically from these that the characteristics of the population will change with each subsequent generation until the population becomes distinctly different from the original; this process is known as evolution.

Why is evolution a theory?

A scientific theory is an explanation of observed phenomena that is supported by the available scientific data. The term "theory" is used as an indication that the explanation will be modified as new data becomes available. For example, the Darwin-Wallace theory was proposed prior to the discovery of the molecular nature of genetics but has since been expanded to encompass that information as well.

Which **scientific disciplines** provide **evidence** for **evolution**?

Although information from any area of natural science is relevant to the study of evolution, there are several in particular that directly support the work of Darwin and Wallace. Paleobiology, geology, and organic chemistry provide insight on how living organisms have evolved. Ecology, genetics, and molecular biology also demonstrate how living species are currently changing in response to their environments and therefore undergoing evolution.

What is **Müllerian mimicry**?

Fritz Müller (1821–1897), a German-born zoologist, described a phenomenon in 1878 in which a group of species with the same adaptations against predation was also of similar appearance. This phenomenon is now called Müllerian mimicry. Müllerian mimics include wasps and bees, all of which have similar yellow-and-black-striped patterns that serve as a warning to potential predators.

What is **Batesian mimicry**?

In 1861 Henry Walter Bates (1825–1892), a British naturalist, proposed that a nontoxic species can evolve (especially in color and color pattern) to look or act like a toxic or unpalatable species in order to avoid being eaten by a predator. The classic example is the viceroy butterfly, which resembles the unpalatable monarch butterfly. In another example, the larva of the hawk moth puffs up its head and thorax when disturbed, looking like the head of a small poisonous snake, complete with eyes. The mimicry even involves behavior; the larva weaves its head back and forth and hisses like a snake. This is called Batesian mimicry.

What is the **Oparin-Haldane hypothesis**?

In the 1920s, while working independently, Alexandr Oparin (1894–1980) and John Haldane (1892–1964) both proposed scenarios for the "prebiotic" conditions on Earth (the conditions that would have allowed organic life to evolve). Although they differed on details, both models described an early Earth with an atmosphere containing ammonia and water vapor. Both also surmised that the assemblage of organic molecules began in the atmosphere and then moved into the seas. The steps of the Oparin-Haldane model are:

1. Organic molecules, including amino acids and nucleotides, are synthesized abiotically (without living cells).

2. Organic building blocks in the prebiotic soup are assembled into polymers of proteins and nucleic acids.

3. Biological polymers are assembled into a self-replicating organism that fed on the existing organic molecules.

Who **verified** the **Oparin-Haldane hypothesis**?

In 1953 (the year that Watson and Crick published their famous paper on DNA structure), Stanley Miller (1930–2007), a graduate student in the lab of Harold Urey (1893–1981), built an apparatus that mimicked what was then thought to be the atmosphere of early Earth, a reducing atmosphere containing methane, ammonia, and hydrogen. In the closed chamber, Miller boiled water, then exposed it to electric shocks and then cooled it. After the apparatus ran for a period of days, Miller tested the water and found several amino acids, which are the building blocks of proteins. Eventually, scientists replicating the Miller-Urey experiment were able to generate other types of amino acids as well as nucleotides (basic units of DNA) and sugars.

What is **punctuated equilibrium**?

Punctuated equilibrium is a model of macroevolution first detailed in 1972 by Niles Eldredge (1943–) and Stephen J. Gould (1941–2002). It can be considered either a rival or supplementary model to the more gradual-moving model of evolution posited by neo-Darwinism. The punctuated equilibrium model essentially asserts that most of geological history shows periods of little evolutionary change, followed by short (geologically speaking, a few million years) periods of time of rapid evolutionary change. Gould and Eldredge's work has been buttressed by the discovery of the Hox genes that control embryonic development. Hox genes are found in all vertebrates and many other species as well; they control the placement of body parts in the developing embryo. Relatively minor mutations in these gene sequences could result in major body changes for species in a short period of time, thereby giving rise to new forms of organisms and therefore new species.

How do **fossils form**?

Fossils are the preserved remains of once-living organisms. Fossils form rarely, since an organism is usually consumed totally or scattered by scavengers after death. If the structures remain intact, fossils can be preserved in amber (hardened tree sap), Siberian permafrost, dry caves, or rock. Rock fossils are the most common. In order to form a rock fossil, three things must happen:

1. The organism must be buried in sediment

2. The hard structures must mineralize

3. The sediment surrounding the fossil must harden and become rock.

What was the Scopes (monkey) trial?

John T. Scopes (1900–1970), a high-school biology teacher, was brought to trial by the State of Tennessee in 1925 for teaching the theory of evolution. He challenged a recent law passed by the Tennessee legislature that made it unlawful to teach in any public school any theory that denied the divine creation of man. He was convicted and sentenced, but the decision was reversed later and the law repealed in 1967.

In the early twenty-first century, pressure against school boards still affects the teaching of evolution. Recent drives by anti-evolutionists either have tried to ban the teaching of evolution or have demanded "equal time" for "special creation" as described in the biblical book of Genesis. This has raised many questions

John T. Scopes (right) with one of his attorneys, Dr. John R. Neal.

about the separation of church and state, the teaching of controversial subjects in public schools, and the ability of scientists to communicate with the public. The gradual improvement of the fossil record, the result of comparative anatomy, and many other developments in biological science have contributed toward making evolutionary thinking more palatable.

Many rock fossils either erode before they can be discovered, or they remain in places inaccessible to scientists. The value of fossils comes not only from the information they give us about the structures of those once-living organisms, but the placement of common fossils in the geologic layers also gives researchers a method for dating other, lesser known, samples.

How did **humans evolve**?

Evolution of the Homo lineage of modern humans (*Homo sapiens*) has been proposed to originate from a hunter of nearly 5 feet (1.5 meters) tall, *Homo habilis*, who is widely presumed to have evolved from an australopithecine ancestor. Near the beginning of the Pleistocene epoch (two million years ago), *Homo habilis* is thought to have transformed into *Homo erectus* (Java Man), who used fire and possessed culture. Middle Pleistocene populations of *Homo erectus* are said to show steady evolution toward the anatomy of *Homo sapiens* (Neanderthals, Cro-Magnons, and modern humans), 120,000 to 40,000 years ago. Pre-modern *Homo sapiens* built huts and made clothing.

What is the **Red Queen hypothesis**?

This hypothesis, also called the law of constant extinction, is named after the Red Queen in Lewis Carroll's (1832–1898) *Through the Looking Glass*, who said, "Now here, you see, it takes all the running you can do to keep in the same place." The idea is that an evolutionary advance by one species represents a deterioration of the environment for all remaining species. This places pressure on those species to advance just to keep up.

CLASSIFICATION, LABORATORY TOOLS, AND TECHNIQUES

How has the **classification of organisms changed** throughout history?

From Aristotle (384–322 B.C.E.) to Carolus Linnaeus (1707–1778), scientists who proposed the earliest classification systems divided living organisms into two kingdoms—plants and animals. During the nineteenth century, Ernst Haeckel (1834–1919) proposed establishing a third kingdom—Protista—for simple organisms that do not appear to fit in either the plant or animal kingdom. In 1969 R. H. Whitaker (1920–1980) proposed a system of classification based on five different kingdoms. The groups Whitaker suggested were the bacteria group Prokaryotae (originally called Monera), Protista, Fungi (for multicellular forms of nonphotosynthetic heterotrophs and single-celled yeasts), Plantae, and Animalia. This classification system is still widely accepted; however a six-kingdom system of classification was proposed in 1977 by Carl Woese (1928–). The groups proposed in the six-kingdom approach are Archaebacteria and Eubacteria (both for bacteria), Protista, Fungi, Plantae, and Animalia. In 1981 Woese proposed a classification system based on three domains (a level of classification higher than kingdom): Bacteria, Archaea, and Eukarya. The domain Eukarya is subdivided into four kingdoms: Protista, Fungi, Plantae, and Animalia.

Who devised one of the **earliest systems** for animal and plant **classification**?

The naming and organizing of the millions of species of plants and animals is frequently called taxonomy; such classifications provide a basis for comparisons and generalizations. Carolus Linnaeus composed a classification system for two major groups: plants (1753) and animals (1758). He categorized the organisms by perceived physical differences and similarities. Every plant and animal was given two scientific names (binomial method) in Latin, one for the species and the other for the group or genus within the species. This system of nomenclature (naming) continues to be used today. Examples of binomical nomenclature include: *Schistocerca americana* (grasshopper) and *Pinus strobus* (white pine).

What is the **difference** between a **domain** and a **kingdom**?

A domain is a taxonomic category above the kingdom level. The three domains are: Bacteria, Archaea, and Eukarya, which are the major categories of life. Essentially, domains are superkingdoms. A kingdom is a taxonomic group that contains one or more phyla. The four traditional kingdoms of Eukarya include: Protista, Fungi, Plantae, and Animalia.

What are the **major characteristics** of each **kingdom** of living organisms?

Kingdom	Cell Type	Characteristics
Monera (Bacterial Kingdoms and Archaean Kingdoms)	Prokaryotic	Single cells lacking distinct nuclei and other membranous organelles
Protista	Eukaryotic	Mainly unicellular or simple multicellular, some containing chloroplasts. Includes protozoa, algae, and slime molds
Fungi	Eukaryotic	Single-celled or multicellular, including yeasts, not capable of photosynthesis
Plantae	Eukaryotic	Single-celled or multicellular, capable of photosynthesis
Animalia	Eukaryotic	Multicellular organisms, many with complex organ systems

How many different **organisms** have been **identified by biologists**?

Approximately 1.5 million different species of plants, animals, and microorganisms have been described and formally named. Some biologists believe this is only a fraction of species that exist; estimating there are more than 10 million species are waiting to be discovered, classified, and named. It is estimated that 15 percent of all species are marine organisms. Most scientists agree that only five percent of bacteria, fungi, nematode, and mite species have been discovered.

What is cell culture?

Cell culture is the cultivation of cells (outside the body) from a multicellular organism. This technique is very important to biotechnology processes because most research programs depend on the ability to grow cells outside the parent animal. Cells grown in culture usually require very special conditions (e.g., specific pH, temperature, nutrients, and growth factors). Cells can be grown in a variety of containers, ranging from a simple Petri dish to large-scale cultures in roller bottles, which are bottles that are rolled gently to keep culture medium flowing over the cells.

How does an *in vivo* study differ from an *in vitro* study?

An *in vivo* study uses living biological organisms and specimens. In contrast, an *in vitro* biological study is carried out in isolation from a living organism, such as in a Petri dish or test tube.

What elements are common to all types of microscopes?

Three elements are needed to form an image: a source of illumination, a specimen to be examined, and a system of lenses that focuses the illumination on the specimen and forms the image.

What distinguishes the different types of microscopes?

Microscopes have played a central role in the development of cell biology, allowing scientists to observe cells and cell structures that are not visible to the human eye. The two basic types of microscopes are light microscopes and electron microscopes. The major differences between light and electron microscopes are the source of illumination and the construction of the lenses. Light microscopes utilize visible light as the source of illumination and a series of glass lenses. Electron microscopes utilize a beam of electrons emitted by a heated tungsten filament as the source of illumination. The lens system consists of a series of electromagnets.

Recent advances using optical techniques have led to the development of specialized light microscopes, including fluorescence microscopy, phase-contrast microscopy, and differential interference contrast microscopy. In fluorescence microscopy, a fluorescent dye is introduced to specific molecules. Both phase-contrast microscopy and differential interference contrast microscopy utilize techniques that enhance and amplify slight changes in the phase of transmitted light as it passes through a structure that has a different refractive index than the surrounding medium.

Who **invented** the *compound microscope*?

The principle of the compound microscope, in which two or more lenses are arranged to form an enlarged image of an object, occurred independently, at about the same time, to more than one person. Certainly many opticians were active in the construction of telescopes at the end of the sixteenth century, especially in Holland, so it is likely that the idea of the microscope may have occurred to several of them independently. In all probability the date may be placed within the period 1590–1609, and the credit should go to three spectacle makers in Holland. Hans Janssen, his son Zacharias (1580–1638), and Hans Lippershey (1570–1619) have all been cited at various times as deserving chief credit. An Englishman, Robert Hooke, was the first to make the best use of a compound microscope, and his book *Micrographia*, published in 1665, contains some of the most beautiful drawings of microscopic observations ever made.

Who **invented** the **electron microscope**?

The theoretical and practical limits to the use of the optical microscope were set by the wavelength of light. When the oscilloscope was developed, it was realized that cathode-ray beams could be used to resolve much finer detail because their wavelength was so much shorter than that of light. In 1928 Ernst Ruska (1906–1988) and Max Knoll (1897–1969), using magnetic fields to "focus" electrons in a cathode-ray beam, produced a crude instrument that gave a magnification of 17, and by 1932 they had developed an electron microscope having a magnification of 400. By 1937 James Hillier (1915–2007) had advanced this magnification to 7,000. The 1939 instrument Vladimir Zworykin (1889–1982) developed gave 50 times more detail than any optical microscope ever could, with a magnification up to two million. The electron microscope revolutionized biological research: for the first time scientists could see the molecules of cell structures, proteins, and viruses.

Which **radioisotope** is most commonly **used** for **biological specimens**?

The most widely used radioisotope in autoradiography is tritium (^{3}H). Tritium allows a resolution of about 1 micrometer with the light microscope and close to 0.1 micrometer with the electron microscope.

Since hydrogen is common in biological molecules, a wide-range of ^{3}H-labeled compounds are potentially available for use in autoradiography. ^{3}H-amino acids are used for locating newly synthesized

Electron microscopes utilize a beam of electrons emitted by a heated tungsten filament as the source of illumination.

proteins, ^{3}H-thymidine is used to monitor DNA synthesis, ribonucleiodies such as ^{3}H-uridine or ^{3}H-cytidine are used to localize newly made RNA molecules, and ^{3}H-glucose is used to study the synthesis of polysaccharides.

How has **centrifugation** been used in **biological applications**?

Centrifugation is the separation of immiscible liquids or solids from liquids by applying centrifugal force. Since the centrifugal force can be very great, it speeds the process of separating these liquids instead of relying on gravity. Biologists primarily use centrifugation to isolate and determine the biological properties and functions of subcellular organelles and large molecules. They study the effects of centrifugal forces on cells, developing embryos, and protozoa. These techniques have allowed scientists to determine certain properties about cells, including surface tension, relative viscosity of the cytoplasm, and the spatial and functional interrelationship of cell organelles when redistributed in intact cells.

What are the **uses** for **chromatography**?

Chromatography is used to separate and identify the chemicals in a mixture. It is useful to: separate and identify the chemicals in a mixture; check the purity of a chemical product; identify impurities in a product; and purify a chemical product (on a laboratory or industrial scale).

How is **chromatography used** to **identify** individual **compounds**?

Chromatography is another technique used to separate mixtures into their individual components. The most common chromatographic techniques are paper chromatography, gas-liquid chromatography (also called gas chromatography), thin layer chromatography, and high pressure (or high performance) liquid chromatography (HPLC). All methods of chromatography share common characteristics. The process is based on the principle that different chemical compounds will stick to a solid surface, or dissolve in a film of liquid, to different degrees. Chromatography involves a sample (or sample extract) being dissolved in a mobile phase (which may be a gas, a liquid, or a supercritical fluid). The mobile phase is then forced through an immobile, immiscible stationary phase. The phases are chosen in such a manner that components of the sample have differing solubilities in each phase. The least soluble component is separated first, and as the separation process continues, the components are separated by increasing solubility.

What is **spectroscopy**?

Spectroscopy includes a range of techniques to study the composition, structure, and bonding of elements and compounds. The different methods of spectroscopy use different wavelengths of the electromagnetic spectrum to study atoms, molecules, ions, and the bonding between them.

Type of Spectroscopy	Wavelength Used
Nuclear magnetic resonance spectroscopy	Radio waves
Infrared spectroscopy	Infrared radiation
Atomic absorption spectroscopy, atomic emission spectroscopy, and ultraviolet spectroscopy	Visible and UV radiation
X-ray spectroscopy	X rays

What is **bioinformatics**?

Bioinformatics is the field of science in which biology, computer science, and information technology merge into a single discipline. The ultimate goal of the field is to enable the discovery of new biological insights as well as to create a global perspective from which unifying principles in biology can be discerned. There are three important sub-disciplines within bioinformatics: 1) the development of new algorithms and statistics with which to assess relationships among members of the large data sets; 2) the analysis and interpretation of various types of data including nucleotide and amino acid sequences, protein domains, and protein structures; and 3) the development and implementation of tools that enable efficient access and management of different types of information.

PLANT WORLD

INTRODUCTION AND HISTORICAL BACKGROUND

What are the **general characteristics** of a plant?

A plant is a multicellular, eukaryotic organism with cellulose-rich cell walls and chloroplasts, which has starch as the primary carbohydrate food reserve. Plants are primarily terrestrial, autotrophic (capable of making their own food) organisms. Most plants contain chlorophylls a and b and xanthophylls (yellow pigments) as well as carotenes (orange pigments).

What are the **major subdivisions** of **botany**?

The major subdivisions of botany are:

Agronomy—The application of plant science to crop production.

Bryology—The study of mosses and liverworts.

Economic botany—The study of the utilization of plants by humans.

Ethnobotany—The study of the use of plants by indigenous peoples.

Forestry—The study of forest management and the utilization of forest products.

Horticulture—The study of ornamental plants, vegetables, and fruit trees.

Paleobotany—The study of fossil plants.

Palynology—The study of pollen and spores.

Phytochemistry—The study of plant chemistry, including the chemical processes that take place in plants.

Plant anatomy—The study of plant cells and tissues.

403

Plant ecology—The study of the role plants play in the environment.

Plant genetics—The study of genetic inheritance in plants.

Plant morphology—The study of plant forms and life cycles.

Plant pathology—The study of plant diseases.

Plant physiology—The study of plant function and development.

Plant systematics—The study of the classification and naming of plants.

A bust of the Greek philosopher Theophrastus, who is considered the "father of botany."

Who is known as the **founder of botany**?

The ancient Greek Theophrastus (c. 372–c. 287 B.C.E.) is known as the father of botany. His two works on botany, *On the History of Plants* and *On the Causes of Plants*, were so comprehensive that 1,800 years passed before any new significant botanical information was discovered. He integrated the practice of agriculture into botany and established theories regarding plant growth and the analysis of plant structure. He related plants to their natural environment and identified, classified, and described 550 different plants.

What is *Gray's Manual*?

Gray's Manual of Botany, first published in 1848 by Asa Gray (1810–1888) under the title *Manual of the Botany of Northern United States*, was one of the first guides to the identification of plants of eastern North America. The publication contained keys and thorough descriptions of plants. The eighth, and centennial, edition was largely rewritten and expanded by Merritt Lyndon Fernald (1873–1950) and published in 1950. This edition was corrected and updated by R. C. Rollins and reprinted in 1987 by Dioscorides Press.

What contributions did **John** and **William Bartram** make to botany?

John Bartram (1699–1777) was the first American-born botanist. He and his son, William Bartram (1739–1823), traveled throughout the American colonies observing the flora and fauna. Although John Bartram never published his observations, he was considered the authority on American plants. In 1791 his son William published his notes on American plants and animals as *Bartram's Travels*.

What was the historical significance of hemp?

During the early years of colonial America, hemp (*Agave sisalana*) was as common as cotton is now. It was an easy crop to grow, requiring little water, no fertilizers, and no pesticides. The fabric looks and feels like linen. It was used for uniforms of soldiers, paper (the first two drafts of the Declaration of Independence were written on hemp paper), and an all-purpose fabric. Betsy Ross's flag was made of red, white, and blue hemp.

How has **plant classification changed** over the years?

The earliest classifications of plants were based on whether the plant was considered medicinal or was shown to have other uses. *De re Rustica* by Cato the Censor (234–149 B.C.E.) lists 125 plants and was one of the earliest catalogs of Roman plants. Gaius Plinius Secundus (23–79 C.E.), known as Pliny the Elder, wrote *Historia naturalis*, which was published in the first century. The book was one of the earliest catalogs of significant plants in the ancient world, describing more than 1,000 plants. Plant classification became more complicated as more and more plants were discovered. One of the earliest plant taxonomists was the Italian botanist Caesalpinus (1519–1603). In 1583 he classified more than 1,500 plants according to various attributes, including leaf formation and the presence of seeds or fruit.

John Ray (1627–1705) was the first botanist to base plant classification on the presence of multiple similarities and features. His *Historia Plantarum Generalis,* published between 1686 and 1704, was a detailed classification of more than 18,000 plants. The book included a distinction between monocotyledon and dicotyledon flowering plants. The French botanist J.P. de Tournefort (1656–1708) was the first to characterize genus as a taxonomic rank that falls between the ranks of family and species. De Tournefort's classification system included 9,000 species in 700 genera. The Swedish naturalist Carolus Linnaeus (1707–1778) published *Species Plantarum* in 1753. It organized plants into 24 classes based on reproductive features. The Linnaean system of binomial nomenclature remains the most widely used system for classifying plants and animals. It is considered an artificial system since it often does not reflect natural relationships.

During the late eighteenth century several natural systems of classification were proposed. The French botanist Antoine Laurent de Jussieu (1686–1758) published *Genera Plantarum*. The tome *Prodromus Systematis Naturalis Regni Vegetabilis* was started in 1824 by the Swiss botanist Augustin Pyrame de Candolle (1778–1841) and completed 50 years later. Another *Genera Plantarum* was published between 1862 and 1883 by the English botanists George Bentham (1800–1884) and Sir Joseph Dalton Hooker (1817–1911).

Charles Darwin's (1809–1882) ideas on evolution began to influence systems of classification during the late nineteenth century. The first major phylogenetic system of plant classification was proposed around the close of the nineteenth century. *Die natürlichen Pflanzenfamilie* (*The Natural Plant Families*), one of the most complete phylogenetic systems of classification and still in use through the twenty-first century, was published between 1887–1915 by the German botanists Adolf Engler (1844–1930) and Karl Prantl (1849–1893). Their system recognizes about 100,000 species of plants, organized by their presumed evolutionary sequence.

Systems of classifications were also developed during the twentieth century. Some works focused on groups of plants, especially flowering plants, rather than all plants. Charles Bessey (1845–1915) was the first American scientist to publish a system of classification in the early twentieth century. Cladistics is one of the newest approaches to classification. It is often defined as a set of concepts and methods for determining cladograms, which portray branching patterns of evolution.

What is the **origin** of **land plants**?

Many scientists believe land plants evolved from green algae. Green algae, especially the charaophytes, share a number of biochemical and metabolic traits with plants. Both contain the same photosynthetic pigments—carotenes, xanthophylls, as well as chlorophylls *a* and *b*. Cellulose is a major component of the cell walls of plants and algae, and both store their excess carbohydrates as starch. In addition, some aspects of cell division, particularly the formation of new cross-walls, only occurs in plants and certain charaophytes, such as species of the genera *Cerara* and *Colechaete*.

How are **tree rings** used to **date** historical **events**?

The study of tree rings is known as dendrochronology. Every year, trees produce an annular ring composed of one wide, light ring and one narrow, dark ring. During spring and early summer, tree stem cells grow rapidly larger, thus producing the wide, light ring. In winter, growth is greatly reduced and cells are much smaller, producing the narrow, dark ring. In the coldest part of winter or the dry heat of summer, no cells are produced. Comparing pieces of dead trees of unknown age with the rings of living trees allows scientists to establish the date when the fragment was part of a living tree. This technique has been used to date the ancient pueblos throughout the southwestern United States. A subfield of dendrochronology is dendroclimatology. Scientists study the tree rings of very old trees to determine climatic conditions of the past. The effects of droughts, pollution, insect infestations, fires, volcanoes, and earthquakes are all visible in tree rings.

Leonardo da Vinci was the first to figure out that you could calculate the age of a tree by counting the number of rings in a cross section of the trunk.

How does a **plant become** a **fossil**?

Fossilization is dependent upon where organisms grow and how quickly they are covered by sediment. Rarely do paleobotanists find the fossil remains of whole plants. Usually only fossilized parts of plants are found. Fossilization occurs in many different ways. Three common methods of fossilization are compression, impression, and molding or casting.

Compression fossils are often formed in water, where heavy sediment flattens leaves or other plant parts. The weight of the sediment squeezes out water present in the plant tissue, leaving only a thin film of tissue. An impression fossil is an imprint of an organism that is left behind when the organism's remains have been completely destroyed, leaving only the contour of the plant. Fossil molds and casts are formed when animal or plant tissues become surrounded by hardened sediment; the tissue then decays. The hollow negative created by the tissue is called a mold. When fossil molds fill with sediment over time, the sediment often conforms to the contours of the mold, resulting in a fossil called a cast.

What was one of the **most-famous criminal cases** involving **forensic botany**?

Forensic botany is the identification of plants or plant products; this form of study can be used to produce evidence for legal trials. One of the first criminal cases to use forensic botany was the famous 1935 trial of Bruno Hauptmann (1899–1936), who was accused, and later convicted, of kidnapping and murdering the son of Charles (1902–1974) and Anne Morrow (1906–2001) Lindbergh. The botanical evidence presented in the case centered on a homemade wooden ladder used during the kidnap-

407

ping and left at the scene of the crime. After extensive investigation, the plant anatomist Arthur Koehler (1885–1967) showed that parts of the ladder were made from wooden planks taken from Hauptmann's attic floor.

How did the **introduction** of the **potato** to Europe **lead to** a devastating **famine** in **Ireland**?

The white potato (*Solanum tuberosum*), native to South America, was first introduced to Spain in the middle of the sixteenth century. It was not widely accepted as a food crop since European relatives of the potato, such as nightshade, mandrake, and henbane, were known to be poisonous or hallucinogenic. In fact, all of the above-ground parts of a potato plant are poisonous and only the tuber is edible. The potato was established as a food crop in Ireland as early as 1625 and became a staple of the diet, especially among the poor, during the eighteenth and early nineteenth centuries. The widespread dependence on potatoes as a main source of food led to massive starvation when the plant pathogen *Phytophthora infestans* destroyed potato fields in the 1840s. Over one million Irish people died from starvation or subsequent disease; another 1.5 million emigrated from Ireland.

How have **dill** and **anise** been **used** throughout history?

Dill (*Anethum graveolen*) has long been used for medicinal purposes. The Egyptians used dill as a soothing medicine. Greeks habitually used the herb to cure the hiccups. During the Middle Ages dill was prized for the protection it purportedly provided against witchcraft. Magicians and alchemists used dill to concoct spells, while a commonly known wives' tale stated that dill added to wine could enhance passion. Colonial settlers brought dill to North America, where it became known as "meetin' seed" because children were given dill seed to chew during long sermons in church.

The Romans brought the licorice-flavored herb, anise (*Pimpinella anisum*) from Egypt to Europe, where they used it as payment for their taxes. It became a popular flavoring for cakes, cookies, bread, and candy.

Is **coffee native** to **Columbia** or **Brazil**?

Although premium coffee is today grown in the mountains of Central and South America, the coffee tree (*Coffea arabica*) is native to Ethiopia. Coffee was widely used in the Arab world before its introduction to European society in the seventeenth century. North and South American coffee plantations were started in the eighteenth century.

Which **plants** have been used to **create dyes**?

Natural materials, including many plants, were the source of all dyes until the late nineteenth century. Blue dye was historically rare and was obtained from the indigo

> ## What is the greatest number of leaves a clover can have?
>
> A 14-leafed white clover (*Trifolium repens*) and a 14-leafed red clover (*Trifolium pretense*) have been found in the United States.

plant (*Indigofera tinctoria*). Another color difficult to obtain for dye was red. The madder plant (*Rubia tinctorum*) was an excellent source of red dye and was used for the famous "red coats" of the British Army. Native Americans painted their faces and dyed their clothes red with the root of the bloodroot wildflower, which is also called redroot, Indian paint, and tetterwort. Bloodroot (*Sanguinaria Canadensis*), found in shady, damp, and woodsy soils, blooms in May and has white flowers that are 2 inches (5 centimeters) wide. Other, more common, natural dyes derived from plant sources are summarized in the following chart:

Common Name	Scientific Name	Part of Plant Used	Color
Black walnut	*Juglans nigra*	Hulls	Dark brown, black
Coreopsis	*Coreopsis*	Flower heads	Orange
Lilac	*Syringa*	Purple flowers	Green
Red cabbage	*Brassica oleracea-capitata*	Outer leaves	Blue, lavender
Turmeric	*Curcuma longa*	Rhizome	Yellow
Yellow onion	*Allium cepa*	Brown, outer leaves	Burnt orange

How much do **plants contribute** to the **human diet?**

In the United States and Western Europe approximately 65 percent of a human being's total caloric intake and 35 percent of consumed protein are obtained from plants or plant products. Soybeans are an example of a plant with high protein content. In developing nations almost 90 percent of calories and more than 80 percent of protein in a person's diet are from plants.

What **plants** are commonly **used** in the **perfume** industry?

Perfumes are made of a mixture of a large variety of scents. Although many perfumes are created synthetically, the expensive designer scents still use natural essential oils extracted from plants. The perfume industry uses all parts of the plant to create a unique blend of scents. Some commonly used plant materials for essential-oil extraction are:

Plant Organ	Source
Bark	Indonesia and Ceylon cinnamons, and cassia
Flowers	Rose, carnation, orange blossoms, ylang-ylang, violet, and lavender

Plant Organ	Source
Gums	Balsam and myrrh
Leaves and stems	Rosemary, geranium, citronella, lemon grass, and a variety of mints
Rhizomes	Ginger
Roots	Sassafras
Seeds and fruits	Orange, lemon, and nutmeg
Wood	Cedar, sandalwood, and pine

Why were **tomatoes** often called **"love apples"** and considered **aphrodisiacs**?

Tomatoes belong to the nightshade family; they were cultivated in Peru and introduced to Europe by Spanish explorers. Tomatoes were introduced to Italy from Morocco, so the Italian name for the fruit was *pomi de Mori* (meaning "apples of the Moors"). The French called the tomato *pommes d'amore* (meaning "apples of love"). This latter name may have referred to the fact that tomatoes were thought to have aphrodisiac powers, or it may have been a corruption of the Italian name. When tomato plants were first introduced to Europe, many people viewed them with suspicion, since poisonous members of the nightshade family were commonly known. Although the tomato is neither poisonous nor an aphrodisiac, it took centuries for it to fully overcome its undeserved reputation.

What **parts** of **plants** are **used** for **spices**?

Spices are aromatic seasonings derived from many different parts of plants including the bark, buds, fruit, roots, seeds, and stems. Some common spices and their sources are:

Spice	Scientific Name	Part Used
Allspice	*Pimenta dioica*	Fruit
Black pepper	*Piper nigrum*	Fruit
Capsicum peppers	*Capsicum annum*	Fruit
	Capsicum baccatum	
	Capsicum chinense	
	Capsicum frutescens	
Cassia	*Cinnamomum cassia*	Bark
Cinnamon	*Cinnamomum zeylanicum*	Inner bark
Cloves	*Eugenia caryophyllata*	Flower
Ginger	*Zingiber officinale*	Rhizome
Mace	*Myristica fragrans*	Seed
Nutmeg	*Myristica fragrans*	Seed
Saffron	*Crocus sativus*	Stigma
Turmeric	*Curcuma longa*	Rhizome
Vanilla	*Vanilla planifolia*	Fruit

> ## How did the search for cinnamon lead to the discovery of North and South America?
>
> Christopher Columbus (1451–1506) was one of many explorers trying to find a direct sea route to Asia, which during the fifteenth century was thought to be rich with spices. Cinnamon and other spices were so valued in Columbus's era that a new, direct route to Asia would have brought untold wealth to the discoverer and his country.

What is the most **expensive spice** in the world?

The world's most expensive spice is saffron. The spice was highly sought after by the ancient civilizations of Egypt, Assyria, Phoenicia, Persia, Crete, Greece, and Rome. The term "saffron" comes from the Arabic word *za'faran*, meaning "yellow."

The spice is obtained from the delicate stigmas of an autumn crocus, *Crocus sativus*, a species native to eastern Mediterranean countries and Asia Minor. Spain is the principal producer of saffron for world markets. *Crocus sativus* is propagated by corms. The blooming period for the crocus is approximately two weeks, after which the flowers must be picked while they are in full bloom and before any signs of wilting. Once picked, the three-part stigmas are removed from the petals before the petals wilt; this is a time-consuming process that can only be done by hand as the stigmas are very fragile. Then the stigmas are roasted and sold either as whole threads (whole stigmas) or powder.

In order to harvest 1 pound (0.45 kilogram) of the spice, between 75,000 and 100,000 flowers must be picked. Approximately 4,000 stigmas yield 1 ounce (28 grams) of the spice. In 2010 it was reported that wholesale price of low-quality saffron was $500 per pound ($1,100 per kilogram). The average retail price in Western countries is $1,000 per pound ($2,200 per kilogram), confirming its place as the world's most costly spice.

What are some common **culinary herbs**?

Herbs are often used to enhance flavors in food. They are usually from the leaves of nonwoody plants.

Common Name	Scientific Name	Part Used
Basil	*Ocimum basilicum*	Leaves
Bay leaves	*Laurus nobilis*	Leaves
Cumin	*Cuminum cyminum*	Fruit
Dill	*Anethum graveolens*	Fruit, leaves

Common Name	Scientific Name	Part Used
Garlic	*Allium satiavum*	Bulbets
Mustard	*Brassica alba; Brassica nigra*	Seed
Onion	*Allium cepa*	Bulb
Oregano	*Origanum vulgare*	Leaves
Parsley	*Petroselinum crispum*	Leaves
Peppermint	*Mentha piperita*	Leaves
Sage	*Salvia officinalis*	Leaves
Tarragon	*Artemesia dracunculus*	Leaves
Thyme	*Thymus vulgaris*	Leaves

What is the **most dangerous poison** produced by a **plant**?

In North America the poisonous hemlock, *Conium maculatum*, is probably the most dangerous plant. The South American lana tree is another dangerous plant. Native Americans used its sap to make curare, a poisonous substance applied to their arrows and spears. It can cause death in a matter of minutes.

What **plants** produce **essential oils** that are commonly used in **aromatherapy**?

Aromatherapy is a holistic approach to healing using essential oils extracted from plants. Holistic medicine looks at the health of the whole individual, and treatments emphasize the connection of mind, body, and spirit. The term "aromatherapy" was first used by Rene Gattefosse (1881–1950), a French perfume chemist. He discovered the healing powers of lavender oil following a laboratory accident during which he burned his hand. Gattefosse began to investigate the properties of lavender oil and other essential oils and published a book on plant extracts. During aromatherapy treatments, essential oils are absorbed through breath or the pores of the skin; this process triggers certain physiological responses. Examples of essential oils and their uses are:

Essential Oil	Common Uses
Cypress	Antiseptic, asthma, coughing, relaxation
Eucalyptus	Anti-inflammatory, arthritis, relaxation
Frankincense	Coughing, bronchitis
Geranium	Dermatitis, relaxation, depression
Ginger	Bronchitis, arthritis, stimulant
Juniper	Antiseptic, aches, pains, relaxation
Lavender	Antiseptic, respiratory infections, relaxation
Marjoram	Respiratory infections, relaxation
Pine	Asthma, arthritis, depression
Roman chamomile	Toothaches, arthritis, tension

What are some specific examples of how plants are economically important?

Materials of plant origin are found in a wide variety of industries including paper, food, textile, and construction. Chocolate is made from cocoa seeds, specifically seeds of the species *Theobroma cacao*. Foxglove (*Digitalis purpurea*) contains cardiac glycosides used to treat congestive heart failure. The berries obtained from the plant *Piper nigrum* produce black pepper. The berries are dried, resulting in black peppercorns, which can then be cracked or ground. Tea can be made from the leaves of *Camellia sinensis*. Fiber taken from the stem of flax plants (*Linum usitatissimum*) has been used to make linen, while the flax seeds are commonly consumed and are a source of linseed oil. Paper money is even made from flax fibers.

Essential Oil	Common Uses
Bulgarian rose	Antiseptic, insomnia, relaxation
Rosemary	Bronchitis, depression, mental alertness
Sandalwood	Acne, bronchitis, depression
Tea tree	Respiratory infections, acne, depression

Who **developed plant breeding** into a modern science?

Luther Burbank (1849–1926) developed plant breeding as a modern science. His breeding techniques included crosses of plant strains native to North America and foreign strains. He obtained seedlings that were then grafted onto fully developed plants for an appraisal of hybrid characteristics. His keen sense of observation allowed him to recognize desirable characteristics, enabling him to select only varieties that would be useful. One of his earliest hybridization successes was the Burbank potato, from which more than 800 new strains and varieties of plants—including 113 varieties of plums and prunes—were developed. More than 20 of these plums and prunes are still commercially important today.

What were some of the **accomplishments** of **Dr. George Washington Carver**?

Because of the work of Dr. George Washington Carver (1864–1943) in plant diseases, soil analysis, and crop management, many southern farmers who adopted his methods increased their crop yields and profits. Carver developed recipes using cowpeas, sweet potatoes, and peanuts. He eventually made 118 products from sweet potatoes, 325 from peanuts, and 75 from pecans. He promoted soil diversification and the adoption of peanuts, soybeans, and other soil-enriching crops. His other work included develop-

George Washington Carver.

ing plastic material from soybeans, which Henry Ford (1863–1947) later used in part of his automobile. Carver also extracted dyes and paints from the Alabama red clay and worked with hybrid cotton. He was a widely talented man who became an almost mythical American folk hero.

When was the **first** practical **greenhouse built**?

French botanist Jules Charles constructed one in 1599 in Leiden, Holland, which housed tropical plants grown for medicinal purposes. The most popular plant there was an Indian date called the tamarind, whose fruit was made into a curative drink.

Who established the **first botanical garden** in the **United States**?

John Bartram (1699–1777) planned and laid out a botanical garden of 5 to 6 acres (2 to 2.5 hectare) in 1728. It is located in Philadelphia, Pennsylvania.

When was the **first plant patent** issued?

Henry F. Bosenberg, a landscape gardener, received U.S. Plant Patent no. 1 on August 18, 1931, for a climbing or trailing rose.

PLANT DIVERSITY

What are the **four major groups** of **plants**?

They are: nonvascular; seedless vascular; flowering, seed-bearing vascular; and nonflowering, seed-bearing vascular. Plants are divided into phyla based on whether they are vascular (containing vascular tissue consisting of cells joined into tubes that transport water and nutrients) or nonvascular. The phyla of vascular plants are then further divided into seedless plants and those that contain seeds. Plants with seeds are divided into flowering and nonflowering groups. Nonvascular plants have traditionally

> ## What feature of liverworts hints to their name?
>
> Liverworts were named during the Middle Ages, when herbalists followed the theoretical approach known as the Doctrine of Signatures. The core philosophy of this perspective was that if a plant part resembled a part of the human body, it would be useful in treating ailments of that organ or part. The thallus of thalloid liverworts resembles a lobed liver. Therefore, in line with the philosophy presented by the doctrine, the plant was used to treat liver ailments. The word "liver" was combined with "wort," which means herb, to form the name "liverwort."

been called bryophytes. Because bryophytes lack a system for conducting water and nutrients, they are restricted in size and live in moist areas close to the ground. Examples of bryophytes are mosses, liverworts, and hornworts. Examples of seedless, vascular plants are ferns, horsetails, and club mosses. The conifers, which are cone-bearing, are seed-bearing, nonflowering vascular plants. The majority of plants are seed-bearing, flowering, vascular plants known as angiosperms.

What are **bryophytes** and where are they found?

Mosses, liverworts, and hornworts—collectively known as bryophytes—are often found in moist environments. However, there are species that inhabit almost all environments, from hot, dry deserts to the coldest regions of Antarctica. They are most noticeable when they grow in a dense mass. They are generally small, compact plants that rarely grow to more than 8 inches (20 centimeters) tall. They have parts that appear leaflike, stemlike, and rootlike, and lack vascular tissue (xylem and phloem).

Why are **mosses important**?

Some mosses are decomposers that break down the substrata and release nutrients for the use of more complex plants. Mosses play an important role in controlling soil erosion. They perform this function by providing ground cover and absorbing water. Mosses are also indicators of air pollution. Under conditions of poor air quality, few mosses will exist. Peat is used as fuel to heat homes and generate electricity. Bryophytes are among the first organisms to invade areas that have been destroyed by a fire or volcanic eruption.

What is unusual about **cave moss**?

Cave moss (*Schistostega pennata*) is a small plant with reflective, sub-spherical cells at its tips. These cells give off an eerie glow that is gold and green in color. In Japan the plant has been the subject of numerous books, television shows, newspaper and magazine articles, and even an opera! There is a national monument to this species near the coast of Hokkaido, where it grows near a small cave.

How can **bryophytes** be used as **bioindicators**?

Bioindicators are physiological, chemical, or behavioral changes that occur in organisms as a result of changes in the environment. Bryophytes of the genus *Hypnum* are particularly sensitive to pollutants, especially sulfur dioxide. As a result, most bryophytes are not found in cities and industrial areas. Mosses and liverworts, especially *Hypnum cupressiforme* and *Homalotecium serieceum*, were used as bioindicators to monitor radioactive fallout from the Chernobyl reactor accident in 1986.

What are the **uses** of **peat moss**?

Peat moss (genus *Sphagnum*) grows mostly in bogs. Peat mosses are favored by gardeners for their ability to increase the water-holding capacity of soils. Due to large, dead cells in their leaf-like parts, they are able to absorb five times as much water as cotton plants. Peat mosses are also used as damp cushions by florists to keep other plants and flowers damp. Species of *Sphagnum* also have medicinal purposes. Certain aboriginal people use peat mosses as disinfectants and, due to their absorbency, as diapers. Peat moss is acidic and is an ideal dressing for wounds. During World War I, the British used more than one million wound dressings made of peat moss. Native North Americans used species of the genera *Mnium* and *Bryum* to treat burns. In Europe species of the genus *Dicranoweisia* have been used to waterproof roofs.

What are the **four groups** of **seedless, vascular plants**?

The seedless, vascular plants include: ferns of the genus *Pterophyta*, which is the largest group; the whisk ferns of the genus *Psilophyta*; the club mosses of the genus *Lycophyta*; and the horsetails of the genus *Arthrophyta*.

What is the **relationship** between **ancient plants** and **coal formation**?

Coal, formed from ancient plant material, is organic. Most of the coal mined today was formed from prehistoric remains of primitive land plants, particularly those of the Carboniferous period, which occurred approximately 300 million years ago. Five main groups of plants contributed to the formation of coal. The first three groups were all seedless, vascular plants: ferns, club mosses, and horsetails. The last two groups were primitive gymnosperms and the now-extinct seed ferns. Forests of these plants were in low-lying, swampy areas that periodically flooded. When these plants died, they decomposed, but as they were covered by water, they did not decompose completely. Over a period of time the decomposed plant material accumulated and consolidated.

> ## What seedless, vascular plant played a role in early photography?
>
> Prior to the invention of flashbulbs, photographers used flash powder that consisted almost entirely of dried spores from club mosses of the genus *Lycopodium*.

Layers of sediment formed over the plant material during each flood cycle. Heat and pressure built up in these accumulated layers and converted the plant material to coal. The various types of coal (lignite, bituminous, and anthracite) were formed as a result of the different temperatures and pressures to which the layers were exposed.

What is a **fiddlehead** and how are they used in **cooking**?

The type of fern typically grown as a houseplant is of the diploid, or sporophyte, generation. It is composed of a rhizome, an underground stem that occurs horizontally, which produces roots and leaves called fronds. As each young frond first emerges from the ground, it is tightly coiled and resembles the top of a violin, hence the name fiddlehead. Fiddleheads have a chewy texture and a flavor that is a cross between asparagus, green beans, and okra. They may be cooked by steaming, simmering, or sautéing, and they are typically served as a side dish. The young fern shoots may also be served raw in salads.

Why are **horsetails** called **"scouring rushes"**?

The epidermal tissue of horsetails contains abrasive particles of silica. Scouring rushes were used by Native Americans to polish bows and arrows. Early North American settlers, who cleaned their pots and pans along stream banks, used horsetails—found in abundance in such areas—to scrub out their dishes.

What are **gymnosperms** and which plants are included in this group?

Gymnosperms (from the Greek terms *gymnos*, meaning "naked," and *sperma*, meaning "seed") produce seeds that are totally exposed or borne on the scales of cones. The four phyla of gymnosperms are: Coniferophyta, conifers including pine, spruce, hemlock and fir; Ginkgophyta, consisting of one species, the ginkgo or maidenhair tree; Cycladophyta, the cycads or ornamental plants; and Gnetophyta, a collection of very unusual vines and trees.

What plant produces the **largest seed cones**?

The largest seed cones are produced by cycads. They may be up to 1 yard (1 meter) in length and weigh more that 3.3 pounds (15 kilograms).

Cycad plants produce the larges seed cones, some weighing more than three pounds.

Do **pine trees** keep their **needles forever?**

Pine needles occur in groups, called fascicles, of two to five needles. A few species have only one needle per fascicle, while others have as many as eight. Regardless of the number of needles, a fascicle forms a cylinder of short shoots that are surrounded at their base by small, scalelike leaves that usually fall off after one year of growth. The needle-bearing fascicles are also shed a few at a time, usually every two to ten years, so that any pine tree, while appearing evergreen, has a complete change of needles every two to four years of less.

In what ways are **gymnosperms economically important?**

Gymnosperms account for approximately 75 percent of the world's timber and a large amount of the wood pulp used to make paper. In North America the white spruce, *Picea glauca*, is the main source of pulpwood used for newsprint and other paper. Other spruce wood is used to manufacture violins and similar string instruments because the wood produces a desired resonance. The Douglas fir, *Pseudotsuga menziesii*, provides more timber than any other North American tree species and produces some of the most desirable lumber in the world. The wood is strong and relatively free of knots. Uses for the wood include house framing, plywood production, structural beams, pulpwood, railroad ties, boxes, and crates. Since most naturally occurring areas of growth have been harvested, the Douglas fir is being grown in managed forests. The wood from the redwood *Sequoia sempervirens* is used for furniture, fences, posts, some construction, and has various garden uses.

In addition to the wood and paper industry, gymnosperms are important in making resin and turpentine. Resin, the sticky substance in the resin canals of conifers, is a combination of turpentine, a solvent, and a waxy substance called rosin. Turpentine is an excellent paint and varnish solvent but is also used to make deodorants, shaving lotions, medications, and limonene—a lemon flavoring used in the food industry. Resin has many uses; it is used by baseball pitchers to improve their grip on the ball and by batters to improve their grip on the bat; violinists apply resin to their bows to increase friction with the strings; dancers apply resin to their shoes to improve their grip on the stage.

What **factors** have **contributed** to the **success** of **seed plants**?

Seed plants do not require water for sperm to swim to an egg during reproduction. Pollen and seeds have allowed them to grow in almost all terrestrial habitats. The sperm of seed plants is carried to eggs in pollen grains by the wind or animal pollinators such as insects. Seeds are fertilized eggs that are protected by a seed coat until conditions are proper for germination and growth.

What are the **two major groups** of **angiosperms** and what are the major differences between the groups?

Angiosperms—made up of the largest number of plant species (240,000)—are classified into two major groups, monocots and dicots. The description of monocots and dicots is based on the first leaves that appear on the plant embryo. Monocots have one seed leaf, while dicots have two seed leaves. There are approximately 65,000 species of monocots and 175,000 species of dicots. Orchids, bamboo, palms, lilies, grains, and many grasses are examples of monocots. Dicots include most trees that are nonconiferous, shrubs, ornamental plants, and many food crops.

What are some **economically important angiosperms**?

Angiosperms produce lumber, ornamental plants, and a variety of foods. Some examples of economically important angiosperms are:

Common Family Name	Genus Name	Economic Importance
Gourd	*Cucurbitaceae*	Food (melons and squashes)
Grass	*Poaceae*	Cereals, forage, ornamentals
Lily	*Liliaceae*	Ornamentals and food (onions)
Maple	*Aceraceae*	Lumber and maple sugar
Mustard	*Brassicaceae*	Food (cabbage and broccoli)
Olive	*Oleaceae*	Lumber, oil, and food

419

What is the most widely cultivated cereal in the world?

Wheat is the most widely cultivated cereal in the world; the grain supplies a major percentage of the nutrients needed by the world's population. Wheat is one of the oldest domesticated plants, and it has been argued that it laid the foundation for Western civilization. Domesticated wheat had its origins in the Near East at least 9,000 years ago. Wheat grows best in temperate grassland biomes that receive 12 to 36 inches (30 to 90 centimeters) of rain per year and have relatively cool temperatures. Some of the top wheat-producing countries are Argentina, Canada, China, India, the Ukraine, and the United States.

Common Family Name	Genus Name	Economic Importance
Palm	*Arecaceae*	Food (coconut), fiber, oils, waxes, furniture
Rose	*Rosaceae*	Fruits (apple and cherry), ornamentals (roses)
Spurge	*Euphorbiaceae*	Rubber, medicinals (castor oil), food (cassava), ornamentals (poinsettia)

Among **angiosperms** what is the **most important** family?

Angiosperms, commonly known as flowering plants, include the grass family. This family is of greater importance than any other family of flowering plants. The edible grains of cultivated grasses, known as cereals, are the basic foods of most civilizations. Wheat, rice, and corn are the most extensively grown of all food crops. Other important cereals are barley, sorghum, oats, millet, and rye.

What is a **weed**?

The dictionary definition is any plant considered undesirable, unattractive, or troublesome, especially a plant growing where it is not wanted in cultivated ground. Some celebrated authors have had their own definition of a weed. Ralph Waldo Emerson (1803–1882) wrote, "What is a weed? A plant whose virtues have not yet been discovered." James Russell Lowell (1819–1891) wrote in 1848, "A weed is no more than a flower in disguise." Ella Wheeler Wilcox (1850–1919), a Wisconsin poet (more famous for her expression "Laugh, and the world laughs with you; Weep, and you weep alone"), wrote in her poem "The Weed" that "a weed is but an unloved flower." Finally, Shakespeare (1564–1616) wrote in *Richard III*, "great weeds do grow apace."

How do **hardwoods** differ from **softwoods**?

"Hardwood" and "softwood" are terms used commercially to distinguish woods. Hardwoods are the woods of dicots, regardless of how hard or soft they are, while softwoods are the woods of conifers. Many hardwoods come from the tropics, while almost all softwoods come from the forests of the northern temperate zone.

PLANT STRUCTURE AND FUNCTION

What are the **major parts** of **vascular plants**?

Vascular plants consist of roots, shoots, and leaves. The root system penetrates the soil and is below ground. The shoot system consists of the stem and the leaves.

The root system is the part of the plant below ground level. It consists of the roots that absorb water and various ions necessary for plant nutrition. The root system anchors the plant in the ground. The shoot system is the part of the plant above ground level. It consists of the stem and leaves. The stem provides the framework for the positioning of the leaves. The leaves are the sites of photosynthesis.

Is there a **relationship** between the **size** of the **root system** and the size of the **shoot system**?

Growing plants maintain a balance between the size of the root system (the surface area available for the absorption of water and minerals) and the shoot system (the photosynthesizing surface). The total water- and mineral-absorbing surface area in young seedlings usually far exceeds the photosynthesizing surface area. As the plant ages, the root-to-shoot ratio decreases. Additionally, if the root system is damaged, reducing the water- and mineral-absorbing surface area, shoot growth is reduced by lack of water, minerals, and root-produced hormones. Similarly, reducing the size of the shoot system limits root growth by decreasing the availability of carbohydrates and shoot-produced hormones to the roots.

What are the **specialized cells** in **plants**?

All plant cells have several common features, such as chloroplasts, a cell wall, and a large vacuole. In addition, a number of specialized cells are found only in vascular plants. They include:

Parenchyma cells—Parenchyma (from the Greek *para*, meaning "beside," and *en + chein*, meaning "to pour in") cells are the most common cells found in leaves, stems, and roots. They are often spherical in shape with only primary cell walls.

Parenchyma cells play a role in food storage, photosynthesis, and aerobic respiration. They are living cells at maturity. Most nutrients in plants such as corn and potatoes are contained in starch-laden parenchyma cells. These cells comprise the photosynthetic tissue of a leaf, the flesh of fruit, and the storage tissue of roots and seeds.

Collenchyma cells—Collenchyma (from the Greek term *kola,* meaning "glue") cells have thickened primary cell walls and lack secondary cell walls. They form strands or continuous cylinders just below the surfaces of stems or leaf stalks. The most common function of collenchyma cells is to provide support for parts of the plant that are still growing, such as the stem. Similar to parenchyma cells, collenchyma cells are living cells once they reach maturity.

Sclerenchyma cells—Sclerenchyma (from the Greek term *skleros,* meaning "hard") cells have tough, rigid, thick secondary cell walls. These secondary cell walls are hardened with lignin, which is the main chemical component of wood. It makes the cell walls more rigid. Sclerenchyma cells provide rigid support for the plant. There are two types of sclerenchyma cells—fiber and sclereid. Fiber cells are long, slender cells that usually form strands or bundles. Sclereid cells, sometimes called stone cells, occur singly or in groups and have various forms. They have a thick, very hard secondary cell wall. Most sclerenchyma cells are dead once they reach maturity.

Xylem—Xylem (from the Greek term *xylos,* meaning "wood") is the main water-conducting tissue of plants and consists of dead, hollow, tubular cells arranged end to end. The water transported in xylem replaces that lost via evaporation through stomata. The two types of water-conducting cells are tracheids and vessel elements. Water flows from the roots of a plant up through the shoot via pits in the secondary walls of the tracheids. Vessel elements have perforations in their end walls to allow the water to flow between cells.

Phloem—The two kinds of cells in the food-conducting tissue of plants, the phloem (from the Greek term *phloios,* meaning "bark") are sieve cells and sieve-tube members. Sieve cells are found in seedless vascular plants and gymnosperms, while sieve-tube members are found in angiosperms. Both types of cells are elongated, slender, tube-like cells arranged end to end with clusters of pores at each cell junction. Sugars (especially sucrose), other compounds, and some mineral ions move between adjacent food-conducting cells. Sieve-tube members have thin primary cell walls but lack secondary cell walls. They are alive once they reach maturity.

Epidermis—Several types of specialized cells occur in the epidermis including guard cells, trichomes, and root hairs. Flattened epidermal cells, one layer thick and coated by a thick layer of cuticle, cover all parts of the primary plant body.

What conditions are necessary for seed germination?

Once the seed is protected and enclosed in a seed coat, it ceases further development and becomes dormant. Seeds remain dormant until the optimum conditions of temperature, oxygen, and moisture are available for germination and further development. In addition to these external factors, some seeds undergo a series of enzymatic and biochemical changes prior to germination.

What is a **seed**?

A seed is a mature, fertilized ovule. It consists of the seed embryo and the nutrient-rich tissue called the endosperm. The embryo consists of a miniature root and shoot.

What are the **functions** of **stems**?

The four main functions of stems are: 1) to support leaves; 2) to produce carbohydrates; 3) to store materials such as water and starch; and 4) to transport water and solutes between roots and leaves. Stems provide the link between the water and dissolved nutrients of the soil and the leaves.

What are the **parts of a leaf** and the major functions of leaves?

Leaves, outgrowths of the shoot apex, are found in a variety of shapes, sizes, and arrangements. Most leaves have a blade, a petiole, stipules, and veins. The blade is the flattened portion of the leaf. The petiole is the slender stalk of the leaf. The stipules, found on some but not all leaves, are located at the base of the petiole where it joins the stem. Stipules may be leaflike and show considerable variation in size. Veins, xylem, and phloem run through the leaf. Leaves are the main photosynthetic organ. However, they are also important in gas exchange and water movement throughout the whole plant.

Why do tree **leaves change color** in the fall?

The carotenoids (pigments in the photosynthesizing cells), which are responsible for the fall colors, are present in the leaves during the growing season. However, the colors are eclipsed by the green chlorophyll. Toward the end of summer, when the chlorophyll production ceases due to declining daylight and a decrease in temperature, the other colors of the carotenoids (such as yellow, orange, red, or purple) become visible. Listed below are the autumn leaf colors of some common trees.

Tree	Color
Sugar maple, sumac	Flame red and orange
Red maple, dogwood, sassafras, scarlet oak	Dark red
Poplar, birch, tulip tree, willow	Yellow
Ash	Plum purple
Oak, beech, larch, elm, hickory, sycamore	Tan or brown
Locust	Stays green until leaves drop
Black walnut, butternut	Drops leaves before they turn color

How **many leaves** are on a **mature tree**?

Leaves are one of the most conspicuous parts of a tree. A maple tree (genus Acer) with a trunk 3 feet (1 meter) wide has approximately 100,000 leaves. Oak (genus *Quercus*) trees have approximately 700,000 leaves. Mature American elm (*Ulmus americana*) trees can produce more than five million leaves per season.

What plant has the **largest leaves**?

The monstera (*Monstera deliciosa*) has dark green, glossy leaves that measure 2 to 3 feet (0.6 to 0.9 meter) long when mature.

How does **water move up** a tree?

Water is carried up a tree through the xylem tissue in a process called transpiration. Constant evaporation from the leaf creates a flow of water from root to shoot. The roots of a tree absorb the vast majority of water that a tree needs. The properties of cohesion

The tropical *Monstera deliciosa* grows the largest leaves of any plant, each measuring up to three feet long.

and adhesion allow the water to move up a tree regardless of its height. Cohesion allows the individual water molecules to stick together in one continuous stream. Adhesion permits the water molecules to adhere to the cellulose molecules in the walls of xylem cells. When the water reaches a leaf, water is evaporated, thus allowing additional water molecules to be drawn up through the tree.

In what ways are **leaves economically important**?

Leaves are used for food and beverages, dyes and fibers, and medicinal and other industrial uses. Certain plants, such as cabbage (*Brassica oleracea*), lettuce (*Lactuca sativa*), spinach (*Spinacia oleracea*), and most herbs—including parsley (*Petroselinum crispum*) and thyme (*Thymus vulgaris*)—are grown for their leaves. Bearberry leaves (*Arctostaphylos uva-ursi*) contain a natural yellow dye, while henna leaves (*Lawsonia inermis*) contain a natural red dye. The leaves of palm trees are used to make clothing, brooms, and thatched huts in tropical climates. Aloe vera leaves are well known for treating burns and are also used in manufacturing medicated soaps and creams.

What are the **functions** of the **root system?**

The major functions of roots are: 1) anchorage in soil; 2) storage of energy resources such as the carrot and sugar beet; 3) absorption of water and minerals from the soil; and 4) conduction of water and minerals to and from the shoot. The roots store the food (energy resources) of the plant. The food is either used by the roots themselves or digested, and the products of digestion are transported back up through the phloem to the above-ground portions of the plant. The roots of some plants are harvested as food for human consumption. Plant hormones are synthesized in the meristematic regions of the roots and transported upward in the xylem to the aerial part of the plant to stimulate growth and development.

How deep does the **root system penetrate** the soil?

The depth to which the root system penetrates the soil is dependent on moisture, temperature, the composition of the soil, and specific plant. Most of the roots actively absorbing water and minerals, the "feeder roots," are found in the upper 3 feet (1 meter) of the soil. The feeder roots of many trees are mainly in the upper 6 inches (15 centimeters) of the soil—the part of the soil richest in organic matter.

Which plant has the **deepest root system?**

Roots of the desert shrub mesquite (*Prosopis juliflora*) have been found growing nearly 175-feet (53.5-meter) deep near Tucson, Arizona.

What is the **economic importance** of **roots?**

Carrots, beets, turnips, radishes, horseradish, sugar beets, and sweet potatoes are all taproots that have been used as food for human consumption for centuries. The spices licorice, sassafras, and sarsaparilla (the flavoring used to make root beer) are derived from roots. The drugs aconite, gentian, ipecac, ginseng, reserpine (a tranquilizer), and protoveratrine (a heart relaxant) are all extracted from the roots of plants.

What are **effective types** of **pollination**?

Effective pollination occurs when viable pollen is transferred to a plant's stigmas, ovule-bearing organs, or ovules (seed precursors). Without pollination, there would be no fertilization. Since plants are immobile organisms, they usually need external agents to transport their pollen from where it is produced in the plant to where fertilization can occur. This situation produces cross-pollination, wherein one plant's pollen is moved by an agent to another plant's stigma. Some plants are able to self-pollinate—transfer their own pollen to their own stigmas. But of the two methods, cross-pollination seems more advantageous, for it allows new genetic material to be introduced.

Cross-pollination agents include insects, wind, birds, mammals, and water. Many times flowers offer one or more "rewards" to attract these agents—sugary nectar, oil, solid food bodies, perfume, a place to sleep, or sometimes the pollen itself. Other times the plant can "trap" the agent into transporting the pollen. Generally, plants use color and fragrances as attractants to lure these agents. For example, a few orchids use a combination of smell and color to mimic the female of certain species of bees and wasps so successfully that the corresponding males will attempt to mate with them. Through this process (pseudocopulation) the orchids achieve pollination. While some plants cater to a variety of agents, other plants are very selective and are pollinated by a single species of insect only. This extreme pollinator specificity tends to maintain the purity of a plant species.

Plant structure can accommodate the type of agent used. For example, plants such as grasses and conifers, whose pollen is carried by the wind, tend to have a simple structure lacking petals, with freely exposed and branched stigmas to catch airborne pollen and dangling anthers (pollen-producing parts) on long filaments. This type of anther allows the light round pollen to be easily caught by the wind. These plants are found in areas such as prairies and mountains, where insect agents are rare. In contrast, semi-enclosed, nonsymmetrical, long-lived flowers such as irises, roses, and snapdragons have a "landing platform" and nectar in the flower base to accommodate insect agents such as the bee. The sticky, abundant pollen can easily become attached to the insect to be borne away to another flower.

What is **tropism**?

Tropism is the movement of a plant in response to a stimulus. The categories include:

Chemotropism—A response to chemicals by plants in which incurling of leaves may occur

Gravitropism—Formerly called geotropism, a response to gravity in which the plant moves in relation to gravity. Shoots of a plant are negatively geotropic (growing upward), while roots are positively geotropic (growing downward)

When was the role of bees in pollination discovered?

The discovery of the role of bees in pollination was discovered by Joseph Gottlieb Kölreuter (1733–1806) in 1761. He was the first to realize that plant fertilization occurs with the help of pollen-carrying insects.

Hydrotropism—A response to water or moisture in which roots grow toward the water source

Paraheliotropism—A response by the plant leaves to avoid exposure to the sun

Phototropism—A response to light in which the plant may be positively phototropic (moving toward the light source) or negatively phototropic (moving away from the light source). Main axes of shoots are usually positively phototropic, whereas roots are generally insensitive to light

Thermotropism—A response to temperature by plants

Thigmotropism or haptotropism—A response to touch by the climbing organs of a plant. For example, the plant's tendrils may curl around a support in a spring-like manner

What are the **major classes** of **plant hormones**?

The five major classes of plant hormones are auxins, gibberellins, cytokinins, ethylene, and abscisic acid.

Hormone	Principal Action	Where Produced or Found in Plant
Auxins	Elongate cells in seedlings, shoot tips, embryos, leaves	Shoot apical meristem
Gibberellins	Elongate and divide cells in seeds, roots, shoots, young leaves	Apical portions of roots and shoots
Cytokinins	Stimulate cell division (cytokinesis) in seeds, roots, young leaves, fruits	Roots
Ethylene	Hastens fruit ripening	Leaves, stems, young fruits
Abscisic acid	Inhibits growth; closes stomata	Mature leaves, fruits, root caps

When were the **major classes** of **plant hormones** identified and **who** is associated with **their identification**?

Auxins—Charles Darwin (1809–1882) and his son, Francis (1845–1925), performed some of the first experiments on growth-regulating substances. They published their

427

results in 1881 in *The Power of Movement in Plants*. In 1926 Frits W. Went (1903–1990) isolated the chemical substance responsible for elongating cells in the tips of oat (genus Avena) seedlings. He named this substance auxin, from the Greek term *auxein*, meaning "to increase."

Gibberellins—In 1926 the Japanese scientist Eiichi Kurosawa discovered a substance produced by a fungus, *Gibberella fujikuroi*, that caused a disease ("foolish seedling disease") in rice (Oryza sativa) seedlings in which the seedlings would grow rapidly but appear sickly and then fall over. The Japanese chemists Teijiro Yabuta (1888–1977) and Yusuke Sumiki (1901–1974) isolated the compound and named it gibberellin in 1938.

Cytokinins—Johannes van Overbeek discovered a potent growth factor in coconut (*Cocos nucifera*) milk in 1941. In the 1950s Folke Skoog (1908–2001) was able to produce a thousand-fold purification of the growth factor but was unable to isolate it. Carlos O. Miller (1923–), Skoog, and their colleagues succeeded in isolating and identifying the chemical nature of the growth factor. They named the substance kinetin and the group of growth regulators to which it belonged cytokinins because of their involvement in cytokinesis or cell division.

Ethylene—Even before the discovery of auxin in 1926, ethylene was known to have effects on plants. In ancient times the Egyptians would use ethylene gas to ripen fruit. During the 1800s shade trees along streets with lamps that burned ethylene, the illuminating gas, would become defoliated from leaking gas. In 1901 Dimitry Neljubov demonstrated that ethylene was the active component of illuminating gas.

Abscisic acid—Philip F. Wareing (1914–1996) discovered large amounts of a growth inhibitor in the dormant buds of ash and potatoes that he called dormin. Several years later in the 1960s, Frederick T. Addicott (1912–2009) reported the discovery in leaves and fruits of a substance capable of accelerating abscission that he called abscisin. It was soon discovered that dormin and abscisin were identical chemically.

What are some **commercial uses** of **plant hormones**?

Plant hormones are used in a variety of applications to control some aspect of plant development. Auxins are used in commercial herbicides as weed killers. Another use of auxins is to stimulate root formation. It is often referred to as the "rooting hormone" and applied to cuttings prior to planting. Some hormones are used to increase fruit production and prevent preharvest fruit drop. Gibberellins are sprayed on Thompson seedless grapes during the flowering stage to thin the flowers on each cluster, thus allowing the remaining flowers to spread out and develop larger fruit. Gibberellins are also used to enhance germination and stimulate the early emergence of seedlings in grapes, citrus fruits, apples, peaches, and cherries. When used on cucumber plants, gibberellins promote the formation of male flowers, which is useful in the production of hybrid seeds.

What is the difference between short-day plants and long-day plants?

Short-day and long-day plants exhibit a response to photoperiodism, or the changes in light and dark in a 24-hour cycle. Short-day plants form flowers when the days become shorter than a critical length, while long-day plants form flowers when the days become longer than a critical length. Short-day plants bloom in late summer or autumn in middle latitudes. Examples of short-day plants are chrysanthemums, goldenrods, poinsettias, soybeans, and ragweed. Long-day plants bloom in spring and early summer. Some examples of long-day plants are clover, irises, and hollyhocks. Florists and commercial plant growers can adjust the amount of light a plant receives to force it to bloom out of season.

Who showed that plant cells were totipotent?

In 1958 Frederick Campion Steward (1904–1993), a botanist at Cornell University, successfully regenerated an entire carrot plant from a tiny piece of phloem. Small pieces of tissue from carrots were grown in a nutrient broth. Cells that broke free from the fragments dedifferentiated, meaning that they reverted to unspecialized cells. However, as these unspecialized cells grew, they divided and redifferentiated back into specialized cell types. Eventually, cell division and redifferentiation produced entire new plants. Each unspecialized cell from the nutrient broth expressed its genetic potential to make all the other cell types in a plant.

Why was Steward successful? Like previous investigators, he supplied the cultured cells with sugars, minerals, and vitamins. In addition, he also added a new ingredient: coconut milk. Coconut milk contains, among other things, a substance that induces cell division. Subsequent research identified this material as cytokinins, a group of plant hormones (growth regulators) that stimulate cell division. Once the cultured cells began dividing, they were transplanted on agar media, where they formed roots and shoots and developed into plants.

FLOWERS AND UNUSUAL PLANTS

What are the parts of a flower?

Sepal—found on the outside of the bud or on the underside of the open flower. It serves to protect the flower bud from drying out. Some sepals ward off predators by their spines or chemicals.

The rose was declared the national flower of the United States in 1986.

Petals—Serve to attract pollinators and are usually dropped shortly after pollination occurs.

Nectar—Contains varying amounts of sugar and proteins that can be secreted by any of the floral organs. It usually collects inside the flower cup near the base of the cup formed by the flower parts.

Stamen—Is the male part of a flower and consists of a filament and anther where pollen is produced.

Pistil—Is the female part, which consists of the stigma, style, and ovary containing ovules. After fertilization the ovules mature into seeds.

What is meant by an **"imperfect" flower**?

An imperfect flower is one that is unisexual, having either stamens (male parts) or pistils (female parts) but not both.

Is there a **national flower** for the **United States**?

The United States adopted the rose as the national flower on October 7, 1986, after many years of deliberation. Other flowers suggested were the dogwood, the mountain laurel, and the columbine.

What is the **national flower** of different **countries**?

Country	National Flower
Argentina	Ceibo
Australia	Wattle
Belgium	Poppy
Bolivia	Cantua buxifolia
Brazil	Cattleya
Canada	Sugar Maple
Chile	Chilean bellflower
China	Narcissus
Costa Rica	Cattleya
Denmark	Clover
Ecuador	Cinchona

Country	National Flower
Egypt	Lotus (water lily)
England	Rose
France	Fleur-de-lis
Germany	Cornflower
Greece	Violet
Holland	Tulip
Honduras	Rose
India	Lotus (Zizyphus)
Ireland	Shamrock
Italy	Lily
Japan	Chrysanthemum
Mexico	Prickly Pear
Newfoundland	Pitcher Plant
New Zealand	Silver fern
Norway	Heather
Persia	Rose
Poland	Poppy
Russia	Sunflower
Scotland	Thistle
South Africa	Protea cynaroides
Spain	Pomegranate
Sweden	Twinflower
Switzerland	Edelweiss
Wales	Leek

What was the **"flower clock"** of **Linnaeus**?

Carolus Linnaeus (1707–1778), who was responsible for the binomial nomenclature classification system of living organisms, invented a floral clock to tell the time of day. He had observed over a number of years that certain plants consistently opened and closed their flowers at particular times of the day, these times varied from species to species. One could deduce the approximate time of day according to which species had opened or closed its flowers. Linnaeus planted a garden displaying local flowers, arranged in sequence of flowering throughout the day, that would flower even on cloudy or cold days. He called it a "horologium florae" or "flower clock."

431

What **flowers** are designated as **symbolic** of each **month** of the year?

Month	Flower
January	Carnation
February	Violet
March	Jonquil
April	Sweet Pea
May	Lily of the Valley
June	Rose
July	Larkspur
August	Gladiola
September	Aster
October	Calendula
November	Chrysanthemum
December	Narcissus

What are the **symbolic meanings** of flowers, herbs, and other plants?

Many cultures and traditions have symbolic meanings to flowers, herbs, and plants. Below is a representative list.

Aloe—Healing, protection, affection

Angelica—Inspiration

Arbor vitae—Unchanging friendship

Bachelor's buttons—Single blessedness

Basil—Good wishes, love

Bay—Glory

Carnation—Alas for my poor heart

Chamomile—Patience

Chives—Usefulness

Clover, White—Think of me

Coriander—Hidden worth

Cumin—Fidelity

Fennel—Flattery

Fern—Sincerity

Geranium, Oak-leaved—True friendship

Goldenrod—Encouragement

Heliotrope—Eternal love

Holly—Hope

Hollyhock—Ambition

Honeysuckle—Bonds of love
Horehound—Health
Hyssop—Sacrifice, cleanliness
Ivy—Friendship, continuity
Lady's mantle—Comforting
Lavender—Devotion, virtue
Lemon balm—Sympathy
Marjoram—Joy, happiness
Mints—Eternal refreshment
Morning glory—Affectation
Nasturtium—Patriotism
Oak—Strength
Oregano—Substance
Pansy—Thoughts
Parsley—Festivity
Pine—Humility
Poppy, Red—Consolation
Rose—Love
Rosemary—Remembrance
Rudbeckia—Justice
Rue—Grace, clear vision
Sage—Wisdom, immortality
Salvia, Blue—I think of you
Salvia, Red—Forever mine
Savory—Spice, interest
Sorrel—Affection
Southernwood—Constancy, jest
Sweet-pea—Pleasures
Sweet woodruff—Humility
Tansy—Hostile thoughts
Tarragon—Lasting interest
Thyme—Courage, strength
Valerian—Readiness
Violet—Loyalty, devotion
Violet, Blue—Faithfulness
Violet, Yellow—Rural happiness
Willow—Sadness
Zinnia—Thoughts of absent friends

What do the different **colors** and **varieties** of **roses symbolize**?

Rose	Meaning
Yellow rose	Jealousy, unfaithfulness
Red rosebud	Youth, beauty
White rose	Silence
Lancaster rose	Union
Burgundy rose	Unconscious beauty
Musk rose	Capricious beauty
Dog rose	Pleasure and pain
Cabbage rose	Ambassador of love
Bridal rose	Happy love
Carolina rose	Dangerous love
May rose	Precocity
Moss rose	Voluptuousness
Christmas rose	Tranquility
Pompon rose	Gentility

Which variety of **orchid** is commonly **used** for **corsages**?

The pale purple Cattelya orchid, named for the English botanist William Cattley, is often used in corsages.

What special **significance** does the **passionflower** have?

Spanish friars of the sixteenth century first gave the name to this flower. They saw in the form of the passionflower (*Passiflora*) a representation of the passion of Christ: the flowers have five petals and five sepals, which was thought to symbolize the ten faithful apostles present at the crucifixion; the corona of five filaments was believed to resemble Christ's crown of thorns; the five stamens represented the five wounds in Christ's body, and the three stigmas stood for the nails driven into his hands and feet. Most species of passionflower are native to the tropical areas of the Western Hemisphere.

Is **hemlock poisonous?**

There are two species known commonly as hemlock: *Conium maculatum* and *Tsuga canadensis*. *Conium maculatum* is a weedy plant. All parts of the plant are poisonous. In ancient times minimal doses of the plant were used to relieve pain with a great risk of poisoning. *Conium maculatum* was used to carry out the death sentence in ancient times. The Greek philosopher Socrates was condemned to death by drinking a potion made from hemlock. It should not be confused with *Tsuga canadensis*, a member of the evergreen family. The leaves of the *Tsuga canadensis* may be used to make tea.

Why do the **leaves** of the **mimosa plant close** in **response** to **touch?**

When a leaf of the mimosa plant—also known as the "sensitive plant"—is touched, a minute electric current is generated that is quickly transmitted to the cells at the base of each leaflet. As soon as the signal arrives to the cells, the water contained in the cells is released. Due to loss of water, the leaves collapse downward.

What is the **fastest-growing land plant?**

Bamboo (*Bambusa* spp.), native to tropical and subtropical regions of Southeast Asia and islands of the Pacific and Indian Oceans, is the plant that gains height most quickly. Bamboo can grow almost 3 feet (1 meter) in 24 hours. This rapid growth is produced partly by cell division and partly by cell enlargement.

What are **carnivorous plants?**

Carnivorous plants are plants that attract, catch, and digest animal prey, absorbing the bodily juices of prey for the nutrient content. There are more than 400 species of carnivorous plants. The species are classified according to the nature of their trapping mechanism. All carnivorous plants have traps made of modified leaves with various incentives or attractants, such as nectar or an enticing color, that can lure prey. Active traps display rapid motion in their capture of prey. The Venus fly trap, *Dionaea muscipula*, and the bladderwort, *Utricularia vulgaris*, have active traps that imprison victims. Each leaf is a two-sided trap with trigger hairs on each side. When the trigger hairs are touched, the trap shuts tightly around the prey.

Semi-active traps employ a two-stage trap in which the prey is caught in the trap's adhesive fluid. As prey struggles in the fluid, the plant is triggered to slowly tighten its grip. The sundew (*Drosera capensis*) and butterwort (*Pinguicula vulgaris*) have semi-active traps.

Passive traps entice insects using nectar. The passive-trap leaf has evolved into a shape resembling a vase or pitcher. Once lured to the leaf, the prey falls into a reservoir of accumulated rainwater and drowns. An example of the passive trap is the pitch-

What is unique about the water lily *Victoria amazonica*?

It is very big! Found only on the Amazon River, this water lily has leaves that can reach up to 6 feet (1.8 meters) in diameter. In a single season, a plant will produce 40 to 50 leaves. The leaves are very buoyant, capable of supporting the weight of a small child. Mature leaves may be able to support 100 pounds (45 kilograms) if evenly distributed on the leaf. The flowers of *Victoria amazonica* reach up to 12 inches (30 centimeters) in height. The flowers open at dusk and are only open on two successive nights. During the first evening, the flower is white. They close during the day and the next evening the flower has changed its color to a pink or purplish-red. Each plant only has one flower at a time, and after the second evening the flower closes up and sinks below the surface of the water.

er plant (*Sarracenia purpurea*). The Green Swamp Nature Preserve in southeastern North Carolina has the most numerous types of carnivorous plants.

How did the **navel orange originate**?

Navel oranges are oranges without seeds. In the early nineteenth century an orange tree in a Brazilian orchard produced seedless fruit although the rest of the trees in the orchard produced seeded oranges. This naturally occurring mutation gave rise to what we now refer to as the navel orange. A bud from the mutant tree was grafted onto another orange tree; branches that resulted were then grafted onto other trees, soon creating orchards of navel orange trees.

How long can an **orange tree produce oranges**?

An average orange tree will produce fruit for 50 years, but 80 years of productivity is not uncommon, and a few trees are known to be still producing fruit after more than a century. An orange tree may attain a height of 20 feet (6.1 meters), but some trees are as much as 30 feet (9.1 meters) high. Orange trees grow well in a variety of soils but prefer subtropical settings.

How are **seedless grapes** grown?

Since seedless grapes cannot reproduce in the manner that grapes usually do (i.e., dropping seeds), growers have to take cuttings from the plants, root them, and then plant the plant cuttings. Seedless grapes come from a naturally occurring mutation in which the hard seed casing fails to develop. Although the exact origin of seedless grapes is unknown, they might have been first cultivated thousands of years ago in present-day Iran or Afghanistan. Currently, 90 percent of all raisins are made from Thompson seedless grapes.

Do **seedless watermelons** occur **naturally**?

Seedless watermelon was first introduced in 1988 after 50 years of research. A seedless watermelon plant requires pollen from a seeded watermelon plant. Farmers frequently plant seeded and seedless plants close together and depend on bees to pollinate the seedless plants. The white "seeds," also known as pods, found in seedless watermelons serve to hold a fertilized egg and embryo. Because a seedless melon is sterile and fertilization cannot take place, pods do not harden and become a black seed, as occurs in seeded watermelons.

What is the **difference** between **poison ivy, oak**, and **sumac**?

These North American woody plants grow in almost any habitat and are quite similar in appearance. Each variety of plant has three-leaf compounds that alternate berry-like fruits and rusty brown stems. Poison ivy (*Rhus radicans*) grows like a vine rather than a shrub and can grow very high, covering tall, stationary items such as trees. The fruit of *R. radicans* is gray in color and is without "hair," and the leaves of the plant are slightly lobed.

Rhus toxicodendron, commonly known as poison oak, usually grows as a shrub, but it can also climb. Its leaflets are lobed and resemble the leaves of oak trees, and its fruit is hairy. Poison sumac (*Rhus vernix*) grows only in acidic, wet swamps of North America. This shrub can grow as high as 12 feet (3.6 meters). The fruit it produces hangs in a cluster and ranges from gray to brown in color. Poison sumac has dark-green leaves that are sharply pointed, compound, and alternating; it also has inconspicuous flowers that are yellowish green. All parts of poison ivy, poison oak, and poison sumac can cause serious dermatitis.

What is **kudzu**?

Kudzu (*Pueraria lobata*) is a vine that was brought from Japan for the 1876 Centennial Exposition in Philadelphia. It was intentionally planted throughout the southern United States during the 1930s in an attempt to control erosion. In fact, the federal government paid farmers as much as eight dollars an acre to plant it. In 1997, however, the government reversed its position on kudzu and referred to it as a "noxious weed." Kudzu grows over everything that it encounters, draping itself across power poles and pine trees like a shawl. The plant is responsible for more than $50 million in lost farm and timber production each year. It grows at a rate of 120,000 acres per year. As of the early twenty-first century it covers between two and four million acres of land throughout the United States, occurring from Connecticut in the East, to Missouri and Oklahoma in the West, and south to Florida and Texas. Kudzu grows as fast as 1 foot (30 centimeters) per day. The latest approach to controlling the growth of kudzu is to have goats chew on it, devouring the leaves, stems, and roots.

What are **succulents**?

A group of more than 30 plant families including the amaryllis, lily, and cactus families forms what is known as the succulents (from the Latin term *succulentis*, meaning "fleshy" or "juicy"). Most members of the group are resistant to droughts as they are dry-weather plants. Even when they live in moist, rainy environments, these plants need very little water.

What **plants** can be used to **determine blood type**?

Lectins—proteins that bind to carbohydrates on cell surfaces—found in lotus plants as well as jack and lima beans can be used to determine a person's blood type. Lectins bind to glycoproteins present on the plasma membrane of red blood cells. Because the cells of different blood types have distinct glycoproteins, cells of each blood type bind to a specific lectin.

What is **wormwood**?

Artemisia absinthium known as wormwood, is a hardy, fragrant perennial that grows to heights of 2 to 4 feet (0.6 to 1.2 meters). Wormwood is native to Europe but has been widely naturalized in North America. Absinthe, a liquor, is distilled and flavored using this plant. Absinthe was banned in the United States in the early 1900s because it was considered habit forming and hazardous to one's health. A revival of absinthe began in the 1990s, when countries in the European Union began to reauthorize its manufacture and sale. It was approved for sale in the United States in 2007, but only if classified as thujone-free. Thujone is a chemical found in wormwood that was reputed to cause hallucinations.

What is the **origin** of the **name "Jimson weed"**?

Jimson weed (*Datura stramonium*) is a corruption of the name "Jamestown weed." The colonists of Jamestown, Virginia, were familiar with this weed. It is also known as thorn apple, mad apple, stinkwort, angel's trumpet, devil's trumpet, stinkweed, dewtry, and white man's weed. Even when consumed in moderate amounts, every part of this plant is poisonous and potentially deadly. Today, extracts are still used to treat asthma, intestinal cramps, and other disorders.

What is **locoweed**?

The legumirous locoweeds or milk vetches (genera *Astragalus* and *Oxytropis*) have been a severe problem for ranchers in the western half of the United Sates and are considered to be some of the most toxic plants for horses, sheep, goats, and cattle. *Loco* is the Spanish word for "crazy" and refers to the staggering and trembling behavior of poisoned animals who walk into things and react to unseen objects. The poisonous compounds are an unusual group of alkaloids that affect certain cells of the central nervous system, explaining the behavioral changes observed.

What are **luffa sponges**?

Luffas are nonwoody vines of the cucumber family. A fibrous skeleton lies inside of the fruit, and this structure is often used as a sponge. The term "loofah" is commonly used when this material is used as a sponge. Dishcloth gourd, rag gourd, and vegetable sponge are other popular names for this sponge.

Loofah sponges are not actually sponges. They come from the fibrous skeleton of the luffa plant.

Which **common houseplants** are **poisonous**?

Philodendron (*Philodendron* and *Monstera*) and dumbcane (*Dieffenbachia*) are some of the most-common poisonous houseplants. All parts of both of these plants are poisonous.

What **plants** are considered to be **child safe**?

These are some plants that are considered to be child safe even if they are swallowed by a child:

African violet	Dandelion	Norfolk Island pine	Tiger lily
Aster	Easter lily	Petunia	Violet
Begonia	Gardenia	Purple passion	Wandering Jew
Boston fern	Impatiens	Rose	Zebra plant
California poppy	Jade plant	Spider Plant	
Coleus	Marigold	Swedish ivy	

TREES AND SHRUBS

What is the **oldest genus** of **living trees**?

The genus *Ginkgo*, commonly known as maidenhair trees, comprises the oldest living trees. This genus is native to China, where it has been cultivated for centuries. It has not been found in the wild and it is likely that it would have become extinct had it not been cultivated. Fossils of 200-million-year-old ginkgoes show that the modern-day ginkgo is nearly identical to its forerunner. As of the early twenty-first century, only one living species of ginkgo remains, *Ginkgo biloba*. The fleshy coverings of the seeds produced by females of the species *G. biloba* have a distinctly foul odor. Horticulturists prefer to cultivate the male plant from shoots to avoid the odor and mess created by the female tree.

Which **tree species** from the **United States** have **lived** the **longest**?

Of the 850 different species of trees in the United States, the oldest species is the bristlecone pine, *Pinus longaeva*. This species grows in the deserts of Nevada and Southern California, particularly in the White Mountains. Some of these trees are believed to be over 4,600 years old. The potential life span of these pines is estimated to be 5,500 years. But potential age of the bristlecone pine is very young when compared to the oldest surviving species in the world, the maidenhair tree (*Ginkgo biloba*) of China. This species of tree first appeared during the Jurassic era, some 200 million years ago. Also called icho, or the ginkyo (meaning "silver apricot"), this species has been cultivated in Japan since 1100 B.C.E.

Longest-Lived Tree Species in the United States

Name of Tree	Scientific Name	Number of Years
Bristlecone pine	*Pinus longaeva*	3,000–4,700
Giant sequoia	*Sequoiadendron giganteum*	2,500
Redwood	*Sequoia sempervirens*	1,000–3,500
Douglas fir	*Pseudotsuga menziesii*	750
Bald cypress	*Taxodium distichum*	600

What are the distinguishing **characteristics** of **pine, spruce**, and **fir trees**?

The best way to tell the difference between the three trees is by their cones and leaves:

What is the origin of the word "sequoia"?

The word "sequoia" was proposed by Austrian botanist Stephen Endlicher (1804–1849) to commemorate the eighteenth-century Cherokee leader Sequoyah (c. 1770–1843), remembered for developing an 83-letter alphabet for the Cherokee language.

Pines

White Pine: Five needles in each bundle; needles soft and 3–5 inches long. Cones can be 4–8 inches long.

Scotch Pine: Two needles in each bundle. Needles are stiff, yellow green, 1.5–3 inches long. Cones are 2–5 inches long.

Spruce

White Spruce: Dark green needles are rigid, but not prickly, grow from all sides of the twig and are less than an inch long. Cones are 1–2.5 inches long and hang downward.

Blue Spruce: Needles are about an inch long, silvery blue, very stiff and prickly; needles grow from all sides of the branch. Cones are 3.5 inches long.

Fir

Balsam Fir: Needles are flat, 1–1.5 inches long and arranged in pairs opposite each other. Cones are upright, cylindrical and 2–4 inches long.

Fraser Fir: Looks like a balsam but needles are smaller and more rounded.

Douglas Fir Single needles, 1–1.5 inches long and very soft. Cone scales have bristles that stick out.

An easy way to identify the various trees is to gently reach out and "shake hands" with a branch remembering that the pine needles come in packages, spruce are sharp and single, and firs are flat and friendly.

Which **conifers** in **North America lose** their **leaves** in winter?

Dawn redwood trees (*Metasequoia*) are deciduous. Their leaves are bright green in summer and turn coppery red in the fall before they drop. Previously known only as a fossil, the tree was found in China in 1941 and has been growing in the United States since the 1940s. The U.S. Department of Agriculture distributed seeds to experimental growers in the United States, and the dawn redwood tree now grows all over the country.

The only native conifers that shed all of their leaves in the fall are the bald cypress (*Taxodium distichum*) and the Larch (*Larix larcina*).

What products does one acre of trees yield when cut and processed?

There are approximately 660 trees on one acre of forest; this number of trees can yield approximately 105,000 feet (32,004 meters) of lumber, more than 30 tons (30,000 kilograms) of paper, or 16 cords of firewood.

What are **Joshua trees**?

Yucca brevifolia, a large shrub found in the southwestern region of the United States, received its common name from Mormon pioneers. They named the tree after the prophet Joshua because its greatly extended branches resemble how Joshua used his outstretched arms to point his spear toward the city of Ai.

What is a **banyan tree**?

The banyan tree, *Ficus benghalensis*, native to tropical regions of Southeast Asia, is a member of the genus *Ficus*. It is a magnificent evergreen that can reach 100 feet (30.48 meters) in height. As the massive limbs spread horizontally, the tree sends down roots that develop into secondary, pillar-like supporting trunks. Over a period of years a single tree may spread to occupy as much as 2,000 feet (610 meters) around its periphery.

What is a **monkey ball tree**?

The osage orange tree, *Maclura pomifera*, produces large, green, orangelike fruits. The fruit is roughly spherical, 3.5 to 5 inches (8.8 to 12.7 centimeters) in diameter, and have a coarse, pebbly surface.

Which **woods** are used for **telephone poles**?

The principal woods used for telephone poles are southern pine, Douglas fir, western red cedar, and lodgepole pine. Ponderosa pine, red pine, jack pine, northern white cedar, and western larch are also used.

Which **woods** are used for **railroad ties**?

Many species of wood are used for railroad ties, but the most common are Douglas fir, hemlock, southern pine, and a variety of oaks and gums.

What **wood** is the favorite for **butcher's blocks**?

Because of its resilience, the preferred wood for making butcher's blocks is derived from the American sycamore (*Platanus occidentalis*), also known as the American planetree, buttonball, buttonwood, and water beech. The wood of Platanus occidentalis is also used as veneers, for decorative surfaces, fence posts, and fuel.

What **wood** is used to make **baseball bats**?

Wooden baseball bats are made from white ash (*Fraxinus Americana*). This wood is ideal for producing bats because it is tough and light, and can thus help drive a ball a great distance. A tree roughly 75 years old and 15.7 inches (40 centimeters) in diameter can produce approximately 60 bats.

From where do **frankincense** and **myrrh originate**?

Frankincense is an aromatic gum resin obtained by tapping the trunks of trees belonging to the genus *Boswellia*. The milky resin hardens when exposed to the air, forming irregularly shaped granules—the form in which frankincense is usually marketed and sold. Also called olibanum, frankincense is used as an ingredient in many different products, including pharmaceuticals, perfumes, fixatives, fumigants, and incense. Myrrh comes from a tree of the genus *Commiphora*, native to the northeastern region of Africa and the Middle East. Myrrh is also a resin obtained from trees; it is used in pharmaceuticals, perfumes, and toothpastes.

Why is the **white birch** known as **"paper birch"**?

The outer layer of the birch tree grows in sheetlike layers. Hence it has been known as "paper birch." It is the same tree that Native Americans used to make paper birch canoes.

Which **trees** are members of the **rose family** (*Rosaceae*)?

The apple, pear, peach, cherry, plum, mountain ash, and hawthorn trees are members of the *Rosaceae* family.

SOIL, GARDENING, AND FARMING

What are the different **types of soil**?

Soil is the weathered outer layer of Earth's crust. It is a mixture of tiny rock

The bark of the white birch tree peels off like sheets of paper, which is why this tree is often called the "paper birch."

fragments and organic matter. There are three broad categories of soils: clay, sandy, and loam. Clay soils are heavy with the particles sticking close together. Most plants have a hard time absorbing the nutrients in clay soil, and the soil tends to become waterlogged. Clay soils can be good for a few deep-rooted plants, such as mint, peas, and broad beans.

Sandy soils are light and have particles that do not stick together. Sandy soil is good for many alpine and arid plants, some herbs such as tarragon and thyme, and vegetables such as onions, carrots, and tomatoes.

Loam soils are a well-balanced mix of smaller and larger particles. They provide nutrients to plant roots easily, they drain well, and loam also retains water very well.

What is the **composition** of **synthetic soil**?

Synthetic soil is composed of a variety of organic and inorganic materials. Inorganic substances used include pumice, calcinated clay, cinders, vermiculite, perlite, and sand. Vermiculite and perlite are used for water retention and drainage. Organic materials used include wood residues, manure, sphagnum moss, plant residues, and peat. Sphagnum peat moss is also helpful for moisture retention and lowers the pH of the mixture. Lime may be added to offset the acidity of peat. Synthetic soil may also be referred to as growing medium, soil mixes, potting mixture, plant substrate, greenhouse soil, potting soil, and amended soil. Most synthetic soils are deficient in important mineral nutrients, which can be added during the mixing process or with water.

What are the essential **nutrient** elements **required** for **plant growth**?

Essential nutrients are chemical elements that are necessary for plant growth. An element is essential for plant growth when: 1) it is required for a plant to complete its life cycle (to produce viable seeds); 2) it is part of a molecule or component of the plant that is itself essential to the plant, such as the magnesium in the chlorophyll molecule; and 3) the plant displays symptoms of deficiency in the absence of the element. Essential nutrients are also referred to as essential minerals and essential inorganic nutrients.

What are the **macronutrients** and **micronutrients** of plants?

The macronutrients of plants are carbon, hydrogen, oxygen, nitrogen, potassium, calcium, phosphorus, magnesium, and sulfur. These all are nearly or in some cases far greater than one percent of the dry weight of a plant. The micronutrients are iron, chlorine, copper, manganese, zinc, molybdenum, and boron. Each of the micronutrients constitutes less than one to several hundred parts per million in plants. Sodium, silicon, cobalt, and selenium are beneficial elements. Research has not shown that these elements are essential for plant growth and development.

What is the **function** of various **plant nutrients**?

Element	Approximate Percent of Dry Weight	Important Functions
Macronutrients		
Carbon	44	Major component of organic molecules
Oxygen	44	Major component of organic molecules
Hydrogen	6	Major component of organic molecules
Nitrogen	1–4	Component of amino acids, proteins, nucleotides, nucleic acids, chlorophyll, coenzymes
Potassium	0.5–6	Component of enzymes, protein synthesis, operation of stomata
Calcium	0.2–3.5	Component of cell walls, maintenance of membrane structure and permeability, activates some enzymes
Magnesium	0.1–0.8	Component of chlorophyll molecule, activates many enzymes
Phosphorus	0.1–0.8	Component of ADP and ATP, nucleic acids, phospholipids, several coenzymes
Sulfur	0.05–1	Components of some amino acids and proteins, coenzyme A

Micronutrients	Concentrations (parts per million)	Important Functions
Chlorine	100–10,000	Osmosis and ionic balance
Iron	25–300	Chlorophyll synthesis, cytochromes, nitrogenase

Micronutrients	Concentrations (parts per million)	Important Functions
Manganese	15–800	Activator of certain enzymes
Zinc	15–100	Activator of many enzymes, active in formation of chlorophyll
Boron	5–75	Possibly involved in carbohydrate transport and nucleic acid synthesis
Copper	4–30	Activator or component of certain enzymes
Molybdenum	0.1–5	Nitrogen fixation, nitrate reduction

What do the **numbers** on a bag of **fertilizer indicate**?

The three numbers, such as 15-20-15, refer to the percentages by weight of macronutrients found in the fertilizer. The first number stands for nitrogen, the second for phosphorus, and the third for potassium. In order to determine the actual amount of each element in the fertilizer, multiply the percentage by the fertilizer's total weight in pounds. For example, in a 50-pound bag of 15-20-15, there are 7.5 pounds of nitrogen, 10 pounds of phosphorus, and 7.5 pounds of potassium. The remaining pounds are filler.

What does **"pH" mean** when applied to **soil**?

Literally, pH stands for "potential of hydrogen" and is the term used by soil scientists to represent the hydrogen ion concentration in a soil sample. The relative alkalinity-acidity is commonly expressed in terms of the symbol pH. The neutral point in the scale is 7: soil testing below 7 is said to be acidic; soil testing above pH 7 is alkaline. The pH values are based on logarithms with a base of ten. Thus, a soil testing pH 5 is ten times as acidic as soil testing pH 5; while a soil testing pH 4 is one hundred times as acidic as soil testing pH 6.

What is the **best soil pH** for **growing plants**?

Nutrients such as phosphorous, calcium, potassium, and magnesium are most available to plants when the soil pH is between 6.0 and 7.5. Under highly acidic (low pH) conditions, these nutrients become insoluble and relatively unavailable for uptake by plants. However, some plants, such as rhododendrons, grow better in acidic soils. High soil pH can also decrease the availability of nutrients. If the soil is more alkaline than pH 8, phosphorous, iron, and many trace elements become insoluble and unavailable for plant uptake.

What is **hydroponics**?

This term refers to growing plants in some medium other than soil; the inorganic plant nutrients (such as potassium, sulfur, magnesium, and nitrogen) are continuously supplied to the plants in solution. Hydroponics is mostly used in areas where there is little soil or unsuitable soil. Since it allows precise control of nutrient levels and oxygenation of the roots, it is often used to grow plants used for research purposes. Julius von Sachs (1832–1897), a researcher in plant nutrition, pioneered modern hydroponics. Research plants have been grown in solution culture since the mid-1800s. William Gericke (1882–1970), a scientist at the University of California, defined the word hydroponics in 1937. In the 50 years that hydroponics has been used on a commercial basis, it has been adapted to many situations. NASA will be using hydroponics in the space station for crop production and to recycle carbon dioxide into oxygen. Although successful for research, hydroponics has many limitations and may prove frustrating for the amateur gardener.

How many **years** can **seeds be kept**?

Seeds stored in an airtight container and kept in a cool, dry place are usable for a long time. The following table indicates how long commonly used seeds can be kept for planting:

Vegetable	Years
Beans	3
Beets	3
Cabbage	4
Carrots	1
Cauliflower	4
Corn, sweet	2
Cucumbers	5
Eggplant	4
Kale	3
Lettuce	4
Melons	4
Onions	1
Peas	1

Vegetable	Years
Peppers	2
Pumpkin	4
Radishes	3
Spinach	3
Squash	4
Swiss chard	3
Tomatoes	3
Turnips	5

What is **Ikebana**?

Ikebana is the Japanese expression for "the arrangement of living material in water." It is the ancient Japanese art of flower arrangement. Ikebana follows certain ancient rules that aim at achieving perfect harmony, beauty, and balance. Some describe Ikebana as sculpture with flowers. In Japan it has been practiced for 1,400 years. Buddhist monks in the sixth century practiced the art using pebbles, rock, and wood with plants and flowers. In Japan Ikebana was evolved and practiced exclusively by men—priests first, then warriors and noblemen. Today, of course, Ikebana is practiced by millions of women as well as men, although the great flower schools in Japan are mostly headed by men.

What was a **victory garden**?

During World War II, U.S. Secretary of Agriculture Claude R. Wickard (1893–1967) encouraged home owners to plant vegetable gardens wherever they could find space. Everyone believed that the produce from such gardens would help lower the price of vegetables needed by the U.S. War Department to feed the troops, thus saving money that could be spent elsewhere on the military. In addition to indirectly aiding the war effort, these gardens were also considered a civil "morale booster"—individual gardeners could feel empowered by their contribution of labor and rewarded by the produce that was grown. This made victory gardens a part of the daily life on the home front. By 1945 there were said to be 20 million victory gardens producing about 40 percent of all American vegetables in many unused scraps of land. Such sites as the strip between a sidewalk and the street, town squares, and the land around Chicago's Cook County jail were used. The term "victory garden" derives from an English book by that title written by Richard Gardner in 1603.

What is the difference between **container-grown, balled-and-burlapped**, and **bare-rooted plants**?

Container-grown plants have been grown in some kind of pot—usually peat, plastic, or clay—for most or all of their lives. Balled-and-burlapped plants have been dug up

Did Johnny Appleseed really plant apple trees?

John Chapman (1774–1845), called Johnny Appleseed, did plant apple orchards in the Midwest. He also encouraged the development of orchards farther west by giving pioneers free seedlings. His depiction as a barefoot tramp roaming the countryside scattering seeds at random from a bag slung over his shoulder, however, is more popular legend than fact. He was a curious figure who often preached from the Bible and from religious philosophy to passersby. At the time of his death in 1845 he was a successful businessman who owned thousands of acres of orchards and nurseries.

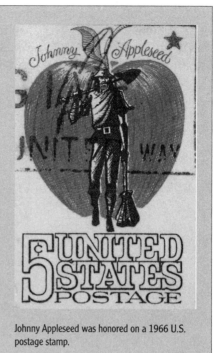

Johnny Appleseed was honored on a 1966 U.S. postage stamp.

with the soil carefully maintained around their roots in burlap. Bare-rooted plants have also been dug from their growing place but without retaining the root ball. Bare-root plants are usually cheaper than the same plants that are sold in containers or burlap wraps. In addition, they are generally available in wider selections and are easy to send through the mail.

What is the **secret of bonsai**, the Japanese art of growing dwarf trees?

These miniature trees with tiny leaves and twisted trunks can be centuries old. To inhibit growth of the plants, they have been carefully deprived of nutrients, pruned of their fastest-growing shoots and buds, and kept in small pots to reduce the root systems. Selective pruning, pinching out terminal buds, and wiring techniques are devices used to control the shape of the trees. Bonsai possibly started during the Chou dynasty (900–250 B.C.E.) in China, when emperors made miniature gardens that were dwarf representations of the provincial lands that they ruled.

What is meant by **xeriscaping**?

A xeriscape, a landscape of low water-use plants, is the modern approach to gardening in areas that experience water shortages. Taken from the Greek word *xeros*, meaning

449

dry, this type of gardening uses drought-resistant plants and low maintenance grasses, which require water only every two to three weeks. Drip irrigation, heavy mulching of plant beds, and organic soil improvements are other xeriscape techniques that allow better water absorption and retention, which in turn decrease garden watering time.

ANIMAL WORLD

INTRODUCTION AND HISTORICAL BACKGROUND

What are the **main characteristics** of **animals**?

Animals are a very diverse group of organisms, but all of them share a number of characteristics. Animals are multicellular eukaryotes that are heterotrophic, ingesting and digesting food inside the body. Animal cells lack the cell walls that provide support in the bodies of plants and fungi. Most animals have muscle systems and nervous systems responsible for movement and rapid response to stimuli in their environment. In addition, most animals reproduce sexually with the diploid stage dominating the life cycle. In most species a large, nonmotile egg is fertilized by a small, flagellated sperm, thus forming a diploid zygote. The transformation of a zygote to an animal of a specific form depends on the controlled expression in the developing embryo of special regulatory genes.

Who is the **"father of zoology"**?

Aristotle (384–322 B.C.E.) is considered the "father of zoology." His contributions to zoology include vast quantities of information about the variety, structure, and behavior of animals, the analysis of the parts of living organisms, and the beginnings of the science of taxonomy.

Who is the **"father of modern zoology"**?

Conrad Gessner (1516–1565), a Swiss naturalist, is often credited as the "father of modern zoology," based on his three-volume *Historia Animalium*, which served as a standard reference work throughout Europe in the sixteenth and seventeenth centuries.

How are **animals classified**?

Animals belong to the kingdom Animalia. Most biologists divide the kingdom into two subkingdoms: 1) Parazoa (from the Greek *para*, meaning "alongside," and *zoa*, meaning "animal"); and 2) Eumetazoa (from the Greek *eu*, meaning "true", *meta*, meaning "later", and *zoa*, meaning "animal"). The only existing animals classified as Parazoa are the sponges (phylum Porifera). Sponges are very different from other animals and function much like colonial, unicellular protoza even though they are multicellular. Cells of sponges can be versatile and change form and function and are not organized into tissues and organs. They also lack symmetry. All other animals have true tissues, are symmetrical, and are classified as eumetazoa.

How can animals be grouped according to **body symmetry**?

Symmetry refers to the arrangement of body structures in relation to the axis of the body. Most animals exhibit either radial or bilateral body symmetry. Animals such as jellyfishes, sea anemones, and starfishes have radial symmetry. In radial symmetry the body has the general form of a wheel or cylinder, and similar structures are arranged as spokes from a central axis. The bodies of all other animals are marked by bilateral symmetry, a design in which the body has right and left halves that are mirror images of each other. A bilaterally symmetrical body plan has a top and a bottom, also known respectively as the dorsal and ventral portions of the body. It also has a front (or anterior) end and a back (or posterior) end.

How many **different animals** are there?

Biologists have described and named more than one million species of animals, and some biologists believe that there are several million more species that remain to be discovered, classified, and named.

When was the **first zoo** in the **United States** established?

The Philadelphia Zoological Garden, chartered in 1859, was the first zoo in the United States. The zoo was delayed by the Civil War, financial difficulties, and restrictions on transporting wild animals. It opened in 1874 on 33 acres (13 hectares), and 282 animals were exhibited.

ANIMAL CHARACTERISTICS AND ACTIVITIES

What is the **difference between** an **invertebrate** and a **vertebrate**?

Invertebrates are animals that lack a backbone. Almost all animals (99 percent) are invertebrates. Of the more than one million identified animals, only 52,000 have a vertebral column; these are referred to as vertebrates. Most biologists believe the millions of species that have yet to be discovered are exclusively invertebrates.

What **invertebrate lives** in both **marine** and **freshwaters** and is one of the most important of all animals?

Copepods, tiny crustaceans, are the link between the photosynthetic life in the ocean or pond and the rest of the aquatic food web. They are primary consumers grazing on algae in the waters of the oceans and ponds. These organisms, among the most abundant multicellular animals on Earth, are then consumed by a variety of small predators, which are eaten by larger predators, and so on. Virtually all animal life in the ocean depends on the copepods, either directly or indirectly. Although humans do not eat copepods directly, our sources of food from the ocean would disappear without the copepods.

What are the major **characteristics** of all **chordates**?

All chordates share a notochord, dorsal nerve cord, and pharyngeal gill pouches. The notochord, a cartilaginous supporting rod, runs along the dorsal part of the body. It is always found in embryos, but in most vertebrates it is replaced during development by a backbone of bony or cartilaginous vertebrae. The tubular dorsal nerve cord, dorsal to the notochord, is formed during development by an infolding of the ectoderm. In vertebrates the nerve cord eventually becomes encased and thus protected by the backbone. The pharyngeal gill pouches appear during embryonic development on both sides of the throat region, the pharynx.

453

What are the three **major groups** of **chordates**?

The chordates are divided into three subphyla: Tunicata, Cephalochordata, and Vertebrata. Tunicates are like little leathery bags that are either free living or attach to pilings, rocks, and seaweeds. They are also called sea squirts because a disturbed animal may contract and shoot streams of water from both of its siphons.

The subphylum Cephalochordata contains the amphioxus or lancelet (*Branchiostoma*), which looks like a small fish and has the three chordate features as an adult. Amphioxus also shows clear serial segmentation or metamerism (from the Greek terms *meta*, meaning "between, among, after," and *meros*, meaning "part"). It is divided lengthwise into a series of muscle segments. Vertebrates, which comprise the third chordate subphylum, retain the same metamerism in internal structures.

What are the **major features** shown by all **vertebrates**?

Animals in the subphylum Vertebrata are distinguished from other chordates by several features. Most prominent is the endoskeleton of bone or cartilage, centering around the vertebral column (spine or backbone). Composed of separate vertebrae (showing internal metamerism), a vertebral column combines flexibility with enough strength to support even a large body. Other vertebrate features include: 1) complex dorsal kidneys; 2) a tail (lost via evolution in some groups) extending by the anus; 3) a closed circulatory systems with a single, well-developed heart; 4) a brain at the anterior end of the spinal cord, with ten or more pairs of cranial nerves; 5) a cranium (skull) protecting the brain; 6) paired sex organs in both males and females; and 7) two pairs of movable appendages—fins in the fishes, which evolved into legs in land vertebrates.

What was the **first group** of **vertebrates**?

The first vertebrates were fishes that appeared 500 million years ago. They were agnathans (from the Greek terms *a*, meaning "without," and *gnath*, meaning "jaw"), small, jawless fishes up to about 8 inches (20 centimeters) long and also known as ostracoderms ("shell skin") because their bodies were covered with bony plates, most notably a head shield protecting the brain.

What group of animals was the first to make a **partial transition** from **water to land**?

Amphibians have made a partial transition to terrestrial life. The living amphibians include newts, salamanders, frogs, and toads. Although lungfish made a partial transition to living out of the water, amphibians were the first to struggle onto land and become adapted to a life of breathing air while not constantly surrounded by water.

The ancestors of amphibians like this frog were the first animals to make the transition from sea to land.

What **group** of **vertebrates** were the **first terrestrial** vertebrates?

Reptiles were the first group of vertebrates that were truly terrestrial. There were a number of adaptations that led to their successful terrestrial life.

What is the **most successful** and diverse group of **terrestrial vertebrates**?

Birds, members of the class Aves, are the most successful of all terrestrial vertebrates. There are 28 orders of living birds with almost 10,000 species distributed over almost the entire earth. The success of birds is basically due to the development of the feather.

What are the **largest** and **smallest vertebrates**?

Largest Vertebrates	Name	Length and Weight
Sea mammal	Blue or sulphur-bottom whale (*Balaenoptera musculus*)	100–110 ft (30.5–33.5 m) long; weighs 135–209 tons (122.4–189.6 tonnes)
Land mammal	African bush elephant (*Loxodonta africana*)	Bull is 10.5 ft (3.2 m) tall at shoulder; weighs 5.25–6.2 tons (4.8–5.6 tonnes)
Living bird	North African ostrich (*Struthio c. camelus*)	8–9 ft (2.4–2.7 m) tall; weighs 345 lb (156.5 kg)

Largest Vertebrates	Name	Length and Weight
Fish	Whale shark (*Rhincodon typus*)	41 ft (12.5 m) long; weighs 16.5 tons (15 tonnes)
Reptile	Saltwater crocodile (*Crocodylus porosus*)	14–16 ft (4.3–4.9 m) long; weighs 900–1,500 lb (408–680 kg)
Rodent	Capybara (*Hydrochoerus hydrochaeris*)	3.25–4.5 ft (1–1.4 m) long; weighs 250 lb (113.4 kg)

Smallest Vertebrates	Name	Length and Weight
Sea mammal	Commerson's dolphin (*Cephalorhynchus commersonii*)	Weighs 50–70 lb (236.7–31.8 kg)
Land mammal	Bumblebee or Kitti's hog-nosed bat (*Craseonycteris thong longyai*) or the pygmy shrew (*Suncus erruscus*)	Bat is 1 in (2.54 cm) long; weighs 0.062–0.07 oz (1.6–2 g); shrew is 1.5–2 in (3.8–5 cm) long; weighs 0.052–0.09 oz (1.5–2.6 g)
Bird	Bee hummingbird (*Mellisuga helenea*)	2.25 in (5.7 cm) long; weighs 0.056 oz (1.6 g)
Fish	Angler fish (male) (*Photocorynus spiniceps*)	0.24 in (6.2 mm) long
Reptile	Gecko (*Spaerodactylus ariasae*)	0.63 in (1.6 cm) long
Rodent	Pygmy mouse (*Baiomys taylori*)	4.3 in (10.9 cm) long; weighs 0.24–0.28 oz (6.8–7.9 g)

What is the **largest group** of **vertebrates**?

The largest group of vertebrates is fishes. They are a diverse group and include almost 25,000 species—almost as many as all the other groups of vertebrates combined. Most members of this group are osteichythes, or "bony fishes," which include basses, trout, and salmon.

What **names** are used for **male** and **female animals**?

Animal	Male Name	Female Name
Alligator	Bull	
Ant		Queen
Ass	Jack, jackass	Jenny
Bear	Boar or he-bear	Sow or she-bear
Bee	Drone	Queen or queen bee

Animal	Male Name	Female Name
Camel	Bull	Cow
Caribou	Bull, stag, or hart	Cow or doe
Cat	Tom, tomcat, gib, gibeat, boarcat, or ramcat	Tabby, grimalkin, malkin, pussy, or queen
Chicken	Rooster, cock, stag, or chanticleer	Hen, partlet, or biddy
Cougar	Tom or lion	Lioness, she-lion, or pantheress
Coyote	Dog	Bitch
Deer	Buck or stag	Doe
Dog	Dog	Bitch
Duck	Drake or stag	Duck
Fox	Fox, dog-fox, stag, reynard, or renard	Vixen, bitch, or she-fox
Giraffe	Bull	Cow
Goat	Buck, billy, billie, billie-goat, or he-goat	She-goat, nanny, nannie, or nannie-goat
Goose	Gander or stag	Goose or dame
Guinea pig	Boar	
Horse	Stallion, stag, horse, stud, slot, stable horse, sire, or rig	Mare or dam
Impala	Ram	Ewe
Kangaroo	Buck	Doe
Leopard	Leopard	Leopardess
Lion (African)	Lion or tom	Lioness or she-lion
Lobster	Cock	Hen
Manatee	Bull	Cow
Mink	Boar	Sow
Moose	Bull	Cow
Mule	Stallion or jackass	She-ass or mare
Ostrich	Cock	Hen
Otter	Dog	Bitch
Owl		Jenny or howlet
Ox	Ox, beef, steer, or bullock	Cow or beef
Partridge	Cock	Hen
Peafowl	Peacock	Peahen
Pigeon	Cock	Hen
Quail	Cock	Hen
Rabbit	Buck	Doe

Animal	Male Name	Female Name
Reindeer	Buck	Doe
Robin	Cock	
Seal	Bull	Cow
Sheep	Buck, ram, male-sheep, or mutton	Ewe or dam
Skunk	Boar	
Swan	Cob	Pen
Termite	King	Queen
Tiger	Tiger	Tigress
Turkey	Gobbler or tom	Hen
Walrus	Bull	Cow
Whale	Bull	Cow
Woodchuck	He-chuck	She-chuck
Wren		Jenny or jennywren
Zebra	Stallion	Mare

What names are used for juvenile animals?

Animal	Name for Young
Ant	Antling
Antelope	Calf, fawn, kid, or yearling
Bear	Cub
Beaver	Kit or kitten
Bird	Nestling
Bobcat	Kitten or cub
Buffalo	Calf, yearling, or spike-bull
Camel	Calf or colt
Canary	Chick
Caribou	Calf or fawn
Cat	Kit, kitten, kitling, kitty, or pussy
Cattle	Calf, stot, or yearling (m. bullcalf or f. heifer)
Chicken	Chick, chicken, poult, cockerel, or pullet
Chimpanzee	Infant
Cicada	Nymph
Clam	Littleneck
Cod	Codling, scrod, or sprag
Condor	Chick
Cougar	Kitten or cub

458

Animal	Name for Young
Cow	Calf (m. bullcalf; f. heifer)
Coyote	Cub, pup, or puppy
Deer	Fawn
Dog	Whelp or puppy
Dove	Pigeon or squab
Duck	Duckling or flapper
Eagle	Eaglet
Eel	Fry or elver
Elephant	Calf
Elk	Calf
Fish	Fry, fingerling, minnow, or spawn
Fly	Grub or maggot
Frog	Polliwog or tadpole
Giraffe	Calf
Goat	Kid
Goose	Gosling
Grouse	Chick, poult, squealer, or cheeper
Horse	Colt (m.), foal, stot, stag, filly (f.), hog-colt, youngster, yearling, or hogget
Kangaroo	Joey
Leopard	Cub
Lion	Shelp, cub, or lionet
Louse	Nit
Mink	Kit or cub
Monkey	Suckling, yearling, or infant
Mosquito	Larva, flapper, wriggler, or wiggler
Muskrat	Kit
Ostrich	Chick
Otter	Pup, kitten, whelp, or cub
Owl	Owlet or howlet
Oyster	Set seed, spat, or brood
Partridge	Cheeper
Pelican	Chick or nestling
Penguin	Fledgling or chick
Pheasant	Chick or poult
Pigeon	Squab, nestling, or squealer
Quail	Cheeper, chick, or squealer
Rabbit	Kitten or bunny

Animal	Name for Young
Raccoon	Kit or cub
Reindeer	Fawn
Rhinoceros	Calf
Sea Lion	Pup
Seal	Whelp, pup, cub, or bachelor
Shark	Cub
Sheep	Lamb, lambkin, shearling, or yearling
Skunk	Kitten
Squirrel	Dray
Swan	Cygnet
Swine	Shoat, trotter, pig, or piglet
Termite	Nymph
Tiger	Whelp or cub
Toad	Tadpole
Turkey	Chick or poult
Turtle	Chicken
Walrus	Cub
Weasel	Kit
Whale	Calf
Wolf	Cub or pup
Woodchuck	Kit or cub
Zebra	Colt or foal

Which animal has the **longest gestation period**?

The animal with the longest gestation period is not a mammal; it is the viviparous amphibian the Alpine black salamander, which can have a gestation period of up to 38 months at altitudes above 4,600 feet (1,402 meters) in the Swiss Alps. It bears two fully metamorphosed young.

How **long** do animals, in particular mammals, **live**?

Of the mammals, humans and fin whales live the longest. Below is the maximum life span for several animal species.

Animal	Latin Name	Maximum Life Span (years)
Marion's tortoise	*Testudo sumeirii*	152+
Quahog	*Venus mercenaria*	c. 150
Common box tortoise	*Terrapene carolina*	138

Animal	Latin Name	Maximum Life Span (years)
European pond tortoise	*Emys orbicularis*	120+
Spur-thighed tortoise	*Testudo graeca*	116+
Fin whale	*Balaenoptera physalus*	116
Human	*Homo sapiens*	110+
Deep-sea clam	*Tindaria callistiformis*	c. 100
Killer whale	*Orcinus orca*	c. 90
European eel	*Anguilla anguilla*	88
Lake sturgeon	*Acipenser fulvescens*	82
Freshwater mussel	*Margaritana margaritifera*	80 to 70
Asiatic elephant	*Elephas maximus*	78
Andean condor	*Vultur gryphus*	72+
Whale shark	*Rhiniodon typus*	c. 70
African elephant	*Loxodonta africana*	c. 70
Great eagle-owl	*Bubo bubo*	68+
American alligator	*Alligator mississipiensis*	66
Blue macaw	*Ara macao*	64
Ostrich	*Struthio camelus*	62.5
Horse	*Equus caballus*	62
Orangutan	*Pongo pygmaeus*	c. 59
Bateleur eagle	*Terathopius ecaudatus*	55
Hippopotamus	*Hippopotamus amphibius*	54.5
Chimpanzee	*Pan troglodytes*	51
White pelican	*Pelecanus onocrotalus*	51
Gorilla	*Gorilla gorilla*	50+
Domestic goose	*Anser a. domesticus*	49.75
Grey parrot	*Psittacus erythacus*	49
Indian rhinoceros	*Rhinoceros unicornis*	49
European brown bear	*Ursus arctos arctos*	47
Grey seal	*Halichoerus gryphus*	46+
Blue whale	*Balaenoptera musculus*	c. 45
Goldfish	*Carassius auratus*	41
Common toad	*Bufo bufo*	40
Roundworm	*Tylenchus polyhyprus*	39
Giraffe	*Giraffa camelopardalis*	36.25
Bactrian camel	*Camelus ferus*	35+
Brazilian tapir	*Tapirus terrestris*	35
Domestic cat	*Felis catus*	34

Animal	Latin Name	Maximum Life Span (years)
Canary	*Serinus caneria*	34
American bison	*Bison bison*	33
Bobcat	*Felis rufus*	32.3
Sperm whale	*Physeter macrocephalus*	32+
American manatee	*Trichechus manatus*	30
Red kangaroo	*Macropus rufus*	c. 30
African buffalo	*Syncerus caffer*	29.5
Domestic dog	*Canis familiaris*	29.5
Lion	*Panthera leo*	c. 29
African civet	*Viverra civetta*	28
Theraphosid spider	*Mygalomorphae*	c. 28
Red deer	*Cervus elaphus*	26.75
Tiger	*Panthera tigris*	26.25
Giant panda	*Ailuropoda melanoleuca*	26
American badger	*Taxidea taxus*	26
Common wombat	*Vombatus ursinus*	26
Bottle-nosed dolphin	*Tursiops truncatus*	25
Domestic chicken	*Gallus g. domesticus*	25
Grey squirrel	*Sciurus carolinensis*	23.5
Aardvark	*Orycteropus afer*	23
Domestic duck	*Anas platyrhynchos domesticus*	23
Coyote	*Canis latrans*	21+
Canadian otter	*Lutra canadensis*	21
Domestic goat	*Capra hircus domesticus*	20.75
Queen ant	*Myrmecina graminicola*	18+
Common rabbit	*Oryctolagus cuniculus*	18+
White or beluga whale	*Delphinapterus leucas*	17.25
Platypus	*Ornithorhynchus anatinus*	17
Walrus	*Odobenus rosmarus*	16.75
Domestic turkey	*Melagris gallapave domesticus*	16
American beaver	*Castor canadensis*	15+
Land snail	*Helix spiriplana*	15
Guinea pig	*Cavia porcellus*	14.8
Hedgehog	*Erinaceus europaeus*	14
Burmeister's armadillo	*Calyptophractus retusus*	12
Capybara	*Hydrochoerus hydrochaeris*	12
Chinchilla	*Chinchilla laniger*	11.3
Giant centipede	*Scolopendra gigantea*	10
Golden hamster	*Mesocricetus auratus*	10

Animal	Latin Name	Maximum Life Span (years)
Segmented worm	*Allolobophora longa*	10
Purse-web spider	*Atypus affinis*	9+
Greater Egyptian gerbil	*Gerbillus pyramidum*	81+
Spiny starfish	*Marthasterias glacialis*	7+
Millipede	*Cylindroiulus landinensis*	7
Coypu	*Myocastor coypus*	6+
House mouse	*Mus musculus*	6
Malagasy brown-tailed mongoose	*Salanoia concolor*	4.75
Cane rat	*Thryonomys swinderianus*	4.3
Siberian flying squirrel	*Pteromys volans*	3.75
Common octopus	*Octopus vulgaris*	2 to 3
Pygmy white-toothed shrew	*Suncus etruscus*	2
Pocket gopher	*Thomomys talpoides*	1.6
Monarch butterfly	*Danaus plexippus*	1.13
Bedbug	*Cimex lectularius*	0.5 (182 days)
Black widow spider	*Latrodectus mactans*	0.27 (100 days)
Common housefly	*Musca domesticus*	0.04 (17 days)

Besides humans, which animals are the most intelligent?

According to Edward O. Wilson (1929–), a behavioral biologist, the ten most intelligent animals are the following:

1. Chimpanzee (two species)
2. Gorilla
3. Orangutan
4. Baboon (seven species, including drill and mandrill)
5. Gibbon (seven species)
6. Monkey (many species, especially macaques, the patas, and the Celebes black ape)
7. Smaller toothed whale (several species, especially killer whale)
8. Dolphin (many of the approximately 80 species)
9. Elephant (two species)
10. Pig

Do animals other than humans have fingerprints?

It is known that gorillas and other primates have fingerprints. Of special interest, however, is that our closest relative, the chimpanzee, does not. Koala bears also have finger-

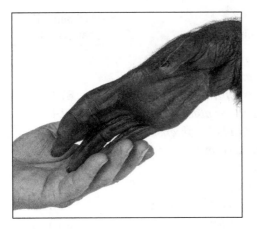

Chimps, gorillas, and other primates have fingerprints just like human beings do.

prints. Researchers in Australia have determined that the fingerprints of koala bears closely resemble those of human fingerprints in size, shape, and pattern.

Do animals have **color vision**?

Most reptiles and birds appear to have a well-developed color sense. Most mammals, however, are color-blind. Apes and monkeys have the ability to tell colors apart. Dogs can distinguish shades of grey and see various shades of blue. Cats seem to be able to distinguish between greens and blues.

Can animals **regenerate parts** of their bodies?

Regeneration does occur in some animals; however, it progressively declines the more complex the animal species becomes. Among primitive invertebrates (lacking a backbone), regeneration frequently occurs. For example, a planarium (flatworm) can split symmetrically, each part becoming a clone of the other. In higher invertebrates regeneration occurs in echinoderms (such as starfish) and arthropods (such as insects and crustaceans). Regeneration of appendages (limbs, wings, and antennae) occurs in insects (such as cockroaches, fruit flies, and locusts) and in crustaceans (such as lobsters, crabs, and crayfish). For example, regeneration of the crayfish's missing claw occurs at its next molt (shedding of its hard cuticle exterior shell/skin for the growing and the subsequent hardening of a new cuticle exterior). However, sometimes the regenerated claw does not achieve the same size of the missing claw. But after every molt (occurring two to three times a year) it grows and will eventually become nearly as large as the original claw. On a very limited basis, some amphibians and reptiles can replace a lost leg or tail.

What is the **frequency** that separates **animal hearing** from human?

The frequency of a sound is the pitch. Frequency is expressed in hertz or Hz. Sounds are classified as infrasounds (below the human range of hearing), sonic range (within the range of human hearing), and ultrasound (above the range of human hearing).

Animal	Frequency range heard (Hz)
Dog	15 to 50,000
Human	20 to 20,000
Cat	60 to 65,000

Do all animals have red blood?

The color of blood is related to the compounds that transport oxygen. Hemoglobin, containing iron, is red and is found in all vertebrates (animals having a backbone) and a few invertebrates (animals lacking a backbone). Annelids (segmented worms) have either a green pigment, chlorocruorin, or a red pigment, hemerythrin. Some crustaceans (arthropods having divided bodies and generally having gills) have a blue pigment, hemocyanin, in their blood.

Animal	Frequency range heard (Hz)
Dolphin	150 to 150,000
Bat	1,000 to 120,000

Do any animals besides people **snore**?

Many animals have been observed snoring occasionally, including dogs, cats, cows, oxen, sheep, buffaloes, elephants, camels, lions, leopards, tigers, gorillas, chimpanzees, horses, mules, zebras, and elands.

Which animals can **run faster** than a **human**?

The cheetah, the fastest mammal, can accelerate from zero to 45 miles (64 kilometers) per hour in two seconds; it has been timed at speeds of 70 miles (112 kilometers) per hour over short distances. In most chases, cheetahs average around 40 miles (63 kilometers) per hour. Humans can run very short distances at almost 28 miles (45 kilometers) per hour maximum. Most of the speeds given in the table below are for distances of one-quarter mile (0.4 kilometer).

Animal	Maximum Speed (mph/kph)
Cheetah	70/112.6
Pronghorn antelope	61/98.1
Wildebeest	50/80.5
Lion	50/80.5
Thomson's gazelle	50/80.5
Quarter horse	47.5/76.4
Elk	45/72.4
Cape hunting dog	45/72.4
Coyote	43/69.2
Gray fox	42/67.6
Hyena	40/64.4

Animal	Maximum Speed (mph/kph)
Zebra	40/64.4
Mongolian wild ass	40/64.4
Greyhound	39.4/63.3
Whippet	35.5/57.1
Rabbit (domestic)	35/56.3
Mule deer	35/56.3
Jackal	35/56.3
Reindeer	32/51.3
Giraffe	32/51.3
White-tailed deer	30/48.3
Wart hog	30/48.3
Grizzly bear	30/48.3
Cat (domestic)	30/48.3
Human	27.9/44.9

SPONGES, COELENTERATES, AND WORMS

What is the **most primitive group** of animals?

Sponges (phylum Porifera, from the Latin terms *porus*, meaning "pore," and *fera*, meaning "bearing") represent the most primitive animals. These organisms are aggregates of specialized cells without true tissues or organs, with little differentiation and integration, and with no body symmetry. A sponge's body is perforated by holes that lead to an inner water chamber. Sponges pump water through those pores and expel it through a large opening at the top of the chamber. While water is passing through the body, nutrients are engulfed, oxygen is absorbed, and waste is eliminated. Sponges are distinctive in possessing choanocytes, special flagellated cells whose beating drives water through the body cavity and that characterize them as suspension feeders (also known as filter feeders).

What is the **basic composition** of a **sponge**?

A sponge is supported by a skeleton made of hard crystals called spicules whose shape and composition are important features in taxonomy. Calcareous sponges have spicules of calcium carbonates, the material of marble and limestone. The silica spicules of the hexactinellid, or glass, sponges are formed into a delicate, glassy network. Demosponges have siliceous spicules and a network of fibrous protein, spongir, that is similar to collagen. The demosponges are the source of natural household sponges, which are made by soaking dead sponges in shallow water until all the cellular material has decayed, leaving the spongin network behind. However, most sponges sold now for household use are plastic and have nothing to do with real sponges.

Are the stings of jellyfishes and Portuguese man-of-war fatal to humans?

The stings of a jellyfish can be very painful and dangerous to humans, but they are generally not fatal. Most stings cause a painful, burning sensation that lasts for several hours. Welts and itchy skin rashes may also appear. Only the sting of the box jelly, or sea wasp (*Chironex fleckeri*), can result in death in humans. The box jelly is the only jellyfish for which a lifesaving, specific antidote exists.

How much **water** does an average **sponge circulate** during a day?

A sponge that is 4 inches (10 centimeters) tall and 0.4 inch (1 centimeter) in diameter pumps about 23 quarts (22.5 liter) of water through its body in one day. To obtain enough food to grow by 3 ounces (100 gram), a sponge must filter about 275 gallons (1,000 kilograms) of seawater.

What animals are members of the **phylum Cnidaria**?

Corals, jellyfishes, sea anemones, and hydras are members of the phylum Cnidaria. The name Cnidaria (from the Greek term *knide*, meaning "nettle," and Latin term *aria*, meaning "like" or "connected with") refers to the stinging structures that are characteristic of some of these animals. These organisms have a digestive cavity with only one opening to the outside; this opening is surrounded by a ring of tentacles used to capture food and defend against predators. Cells in the tentacles and outer body surface contain stinging, harpoonlike structures called nematocysts. Cnidarians are the first group in the animal hierarchy to have their cells organized into tissues.

What are some **interesting features** of **jellyfishes**?

Jellyfishes live close to the shores of most oceans and spend most of their time floating near the surface. Jellyfishes have bell-shaped bodies that are between 95 percent and 96 percent water. They have a muscular ring around the margin of the bell that contracts rhythmically to propel them through the water. Jellyfishes are carnivores, subduing their prey with stinging tentacles and drawing the paralyzed animal into the digestive cavity. Jellyfishes are gelatinous—you can see through their bodies.

How does a **nematocyst** work?

A nematocyst is a specialized organelle found in all cnidarians. Each nematocyst features a coiled, threadlike tube lined with a series of barbed spines. The nematocyst is used to capture prey and may also be used for defense purposes. When it is triggered to

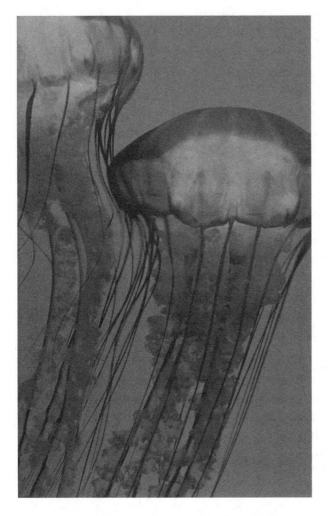

discharge, the extremely high osmotic pressure within the nematocyst (140 atmospheres) causes water to rush into the capsule, increasing the hydrostatic pressure and expelling the thread with great force. The barb instantly penetrates the prey, stinging it with a highly toxic protein.

What members of **Cnidaria** are **economically important**?

Reef-building corals are among the most important members of Cnidaria. Coral reefs are among the most productive of all ecosystems. They are large formations of calcium carbonate (limestone) in tropical seas laid down by living organisms over thousands of years. Fishes and other animals associated with reefs provide an important source of food for humans, and reefs serve as tourist attractions. Many terrestrial organisms also benefit from coral reefs, which form and maintain the

Jellyfish are delicate creatures that are also carnivorous. They use a poisonous sting to subdue their prey. It makes sense that they use strong poisons to kill prey quickly before it can damage the jellyfish's delicate body.

foundation of thousands of islands. By providing a barrier against waves, reefs also protect shorelines against storms and erosion.

How are **coral reefs formed** and how fast are they built?

Coral reefs grow only in warm, shallow water. The calcium carbonate skeletons of dead corals serve as a framework upon which layers of successively younger animals attach themselves. Such accumulations, combined with rising water levels, slowly lead to the formation of reefs that can be hundreds of meters deep and long. The coral animal, or polyp, has a columnar form; its lower end is attached to the hard floor of

> ## What gives coral their colors?
>
> Coral have a symbiotic relationship with zooxanthellae. Zooxanthellae are photosynthetic dinoflagellates (one-celled animals) that give coral their characteristic colors of pink, purple, and green. Coral that have expelled the zooxanthellae appear white.

the reef, while the upper end is free to extend into the water. A whole colony consists of thousands of individuals. There are two kinds of corals, hard and soft, depending on the type of skeleton secreted. The polyps of hard corals deposit around themselves a solid skeleton of calcium carbonate (chalk), so most swimmers see only the skeleton of the coral; the animal is in a cuplike formation into which it withdraws during the daytime. The major reef builder in Florida and Caribbean waters, *Montastrea annularis* (star coral), requires about 100 years to form a reef just 3 feet (1 meter) high.

Is **coral bleaching** related to changes in the **environment**?

Although corals can capture prey, many tropical species are dependent on photosynthetic algae (zooxanthellae) for nutrition. These algae live within the cells that line the digestive cavity of the coral. The symbiotic relationship between coral and zooxanthellae is mutually beneficial. The algae provide the coral with oxygen and carbon and nitrogen compounds. The coral supplies the algae with ammonia (waste product), from which the algae make nitrogenous compounds for both partners. Coral bleaching is the stress-induced loss of zooxanthellae that live in coral cells. In coral bleaching the algae lose their pigmentation or are expelled from coral cells. Without the algae, coral become malnourished and die. The causes of coral bleaching are not completely understood, but it is believed that environmental factors are involved. Pollution, invasive bacteria such as Vibrio, salinity changes, temperature changes, and high concentrations of ultraviolet radiation (associated with the destruction of the ozone layer) all contribute to coral bleaching.

What three groups are included in the **flatworms**?

Flatworms belong to the phylum Platyhelminthes. They are flat, elongated, acoelomate animals that exhibit bilateral symmetry and have primitive organs. The members of the flatworms are: 1) planarians, 2) flukes, and 3) tapeworms.

What are the **most common tapeworm infections** in humans?

Tapeworms, members of the class Cestoda, have long, flat bodies in which there is a linear series of sets of reproductive organs. Each set or segment is called a proglottid.

What is the most famous roundworm?

One soil nematode, *Caenorhabditis elegans*, is widely cultured and has become a model research organism in developmental biology. The study of this animal was begun in 1963 by Sydney Brenner (1927–), who received the Nobel Prize in Physiology or Medicine in 2002. The species normally lives in soil but is easily grown in the laboratory in Petri dishes. It is only about 0.06 inch (1.5 millimeters) long, has a simple, transparent body consisting of only 959 cells, and grows from zygote to mature adult in only three and a half days. The genome (genetic material) of *C. elegans*, consisting of 14,000 genes, was the first animal genome to be completely mapped and sequenced.

The small transparent body of this nematode allows researchers to locate cells in which a specific, developmentally important gene is active. These cells show up as bright green spots in a photograph because they have been genetically engineered to produce a green fluorescent protein known as GFP. The complete "wiring diagram" of its nervous system is known, including all the neurons and all connections between them. Much of the knowledge of nematode genetics and development gained from the study of *C. elegans* is transferable to the study of other animals.

Tapeworm	Means of Infection
Beef tapeworm (*Taenia saginata*)	Eating rare beef; most common of all tapeworms in humans
Pork tapeworm (*Taenia solium*)	Eating rare pork; less common than beef tapeworm
Fish tapeworm (*Diphyllobothrium latum*)	Eating rare or poorly cooked fish; fairly common in the Great Lakes region of the United States

How numerous are roundworms?

Roundworms, or nematodes, are members of the phylum Nematoda (from the Greek term *nematos*, meaning "thread") and are numerous in two respects: 1) number of known and potential species; and 2) the total number of these organisms in a habitat. Approximately 12,000 species of nematodes have been named, but it has been estimated that if all species were known, the number would be closer to 500,000. Nematodes live in a variety of habitats ranging from the sea to soil. Six cubic inches (100 cubic centimeters) of soil may contain several thousand nematodes. A square yard (0.85 square meter) of woodland or agricultural soil may contain several million of them. Good topsoil may contain billions of nematodes per acre.

What are the **most common roundworm infections** in humans in the United States?

Roundworm	Means of Infection
Hookworm (*Ancylostoma duodenale* and *Necator americanus*)	Contact with soil-based juveniles; common in southern states
Pinworm (*Enterobius vermicularis*)	Inhalation of dust that contains ova and by contamination with fingers; most common worm parasite in United States
Intestinal roundworm (*Ascaris lumbricoides*)	Ingestion of embryonated ova in contaminated food; common in rural Appalachia and southeastern states
Trichina worm (*Trichinella spiralis*)	Ingestion of infected meat; occurs occasionally in humans throughout North America

What are the major groups of **segmented worms**?

Members of the phylum Annelida, the segmented worms, have bilateral symmetry and a tubular body that may have 100 to 175 ringlike segments. The three classes of segmented worms are: 1) Polychaeta, the sandworms and tubeworms; 2) Oligochaeta, the earthworms; and 3) Hirudinea, the leeches.

In what ways are **earthworms beneficial**?

Earthworms help maintain fertile soil. An earthworm literally eats its way through soil and decaying vegetation. As it moves about, the soil is turned, aerated, and enriched by nitrogenous wastes. Charles Darwin (1809–1882) calculated that a single earthworm could eat its own weight in soil every day. Much of what is eaten is then excreted on Earth's surface in the form of "casts." The worms then re-bury these casts with their burrowing process. In addition, Darwin claimed that 2.5 acres (1 hectare) of soil might contain 155,000 earthworms, which in one year would bring 18 tons of soil to the surface and in 20 years might build a new layer 3 inches (11 centimeters) thick.

What is the **largest leech**?

Most leeches are between 0.75 inch and 2 inches (2 and 6 centimeters) in length, but some "medicinal" leeches reach 8 inches (20 centimeters). The giant of all leeches is the Amazonian *Haementeriam ghilanii* (from the Greek term *haimateros*, meaning "bloody"), which reaches 12 inches (30 centimeters) in length.

Why are **leeches** important in the field of **medicine**?

Leeches have been used in the practice of medicine since ancient times. During the 1800s leeches were widely used for bloodletting because of the mistaken idea that body disorders and fevers were caused by an excess of blood. Leech collecting and culture were practiced on a commercial scale during this time. William Wadsworth's (1770–1850) poem "The Leech-Gatherer" was based on this use of leeches.

The medical leech, *Hirudo medicinalis*, is used to remove blood that has accumulated within tissues as a result of injury or disease. Leeches have also been applied to fingers or toes that have been surgically reattached to the body. The sucking by the leech unclogs small blood vessels, permitting blood to flow normally again through the body part. The leech releases hirudin, secreted by the salivary glands, which is an anticoagulant that prevents blood from clotting and dissolves preexisting clots. Other salivary ingredients dilate blood vessels and act as an anesthetic. A medicinal leech can absorb as much as five to ten times its body weight in blood. Complete digestion of this blood takes a long time, and these leeches feed only once or twice a year in this manner.

MOLLUSKS AND ECHINODERMS

What are the major groups of **mollusks**?

There are four major groups of mollusks: 1) chitons; 2) gastropods, which include snails, slugs, and nudibranches; 3) bivalves, which include clams, oysters, and mussels; and 4) cephalopods, which include squids and octopods. Although mollusks vary widely in external appearance, most share the following body plan: 1) a muscular foot, usually used for movement; 2) a visceral mass containing most of the internal organs; and 3) a mantle—a fold of tissue that drapes over the visceral mass and secretes a shell (in organisms that have a shell).

How fast does a **snail move**?

Many snails move at a speed of less than 3 inches (8 centimeters) per minute. This means that if a snail did not stop to rest or eat, it could travel 16 feet (4.8 meters) per hour.

How many **tentacles** do the **cephalopods** have?

Octopods have eight tentacles or arms, squids have ten tentacles, and there are as many as 90 in the chambered nautilus.

Are freshwater **clams** an **endangered group**?

Although freshwater clams are found on every continent except Antarctica, they are now considered one of the most jeopardized groups of animals in the world. Approximately 270 species belong to the family Unionidae, found in North America. A total of
72 percent of our 270 native mussel species are listed as recently extinct, endangered,

threatened, or of special concern due to human impact on aquatic habitat, commercial harvesting, the introduction of carp, water pollution, and the invasion of zebra mussels.

What has been the impact of **zebra mussles** on North American waterways?

Zebra mussels (*Dreissena polymorpha*) are black-and-white-striped bivalve mollusks. They are hard-shelled species that adhere to hard surfaces with byssal threads. They were probably introduced to North America in 1985 or 1986 via a discharge of a foreign ship's ballast water into Lake St. Clair. They have spread throughout the Great Lakes, the Mississippi River, and as far east as the Hudson River. High densities of zebra mussels have been found in the intakes, pipes, and heat exchangers of waterways throughout the world. They can clog the water intakes of power plants, industrial sites, and public drinking water systems; foul boat hulls and engine cooling water systems; and disrupt aquatic ecosystems. Water-processing facilities must be cleaned manually to rid the systems of the mussels. Zebra mussels are a threat to surface water resources because they reproduce quickly, have free-swimming larva and rapid growth, lack competitors for space or food, and have no predators.

An octopus has eight tentacles while a squid has ten; both animals are cephalopods.

How are **pearls** created?

Pearls are formed in saltwater oysters and freshwater clams. There is a curtainlike tissue called the mantle within the body of these mollusks. Certain cells on the side of the mantle toward the shell secrete nacre, also known as mother-of-pearl, during a specific stage of the shell-building process. A pearl is the result of an oyster's reaction to a foreign body, such as a piece of sand or a parasite, within the oyster's shell. The oyster neutralizes the invader by secreting thin layers of nacre around the foreign body, eventually building it into a pearl. The thin layers are alternately composed of calcium carbonate, argonite, and conchiolin. Irritants intentionally placed within an oyster result in the production of what are called cultured pearls.

473

Which **mollusks** produce the most **cultured pearls**?

Cultured pearls are produced by both freshwater and marine mollusks. Most of the world's cultured pearls (known as freshwater pearls) are produced by freshwater mussels belonging to the family Unionidae. Most saltwater pearls are produced by three species of oysters belonging to the genus Pinctada, including *Pinctada imbricata*, *Pinctada maxima*, and *Pinctada margaritifera*.

ARTHROPODS: CRUSTACEANS, INSECTS, AND SPIDERS

Why are **arthropods** considered the **most biologically successful phylum** of animals?

Members of the phylum Arthropoda are characterized by jointed appendages and an exoskeleton of chitin. There are more than one million species of arthropods currently known to science, and many biologists believe there are millions more to be identified. Arthropods are the most biologically successful group of animals because they are the most diverse and live in a greater range of habitats than do the members of any other phylum of animals.

How large is the **arthropod population**?

Zoologists estimate that the arthropod population of the world, including crustaceans, spiders, and insects, numbers about a billion million (10^{18}) individuals. More than one million arthropod species have been described, with insects making up the vast majority of them. In fact, two out of every three organisms known on Earth are arthropods, and the phylum is represented in nearly all habitats of the biosphere. About 90 percent of all arthropods are insects, and about half of the named species of insects are beetles.

How **many species** of **insects** are there?

Estimates of the number of recognized insect species range from about 750,000 to upward of one million—but some experts think that this represents less than half of the number that exists in the world. About 7,000 new insect species are described each year, but unknown numbers are lost annually from the destruction of their habitats, mainly tropical forests.

Do **centipedes** actually have **one hundred legs** and **millipedes** have more than **one thousand legs**?

Centipedes (class Chilopoda) always have an uneven number of pairs of walking legs, varying from 15 to more than 171. The true centipedes (order Scolopendromorpha)

> ### How rare are blue lobsters?
>
> Just one in 30 million lobsters is blue. Other mutant colors include yellow and orange. Strangely colored lobsters are mutations of the brown lobster.

have 21 or 23 pairs of legs. Common house centipedes (*Scutigera coleoptrato*) have 15 pairs of legs. Centipedes are all carnivorous and feed mainly on insects. Millipedes (class Diplopoda) have thirty or more pairs of legs. They are herbivores, feeding mainly on decaying vegetation.

What **arthropods** are of **medical importance** in the United States?

Arthropod	Effect on Human Health
Black widow spider (*Latrodectus mactans*)	Venomous bite
Brown recluse or violin spider (*Loxosceles reclusa*)	Venomous bite
Scorpion (*Centruroides exilicauda*)	Venomous bite
Chiggers (*Trombiculid mites*)	Dermatitis
Itch mite (*Sarcoptes scabiei*)	Scabies
Deer tick (*Ixodes dammini*)	Bite transmits Lyme disease
Dog tick, wood tick (*Dermacentor* species)	Bite transmits Rocky Mountain spotted fever
Mosquitoes	Bite transmits disease (West Nile virus, encephalitis, filarial worms)
Horseflies, deerflies	Female has painful bite
Houseflies	Many transmit bacteria and viruses
Fleas	Dermatitis
Bees, wasps, ants	Venomous stings (single stings not dangerous unless person is allergic)

How long does it take the average **spider** to **weave a complete web**?

The average orb-weaver spider takes 30 to 60 minutes to completely spin its web. These species of spiders (order Araneae) use silk to capture their food in a variety of ways, ranging from the simple trip wires used by large bird-eating spiders to the complicated and beautiful webs spun by orb spiders. Some species produce funnel-shaped webs, and other communities of spiders build communal webs.

A completed web features several spokes leading from the initial structure. The number and nature of the spokes depend on the species. The spider replaces any damaged threads by gathering up the thread in front of it and producing a new one behind it. The orb web must be replaced every few days because it loses its stickiness (and its ability to entrap food).

What are the **largest** and **smallest aerial spiderwebs**?

The largest aerial webs are spun by the tropical orb weavers of the genus *Nephila*, which produce webs that measure up to 18.9 feet (6 meters) in circumference. The smallest webs are produced by the species *Glyphesis cottonae*; their webs cover an area of about 0.75 square inch (4.84 square centimeters).

How **many eggs** does a **spider lay**?

The number of eggs varies according to the species. Some larger spiders lay over 2,000 eggs, but many tiny spiders lay one or two and perhaps no more than a dozen during their lifetime. Spiders of average size probably lay a hundred or so. Most spiders lay all their eggs at one time and enclose them in a single egg sac; others lay eggs over a period of time and enclose them in a number of egg sacs.

Are **spiders** really **dangerous**?

Most spiders are harmless organisms that, rather than being dangerous to humans, are actually allies in the continuing battle to control insects. Most venom produced by spiders to kill prey is usually harmless to humans. However, there are two spiders in the United States that can produce severe or even fatal bites. They are the black widow spider (*Latrodectus mactans*) and the brown recluse spider (*Loxosceles reclusa*). Black widows are shiny black, with a bright red "hourglass" on the underside of the abdomen. The venom of the black widow is neurotoxic and affects the nervous system. About 4 out of 1,000 black widow bites have been reported as fatal. Brown recluse spiders have a violin-shaped strip on their back. The venom of the brown recluse is hemolytic and causes the death of tissues and skin surrounding the bite. Their bite can be mild to serious and sometimes fatal.

The black widow spider has a distinctive red hourglass marking on its abdomen. While the venomous bite of these spiders causes extreme pain, fewer than one percent of people injured will actually die.

Do male mosquitoes bite humans?

No. Male mosquitoes live on plant juices, sugary saps, and liquids arising from decomposition. They do not have a biting mouth that can penetrate human skin as female mosquitoes do. In some species the females, who lay as many as 200 eggs, need blood to lay their eggs. These are the species that bite humans and other animals.

What **first aid measures** can be used for a bite by a **black widow spider**?

The black widow spider (*Latrodectus mactans*) is common throughout the United States. Its bite is severely poisonous, but no first aid measures are of value. Age, body size, and degree of sensitivity determine the severity of symptoms, which include an initial pin-prick with a dull numbing pain, followed by swelling. An ice cube may be placed over the bite to relieve pain. Between 10 and 40 minutes after the bite, severe abdominal pain and rigidity of stomach muscles develop. Muscle spasms in the extremities, ascending paralysis, and difficulty in swallowing and breathing may follow. The mortality rate is less than one percent, but anyone who has been bitten should see a doctor; the elderly, infants, and those with allergies are most at risk and may require hospitalization.

Which is **stronger—steel** or the **silk** from a **spider's web**?

Spider silk is stronger. Well known for its strength and elasticity, the strongest spider silk has tensile strength second only to fused quartz fibers and five times greater than that of steel of equivalent weight. Tensile strength is the longitudinal stress that a substance can bear without tearing apart.

What is a **"bug,"** biologically speaking?

The biological meaning of the word "bug" is significantly more restrictive than in common usage. People often refer to all insects as "bugs," even using the word to include such organisms as bacteria and viruses as well as glitches in computer programs. In the strictest biological sense, a "bug" is a member of the order Hemiptera, also called true bugs. Members of Hemiptera include bedbugs, squash bugs, clinch bugs, stink bugs, and water striders.

What is the **largest group** of **insects** that has been identified and classified?

The largest group of insects that has been identified and classified is the order Coleoptera (beetles, weevils, and fireflies) with some 350,000 to 400,000 species. Beetles are the dominant form of life on Earth, as one of every five living species is a beetle.

Which **insect** has the **best sense of smell**?

Giant male silk moths may have the best sense of smell in the world. They can smell a female's perfume nearly 7 miles (11 kilometers) away.

What are the **stages** of **insect metamorphosis**?

There are two types of metamorphoses (marked structural changes in the growth processes): complete and incomplete. In complete metamorphosis, the insect (such as the ant, moth, butterfly, termite, wasp, or beetle) goes through all the distinct stages of growth to reach adulthood. In incomplete metamorphosis, the insect (such as the grasshopper, cricket, or louse) does not go through all the stages of complete metamorphoses.

Complete Metamorphosis

Egg—One egg is laid at a time or many (as much as 10,000).

Larva—What hatches from the eggs is called a larva. A larva can look like a worm.

Pupa—After reaching its full growth, the larva hibernates, developing a shell or pupal case for protection. A few insects (e.g., the moth) spin a hard covering called a "cocoon." The resting insect is called a pupa (except the butterfly, which is called a chrysalis), and remains in the hibernation state for several weeks or months.

Adult—During hibernation, the insect develops its adult body parts. When it has matured physically, the fully grown insect emerges from its case or cocoon.

Incomplete Metamorphosis

Egg—One egg or many eggs are laid.

Early stage nymph—Hatched insect resembles an adult, but smaller in size. However, those insects that would normally have wings have not yet developed them.

Late-stage nymph—At this time, the skin begins to molt (shed), and the wings begin to bud.

Adult—The insect is now fully grown.

What are some **beneficial insects**?

Beneficial insects include bees, wasps, flies, butterflies, moths, and others that pollinate plants. Many fruits and vegetables depend on insect pollinators for the production of seeds. Insects are an important source of food for birds, fish, and many animals. In some countries such insects as termites, caterpillars, ants, and bees are eaten as food by people. Products derived from insects include honey and beeswax, shellac, and silk. Some predators such as mantises, ladybugs or lady beetles, and lacewings feed on other harmful insects. Other helpful insects are parasites that live on or in the body of harmful insects. For example, some wasps lay their eggs in caterpillars that damage tomato plants.

How is the **light** in **fireflies produced**?

The light produced by fireflies (*Photinus pyroles*), or lightning bugs, is a kind of heat-less light called bioluminescence. It is caused by a chemical reaction in which the substance luciferin undergoes oxidation when the enzyme luciferase is present. The flash is a photon of visible light that radiates when the oxidating chemicals produce a high-energy state, which then revert back to their normal state. The flashing is controlled by the nervous system and takes place in special cells called photocytes. The nervous system, photocytes, and the tracheal end organs control the flashing rate. The air temperature also seems to be correlated with the flashing rate. The higher the temperature, the shorter the interval between flashes—eight seconds at 65°F (18.3°C) and four seconds at 82°F (27.7°C). Scientists are uncertain as to why this flashing occurs. The rhythmic flashes could be a means of attracting prey or enabling mating fireflies to signal in heliographic codes (that differ from one species to another), or they could serve as a warning signal.

What is the **most destructive insect** in the world?

The most destructive insect is the desert locust (*Schistocera gregaria*), the locust of the Bible, whose habitat ranges from the dry and semi-arid regions of Africa and the Middle East, through Pakistan and northern India. This short-horn grasshopper can eat its own weight in food a day, and during long migratory flights a large swarm can consume 20,000 tons (18,144,000 kilograms) of grain and vegetation a day.

Who **introduced** the **gypsy moth** into the **United States**?

In 1869, Professor Leopold Trouvelot (1827–1895) brought gypsy moth egg masses from France to Medford, Massachusetts. His intention was to breed the gypsy moth (*Porthetria dispar*) with the silkworm to overcome a wilt disease of the silkworm. He placed the egg masses on a window ledge, and evidently the wind blew them away. About ten years later these caterpillars were numerous on trees in that vicinity, and in 20 years, trees in eastern Massachusetts were being defoliated. In 1911, a contaminated plant shipment from Holland also introduced the gypsy moth to Massachusetts.

Gypsy moths are now found throughout the entire northeastern United States and portions of Virginia, North Carolina, Ohio, and Michigan.

The gypsy moth lays its eggs on the leaves of oaks, birches, maples, and other hardwood trees. When the yellow hairy caterpillars hatch from the eggs, they devour the leaves in such quantities that the tree becomes temporarily defoliated. Sometimes this causes the tree to die. The caterpillars grow from 0.5 inch (3 millimeters) to about 2 inches (5.1 centimeters) before they spin a pupa, in which they will metamorphose into adult moths.

Are there any **natural predators** of **gypsy moth caterpillars**?

About 45 kinds of birds, squirrels, chipmunks, and white-footed mice eat this serious insect pest. Among the 13 imported natural enemies of the moth, two flies, *Compislura concinnata* (a tachnid fly) and *Sturnia scutellata*, parasitize the caterpillar. Other parasites and various wasps have also been tried as controls, as well as spraying and male sterilization.

How does a **butterfly differ** from a **moth**?

Characteristic	Butterflies	Moths
Antennae	Knobbed	Unknobbed
Active-time of day	Day	Night
Coloration	Bright	Dull
Resting wing position	Vertically above body	Horizontally beside body

While these guidelines generally hold true, there are exceptions. Moths have hairy bodies, and most have tiny hooks or bristles linking the fore-wing to the hind-wing; butterflies do not have either characteristic.

Do **butterflies** see **color**?

Butterflies have highly developed sensory capabilities. They have the widest visual spectrum of any animal and are able to see from the red end of the spectrum all the way to near ultraviolet. They are therefore able to distinguish colors that humans are unable to see.

Has the United States selected a **national insect**?

No, the United States does not have a national insect. Congress did consider naming the monarch butterfly as the national insect, but the legislation did not pass.

How **many bees** are in a **bee colony**?

On average, a bee colony contains from 50,000 to 70,000 bees, which produce a harvest of from 60 to 100 pounds (27 to 45 kilograms) of honey per year. A little more than one third of the honey produced by the bees is retained in the hive to sustain the population.

How **many flowers** need to be tapped for **bees** to gather enough **nectar** to produce **one pound of honey**?

Bees must gather 4 pounds (1.8 grams) of nectar, which requires the bees to tap about two million flowers, in order to produce 1 pound (454 grams) of honey. The honey is gathered by worker bees, whose life span is three to six weeks, long enough to collect about a teaspoon of nectar.

Who discovered the **"dance of the bees"**?

In 1943, Karl von Frisch (1886–1982) published his study on the dance of the bees. It is a precise pattern of movements performed by returning forager (worker) honeybees in order to communicate the direction and distance of a food source to the other workers in the hive. The dance is performed on the vertical surface of the hive and two kinds of dances have been recognized: the round dance (performed when food is nearby) and the waggle dance (done when food is farther away).

What are **"killer bees"**?

Africanized honeybees—the term entomologists prefer rather than killer bees—are a hybrid originating in Brazil, where African honeybees were imported in 1956. The breeders, hoping to produce a bee better suited to producing more honey in the tropics, instead found that African bees soon hybridized with and mostly displaced the familiar European honeybees. Although they produce more honey, Africanized honeybees (*Apis mellifera scutellata*) also are more dangerous than European bees because they attack intruders in greater numbers. Since their introduction, they have been responsible for approximately 1,000 human deaths. In addition to such safety issues, concern is growing regarding the effect of possible hybridization on the U.S. beekeeping industry.

In October 1990, the bees crossed the Mexican border into the United States; they reached Arizona in 1993. In 1996, six years after their arrival in the United States, Africanized honeybees could be found in parts of Texas, Arizona, New Mexico, and California. As of 2009, Africanized honeybees are also found in Nevada, Utah, Oklahoma, Arkansas, and Louisiana. Their migration northward has slowed partially because they are a tropical insect and cannot live in colder climates. Experts have suggested two possible ways of limiting the spread of the Africanized honeybees. The first is drone-flooding, a process by which large numbers of European drones are kept in areas where commercially reared European queen bees mate, thereby ensuring that only limited mating occurs between Africanized drones and European queens. The second method is frequent re-queening, in which a beekeeper replaces a colony's queen with

How do fleas jump so far?

The jumping power of fleas comes both from strong leg muscles and from pads of a rubber-like protein called resilin. The resilin is located above the flea's hind legs. To jump, the flea crouches, squeezing the resilin, and then it relaxes certain muscles. Stored energy from the resilin works like a spring, launching the flea. A flea can jump well both vertically and horizontally. Some species can jump 150 times their own length. To match that record, a human would have to spring over the length of two and a quarter football fields—or the height of a 100-story building—in a single bound. The common flea (*Pulex irritans*) has been known to jump 13 inches (33 centimeters) in length and 7.25 inches (18.4 centimeters) in height.

one of his or her own choosing. The beekeeper can then be assured that the queens are European and that they have already mated with European drones.

How much **weight** can an **ant carry**?

Ants are the "superweight lifters" of the animal kingdom. They are strong in relation to their size and can carry objects 10 to 20 times their own weight—some species can carry objects up to 50 times their own weight. Ants are able to carry these objects great distances and even climb trees while carrying them. This is comparable to a 100-pound person picking up a small car, carrying it seven to eight miles on his back and then climbing the tallest mountain while still carrying the car!

How has **flight** contributed to the **success of insects**?

Flight is one key to the great success of insects. An animal that can fly can escape many predators, find food and mates, and disperse to new habitats much faster than an animal that must crawl about on the ground.

Why do some biologists consider the **insects** the **most successful** group of **animals**?

With more than one million described species (and perhaps millions more not yet identified), class Insecta is the most successful group of animals on Earth in terms of diversity, geographic distribution, number of species, and number of individuals. More species of insects have been identified than of all other groups of animals combined. What insects lack in size, they make up for in sheer numbers. If we could weigh all the insects in the world, their weight would exceed that of all the remaining terrestrial animals. About 200 million insects are alive at any one time for each human.

What **names** are used for **groups of animals**?

Animal	Group Name
Ants	Nest, army, colony, state, or swarm
Bees	Swarm, cluster, nest, hive, or erst
Caterpillars	Army
Eels	Swarm or bed
Fish	School, shoal, haul, draught, run, or catch
Flies	Business, hatch, grist, swarm, or cloud
Frogs	Arm
Gnats	Swarm, cloud, or horde
Goldfish	Troubling
Grasshoppers	Cloud
Hornets	Nest
Jellyfish	Smuck or brood
Lice	Flock
Locusts	Swarm, cloud, or plague
Minnows	Shoal, steam, or swarm
Oysters	Bed
Sardines	Family
Sharks	School or shoal
Snakes	Bed, knot, den, or pit
Termites	Colony, nest, swarm, or brood
Toads	Nest, knot, or knab
Trout	Hover
Turtles	Bale or dole
Wasps	Nest, herd, or pladge

FISH, AMPHIBIANS, AND REPTILES

What are **chondrichthyes**?

Chondrichthyes are fishes that have a cartilaginous skeleton rather than a bony skeleton; they include such organisms as sharks, skates, and rays.

How many **kinds of sharks** are there and how many are **dangerous**?

The United Nations's Food and Agricultural Organization lists 354 species of sharks, ranging in length from 6 inches (15 centimeters) to 49 feet (15 meters). While 35 species are known to have attacked humans at least once, only a dozen do so on a regular basis. The relatively rare great white shark (*Carcharodan carcharias*) is the largest predatory fish. The largest specimen accurately measured was 20 feet, 4 inches (6.2 meters) long and weighed 5,000 pounds (2,270 kilograms).

How **far** from **shore** do **shark attacks** occur?

In a study of 570 shark attacks, it was found that most shark attacks occur near shore. These data are not surprising since most people who enter the water stay close to the shore.

Distance from Shore	Percentage of Shark Attacks	Percentage of People Who Swim that Distance
50 ft (15 m)	31	39
100 ft (30 m)	11	15
200 ft (60 m)	9	12
300 ft (90 m)	8	11
400 ft (120 m)	2	2
500 ft (150 m)	3	5
1,000 ft (300 m)	6	9
1 mile (1.6 km)	8	6
> 1 mile (> 1.6 km)	22	1

What is unusual about the **teeth** of **sharks**?

Sharks were among the first vertebrates to develop teeth. The teeth are not set into the jaw but rather sit atop it. They are not firmly anchored and are easily lost. The teeth are arranged in 6 to 20 rows, with the ones in front doing the biting and cutting. Behind these teeth, others grow. When a tooth breaks or is worn down, a replacement moves forward. One shark may eventually develop and use more than 20,000 teeth in a lifetime.

A blacktip shark (left) and a nurse shark. The blacktip is one of the most dangerous sharks to humans, while nurse sharks are relatively docile.

Is the **whale shark** a **mammal** or a **fish**?

The whale shark (*Rhincodon typus*) is a shark, not a whale. It is, therefore, a fish. This species' name merely indicates that it is the largest of all shark species (weighing 40,000 pounds [18,144 kilograms] or more and growing to lengths of 49 feet [15 meters] or more) and the largest fish species in the world. However, it is completely harmless to humans.

What **general characteristics** do all **fishes** have in common?

All fishes have the following characteristics: 1) gills that extract oxygen from water; 2) an internal skeleton with a skin that surrounds the dorsal nerve cord; 3) single-loop blood circulation in which the blood is pumped from the heart to the gills and then to the rest of the body before returning to the heart; 4) nutritional deficiencies, particularly some amino acids that must be consumed and cannot be synthesized.

How is the **age** of a **fish determined**?

One way to determine the age of a fish is by its scales, which have growth rings just as trees do. Scales have concentric bony ridges or "circuli," which reflect the growth patterns of the individual fish. The portion of the scale that is embedded in the skin contains clusters of these ridges (called "annuli"); each cluster marks one year's growth cycle.

485

The movement, which confuses predators, happens because fish detect pressure changes in the water. The detection system, called the lateral line, is found along each side of the fish's body. Along the line are clusters of tiny hairs inside cups filled with a jellylike substance. If a fish becomes alarmed and turns sharply, it causes a pressure wave in the water around it. This wave pressure deforms the "jelly" in the lateral line of nearby fish. This moves the hairs that trigger nerves, and a signal is sent to the brain telling the fish to turn.

At what **speeds** do **fishes swim**?

The maximum swimming speed of a fish is somewhat determined by the shape of its body and tail and by its internal temperature. The cosmopolitan sailfish (*Istiophorus platypterus*) is considered to be the fastest fish species, at least for short distances, swimming at greater than 60 miles (95 kilometers) per hour. Some American fishermen believe, however, that the bluefin tuna (*Thunnus thynnus*) is the fastest, but the fastest speed recorded for them so far is 43.4 miles (69.8 kilometers) per hour. Data is extremely difficult to secure because of the practical difficulties in measuring the speeds. The yellowfin tuna (*Thunnus albacares*) and the wahoe (*Acanthocybium solandri*) are also fast, timed at 46.35 miles (74.5 kilometers) per hour and 47.88 miles (77 kilometers) per hour during 10- to 20-second sprints. Flying fish swim at more than 40 miles (over 64 kilometers) per hour, dolphins at 37 miles (60 kilometers) per hour, trout at 15 miles (24 kilometers) per hour, and blenny at 5 miles (8 kilometers) per hour. Humans can swim 5.19 miles (8.3 kilometers) per hour.

How much **electricity** does an **electric eel generate**?

An electric eel (*Electrophorus electricus*) has current-producing organs made up of electric plates on both sides of its vertebral column running almost its entire body length. The charge—on the average of 350 volts, but as great as 550 volts—is released by the central nervous system. The shock consists of four to eight separate charges, which last only two- to three-thousandths of a second each. These shocks, used as a defense mechanism, can be repeated up to 150 times per hour without any visible fatigue to the eel. The most powerful electric eel, found in the rivers of Brazil, Colombia, Venezuela, and Peru, produces a shock of 400 to 650 volts.

What does the word **"amphibian" mean**?

The word "amphibian," from the Greek term *amphibia*, means "both lives" and refers to the animals' double life on land and in water. The usual life cycle of amphibians begins with eggs laid in water, which develop into aquatic larvae with external gills; in a development that recapitulates its evolution, the fishlike larva develops lungs and limbs and becomes an adult.

What are the **major groups** of **amphibians**?

The following chart illustrates the three major groups of amphibians.

Examples	Order	Number of Living Species
Frogs and toads	Anura (Salientia)	3,450
Salamanders and newts	Caudata (Urodela)	360
Caecilians	Apoda (Gymnophiona)	160

What is the **difference** between a **reptile** and an **amphibian**?

Reptiles are clad in scales, shields, or plates, and their toes have claws; amphibians have moist, glandular skins, and their toes lack claws. Reptile eggs have a thick, hard, or parchmentlike shell that protects the developing embryo from moisture loss, even on dry land. The eggs of amphibians lack this protective outer covering and are always laid in water or in damp places. Young reptiles are miniature replicas of their parents in general appearance if not always in coloration and pattern. Juvenile amphibians pass through a larval, usually aquatic, stage before they metamorphose (change in form and structure) into the adult form. Reptiles include alligators, crocodiles, turtles, and snakes. Amphibians include salamanders, toads, and frogs.

What **features of reptiles** enabled them to become **true land vertebrates**?

Legs were arranged to support the body's weight more effectively than in amphibians, allowing reptile bodies to be larger and to run. Reptilian lungs were more developed with a greatly increased surface area for gas exchange than the saclike lungs of amphibians. The three-chambered heart of reptiles was more efficient than the three-chambered amphibian heart. In addition, the skin was covered with hard, dry scales to minimize water loss. However, the most important evolutionary adaptation was the amniotic egg, in which an embryo could survive and develop on land. The eggs were surrounded by a protective shell that prevented the developing embryo from drying out.

What groups of **reptiles** are **living today**?

The three orders of reptiles that are alive today are: 1) Chelonia, which includes turtles, terrapins, and tortoises; 2) Squamata, which includes lizards and snakes; and 3) Crocodilia, which includes crocodiles and alligators.

Which **venomous snakes** are **native** to the **United States**?

Snake	Average Length
Rattlesnakes	
Eastern diamondback (*Crotalus adamateus*)	33–65 in (84–165 cm)
Western diamondback (*Crotalus atrox*)	30–65 in (76–419 cm)
Timber rattlesnake (*Crotalus horridus horridus*)	32–54 in (81–137 cm)
Prairie rattlesnake (*Crotalus viridis viridis*)	32–46 in (81–117 cm)
Great Basin rattlesnake (*Crotalus viridis lutosus*)	32–46 in (81–117 cm)
Southern Pacific rattlesnake (*Crotalus viridis helleri*)	30–48 in (76–122 cm)
Red diamond rattlesnake (*Crotalus ruber ruber*)	30–52 in (76–132 cm)
Mojave rattlesnake (*Crotalus scutulatus*)	22–40 in (56–102 cm)
Sidewinder (*Crotalus cerastes*)	18–30 in (46–76 cm)
Moccasins	
Cottonmouth (*Agkistrodon piscivorus*)	30–50 in (76–127 cm)
Copperhead (*Agkistrodon contortrix*)	24–36 in (61–91 cm)
Cantil (*Agkistrodon bilineatus*)	30–42 in (76–107 cm)
Coral snakes	
Eastern coral snake (*Micrurus fulvius*)	16–28 in (41–71 cm)

What is the **fastest snake** on **land**?

The black mamba (*Dendroaspis polylepis*), a deadly poisonous African snake that can grow up to 13 feet (4 meters) in length, has been recorded reaching a speed of 7 miles (11 kilometers) per hour. A particularly aggressive snake, it chases animals at high speeds holding the front of its body above the ground.

Are **tortoises** and **terrapins** the same as **turtles**?

The terms "turtle," "tortoise," and "terrapin" are used for various members of the order Testudines (from the Latin term *testudo*, meaning "tortoise"). In North American usage they are all correctly called turtles. The term "tortoise" is often used for land turtles. In British usage the term "tortoise" is the inclusive term, and "turtle" is only applied to aquatic members of the order.

What are the **upper** and **lower shells** of a turtle called?

The turtle (order Testudines) uses its shell as a protective device. The upper shell is called the dorsal carapace and the lower shell is called the ventral plastron. The shell's sections are referred to as the scutes. The carapace and the plastron are joined at the sides.

The upper shell of this wood turtle is called the "carapace," while the bottom shell is called the "plastron."

How **fast** can a **crocodile run** on land?

In smaller crocodiles, the running gait can change into a bounding gallop that can achieve speeds of 2 to 10 miles (3 to 17 kilometers) per hour.

BIRDS

What **names** are used for **groups of birds**?

A group of birds in general is called a congregation, flight, flock, volery, or volley. Below is a list specific to types of birds.

Bird	Group Name
Bitterns	Siege or sedge
Budgerigars	Chatter
Chickens	Flock, run, brood, or clutch
Coots	Fleet or pod
Cormorants	Flight

489

Do all birds fly?

No. Among the flightless birds, the penguins and the ratites are the best known. Ratites include emus, kiwis, ostriches, rheas, and cassowaries. They are called ratite because they lack a keel on the breastbone. All of these birds have wings but lost their power to fly millions of years ago. Many birds that live isolated on oceanic islands (for example, the great auk) apparently became flightless in the absence of predators and the consequent gradual disuse of their wings for escape.

Bird	Group Name
Cranes	Herd or siege
Crows	Murder, clan, or hover
Curlews	Herd
Doves	Flight, flock, or dole
Ducks	Paddling, bed, brace, flock, flight, or raft
Eagles	Convocation
Geese	Gaggle or plump (on water), flock (on land), skein (in flight), or covert
Goldfinches	Charm, chattering, chirp, or drum
Grouses	Pack or brood
Gulls	Colony
Hawks	Cast
Hens	Brood or flock
Herons	Siege, sege, scattering, or sedge
Jays	Band
Larks	Exaltation, flight, or ascension
Magpies	Tiding or tittering
Mallards	Flush, sord, or sute
Nightingales	Watch
Partridges	Covey
Peacocks	Muster, ostentation, or pride
Penguins	Colony
Pheasants	Nye, brood, or nide
Pigeons	Flock or flight
Plovers	Stand, congregation, flock, or flight
Quails	Covey or bevy
Sparrows	Host
Starlings	Chattering or murmuration

Bird	Group Name
Storks	Mustering
Swallows	Flight
Swans	Herd, team, bank, wedge, or bevy
Teals	Spring
Turkeys	Rafter
Turtle doves	Dule
Woodpeckers	Descent
Wrens	Herd

What accounts for the **different colors** of **bird feathers**?

The vivid color of feathers is of two kinds: pigmentary and structural. Red, orange and yellow feathers are colored by pigments called lipochromes deposited in the feather barbules as they are formed. Black, brown, and gray colors are from another pigment, melanin. Blue feathers depend not on pigment but on scattering of shorter wavelengths of light by particles within the feather. These are structural feathers. Green colors are almost always a combination of yellow pigment and blue feather structure. Another kind of structural color is the beautiful iridescent color of many birds, which ranges from red, orange, copper, and gold to green, blue, and violet. Iridescent color is based on interference that causes light waves to reinforce, weaken, or eliminate each other. Iridescent colors may change with the angle of view.

What bird has the **biggest wing span**?

Three members of the albatross family—the wandering albatross (*Diomedea exculans*), the royal albatross (*Diomedea epomophora*), and the Amsterdam Island albatross (*Diomeda amsterdiamensis*)—have the greatest wingspan of any bird species with a spread of 8 to 11 feet (2.5 to 3.3 meters).

How **fast** do a **hummingbird's wings move**?

Hummingbirds are the only family of birds that can truly hover in still air for any length of time. They need to do so in order to hang in front of a flower while they perform the delicate task of inserting their slim, sharp bills into its depths to drink nectar. Their thin wings are not contoured into the shape of aerofoils and do not generate lift in this way. Their paddle-shaped wings are, in effect, hands that swivel at the shoulder. They beat them in such a way that the tip of each wing follows the line of a figure eight lying on its side. The wing moves forward and downwards into the front loop of the eight, creating lift. As it begins to come up and go back, the wing twists through 180 degrees so that once again it creates a downward thrust.

The hummingbird's method of flying does have one major limitation: the smaller the wing, the faster it has to beat in order to produce sufficient downward thrust. An average-sized hummingbird beats its wings 25 times per second. Small species beat their wings 50 to 80 times per second, and even faster during courtship displays. The bee hummingbird, native to Cuba, is only 2 inches (5 centimeters) long and beats its wings at an astonishing 200 times per second.

How **fast** does a **hummingbird fly** and how far does the hummingbird migrate?

Hummingbirds fly at speeds up to 71 miles (80 kilometers) per hour. The longest migratory flight of a hummingbird documented to date is the flight of a rufous hummingbird from Ramsey Canyon, Arizona, to near Mt. Saint Helens, Washington, a distance of 1,414 miles (2,277 kilometers). Bird-banding studies are now in progress to verify that a few rufous hummingbirds do make a 11,000 to 11,500 mile (17,699 to 18,503 kilometer) journey along a super Great Basin High route, a circuit that could take a year to complete. Hummingbird studies, however, are difficult to complete because so few banded birds are recovered.

How **fast** do **birds fly**?

Different species of birds fly at different speeds. The following table lists the flight speeds of some birds:

Bird	Speed (mph / kph)
Peregrine falcon	168–217 / 270.3–349.1
Swift	105.6 / 169.9
Merganser	65 / 104.6
Golden plover	50–70 / 80.5–112.6
Mallard	40.6 / 65.3
Wandering albatross	33.6 / 54.1
Carrion crow	31.3 / 50.4
Herring gull	22.3–24.6 / 35.9–39.6
House sparrow	17.9–31.3 / 28.8–50.4
Woodcock	5 / 8

Why do **birds migrate** annually?

Migratory behavior in birds is inherited; however, birds will not migrate without certain physiological and environmental stimuli. In the late summer, the decrease in sunlight stimulates the pituitary gland and the adrenal gland of migrating birds, causing them to produce the hormones prolactin and corticosterone respectively. These hormones in turn cause the birds to accumulate large amounts of fat just under the skin,

Which bird migrates the greatest distance?

The arctic tern (*Sterna paradisaea*) migrates the longest distance of any bird. They breed from subarctic regions to the very limits of land in the arctic of North America and Eurasia. At the end of the northern summer, the arctic tern leaves the north on a migration of more than 11,000 miles (17,699 kilometers) to its southern home in Antarctica. A tern tagged in July on the arctic coast of Russia was recovered the next May near Fremantle, Australia, a record 14,000 miles (22,526 kilometers) away.

providing them with enough energy for the long migratory flights. The hormones also cause the birds to become restless just prior to migration. The exact time of departure, however, is dictated not only by the decreasing sunlight and hormonal changes, but also by such conditions as the availability of food and the onset of cold weather.

The major wintering areas for North American migrating birds are the southern United States and Central America. Migrating ducks follow four major flyways south: the Atlantic flyway, the Mississippi flyway, the central flyway, and the Pacific flyway. Some bird experts propose that the birds return north to breed for several reasons: (1) Birds return to nest because there is a huge insect supply for their young; (2) The higher Earth's latitude in the summer in the Northern Hemisphere, the longer the daylight available to the parents to find food for their young; (3) Less competition exists for food and nesting sites in the north; (4) In the north, there are fewer mammal predators for nesting birds (which are particularly vulnerable during the nesting stage); (5) Birds migrate south to escape the cold weather, and they return north when the weather is more temperate.

Why do **geese fly** in **formation**?

Aerodynamicists have suspected that long-distance migratory birds, such as geese and swans, adapt the "V" formation in order to reduce the amount of energy needed for such long flights. According to theoretical calculations, birds flying in a "V" formation can fly some ten percent farther than a lone bird can. Formation flying lessens the drag (the air pressure that pushes against the wings). The effect is similar to flying in a thermal upcurrent, where less total lift power is needed. In addition, when flying, each bird creates behind it a small area of disturbed air. Any bird flying directly behind it would be caught in this turbulence. In the "V" formation of Canada geese, each bird flies not directly behind the other, but to one side or above the bird in front.

How are **birds related** to **dinosaurs**?

Birds are essentially modified dinosaurs with feathers. Robert T. Bakker (1945–) and John H. Ostrom (1928–2005) did extensive research on the relationship between birds and dinosaurs in the 1970s and concluded that the bony structure of small dinosaurs was very similar to *Archaeopteryx*, the first animal classified as a bird, but that dinosaur fossils showed no evidence of feathers. They proposed that birds and dinosaurs evolved from the same source.

Why is *Archaeopteryx* important?

Archaeopteryx is the first known bird. It had true feathers that provided insulation and allowed this animal to form scoops with its wings for catching prey.

How sensitive is the **hearing** of **birds**?

In most species of birds, the most important sense after sight is hearing. Birds' ears are close to their bodies and are covered by feathers. The

The discovery of this fossil of an *archaeopteryx* spurred the theory that birds are the evolutionary descendants of dinosaurs.

feathers, however, do not have barbules, which would obstruct sound. Nocturnal raptors, such as the great horned owl, have a very well-developed sense of hearing in order to be able to capture their prey in total darkness.

How do **birds learn to sing** the distinctive melody of their respective species?

The ability to learn the proper song appears to be influenced by both heredity and experience. Scientists have speculated that a bird is genetically programmed with the ability to recognize the song of its own species and with the tendency to learn its own song. As a bird begins to sing, it goes through a stage of practice (which closely resembles the babbling of human infants) through which it perfects the notes and structure of its distinctive song. In order to produce a perfect imitation, the bird must apparently hear the song from an adult during its first months of life.

What is unusual about the way the emperor penguin's eggs are incubated?

Each female emperor penguin (*Aptenodytes forsteri*) lays one large egg. Initially, both sexes share in incubating the egg by carrying it on his or her feet and covering it with a fold of skin. After a few days of passing the egg back and forth, the female leaves to feed in the open water of the Arctic Ocean. Balancing their eggs on their feet, the male penguins shuffle about the rookery, periodically huddling together for warmth during blizzards and frigid weather. If an egg is inadvertently orphaned, a male with no egg will quickly adopt it. Two months after the female's departure, the chick hatches. The male feeds it with a milky substance he regurgitates until the female returns. Now padded with blubber, the females take over feeding the chicks with fish they have stored in their crops. The females do not return to their mate, however, but wander from male to male until one allows her to take his chick. It is then the males' turn to feed in open water and restore the fat layer they lost while incubating.

Which birds lay the **largest** and **smallest eggs**?

The elephant bird (*Aepyornis maximus*), an extinct flightless bird of Madagascar, also known as the giant bird or roc, laid the largest known bird eggs. Some of these eggs measured as much as 13.5 inches (34 centimeters) in length and 9.5 inches (24 centimeters) in diameter. The largest egg produced by any living bird is that of the North African ostrich (*Struthio camelus*). The average size is 6 to 8 inches (15 to 20.5 centimeters) in length and 4 to 6 inches (five to 15 centimeters) in diameter.

The smallest mature egg, measuring less than 0.39 inch (1 centimeter) in length, is that of the vervain hummingbird (*Mellisuga minima*) of Jamaica.

Generally speaking, the larger the bird, the larger the egg. However, when compared with the bird's body size, the ostrich egg is one of the smallest eggs, while the hummingbird's egg is one of the largest. The Kiwi bird of New Zealand lays the largest egg, relative to body size, of any living bird. Kiwis are comparable to chickens in size, but their eggs are comparable to ostrich eggs in size. The egg of a Brown Kiwi is 14 to 20 percent of the female's body weight. Its egg weighs up to 1 pound (0.5 kilogram).

What are the **natural predators** of the **penguin**?

The leopard seal (*Hydrurga leptonyx*) is the principal predator of both the adult and juvenile penguin. The penguin may also be caught by a killer whale while swimming in open water. Eggs and chicks that are not properly guarded by adults are often devoured by skuas and sheathbills.

Why don't woodpeckers get headaches?

Woodpeckers' skulls are particularly sturdy to withstand the force of the blows as they hammer with their beaks. They are further aided by strong neck muscles to support their heads.

How does a **homing pigeon** find its **way home**?

Scientists currently have two hypotheses to explain the homing flight of pigeons. Neither has been proved to the satisfaction of all the experts. The first hypothesis involves an "odor map." This theory proposes that young pigeons learn how to return to their original point of departure by smelling different odors that reach their home in the winds from varying directions. They would, for example, learn that a certain odor is carried on winds blowing from the east. If a pigeon were transported eastward, the odor would tell it to fly westward to return home. The second hypothesis proposes that a bird may be able to extract its home's latitude and longitude from Earth's magnetic field. It may be proven in the future that neither theory explains the pigeon's navigational abilities or that some synthesis of the two theories is plausible.

What is the name of the **bird** that **perches** on the **black rhinoceros's back**?

The bird, a relative of the starling, is called an oxpecker (a member of the Sturnidae family). Found only in Africa, the yellow-billed oxpecker (*Buphagus africanus*) is widespread over much of western and central Africa, while the red-billed oxpecker (*Buphagus erythrorhynchus*) lives in eastern Africa from the Red Sea to Natal.

Seven to eight inches (17–20 centimeters) long with a coffee-brown body, the oxpecker feeds on more than 20 species of ticks that live in the hide of the black rhinoceros (*Diceros bicornis*), also called the hook-lipped rhino. The bird spends most of its time on the rhinoceros or on other animals, such as the antelope, zebra, giraffe, or buffalo. The bird has even been known to roost on the body of its host.

The relationship between the oxpecker and the rhinoceros is a type of symbiosis (a close association between two organisms in which at least one of them benefits) called mutualism. The rhinoceros' relief of its ticks and the bird's feeding clearly demonstrates mutualism (a condition in which both organisms benefit). In addition, the oxpecker, having much better eyesight than the nearsighted rhinoceros, alerts its host with its shrill cries and flight when danger approaches.

Why **don't birds** get **electrocuted** when they sit on **wires**?

In general, birds do not get electrocuted while just sitting on power transmission wires. Most electrocutions happen when a bird opens its wingspan and completes a circuit by bridging the gap between two live wires or a live wire and a grounded wire, or other parts such as transformers and grounded metal crossarms.

Which state was the first to "officially" name a state bird?

In 1926, Kentucky officially named the cardinal as its state bird.

When was the bald eagle adopted as the national bird of the United States?

On June 20, 1782, the citizens of the newly independent United States of America adopted the bald or "American" eagle as their national emblem. At first the heraldic artists depicted a bird that could have been a member of any of the larger species, but by 1902, the bird portrayed on the seal of the United States of America had assumed its proper white plumage on the head and tail. The choice of the bald eagle was not unanimous; Benjamin Franklin (1706–1790) preferred the wild turkey. Oftentimes a tongue-in-cheek humorist, Franklin thought the turkey a wily but brave, intelligent, and prudent bird. He viewed the eagle on the other hand as having "a bad moral character" and "not getting his living honestly," preferring instead to steal fish from hardworking fishhawks. He also found the eagle a coward that readily flees from the irritating attacks of the much smaller kingbird.

MAMMALS

What names are used for groups or companies of mammals?

Mammal	Group Name
Antelopes	Herd
Apes	Shrewdness
Asses	Pace, drove, or herd
Baboons	Troop
Bears	Sloth
Beavers	Family or colony
Boars	Sounder
Buffaloes	Troop, herd, or gang
Camels	Flock, train, or caravan
Caribou	Herd
Cattle	Drove or herd
Deer	Herd or leash
Elephants	Herd

Mammal	Group Name
Elks	Gang or herd
Foxes	Cloud, skulk, or troop
Giraffes	Herd, corps, or troop
Goats	Flock, trip, herd, or tribe
Gorillas	Band
Horses	Haras, stable, remuda, stud, herd, string, field, set, team, or stable
Jackrabbits	Husk
Kangaroos	Troop, mob, or herd
Leopards	Leap
Lions	Pride, troop, flock, sawt, or souse
Mice	Nest
Monkeys	Troop or cartload
Moose	Herd
Mules	Barren or span
Oxen	Team, yoke, drove, or herd
Porpoises	School, crowd, herd, shoal, or gam
Reindeer	Herd
Rhinoceri	Crash
Seals	Pod, herd, trip, rookery, or harem
Sheep	Flock, hirsel, drove, trip, or pack
Squirrels	Dray
Swine	Sounder, drift, herd, or trip
Walruses	Pod or herd
Weasels	Pack, colony, gam, herd, pod, or school
Whales	School, gam, mob, pod, or herd
Wolves	Rout, route, or pack
Zebras	Herd

Which **mammals** have the **shortest gestation periods**?

Gestation is the period of time between fertilization and birth in oviparous animals. The shortest gestation period known is 12 to 13 days, shared by three marsupials: the American or Virginian opossum (*Didelphis marsupialis*); the rare water opossum, or yapok (*Chironectes minimus*) of central and northern South America; and the eastern native cat (*Dasyurus viverrinus*) of Australia. The young of each of these marsupials are born while still immature and complete their development in the ventral pouch of their mother. While 12 to 13 days is the average, the gestation period is sometimes as short as eight days. The longest gestation period for a mammal is that of the African elephant (*Loxodonta africana*) with an average of 660 days, and a maximum of 760 days.

Though some mammals can glide, the only mammals that can truly fly are bats, like these fruit bats.

Do any **mammals fly**?

Bats (order Chiroptera with 986 species) are the only truly flying mammals, although several gliding mammals are referred to as "flying" (such as the flying squirrel and flying lemur). The "wings" of bats are double membranes of skin stretching from the sides of the body to the hind legs and tail, and are actually skin extensions of the back and belly. The wing membranes are supported by the elongated fingers of the forelimbs (or arms). Nocturnal (active at night), ranging in length from 1.5 inches (25 millimeters) to 1.3 feet (40.6 centimeters), and living in caves or crevices, bats inhabit most of the temperate and tropical regions of both hemispheres. The majority of species feed on insects and fruit, while some tropical species eat pollen and nectar of flowers, and insects found inside them. Moderate-sized species usually prey on small mammals, birds, lizards, and frogs, and some eat fish. But true vampire bats (three species) eat the blood of animals by making an incision in the animal's skin—from these bats, animals can contract rabies.

Most bats do not find their way around by sight but have evolved a sonar system, called "echolocation," for locating solid objects. Bats emit vocal sounds through the nose or mouth while flying. These sounds, usually above the human hearing range, are reflected back as echoes. This method enables bats, when flying in darkness, to avoid solid objects and to locate the position of flying insects. Bats have the most acute sense of hearing of any land animal, hearing frequencies as high as 120 to 210 kilohertz. The highest frequency humans can hear is 20 kilohertz.

What are the **only mammals** that **cannot jump**?

It might not be surprising to learn that neither the rhinoceros nor the elephant can jump, since their enormous weight makes the feat difficult. However, the third mammal that cannot jump is the pronghorn sheep, which was called an "antelope" in the famous song "Home on the Range." The pronghorn sheep's inability to jump has been a particular disadvantage in its North American home, where fences have prevented populations from migrating and hindered the pronghorn's ability to find mates and breed.

How does the **breath-holding capability** of a **human compare** with other mammals?

Mammal	Average Time (minutes)
Human	1
Polar bear	1.5
Pearl diver (human)	2.5
Sea otter	5
Porpoise	6
Platypus	10
Muskrat	12
Hippopotamus	15
Seal	15–28
Sea cow	16
Beaver	20
Greenland whale	60
Sperm whale	90
Bottle-nosed whale	120

How **deep** do **marine mammals dive**?

The maximum depths and the longest durations of time underwater by various aquatic mammals are listed below:

Mammal	Maximum Depth (feet/meters)	Maximum Time Underwater
Porpoise	984 / 300	15 minutes
Fin whale	1,148 / 350	20 minutes
Bottle-nosed whale	1,476 / 450	120 minutes
Weddell seal	1,968 / 600	70 minutes
Sperm whale	> 6,562 / > 2,000	90 minutes

Is there any truth to the saying "blind as a bat"?

The saying "blind as a bat" is not true. Although bats rely on sound to navigate and find food, they have all the elements found in a normal mammalian eye and they do see.

How does a **human's heartbeat compare** with those of other mammals?

Mammal	Resting Heart Rate (beats per minute)
Human	75
Horse	48
Cow	45–60
Dog	90–100
Cat	110–140
Rat	360
Mouse	498

How does a **bat catch flying insects** in total darkness?

Bats use sound waves for communication and navigation. They emit supersonic radiation ranging from as low as 200 hertz to as high as 30,000 hertz. The sounds are emitted through the bat's nostrils or mouth and are aided by a complex flap structure to provide precise directivity to the radiation. Echo returns from the emissions allow a bat to pick out a tiny flying insect some distance ahead. Highly sensitive ears and an ability to maneuver with great agility enables many bats to fly around in a darkened cave, catching insects without fear of collision.

What are some **animals** that have **pouches**?

Marsupials (meaning "pouched" animals) differ from all other living mammals in their anatomical and physiological features of reproduction. Most female marsupials, including kangaroos, bandicoots, wombats, banded anteaters, koalas, opossums, wallabies, and tasmanian devils, possess an abdominal pouch (called a marsupium), in which their young are carried. In some small terrestrial marsupials, however, the marsupium is not a true pouch but merely a fold of skin around the mammae (milk nipples).

The short gestation period in marsupials (in comparison to other similarly sized mammals) allows their young to be born in an "undeveloped" state. Consequently, these animals have been viewed as "primitive" or second-class mammals. However, some now see that the reproductive process of marsupials has an advantage over that of placental mammals. A female marsupial invests relatively few resources during the brief gestation period, more so during the lactation (nursing period) when the young are in the marsupium. If the female marsupial loses its young, it can conceive again sooner than a placental mammal in a comparable situation.

Which **mammals lay eggs** and suckle their young?

The duck-billed platypus (*Ornithorhynchus anatinus*), the short-nosed echidna or spiny anteater (*Tachyglossus aculeatus*), and the long-nosed echidna (*Zaglossus bruijni*), indigenous to Australia, Tasmania, and New Guinea, are the only three species of mammals that lay eggs (a non-mammalian feature) but suckle their young (a mammalian feature). These mammals (order Monotremata) resemble reptiles in that they lay rubbery shell-covered eggs that are incubated and hatched outside the mother's body. In addition, they resemble reptiles in their digestive, reproductive, and excretory systems, and in a number of anatomical details (eye structure, presence of certain skull bones, pectoral [shoulder] girdle, and rib and vertebral structures). They are, however, classed as mammals because they have fur and a four-chambered heart, nurse their young from gland milk, are warm-blooded, and have some mammalian skeletal features.

What **freshwater mammal** is **venomous**?

The male duck-billed platypus (*Ornithorhynchus anatinus*) has venomous spurs located on its hind legs. When threatened, the animal will drive them into the skin of a potential enemy, inflicting a painful sting. The venom this action releases is relatively mild and generally not harmful to humans.

What is the **difference** between **porpoises** and **dolphins**?

Marine dolphins (family Delphinidae) and porpoises (family Phocoenidae) together comprise about 40 species. The chief differences between dolphins and porpoises occur in the snout and teeth. True dolphins have a beaklike snout and cone-shaped teeth. True porpoises have a rounded snout and flat or spade-shaped teeth.

How do the **great whales compare** in weight and length?

Whale	Average Weight (tons/kg)	Greatest Length (ft/meters)
Sperm	35 / 31,752	59 / 18
Blue	84 / 76,204	98.4 / 30
Finback	50 / 45,360	82 / 25
Humpback	33 / 29,937	49.2 / 15
Right	50 (est.) / 45,360 (est.)	55.7 / 17

What fictional character was inspired by manatees?

Manatees and their close relatives, sea cows and dugongs, may have been the inspiration for the mermaid legend. The scientific family name for sea cows is *Sirenus*, which is derived from the word "siren"—the original name for the beautiful mermaids that lured love-struck sailors to their deaths in ancient legends.

Whale	Average Weight (tons/kg)	Greatest Length (ft/meters)
Sei	17 / 15,422	49.2 / 15
Gray	20 / 18,144	39.3 / 12
Bowhead	50 / 45,360	59 / 18
Bryde's	17 / 15,422	49.2 / 15
Minke	10 / 9,072	29.5 / 9

What is the **fastest swimming whale**?

The orca or killer whale (*Orcinus orca*) is the fastest swimming whale. In fact, it is the fastest swimming marine mammal with speeds that reach 31 miles per hour (50 kilometers per hour).

What is the name of the **seal-like animal** in **Florida**?

The West Indian manatee (*Trichechus manatus*), in the winter, moves to more temperate parts of Florida, such as the warm headwaters of the Crystal and Homosassa Rivers in central Florida or the tropical waters of southern Florida. When the air temperature rises to 50°F (10°C), it will wander back along the Gulf coast and up the Atlantic coast as far as Virginia. Long-range offshore migrations to the coast of Guyana and South America have been documented. In 1893, when the population of manatees in Florida was reduced to several thousand, the state gave it legal protection from being hunted or commercially exploited. However, many animals continue to be killed or injured by the encroachment of humans. Entrapment in locks and dams, collisions with barges and power boat propellers, and other man-made objects, cause at least 30 percent of the manatee deaths, which total 125 to 130 annually.

What is the **only four-horned animal** in the world?

The four-horned antelope (*Tetracerus quadricornis*) is a native of central India. The males have two short horns, usually 4 inches (10 centimeters) in length, between their ears, and an even shorter pair, 1 to 2 inches (2.5 to five centimeters) long, between the brow ridges over their eyes. Not all males have four horns, and in some the second pair eventually falls off. The females have no horns at all.

Is there a **cat** that lives in the **desert**?

The sand cat (*Felis margarita*) is the only member of the cat family tied directly to desert regions. Found in North Africa, the Arabian Peninsula, Turkmenistan, Uzbekistan, and western Pakistan, the sand cat has adapted to extremely arid desert areas. The padding on the soles of its feet is well suited to the loose sandy soil, and it can live without drinking free-standing water. Having sandy or grayish-ochre dense fur, its body length is 17.5 to 22 inches (45 to 57 centimeters). Mainly nocturnal (active at night), the cat feeds on rodents, hares, birds, and reptiles.

The Chinese desert cat (*Felis bieti*) does not live in the desert as its name implies, but inhabits the steppe country and mountains. Likewise, the Asiatic desert cat (*Felis silvestris ornata*) inhabits the open plains of India, Pakistan, Iran, and Asiatic Russia.

What is the only **American canine** that can **climb trees**?

The gray fox (*Urocyon cinereoargenteus*) is the only American canine that can climb trees.

Which **bear** lives in a **tropical rain forest**?

The Malayan sun bear (*Ursus malayanus*) is one of the rarest animals in the tropical forests of Sumatra, the Malay Peninsula, Borneo, Burma, Thailand, and southern China. The smallest bear species, with a length of 3.3 to 4.6 feet (1 to 1.4 meters) and weighing 60 to 143 pounds (27 to 65 kilograms), it has a strong, stocky body. Against its black, short fur it has a characteristic orange-yellow-colored crescent across its chest, which according to legend represents the rising sun. With powerful paws having long, curved claws to help it climb trees in the dense forests, it is an expert tree climber. The sun bear tears at tree bark to expose insects, larvae, and the nests of bees and termites. Fruit, coconut palms, and small rodents are also part of its diet. Sleeping and sunbathing during the day, it is active at night. Unusually shy and retiring, cautious and intelligent, the sun bear is declining in population as its native forests are being destroyed.

What is the **largest terrestrial mammal** in North America?

The bison (*Bison bison*) is the largest terrestrial mammal in North America. It weighs 3,100 pounds (1,406 liograms) and is 6 feet (1.8 meters) high.

Do **camels store water** in their **humps**?

The hump or humps do not store water, since they are fat reservoirs. The ability to go long periods without drinking water, up to ten months if there is plenty of green vegetation and dew to feed on, results from a number of physiological adaptations. One major factor is that camels can lose up to 40 percent of their body weight with no ill effects. A camel can also withstand a variation of its body temperature by as much as 14 degrees. A camel can drink 30 gallons of water in ten minutes and up to 50 gallons over several hours. A one-humped camel is called a dromedary or Arabian camel; a Bactrian camel has two humps and lives in the wild on the Gobi Desert. Today, the Bactrian is confined to Asia, while most of the Arabian camels are on African soil.

How many **quills** does a **porcupine** have?

For its defensive weapon, the average North American porcupine has about 30,000 quills or specialized hairs, comparable in hardness and flexibility to slivers of celluloid and so sharply pointed that they can penetrate any hide. The quills that do the most damage are the short ones that stud the porcupine's muscular tail. With a few lashes, the porcupine can send a rain of quills that have tiny scalelike barbs into the skin of its adversary. The quills work their way inward because of their barbs and the involuntary muscular action of the victim. Sometimes the quills can work themselves out, but other times the quills pierce vital organs, and the victim dies.

Slow-footed and stocky, porcupines spend much of their time in the trees, using their formidable incisors to strip off bark and foliage for their food, and supplement their diets with fruits and grasses. Porcupines have a ravenous appetite for salt; as herbivores (plant-eating animals), their diets have insufficient salt. So natural salt licks, animal bones left by carnivores (meat-eating animals), yellow pond lilies, and other items having a high salt content (including paints, plywood adhesives, and human clothing that bears traces of sweat) have a strong appeal to porcupines.

What is the **difference** between an **African elephant** and an **Indian elephant**?

The African elephant (*Loxodonta africana*) is the largest living land animal, weighing up to 8.25 tons (7,500 kilograms) and standing 10 to 13 feet (3 to 4 meters) at the shoulder. The Indian elephant (*Elephas maximus*) weighs about 6 tons (5,500 kilograms) with a shoulder height of 10 feet (3 meters). Other differences are:

African Elephant	Indian Elephant
Larger ears	Smaller ears
Gestation period of about 670 days	Gestation period of about 610 days
Ear tops turn backwards	Ear tops turn forward
Concave back	Convex back
Three toenails on hind feet	Four toenails on hind feet
Larger tusks	Smaller tusks
Two finger-like lips at tips of trunk	One lip at tip of trunk

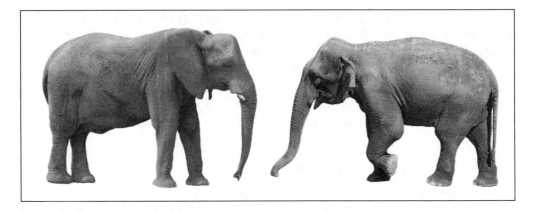

An African elephant (left) and Asian elephant. It is easy to see the differences between these two animals when they are next to each other!

Why do **cows** have **four stomachs**?

The stomachs of cows, as well as all ruminants, are divided into four sections—the rumen, reticulum, omasum, and abomasums. Ruminants eat rapidly and do not chew much of their food completely before they swallow it. The liquid part of their food enters the reticulum first, while the solid part of their food enters the rumen where it softens. Bacteria in the rumen initially break it down as a first step in digestion. Ruminants later regurgitate it into the mouth where they chew their cud. Cows chew their cud about six to eight times per day, spending a total of five to seven hours in rumination. The chewed cud goes directly into the other chambers of the stomach, where various microorganisms assist in further digestion.

Why were **Clydesdale horses** used as **war horses**?

The Clydesdales were among a group of European horses referred to as the "Great Horses," which were specifically bred to carry the massively armored knights of the Middle Ages. These animals had to be strong enough to carry a man wearing as much as 100 pounds (45 kilograms) of armor as well as up to 80 pounds (36 kilograms) of armor on their own bodies. However, the invention of the musket quickly ended the use of Clydesdales and other Great Horses on the battlefield as speed and maneuverability became more important than strength.

Why are **Dalmatians "firehouse dogs"**?

Before automobiles, coaches and carriages were often accompanied by dogs that kept horses company and guarded them from theft. Dalmatians were particularly well known for the strong bond they formed with horses, and firemen, who often owned the strongest and speediest horses in the area, kept the dogs at the station to deter

> ## How many vertebrae are in the neck of a giraffe?
>
> **A** giraffe neck has seven vertebrae, the same as other mammals, but the vertebrae are greatly elongated.

horse thieves. Although fire engines have replaced horses, Dalmatians have remained a part of firehouse life, both for the appeal of these beautiful dogs and for their nostalgic tie to the past.

What is the **chemical composition** of a **skunk's spray**?

The chief odorous components of the spray have been identified as crotyl mercaptan, isopentyl mercaptan, and methyl crotyl disulfide in the ratio of 4:4:3. The liquid is an oily, pale-yellow, foul-smelling spray that can cause severe eye inflammation. This defensive weapon is discharged from two tiny nipples located just inside the skunk's anus—either as a fine spray or a short stream of rain-sized drops. Although the liquid's range is 6.5 to 10 feet (2 to 3 meters), its smell can be detected 1.5 miles (2.5 kilometers) downwind.

PETS

What is the **oldest breed** of **dog**?

Dogs are the oldest domestic animal, originating 12,000 to 14,000 years ago. They are believed to be descendants of wild canines, most likely wolves, which began to frequent human settlements where food was more readily available. The more aggressive canines were probably driven off or killed, while the less dangerous ones were kept to guard, hunt, and later herd other domesticated animals, such as sheep. Attempts at selectively breeding desirable traits likely began soon after.

The oldest purebred dog is believed to be the saluki. Sumerian rock carvings in Mesopotamia that date to about 7000 B.C.E. depict dogs bearing a striking resemblance to the saluki. The dogs are 23 to 28 inches (58 to 71 centimeters) tall with a long, narrow head. The coat is smooth and silky and can be white, cream, fawn, gold, red, grizzle (bluish-gray) and tan, black and tan, or tricolor (white, black, and tan). The tail is long and feathered. The saluki has remarkable sight and tremendous speed, which makes it an excellent hunter.

The oldest American purebred dog is the American Foxhound. It descends from a pack of foxhounds belonging to an Englishman named Robert Brooke who settled in

Maryland in 1650. These dogs were crossed with other strains imported from England, Ireland, and France to develop the American Foxhound. This dog stands 22 to 25 inches (56 to 63.5 centimeters) tall. It has a long, slightly domed head, with a straight, squared-out muzzle. The coat is of medium length and can be any color. They are used primarily for hunting.

Which **breeds** of **dogs** are best for **families** with **young children**?

Research has shown that golden retriever, Labrador retriever, beagle, collie, bichon frise, cairn terrier, pug, coonhound, boxer, basset hound or mixes of these breeds are best for families with young children.

How do the **bones** of an **adult human compare** to those of an adult **dog**?

	Adult Human	Adult Dog
Number of bones	206	321
Number of vertebrae	33	50
Number of joints	over 200	over 300
Age of maturity	18	2
Longest bone	femur (thighbone)	ulna (forearm bone)
Smallest bone	ossicles (ear bones)	ossicles (ear bones)
Number of ribs	12 on each side	18 on each side

What are the **different classifications** of **dogs**?

Dogs are divided into groups according to the purpose for which they have been bred.

Group	Purpose	Representative Breeds
Sporting dogs	Retrieving game birds	Cocker spaniel, English setter, English and water fowl springer spaniel, golden retriever, Irish setter, Labrador retriever, pointer
Hounds	Hunting	Basenji, beagle, dachshund, foxhound, greyhound, saluki, Rhodesian ridgeback
Terriers	Hunting small animals such as rats and foxes	Airedale terrier, Bedlington terrier, bull terrier, fox terrier, miniature schnauzer, Scottish terrier, Skye terrier, West Highland white terrier
Toy dogs	Small companions or lap dogs	Chihuahua, Maltese, Pekingese, Pomeranian, pug, Shih Tzu, Yorkshire terrier

Group	Purpose	Representative Breeds
Herding dogs	Protect sheep and other livestock	Australian cattle dog, bouviers des Flandres, collie, German shepherd, Hungarian puli, Old English sheepdog, Welsh corgi
Working dogs	Herding, rescue, and sled dogs	Alaskan malamute, boxer, Doberman pinscher, Great Dane, mastiff, St. Bernard, Siberian husky
Non-sporting dogs	No specific purpose, not toys	Boston terrier, bulldog, Dalmatian, Japanese akita, keeshond, Lhasa apso, poodle

Which **breeds of dogs** are the most **dangerous** to humans?

Data collected during the 20-year period, 1979 to 1998, reported 238 dog-bite related fatalities. The breeds most associated with fatalities in this study were:

Breed	Known Fatal Attacks
Pit bull	66
Rottweiler	39
German shepherd	17
Husky	15
Malamute	12
Doberman pinscher	9
Chow chow	8
Great Dane	7
St. Bernard	7
Akita	4

A more recent study for the three-year period of January 2006 to December 2008 continues to show similar results.

Breed	Known Fatal Attacks	Percentage of Total Fatal Attacks
Pit bull	52	59
Rottweiler	12	14
American bulldog	4	5
Husky	4	5
German shepherd	3	3
Doberman pinscher	2	2
Chow chow	2	2
Wolf hybrid	2	2
Labrador	2	2

Which **dogs** are the **easiest to train**?

In a study of 56 popular dog breeds the top breeds to train were Shetland sheepdogs, Shih Tzus, miniature toy and standard poodles, Bichons Frises, English Springer Spaniels, and Welsh Corgis.

What are the **top ten dog names**?

In one recent survey, the top ten dog names in 2009 were:

1. Bella
2. Max
3. Daisy
4. Sadie
5. Charlie
6. Buddy
7. Bailey
8. Molly
9. Ginger
10. Coco

Why do **dogs hear more** than humans?

A dog's ears are highly mobile, allowing it to scan its environment for sounds. The ears capture the sounds and funnel them down to the eardrum. Dogs can hear sounds from four times farther away than humans.

Why do **dogs howl** at **sirens**?

The high pitch of a siren is very similar to the pitch of a dog's howl. A dog's howl is a way of communicating with other dogs—either to indicate location or to define territory. When a dog responds to an ambulance or fire engine siren, he is "returning the call of the wild."

What **breeds** of dogs **do not shed**?

Poodles, Kerry blue terriers, and schnauzers do not shed.

Which breed is known as the **wrinkled dog**?

The shar-pei, or Chinese fighting dog, is covered with folds of loose skin. It stands 18 to 20 inches (46 to 51 centimeters) tall and weighs up to 50 pounds (22.5 kilograms). Its solid-colored coat can be black, red, fawn, or cream. The dog originated in Tibet or the northern provinces of China some 2,000 years ago. The People's Republic of China put

> ## What was the contribution to medical science of a dog named Marjorie?
>
> **M**arjorie was a diabetic black-and-white mongrel that was the first creature to be kept alive by insulin, a substance that controls the level of sugar in the blood.

such a high tax on shar-peis, however, that few people could afford to keep them, and the dog was in danger of extinction. But a few specimens were smuggled out of China, and the breed has made a comeback in the United States, Canada, and the United Kingdom. Although bred as a fighting dog, the shar-pei is generally an amiable companion.

Which breed is known as the **voiceless dog**?

The basenji dog does not bark. When happy, it will make an appealing sound described as something between a chortle and a yodel. It also snarls and growls on occasion. One of the oldest breeds of dogs, and originating in central Africa, the basenji was often given as a present to the pharaohs of ancient Egypt. Following the decline of the Egyptian civilization, the basenji was still valued in central Africa for its hunting prowess and its silence. The dog was rediscovered by English explorers in the nineteenth century, although it was not widely bred until the 1940s.

The basenji is a small, lightly built dog with a flat skull and a long, rounded muzzle. It measures 16 to 17 inches (40 to 43 centimeters) in height at the shoulder and weighs 22 to 24 pounds (10 to 11 kilograms). The coat is short and silky in texture. The feet, chest, and tail tip are white; the rest of the coat is chestnut red, black, or black and tan.

What **food odors** do **dogs like** best?

In a study of different foods, researchers found that liver and chicken ranked higher than everything else, including hamburgers, fish, vegetables, and fresh fruit.

What is the **rarest breed of dog**?

The Tahltan bear dog, of which only a few remain, is thought to be the rarest dog. In danger of extinction, this breed was once used by the Tahltan Indians of western Canada to hunt bear, lynx, and porcupine.

What is the **newest method** of **tagging a dog**?

There is now a computer-age dog tag. A microchip is implanted painlessly between the dog's shoulder blades. The semiconductor carries a 10-digit code, which can be read by a scanner. When the pet is found, the code can be phoned into a national database to locate the owner. The microchip can store license number, medical condition, and the owner's address and phone number.

How is the **age** of a **dog or cat** computed in **human years**?

When a cat is one year old, it is about 20 years old in human years. Each additional year is multiplied by four. Another source counts the age of a cat slightly differently. At age one, a cat's age equals 16 human years. At age two, a cat's age is 24 human years. Each additional year is multiplied by four.

When a dog is one year old, it is about 15 years old in human years. At age two it is about 24; after age two, each additional year is multiplied by four.

Do dogs and cats have **good memories**?

Dogs do have long-term memories, especially for those whom they love. Cats have a memory for things that are important to their lives. Some cats seem to have extraordinary "memories" for finding places. Taken away from their homes, they seem able to remember where they live. The key to this "homing" ability could be a built-in celestial navigation, similar to that used by birds, or the cats' navigational ability could be attributed to the cats' sensitivity to Earth's magnetic fields. When magnets are attached to cats, their normal navigational skills are disrupted.

What is the **original breed** of **domestic cat** in the United States?

The American shorthair is believed by some naturalists to be the original domestic cat in America. It is descended from cats brought to the New World from Europe by the early settlers. The cats readily adapted to their new environment. Selective breeding to enhance the best traits began early in the twentieth century.

The American shorthair is a very athletic cat with a lithe, powerful body, excellent for stalking and killing prey. Its legs are long, heavy, and muscular, ideal for leaping and for coping with all kinds of terrain. The fur, in a wide variety of color and coat patterns, is thick enough to protect the animal from moisture and cold, but short enough to resist matting and snagging.

Although this cat makes an excellent house pet and companion, it remains very self-sufficient. Its hunting instinct is so strong that it exercises the skill even when well-provided with food. The American shorthair is the only true "working cat" in the United States.

What is a **tabby cat**?

"Tabby," the basic feline coat pattern, dates back to the time before cats were domesticated. The tabby coat is an excellent form of camouflage. Each hair has two or three dark and light bands, with the tip always dark. There are four variations on the basic tabby pattern.

The mackerel (also called striped or tiger) tabby has a dark line running down the back from the head to the base of the tail, with several stripes branching down the sides. The legs have stripes, and the tail has even rings with a dark tip. There are two rows of dark spots on the stomach. Above the eyes is a mark shaped like an "M" and dark lines run back to the ears. Two dark necklace-like bands appear on the chest.

The blotched, or classic, tabby markings seem to be the closest to those found in the wild. The markings on the head, legs, tail, and stomach are the same as the mackerel tabby. The major difference is that the blotched tabby has dark patches on the shoulder and side, rimmed by one or several lines.

The spotted tabby has uniformly shaped round or oval dark spots all over the body and legs. The forehead has an "M" on it, and a narrow, dark line runs down the back.

The Abyssinian tabby has almost no dark markings on its body; they appear only on the forelegs, the flanks, and the tail. The hairs are banded except on the stomach, where they are light and unicolored.

What controls the **formation** of the **color points** in a **Siamese cat**?

The color points are due to the presence of a recessive gene that operates at cooler temperatures, limiting the color to well-defined areas—the mask, ears, tail, lower legs, and paws—the places at the far reaches of the cardiovascular system of the cat.

There are four classic varieties of Siamese cats. Seal-points have a pale fawn to cream colored coat with seal-brown markings. Blue-points are bluish-white with slate blue markings. Chocolate-points are ivory colored with milk-chocolate brown colored markings. Lilac-points have a white coat and pinkish-gray markings. There are also some newer varieties with red, cream, and tabby points.

The Siamese originated in Thailand (once called Siam) and arrived in England in the 1880s. They are medium-sized

The color points on a Siamese cat are the result of a recessive gene that affects fur color at different temperatures.

Why do cats have whiskers?

The function of a cat's whiskers is not fully understood. They are thought to have something to do with the sense of touch. Removing them can disturb a cat for some time. Some people believe that the whiskers act as antennae in the dark, enabling the cat to identify things it cannot see. The whiskers may help the cat to pinpoint the direction from which an odor is coming. In addition, the cat is thought to point some of its whiskers downwards to guide it when jumping or running over uneven terrain at night.

and have long, slender, lithe bodies, with long heads and long, tapering tails. Extroverted and affectionate, Siamese are known for their loud, distinctive voices, which are impossible to ignore.

Why do **cats' eyes shine** in the dark?

A cat's eyes contain a special light-conserving mechanism called the *tapetum lucidum*, which reflects any light not absorbed as it passes through the retina of each eye. The retina gets a second chance (so to speak) to receive the light, aiding the cat's vision even more. In dim light, when the pupils of the cat's eyes are opened the widest, this glowing or shining effect occurs when light hits them at certain angles. The *tapetum lucidum*, located behind the retina, is a membrane composed of 15 layers of special, glittering cells that all together act as a mirror. The color of the glow is usually greenish or golden, but the eyes of the Siamese cat reflect a luminous ruby red.

Why and how do **cats purr**?

Experts cannot agree on how or why cats purr, or on where the sound originates. Some think that the purr is produced by the vibration of blood in a large vein in the chest cavity. Where the vein passes through the diaphragm, the muscles around the vein contract, nipping the blood flow and setting up oscillations. These sounds are magnified by the air in the bronchial tubes and the windpipe. Others think that purring is the vibrations of membranes, called false vocal cords, located near the vocal cords. No one knows for sure why a cat purrs, but many people interpret the sound as one of contentment. Cats are also known to purr while in pain, such as while giving birth or dying, possibly as a way to soothe themselves.

What are the **top ten cat names**?

In one recent survey, the top ten cat names in 2009 were:

1. Bella
2. Sassy
3. Angel
4. Sugar
5. Max
6. Patches
7. Tigger
8. Princess
9. Lucy
10. Kitty

Which **plants** are **poisonous** to **cats**?

Certain common houseplants are poisonous to cats:

- Caladium (Elephant's ears)
- Dieffenbachia (Dumb cane)
- *Euphorbia pulcherrima* (Poinsettia)
- Hedera (True ivy)
- Mistletoe
- Oleander
- Philodendron
- *Prunus laurocerasus* (Common or cherry laurel)
- Rhododendron (Azalea)
- *Solanum capiscastrum* (Winter or False Jerusalem cherry)

How can pets be treated to **remove skunk odor**?

From a pet store, purchase one of the products specifically designed to counteract skunk odor. Most of these are of the enzyme or bacterial enzyme variety and can be used without washing the pet first. A dog may also be given a bath with tomato juice, diluted vinegar, or neuthroleum-alpha, or you could try mint mouthwash, aftershave, or soap and water.

Which types of **birds** make the **best pets**?

There are several birds that make good house pets and have a reasonable life expectancy:

Bird	Life Expectancy (years)	Considerations
Finch	2–3	Easy care
Canary	8–10	Easy care; males sing

Bird	Life Expectancy (years)	Considerations
Budgerigar (parakeet)	8–15	Easy care
Cockatiel	15–20	Easy care; easy to train
Lovebird	15–20	Cute, but not easy to care for or train
Amazon parrot	50–60	Good talkers, but can be screamers
African grey parrot	50–60	Talkers; never scream

What are some **unusual animals** that have been **White House pets**?

Several unusual animals have resided at the White House. In 1825, the Marquis de Lafayette (1757–1834) toured America and was given an alligator by a grateful citizen. While Lafayette was the guest of President John Quincy Adams (1767–1848), the alligator took up residence in the East Room of the White House for several months. When Lafayette departed, he took his alligator with him. Mrs. John Quincy Adams (1775–1852) also kept unusual pets: silkworms that feasted on mulberry leaves. Other residents kept a horned-toad, another a green snake, and still another a kangaroo rat. Theodore Roosevelt (1858–1919) brought home a badger that was presented to him as he campaigned in Kansas. The Abraham Lincoln (1809–1865) household contained rabbits and a pair of goats named Nanny and Nanko. President Calvin Coolidge (1872–1933) kept a raccoon as a pet instead of eating it for Thanksgiving dinner, as was intended by the donors from the State of Mississippi. Given the name Rebecca, the raccoon was kept in a large pen near the President's office.

Other unusual White House pets were:

U.S. President	Pets
Martin Van Buren (1782–1862)	Two tiger cubs
William Henry Harrison (1773–1841)	Billy goat; Durham cow
Andrew Johnson (1808–1875)	Pet mice
Theodore Roosevelt (1858–1919)	Lion, hyena, wildcat, coyote, five bears, zebra, barn owl, snakes, lizards, roosters, raccoon
William Howard Taft (1857–1930)	Cow
Calvin Coolidge (1872–1933)	Raccoons, donkey, bobcat, lion cubs, wallaby, pigmy hippo, bear

HUMAN BODY

INTRODUCTION AND HISTORICAL BACKGROUND

Which **scientific disciplines** study the **human body**?

The scientific disciplines of anatomy and physiology study the human body. Anatomy (from the Greek *ana* and *temnein*, meaning "to cut up") is the study of the structure of the body parts, including their form and organization. Physiology (from the Latin, meaning "the study of nature") is the study of the function of the various body parts and organs. Anatomy and physiology are usually studied together to achieve a complete understanding of the human body.

What were **Aristotle's contributions** to **anatomy**?

Aristotle (384–322 B.C.E.) wrote several works laying the foundations for comparative anatomy, taxonomy, and embryology. He investigated carefully all kinds of animals, including humans. His works on life sciences, *On Sense and Sensible Objects, On Memory and Recollection, On Sleep and Waking, On Dreams, On Divination by Dreams, On Length and Shortness of Life, On Youth and Age,* and *On Respiration,* are collectively called *Parva Naturalia.*

Whose work during the **Roman era** became the **authority on anatomy**?

Galen (130–200), a Greek physician, anatomist, and physiologist living during the time of the Roman Empire, was one of the most influential and authoritative authors on medical subjects. His writings include *On Anatomical Procedures, On the Usefulness of the Parts of the Body, On the Natural Faculties,* and hundreds of other treatis-

517

es. Since human dissection was forbidden, Galen made most of his observations on different animals. He correctly described bones and muscles and observed muscles working in contracting pairs. He was also able to describe heart valves and structural differences between arteries and veins. While his work contained many errors, he provided many accurate anatomical details that are still regarded as classics. Galen's writings were the accepted standard text for anatomical studies for 1,400 years.

Who is considered the **founder of physiology**?

As an experimenter, Claude Bernard (1813–1878) enriched physiology by his introduction of numerous new concepts into the field. The most famous of these concepts is that of the *milieu intérieur* or internal environment. The complex functions of the various organs are closely interrelated and are all directed to maintaining the constancy of internal conditions despite external changes. All cells exist in this aqueous (blood and lymph) internal environment, which bathes the cells and provides a medium for the elementary exchange of nutrients and waste material.

Who **coined** the **term homeostasis**?

Walter Bradford Cannon (1871–1945), who elaborated on Claude Bernard's concept of the *milieu intérieur* (interior environment), used the term homeostasis to describe the body's ability to maintain a relative constancy in its internal environment.

Which **chemicals** constitute the **human body**?

About 24 elements are used by the body in its functions and processes.

Major Elements in the Human Body

Element	Percentage	Function
Oxygen production	65.0	Part of all major nutrients of tissues; vital to energy
Carbon	18.5	Essential life element of proteins, carbohydrates, and fats; building blocks of cells
Hydrogen	9.5	Part of major nutrients; building blocks of cells
Nitrogen	3.3	Essential part of proteins, DNA, RNA; essential to most body functions
Calcium	1.5	Forms nonliving bone parts; a messenger between cells
Phosphorous	1.0	Important to bone building; essential to cell energy

Potassium, sulfur, sodium, chlorine, and magnesium each occur at 0.35 percent or less. There are also traces of iron, cobalt, copper, manganese, iodine, zinc, fluorine, boron, aluminum, molybdenum, silicon, chromium, and selenium.

What is the **average life span** of **cells** in the human body?

The human body is self-repairing and self-replenishing. According to one estimate, almost 200 billion cells die each hour. In a healthy body, dying cells are simultaneously replaced by new cells.

Cell Type	Length of Time
Red blood cells	120 days
Lymphocytes	Over 1 year
Other white blood cells	10 hours
Platelets	10 days
Bone cells	25–30 years
Brain cells*	Lifetime
Colon cells	3–4 days
Liver cells	500 days
Skin cells	19–34 days
Spermatozoa	2–3 days
Stomach cells	2 days

*Brain cells are the only cells that do not divide further during a person's lifetime. They either last the entire lifetime or, if a cell in the nervous system dies, it is not replaced.

What are the **types** of human **body shapes**?

The best known example of body typing (classifying body shape in terms of physiological functioning, behavior, and disease resistance) was devised by American psychologist William Herbert Sheldon (1898–1977). Sheldon's system, known as somatotyping, distinguishes three types of body shapes, ignoring overall size: endomorph, mesomorph, and ectomorph. The extreme endomorph tends to be spherical: a round head, a large, fat abdomen, weak penguin-like arms and legs, with heavy upper arms and thighs but slender wrists and ankles. The extreme mesomorph is characterized by a massive cubical head, broad shoulders and chest, and heavy muscular arms and legs. The extreme ectomorph has a thin face, receding chin, high forehead, a thin, narrow chest and abdomen, and spindly arms and legs. In Sheldon's system there are mixed body types, determined by component ratings. Sheldon assumed a close relationship between body build and behavior and temperament. This system of body typing has many critics.

What **percent** of human **body weight** is **water**?

The human body is 61.8 percent water by weight. Water is found is every tissue.

Tissue	Percentage Body Weight	Percentage Water	Water (quarts/liters)
Muscle	41.7	75.6	23.35/22.1
Skin	18	72	9.58/9.07
Blood	8	83	4.91/4.65
Skeletal	15.9	22	2.59/2.45
Brain	2	74.8	1.12/1.05
Liver	2.3	68.3	1.16/1.1
Intestines	1.8	74.5	0.99/0.94
Fat tissue	8.5	10	0.74/0.7
Lungs	0.7	79	0.41/0.39
Heart	0.5	79.2	0.3/0.28
Kidneys	0.4	82.7	0.24/0.23
Spleen	0.2	7.8	0.12/0.11

TISSUES, ORGANS, AND GLANDS

What are the **levels** of **structural organization** in vertebrate animals, including humans?

Every vertebrate animal has four major levels of hierarchical organization: cell, tissue, organ, and organ system. Each level in the hierarchy is of increasing complexity, and all organ systems work together to maintain life.

What are the **four major types** of **tissue**?

A tissue (from the Latin *texere*, meaning "to weave") is a group of similar cells that perform a specific function. The four major types of tissue are epithelial, connective, muscle, and nerve. Each type of tissue performs different functions, is located in different parts of the body, and has certain distinguishing features. The table below explains these differences.

Characteristics of Tissues

Tissue	Function	Location	Distinguishing Features
Epithelial	Protection, secretion, absorption, excretion	Covers body surfaces, covers and lines internal organs compose glands	Lacks blood vessels
Connective	Bind, support, protect, fill spaces, store fat, produce blood cells	Widely distributed throughout the body	Matrix between cells, good blood supply
Muscle	Movement	Attached to bones, in the walls of hollow internal organs, heart	Contractile
Nervous	Transmit impulses for coordination	Regulation, integration, and sensory reception	
Brain, spinal cord nerves	Cells connect to each other and other body parts		

What is the **matrix in blood**?

Blood is a loose connective tissue whose matrix is a liquid called plasma. Blood consists of red blood cells (erythrocytes), white blood cells (leukocytes), and platelets (thrombocytes), which are tiny pieces of bone marrow cell. Plasma also contains water, salts, sugars, lipids, and amino acids. Blood is approximately 55 percent plasma and 45 percent formed elements.

How **strong** is **bone**?

Bone is one of the strongest materials found in nature. It is a rigid connective tissue that has a matrix of collagen fibers embedded in calcium salts. Most of the skeletal system is composed of bone, which provides support for muscle attachment and protects the internal organs. One cubic inch of bone can withstand loads of at least 19,000 pounds (8,626 kilograms), which is approximately the weight of five standard-size pickup trucks. This is roughly four times the strength of concrete. Bone's resistance to load is equal to that of aluminum and light steel. Ounce for ounce, bone is actually stronger than steel and reinforced concrete since steel bars of comparable size would weigh four or five times as much as bone.

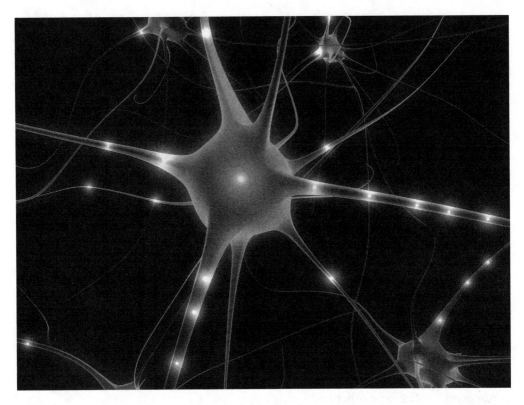

Neurons are nerve cells that communicate with other nerve cells by sending electrical impulses through dendrites and axons.

What is the **hardest substance** in the **body**?

Tooth enamel is the hardest substance in the body. It is composed of 96 percent mineral salts and 4 percent organic matter and water.

What are the **three types** of **muscle tissue**?

There are three types of muscle tissue in the body: 1) smooth muscle; 2) skeletal muscle; and 3) cardiac muscle. Muscle tissue, consisting of bundles of long cells called muscle fibers, is specialized for contraction. It enables body movements, as well as the movement of substances within the body.

What type of **cell** is found in **nerve tissue**?

Neurons are specialized cells that produce and conduct "impulses," or nerve signals. Neurons consist of a cell body, which contains a nucleus and two types of cytoplasmic extensions: dendrites and axons. Dendrites are thin, highly branched extensions that receive signals. Axons are tubular extensions that transmit nerve impulses away from the cell body, often to another neuron. Nerve tissue also has supporting cells, called neuroglia or glial cells, which nourish the neurons, insulate the dendrites and axons, and promote quicker transmission of signals.

What is the **largest nerve** in the body?

The sciatic nerve is the largest in the human body—about as thick as a lead pencil—0.78 inch (1.98 centimeters). It is a broad, flat nerve composed of fibers that run from the spinal cord to the back of each leg.

How many **different types** of **neurons** are found in nerve tissue?

There are three main types of neurons: 1) sensory neurons; 2) motor neurons; and 3) interneurons (also called association neurons). Sensory neurons conduct impulses from sensory organs (eyes, ears, and the surface of the skin) into the central nervous system. Motor neurons conduct impulses from the central nervous system to muscles or glands. Interneurons are neither sensory neurons nor motor neurons. They permit elaborate processing of information to generate complex behaviors. Interneurons comprise the majority of neurons in the central nervous system.

How **many neurons** are in the **nervous system**?

There are approximately 20 billion neurons in the nervous system.

What is an **organ**?

An organ is a group of several different tissues working together as a unit to perform a specific function or functions. Each organ performs functions that none of the component tissues can perform alone. This cooperative interaction of different tissues is a basic feature of animals, including humans. The heart is an example of an organ. It consists of cardiac muscle wrapped in connective tissue. The heart chambers are lined with epithelium. Nerve tissue controls the rhythmic contractions of the cardiac muscles.

In an infant, the heart is about one thirtieth of total body weight. In an average adult, the heart is about one three-hundredth of total body weight; this equals about 11 ounces (310 grams) in males and 8 ounces (225 grams) in females.

What is the **size and location** of the **heart**?

Heart size varies with body size. The average adult's heart is about 5.5 inches (14 centimeters) long and 3.5 inches (9 centimeters) wide, or approximately the size of one's fist. The heart is located just above the diaphragm, between the right and left lungs. One-third of the heart is located on the right size of the chest, while two-thirds are located on the left side of the chest.

What is the **largest organ** in the human body?

The largest and heaviest human organ is the skin, with a total surface area of about 20 square feet (1.9 square meters) for an average person or 25 square feet (2.3 square meters) for a large person; it weighs 5.6 pounds (2.7 kilograms) on average representing four percent of the average weight of the human body. Although generally it is not thought of as an organ, anatomically it is an organ.

Are both of the **lungs identical**?

The lungs are cone-shaped organs in the thoracic cavity. The right lung consists of three lobes (right superior lobe, right middle lobe, and right inferior lobe) while the left lung has only two lobes (left superior lobe and left inferior lobe) and is slightly smaller than the right lung. Although relatively large, each lung weighs only 1 pound (2.2 kilograms).

What is the **basic unit** of the **brain**?

Neurons are the nerve cells that are the major constituent of the brain. At birth the brain has the maximum number of neurons—20 billion to 200 billion neurons. Thousands are lost daily, never to be replaced and apparently not missed, until the cumulative loss builds up in very old age.

How **large** is the **brain**?

The brain weighs about 3 pounds (1.4 kilograms). The average brain has a volume of 71 cubic inches (1,200 cubic centimeters). In general, the brain of males averages about ten percent larger than those of females due to overall differences in average body size. The brain contains approximately 100 billion neurons and 1 trillion neuroglia. There is no correlation between brain size and intelligence.

What are the **major divisions** of the **brain**?

The brain has four major divisions: 1) brainstem, including the medulla oblongata, pons, and midbrain; 2) cerebellum; 3) cerebrum; and 4) diencephalon. The diencephalon is further divided into the thalamus, hypothalamus, epithalamus, and ventral thalamus or subthalamus.

Brain Area	General Functions
Brainstem	
Medulla oblongata	Relays messages between spinal cord and brain and to cerebrum; center for control and regulation of cardiac, respiratory, and digestive activities
Pons	Relays information from medulla and other areas of the brain; controls certain respiratory functions
Midbrain	Involved with the processing of visual information, including visual reflexes, movement of eyes, focusing of lens and dilation of pupils
Cerebellum	Processing center involved with coordination of movements, balance and equilibrium, posture; processes sensory information used by motor systems
Cerebrum	Center for conscious thought processes and intellectual functions, memory, sensory perception, and emotions
Diencephalon	
Thalamus	Relay and processing center for sensory information
Hypothalamus	Regulates body temperature, water balance, sleep-wake cycles, appetite, emotions, and hormone production

What **regulates body temperature** in humans?

The hypothalamus controls internal body temperature by responding to sensory impulses from temperature receptors in the skin and in the deep body regions. The hypothalamus establishes a "set point" for the internal body temperature, then constantly compares this with its own actual temperature. If the two do not match, the hypothalamus activates either temperature-decreasing or temperature-increasing procedures to bring them into alignment.

What is an **organ system**?

An organ system is a group of organs working together to perform a vital body function. There are 12 major organ systems in the human body.

Organ Systems and Their Functions

Organ System	Components	Functions
Cardiovascular and circulatory	Heart, blood, and blood vessels	Transports blood throughout the body, supplying nutrients and carrying oxygen to the lungs and wastes to kidneys
Digestive	Mouth, esophagus, stomach, intestines, liver, and pancreas	Ingests food and breaks it down into smaller chemical units
Endocrine	Pituitary, adrenal, thyroid, and other ductless glands	Coordinates and regulates the activities of the body
Excretory	Kidneys, bladder, and urethra	Removes wastes from the bloodstream
Immune	Lymphocytes, macrophages, and antibodies	Removes foreign substances
Integumentary	Skin, hair, nails, and sweat glands	Protects the body
Lymphatic	Lymph nodes, lymphatic capillaries, lymphatic vessels, spleen, and thymus	Captures fluid and returns it to the cardiovascular system
Muscular	Skeletal muscle, cardiac muscle, and smooth muscle	Allows body movements
Nervous	Nerves, sense organs, brain, and spinal cord	Receives external stimuli, processes information, and directs activities
Reproductive	Testes, ovaries, and related organs	Carries out reproduction
Respiratory	Lungs, trachea, and other air passageways	Exchanges gases—captures oxygen (O_2) and disposes of carbon dioxide (CO_2)
Skeletal	Bones, cartilage, and ligaments	Protects the body and provides support for locomotion and movement

What is a **gland**?

Glands are secretory cells or multicellular structures that are derived from epithelium and often stay connected to it. They are specialized for the synthesis, storage, and secretion of chemical substances. Glands are classified as either endocrine or exocrine glands. Endocrine glands do not have ducts, but release their secretions directly into the extracellular fluid. The secretions pass into capillaries and are then transported by the bloodstream to target cells elsewhere in the body. Exocrine glands have ducts that carry the secretions to some body surface. Mucus, saliva, perspiration, earwax, oil, milk, and digestive enzymes are examples of exocrine secretions.

Which gland is the largest?

The liver is the largest gland and the second largest organ after the skin. At 2.5 to 3.3 pounds (1.1 to 1.5 kilograms) the liver is seven times larger than it needs to be to perform its estimated 500 functions. It is the main chemical factory of the body. A ducted gland that produces bile to break down fats and reduce acidity in the digestive process, the liver is also a part of the circulatory system. It cleans poisons from the blood and regulates blood composition.

How many sweat glands are present in the body?

Sweat glands are present on all regions of the skin. There can be as many as 90 glands per square centimeter on the leg, 400 glands per cubic centimeter on the palms and soles, and an even greater number on the fingertips. Collectively, there are over two million sweat glands in the adult human body.

What are the seven endocrine glands?

The major endocrine glands include the pituitary, thyroid, parathyroids, adrenals, pancreas, testes, and ovaries. These glands secrete hormones into the blood system, which generally stimulate some change in metabolic activity:

Pituitary—secretes ACTH to stimulate the adrenal cortex, which produces aldosterone to control sodium and potassium reabsorption by the kidneys; FSH to stimulate gonad function and prolactin to stimulate milk secretion of breasts; TSH to stimulate thyroid gland to produce thyroxin; LH to stimulate ovulation in females and testerone production in males; GH to stimulate general growth. Stores oxytocin for uterine contraction.

Thyroid—secretes triiodothyronine (T3) and thyroxine (T4) to stimulate metabolic rate, especially in growth and development, and secretes calcitonin to lower blood-calcium levels.

Parathyroids—secrete hormone PTH to increase blood-calcium levels, and stimulates calcium reabsorption in kidneys.

Adrenals—secrete epinephrine and norepinephrine to help the body cope with stress, and raise blood pressure, heart rate, metabolic rate, blood sugar levels, etc. Aldosterone secreted by the adrenal cortex maintains sodium-potassium balance in kidneys and cortisol helps the body adapt to stress, mobilizes fat, and raises blood sugar level.

Pancreas—secretes insulin to control blood sugar levels, stimulates glycogen production, fat storage, and protein synthesis. Glucagon secretion raises blood sugar level and mobilizes fat.

Ovaries and testes—secrete estrogens, progesterone, or testosterone to stimulate growth and reproductive processes.

BONES AND MUSCLES

How **many bones** are in the human body?

Babies are born with about 300 to 350 bones, but many of these fuse together between birth and maturity to produce an average adult total of 206. Bone counts vary according to the method used to count them, because a structure may be treated as either multiple bones or as a single bone with multiple parts.

Location	Number
Skull	22
Ears (pair)	6
Vertebrae	26
Sternum	3
Ribs	24
Throat	1
Pectoral girdle	4
Arms (pair)	60
Hip bones	2
Legs (pair)	58
TOTAL	*206*

What are the **major divisions** of the human **skeleton**?

The human skeleton has two major divisions: the axial skeleton and the appendicular skeleton. The axial skeleton includes the bones of the center or axis of the body. The appendicular skeleton consists of the bones of the upper and lower extremities.

What is the **smallest bone** in the body?

The stapes (stirrup) in the middle ear is the smallest bone in the body. It measures 1.02 to 1.34 inches (2.6 to 3.4 centimeters) and weighs 0.00071 to 0.0015 ounce (0.002 to 0.004 gram).

How does **exercise affect bone** tissue?

Bone adapts to changing stresses and forces. When muscles increase and become more powerful due to exercise, the corresponding bones also become thicker and

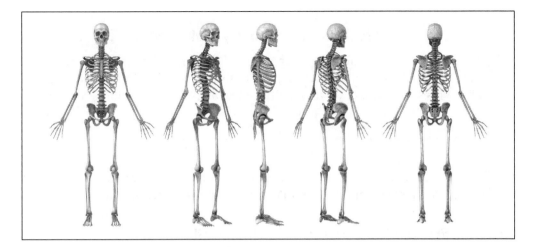

There are 206 bones in the human skeleton.

stronger through stimulation of osteoblasts. Regular exercise (especially weight-bearing exercise) maintains normal bone structure. Bones that are not subjected to normal stresses, such as an injured leg immobilized in a cast, quickly degenerate. It is estimated that unstressed bones lose up to a third of their mass after a few weeks. The adaptability of bones allows them to rebuild just as quickly when regular, normal weight-bearing activity is resumed.

Which is the **only bone** that **does not touch** another bone?

The hyoid bone is the only bone that does not touch another bone. Located above the larynx, it supports the tongue and provides attachment sites for the muscles of the neck and pharynx used in speaking and swallowing. The hyoid is carefully examined when there is a suspicion of strangulation, since it is often fractured from such trauma.

How long does it take for the **"soft spots"** on a **baby's skull** to **disappear**?

The "soft spots" of a baby's skull are areas of incompletely ossified bones called fontanels. The bones of the skull are connected by fibrous, pliable, connective tissue at birth. The flexibility of these connections allows the skull bones to move and overlap as the infant passes through the birth canal. The fontanels begin to close about two months after birth. The largest of the fontanels, the frontal fontanel located on the top of the skull, does not close until 18 to 24 months of age.

Which **two cervical vertebrae** allow the **head to move**?

The first two cervical vertebrae, C1 and C2, allow the head to move. The first cervical vertebra, the C1 or atlas, articulates with the occipital bone of the skull and makes it 529 possible for a person to nod his or her head. The second cervical vertebra, C2, known as the axis, forms a pivot point for the atlas to move the skull in a side-to-side rotation.

Why is the **first cervical vertebra**, C1, called the **atlas**?

The first cervical vertebra is called the atlas after the Greek god Atlas, who was condemned to carry the earth and heavens on his shoulders. This vertebra has a ringlike structure with a large central opening and supports the head.

How does the **anatomical usage** of the word **"arm"** differ **from the** common usage?

Anatomically, the word arm refers only to the humerus, the long bone between the shoulder and the elbow. Common usage refers to the entire length of the limb from the shoulder to the wrist as the arm.

What makes **knuckles crack**?

When a person pulls quickly on his or her finger, a vacuum is created in the joint space between the bones, displacing the fluid normally found in the space. The popping sound occurs when the fluids rush back into the empty gap.

Is it **harmful** to **crack** one's **knuckles**?

A study of 300 knuckle crackers found no apparent connection between joint cracking and arthritis. Other damage was observed, including soft tissue damage to the joint capsule and a decrease in grip strength. The rapid, repeated stretching of the ligaments surrounding the joint is most likely the cause of damage to the soft tissue.

What are the **functions** of the **muscular system**?

Muscles are identified as being voluntary muscles or involuntary muscles. The skeletal muscles are called voluntary muscles because the person controls their use. They are used to move the various parts of the body. Smooth muscles, found in the stomach

> ## What is the funny bone?
>
> The funny bone is not a bone but part of the ulnar nerve located at the back of the elbow. A bump in this area can cause a tingling sensation or produce a temporary numbness and paralysis of muscles on the forearm.

and intestinal walls, vein and artery walls, and in various internal organs, are called involuntary muscles, because they are not generally controlled by the person. The cardiac muscles, or the heart muscles, are also involuntary muscles. The major functions of the muscular system are:

1. Body movement due to the contraction of skeletal muscles
2. Maintenance of posture also due to skeletal muscles
3. Respiration due to movements of the muscles of the thorax
4. Production of body heat, which is necessary for the maintenance of body temperature, as a by-product of muscle contraction
5. Communication, such as speaking and writing, which involve skeletal muscles
6. Constriction of organs and vessels, especially smoother muscles that can move solids and liquids in the digestive tract and other secretions, including urine, from organs
7. Heartbeat caused by the contraction of cardiac muscle that propels blood to all parts of the body

How **many muscles** are in the human body?

There are about 650 muscles in the body, although some authorities believe there are as many as 850 muscles. No exact figure is available because experts disagree about which are separate muscles and which ones branch off larger ones. Also, there is some variability from one person to another, though the general musculature remains the same.

Which are the **largest** and **smallest muscles** in the human body?

The largest muscle is the gluteus maximus (buttock muscle), which moves the thighbone away from the body and straightens out the hip joint. It is also one of the stronger muscles in the body. The smallest muscle is the stapedius in the middle ear. It is thinner than a thread and 0.05 inches (0.127 centimeters) in length. It activates the stirrup that sends vibrations from the eardrum to the inner ear.

What is the **longest muscle** in the human body?

The longest muscle is the sartorius, which runs from the waist to the knee. Its purpose is to flex the hip and knee.

What are the hamstring muscles?

There are three hamstring muscles, located at the back of the thigh. They flex the leg on the thigh, such as when one kneels. They also extend the hip whenever one, for example, sits in a chair. Hamstring injuries are probably the most common muscle injury among runners. Maintaining flexibility and strengthening the muscle helps to prevent injury. Hamstring muscles are also prone to reinjury.

Which **muscle** is the **most variable** among humans?

The platysma muscle in the side of the neck is probably the most variable. It can cover the whole region in some people while in others it is straplike or in a few situations it is missing completely.

How many **muscles** does it take to produce a **smile** and a **frown**?

Seventeen muscles are used in smiling while the average frown uses 43.

Why does **excessive exercise** cause muscles to become **stiff and sore**?

During vigorous exercise, the circulatory system cannot supply oxygen to muscle fibers quickly enough. In the absence of oxygen, the muscle cells begin to produce lactic acid, which accumulates in the muscle. It is this buildup of lactic acid that causes soreness and stiffness.

SKIN, HAIR, AND NAILS

What **organs** are included in the **integumentary system**?

The integumentary system (from the Latin *integere*, meaning "to cover") includes skin, hair, glands, and nails. The main function of the integumentary system is to provide the body with a protective barrier between the organs inside the body and the changing environment outside.

How much **skin** does a person **shed** in one year?

An average man or woman sheds about 600,000 particles of skin per hour, which is approximately 1.5 pounds (680 grams) per year. Using this figure, by the age of 70, a person will have lost 105 pounds (47.6 kilograms) of skin which is equivalent to two-thirds of their entire body weight.

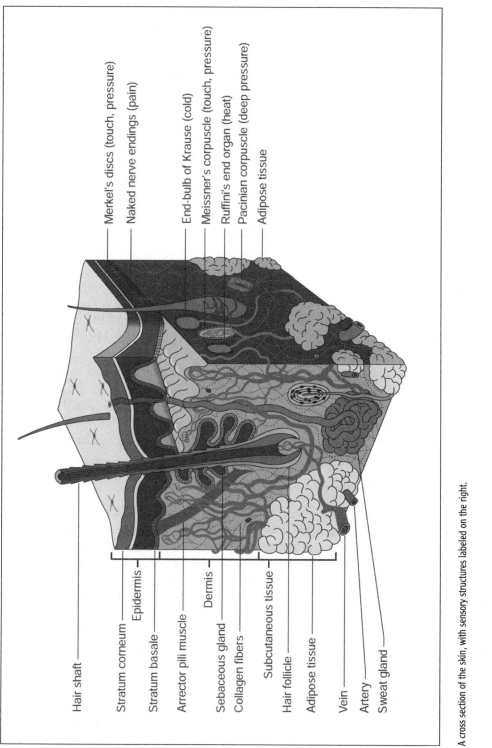

Merkel's discs (touch, pressure)
Naked nerve endings (pain)
End-bulb of Krause (cold)
Meissner's corpuscle (touch, pressure)
Ruffini's end organ (heat)
Pacinian corpuscle (deep pressure)
Adipose tissue

Hair shaft
Stratum corneum
Epidermis
Stratum basale
Arrector pili muscle
Dermis
Sebaceous gland
Collagen fibers
Subcutaneous tissue
Hair follicle
Adipose tissue
Vein
Artery
Sweat gland

A cross section of the skin, with sensory structures labeled on the right.

533

What is the purpose of goose bumps?

The puckering of the skin that takes place when goose-flesh is formed is the result of contraction of the muscle fibers in the skin. This muscular activity will produce more heat, and raises the temperature of the body.

How **thick** is **skin**?

The thickness of skin varies, depending on where it is found on the body. Skin averages 0.05 inches (0.127 centimeters) in thickness. The thinnest skin is found in the eyelids and is less than 0.002 inches (0.005 centimeters) thick, while the thickest skin is on the upper back (0.2 inches or 0.5 centimeters).

What are the various **layers** of the **skin**?

Skin is a tissue membrane that consists of layers of epithelial and connective tissues. The outer layer of the skin's epithelial tissue is the epidermis and the inner layer of connective tissue is the dermis. The epidermis layer is replaced continually as new cells, produced in the stratum basale, mature, and are pushed to the surface by the newer cells beneath; the entire epidermis is replaced in about 27 days. The dermis, the lower layer, contains nerve endings, sweat glands, hair follicles, and blood vessels. The upper portion of the dermis has small, fingerlike projections called "papillae," which extend into the upper layer. The patterns of ridges and grooves visible on the skin of the soles, palms, and fingertips are formed from the tops of the dermal papillae.

What determines **skin color**?

Three factors contribute to skin color: 1) the amount and kind (yellow, reddish brown, or black) of melanin in the epidermis; 2) the amount of carotene (yellow) in the epidermis and subcutaneous tissue; and 3) the amount of oxygen bound to hemoglobin (red blood cell pigment) in the dermal blood cells. Skin color is genetically determined, for the most part. Differences in skin color result not from the number of melanocytes an individual has, but rather from the amount of melanin produced by the melanocytes and the size and distribution of the pigment granules. Although darker-skinned people have slightly more melanocytes than those who are light-skinned, the distribution of melanin in the higher levels of the epidermis contributes to their skin color.

Are **freckles dangerous**?

Freckles, those tan or brown spots on the skin, are small areas of increased skin pigment or melanin. There is a genetic tendency to develop freckles, and parents with freckles often pass this trait down to their children. Freckles usually occur on the face, arms, and other parts of the body that are exposed to the sun. Freckles themselves pose no health risks, but individuals who freckle easily are at an increased risk for skin cancer.

What is **dermatoglyphics**?

Dermatoglyphics, the study of fingerprints, recognizes three basic patterns of fingerprints. They are arches, loops, and whorls. The lines or ridges of an arch run from one side of the finger to the other with an upward curve in the center. In a loop, the ridges begin on one side, loop around the center, and return to the same side. The ridges of a whorl form a circular pattern. Dermatoglyphics is of interest in such diverse fields as medicine, anthropology, and criminology.

Who first used **fingerprints** as a means of **identification**?

It is generally acknowledged that Francis Galton (1822–1911) was the first to classify fingerprints. However, his basic ideas were further developed by Sir Edward Henry (1850–1931), who devised a system based on the pattern of the thumb print. In 1901 in England, Henry established the first fingerprint bureau with Scotland Yard called the Fingerprint Branch.

Do **identical twins** have the **same fingerprints**?

No. Even identical twins have differences in their fingerprints, which, though subtle, can be discerned by experts.

How fast do **fingernails grow**?

Healthy nails grow about 0.12 inch (3 milimeters) each month or 1.4 inches (3.5 centimeters) each year. It takes approximately three months for a whole fingernail to be replaced. The middle fingernail grows the fastest, because the longer the finger the faster its nail growth. The thumbnail grows the slowest.

Do **fingernails** and **toenails grow** at the same **rate**?

Fingernails tend to grow a little faster than toenails.

Are **fingernails and toenails** the same **thickness**?

No. Toenails are approximately twice as thick as fingernails.

How much does **human hair grow** in a **year**?

Each hair grows about 9 inches (23 centimeters) every year.

How **fast** do **eyelashes grow**?

Eyelashes are replaced every three months. An individual will grow about 600 complete eyelashes in a lifetime.

Does human **hair grow faster** in **summer** or **winter**?

During the summertime, the rate of human hair growth increases by about 10 percent to 15 percent. This is because warm weather enhances blood circulation to the skin and scalp, which in turn nourishes hair cells and stimulates growth. In cold weather, when blood is needed to warm internal organs, circulation to the body surface slows and hair cells grow less quickly.

How many hairs are on the human body?

On the average human body, there are approximately five million hairs.

How many hairs does the average person have on his or her **head**?

The amount of hair on the head varies from one individual to another. An average person has about 100,000 hairs on their scalp (blonds 140,000, brunettes 155,000, and redheads only 85,000). Most people shed between 50 to 100 hairs daily.

How is **hair color determined**?

Genes determine hair color by directing the type and amount of pigment that epidermal melanocytes produce. If these cells produce an abundance of melanin, the hair is dark. If an intermediate quantity of pigment is produced, the hair is blond. If no pigment is produced, the hair appears white. A mixture of pigmented and unpigmented hair is usually gray. Another pigment, trichosiderin, is found only in red hair.

Why does **hair turn gray** as part of the aging process?

The pigment in hair, as well as in the skin, is called melanin. There are two types of melanin: eumelanin, which is dark brown or black, and pheomelanin, which is reddish yellow. Both are made by a type of cell called a melanocyte that resides in the hair bulb and along the bottom of the outer layer of skin, or epidermis. The melanocytes pass this pigment to adjoining epidermal cells called keratinocytes, which produce the protein keratin—hair's chief component. When the keratinocytes undergo their scheduled death, they retain the melanin. Thus, the pigment that is visible in the hair and in the skin lies in these dead keratinocyte bodies. Gray hair is simply hair with less melanin, and white hair has no melanin at all. It remains unclear as to how hair loses its pigment. In the early stages of graying, the melanocytes are still present but inactive. Later they seem to decrease in number. Genes control this lack of deposition of melanin. In some families, many members' hair turns white while they are still in their twenties. Generally speaking, among Caucasians 50 percent are gray by age 50. There is, however, wide variation.

Why is some **hair curly** while some hair is **straight**?

The shape of the hair follicle determines how wavy a hair will be. Round follicles produce straight hair. Oval follicles produce wavy hair. Flat follies produce curly hair.

What information can a **forensic scientist** determine from a **human hair**?

A single strand of human hair can identify the age and sex of the owner, drugs and narcotics the individual has taken or used (up to 90 days previously), and, through DNA evaluation and sample comparisons, from whose head the hair came.

BLOOD AND CIRCULATION

What are the **functions** of the **circulatory system**?

The cardiovascular system provides a transport system between the heart, lungs, and tissue cells. The most important function is to supply nutrients to tissues and remove waste products.

What is the difference between the **cardiovascular system** and the **circulatory system**?

The cardiovascular system refers to the heart (cardio) and blood vessels (vascular). The circulatory system is a more general term encompassing the blood, blood vessels, heart, lymph, and lymph vessels.

Which **structures** and **organs** constitute the **cardiovascular system**?

Technically speaking, the structures of the cardiovascular system are the heart and blood vessels. Blood, a connective tissue, plays a major role in the cardiovascular system and is usually discussed within the context of the cardiovascular system.

How hard does the **heart work**?

The heart squeezes out about 2 ounces (71 grams) of blood at every beat and daily pumps at least 2,500 gallons (9,450 liters) of blood. On the average, the adult heart beats 70 to 75 times a minute. The rate of the heartbeat is determined in part by the

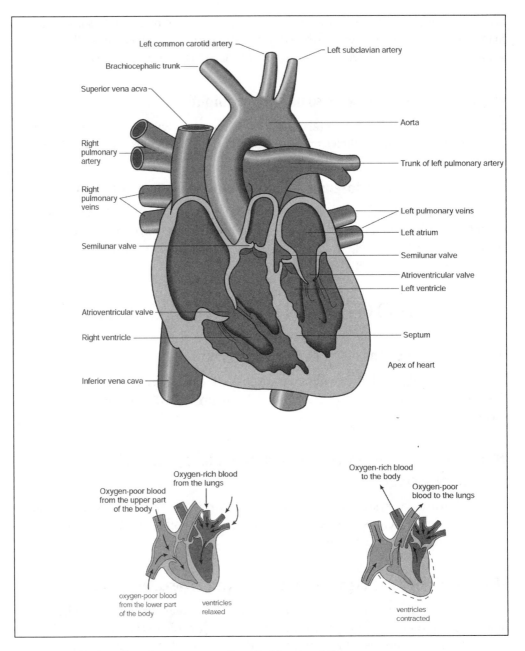

Left common carotid artery
Left subclavian artery
Brachiocephalic trunk
Superior vena acva
Aorta
Right pulmonary artery
Trunk of left pulmonary artery
Right pulmonary veins
Left pulmonary veins
Left atrium
Semilunar valve
Semilunar valve
Atrioventricular valve
Left ventricle
Atrioventricular valve
Right ventricle
Septum
Apex of heart
Inferior vena cava

Oxygen-rich blood from the lungs
Oxygen-poor blood from the upper part of the body
Oxygen-rich blood to the body
Oxygen-poor blood to the lungs
oxygen-poor blood from the lower part of the body
ventricles relaxed
ventricles contracted

This illustration of the human heart shows its structure and how blood flows through the organ.

size of the organism. Generally, the smaller the size, the faster the heartbeat. Thus women's hearts beat six to eight beats per minute faster than men's hearts do. At birth the heart of a baby can beat as fast as 130 times per minute.

Does your **heart stop beating** when you **sneeze**?

The heart does not stop when you sneeze. Sneezing, however, does affect the cardio-vascular system. It causes a change in pressure inside the chest. This change in pressure affects the blood flow to the heart, which in turn affects the heart's rhythm. Therefore, a sneeze does produce a harmless delay between one heartbeat and the next, often misinterpreted as a "skipped beat."

What are the **functions** of **blood**?

The functions of blood can be divided into three general categories: transportation, regulation, and protection.

Functions of Blood

Function	Examples
Transportation	Gases (oxygen and carbon dioxide), nutrients, metabolic waste
Regulation	Body temperature, normal pH, fluid volume/pressure
Protection	Against blood loss, against infection

What is the **normal pH** of **blood**?

The normal pH of arterial blood is 7.4 while the pH of venous blood is about 7.35. Arterial blood has a slightly higher pH because it has less carbon dioxide.

What is the **amount** of **carbon dioxide** found in **normal blood**?

Carbon dioxide normally ranges from 19 to 50 millimeters per liter in arterial blood and 22 to 30 millimeters per liter in venous blood.

How much blood does the average-sized **adult** human have?

An adult man has 5.3 to 6.4 quarts, or 1.5 gallons (5 to 6 liters), of blood, while an adult woman has 4.5 to 5.3 quarts, or 1.2 gallons (4 to 5 liters). Differences are due to the sex of the individual, body size, fluid and electrolyte concentrations, and amount of body fat.

How many **miles of blood vessels** are contained in the body?

If they could be laid end to end, the blood vessels—arteries, arterioles, capillaries, venules, and veins—would span about 60,000 miles (96,500 kilometers). This would be enough to encircle the earth more than two times.

Why is blood sticky?

Blood is sticky because it is denser than water and about five times more viscous than water. Blood is viscous mainly due to the red blood cells. When the number of these cells increases, the blood becomes thicker and flows slower. Conversely, if the number of red blood cells decreases, blood thins and flows faster.

What is the **largest artery** in the human body?

The aorta is the largest artery in the human body. In adults, it is approximately the size of a garden hose. Its internal diameter is 1 inch (2.5 centimeters) thick.

What is the **largest vein** in the human body?

The largest vein in the human body is the inferior vena cava, the vein that returns blood from the lower half of the body back to the heart.

Why does blood in the **veins look blue**?

Since venous blood is oxygen-poor blood, it is not as bright red as arterial blood. It appears as a deep, dark-red, almost purplish color. Seeing "blue blood" in veins through the skin is a combination of light passing through the skin and the oxygen-poor blood.

How **big** are **capillaries**?

The diameter of a capillary is about 0.0003 inches (0.0076 millimeters), which is just about the same as a single red blood cell. A capillary is only about 0.04 inches (1 millimeter) long. If all the capillaries in a human body were placed end to end, the collective length would be approximately 25,000 miles (46,325 kilometers), which is slightly more than the circumference of the earth at the equator: 24,900 miles (46,139 kilometers).

Who **discovered** the **ABO system** of typing blood?

The Austrian physician Karl Landsteiner (1868–1943) discovered the ABO system of blood types in 1909. Landsteiner had investigated why blood transfused from one individual was sometimes successful and other times resulted in the death of the patient. He theorized that there must be several different blood types. A person with one type of blood will have antibodies to the antigens in the blood type they do not have. If a transfusion occurs between two individuals with different blood types, the red blood cells will clump together, blocking the blood vessels.

Which of the **major blood types** are the most common in the United States?

The following table lists the blood types and their rate of occurrence in the United States.

Distribution of Blood Types in the United States

Blood Type	Frequency
O+	37.4%
O–	6.6%
A+	35.7%
A–	6.3%
B+	8.5%
B–	1.5%
AB+	2.4%
AB–	0.6%

What are the **preferred** and permissible **blood types** for **transfusions**?

The table below lists the blood types that are best matched with other blood types.

Blood Type of Recipient	Preferred Blood Type of Donor	Permissible Blood Type(s) of Donor in an Emergency
A	A	A, O
B	B	B, O
AB	AB	AB, A, B, O
O	O	O

Which **blood type** is the **universal donor** and which is the universal recipient?

Persons with blood type O– are universal donors. They are able to donate blood to anyone. Persons with blood type AB+ are universal recipients. They are able to receive blood from any donor.

What is the **Rh factor**?

In addition to the ABO system of blood types, blood types can also be grouped by the Rhesus factor, or Rh factor, an inherited blood characteristic. Discovered independently in 1939 by Philip Levine (1900–1987) and R.E. Stetson and in 1940 by Karl Landsteiner and A.S. Weiner, the Rh system classifies blood as either having the Rh factor or lacking it. Pregnant women are carefully screened for the Rh factor. If a mother is found to be Rh-negative, the father is also screened. Parents with incompatible Rh factors can lead to potentially fatal blood problems in newborn infants. The condition can be treated with a series of blood transfusions.

What are the **blood group combinations** that can normally be used to **prove** that a man is **not the father** of a particular child?

If the Mother Is	And the Child Is	The father Can Be	But Not
O	O	O, A, or B	AB
O	A	A or AB	O or B
O	B	B or AB	O or A
A	O	O, A, or B	AB
A	A	any group	
A	B	B or AB	O or A
A	AB	B or AB	O or A
B	O	O, A, or B	AB
B	B	any group	
B	A	A or AB	O or B
B	AB	A or AB	O or B
AB	AB	A, B, or AB	O

No child can acquire a gene, and consequently a blood grouping, if it is not possessed by either parent.

What are the primary **functions** of the **lymphatic system**?

The lymphatic system consists of the lymphatic vessels, lymph, and lymphoid organs. It is responsible for maintaining proper fluid balance in tissues and blood, in addition to its role defending the body against disease-causing agents. The primary functions of the lymphatic system are: 1) to collect the interstitial fluid that consists of excess water and proteins and return it to the blood; 2) to transport lipids and other nutrients that are unable to enter the bloodstream directly; and 3) to protect the body from foreign cells and microorganisms.

What are the **functions** of the **spleen**?

The primary function of the spleen is the filtering of blood and removal of abnormal blood cells by phagocytosis. The spleen also stores iron from worn-out blood cells, which is then returned to the circulation and used by the bone marrow to produce new blood cells. The immune reaction begins in the spleen with the activation of immune response by B cells and T cells in response to antigens in the blood.

How serious is **damage** to the **spleen**?

Damage to the spleen, which is often the result of an injury caused by a blow to the left side of the abdomen, may be life-threatening. Since the spleen is a fragile organ, an injury can easily rupture it, resulting in serious internal bleeding, hemorrhaging, circulatory shock, and even death. Once the spleen ruptures, the only remedy is to surgically remove it in a procedure called a splenectomy.

How do **T cells differ** from **B lymphocytes**?

Lymphocytes are one variety of white blood cells and are the primary cells of the lymphatic system—the body's immune system. The immune system fights invading organisms that have penetrated the body's general defenses. T cells, responsible for dealing with most viruses, for handling some bacteria and fungi, and for cancer surveillance, are one of the two main classes of lymphocytes. T lymphocytes, or T cells, compose about 60 to 80 percent of the lymphocytes circulating in the blood. They have been "educated" in the thymus to perform particular functions.

Killer T cells are sensitized to multiply when they come into contact with antigens (foreign proteins) on abnormal body cells (cells that have been invaded by viruses, cells in transplanted tissue, or tumor cells). These killer T cells attach themselves to the abnormal cells and release chemicals (lymphokines) to destroy them.

Helper T cells assist killer cells in their activities and control other aspects of the immune response. When B lymphocytes, which compose approximately 10–15 per-

543

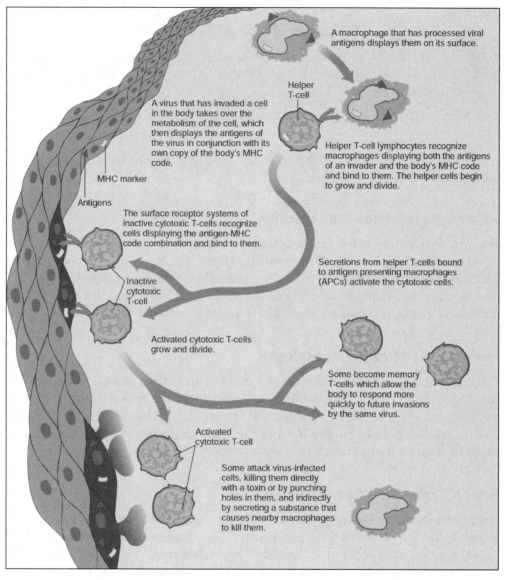

A macrophage that has processed viral antigens displays them on its surface.

Helper T-cell

A virus that has invaded a cell in the body takes over the metabolism of the cell, which then displays the antigens of the virus in conjunction with its own copy of the body's MHC code.

MHC marker

Antigens

Helper T-cell lymphocytes recognize macrophages displaying both the antigens of an invader and the body's MHC code and bind to them. The helper cells begin to grow and divide.

The surface receptor systems of inactive cytotoxic T-cells recognize cells displaying the antigen-MHC code combination and bind to them.

Inactive cytotoxic T-cell

Secretions from helper T-cells bound to antigen presenting macrophages (APCs) activate the cytotoxic cells.

Activated cytotoxic T-cells grow and divide.

Some become memory T-cells which allow the body to respond more quickly to future invasions by the same virus.

Activated cytotoxic T-cell

Some attack virus-infected cells, killing them directly with a toxin or by punching holes in them, and indirectly by secreting a substance that causes nearby macrophages to kill them.

How the immune system works.

cent of total lymphocytes, contact the antigens on abnormal cells, the lymphocytes enlarge and divide to become plasma cells. Then the plasma cells secrete vast numbers of immunoglobulins or antibodies into the blood, which attach themselves to the surfaces of the abnormal cells, to begin a process that will lead to the destruction of the invaders. NK (natural killer) cells account for the remaining five to ten percent of the circulating lymphocytes. They attack foreign cells, normal cells infected with viruses, and cancer cells that appear in normal tissues.

How does the **immune system work**?

The immune system has two main components: white blood cells and antibodies circulating in the blood. The antigen-antibody reaction forms the basis for this immunity. When an antigen (*anti*body *gen*erator)—such as a harmful bacterium, virus, fungus, parasite, or other foreign substance—invades the body, a specific antibody is generated to attack the antigen. The antibody is produced by B lymphocytes (B cells) in the spleen or lymph nodes. An antibody may either destroy the antigen directly or it may "label" it so that a white blood cell (called a macrophage, or scavenger cell) can engulf the foreign intruder.

After a human has been exposed to an antigen, a later exposure to the same antigen will produce a faster immune system reaction. The necessary antibodies will be produced more rapidly and in larger amounts. Artificial immunization uses this antigen-antibody reaction to protect the human body from certain diseases by exposing the body to a safe dose of antigen to produce effective antibodies as well as a "readiness" for any future attacks of the harmful antigen.

What is the **composition** of **lymph**?

Lymph is similar in composition to blood plasma. The main chemical difference is that lymph does not contain erythrocytes. It also contains a much lower concentration of protein than plasma since most protein molecules are too large to filter through the capillary wall. Lymph contains water, some plasma proteins, electrolytes, lipids, leukocytes, coagulation factors, antibodies, enzymes, sugars, urea, and amino acids.

What is the **route** the **fluid travels** in the body?

Blood flows from the arteries to the capillaries, with a portion leaking into the interstitial spaces. Once the fluid leaves the interstitial spaces and enters the lymphatic capillaries, it is called lymph. It flows from the lymphatic capillaries through the lymphatic vessels and lymph nodes to the lymphatic ducts, eventually entering either the right lymphatic duct or the thoracic duct. It finally drains into the subclavian veins and is returned to the blood.

What are the **major functions** of the **respiratory system**?

The major functions of the respiratory system are:

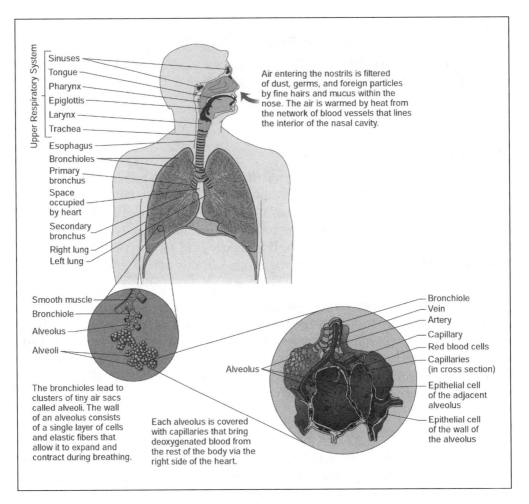

The human respiratory system.

1. Gas exchange: the respiratory system allows oxygen in the air to enter the blood and carbon dioxide to leave the blood and enter the air. The cardiovascular system transports oxygen from the lungs to the cells of the body and carbon dioxide from the cells of the body to the lungs.

2. Regulation of blood pH, which can be altered by changes in blood carbon dioxide levels.

3. Voice production as air moves past the vocal cords to make sound and speech.

4. Olfaction, or the sense of smell, occurs when airborne molecules are drawn into the nasal cavity.

5. Innate immunity, providing protection against some microorganisms by preventing their entry into the body or by removing them from respiratory surfaces.

What are the **two divisions** of the **respiratory system**?

The respiratory system is divided into the upper respiratory system and the lower respiratory system. The upper respiratory system includes the nose, nasal cavity, and sinuses. The lower respiratory system includes the larynx, trachea, bronchi, bronchioles, and alveoli.

What is the essential **relationship** between the **heart** and **lungs**?

The teamwork of the heart and lungs ensures that the body has a constant supply of oxygen for metabolic activities and that the major waste product of metabolism, carbon dioxide, is continuously removed. This occurs through the pulmonary circulation, with the heart supplying blood that has moved through the body to the lungs. The lungs connect to the heart through blood vessels. The pulmonary artery delivers deoxygenated blood to the lungs from the right ventricle and the pulmonary vein delivers oxygenated blood to the left atrium of the heart.

How does the **body introduce oxygen** to the blood and where does this happen?

Blood entering the right side of the heart (right auricle or atrium) contains carbon dioxide, a waste product of the body. The blood travels to the right ventricle, which pushes it through the pulmonary artery to the lungs. In the lungs, the carbon dioxide is removed and oxygen is added to the blood. Then the blood travels through the pulmonary vein carrying the fresh oxygen to the left side of the heart, first to the left auricle, where it goes through a one-way valve into the left ventricle, which must push the oxygenated blood to all portions of the body (except the lungs) through a network of arteries and capillaries. The left ventricle must contract with six times the force of the right ventricle, so its muscle wall is twice as thick as the right.

What are the **two phases** of **breathing**?

Breathing, or ventilation, is the process of moving air into and out of the lungs. The two phases are: 1) inspiration, or inhalation; and 2) expiration, or exhalation. Inspiration is the movement of air into the lungs, while expiration is the movement of air out of the lungs. The respiratory cycle consists of one inspiration followed by one expiration. The volume of air that enters or leaves during a single respiratory cycle is called the tidal volume. Tidal volume is typically 500 milliliters, meaning that 500 milliliters of air enters during inspiration and the same amount leaves during expiration.

How **much air** does a person **breathe** in a **lifetime**?

During his or her life, the average person will breathe about 75 million gallons (284 million liters) of air. Per minute, the human body needs 2 gallons (7.5 liters) of air   547 when lying down, 4 gallons (15 liters) when sitting, 6 gallons (23 liters) when walking, and 12 gallons (45 liters) or more when running.

Why is it more **difficult** to **breathe** at **high altitudes**?

It is difficult to breathe at high altitudes because there is less oxygen available in the atmosphere. If the concentration of oxygen in the alveoli drops, the amount of oxygen in the blood drops. At altitudes of 9,843 feet (3,000 meters) or more, people often feel lightheaded, especially if they are exercising and placing extra demands on the cardiovascular and respiratory systems.

What **causes hiccups** and how can they be **cured**?

Hiccupping is the involuntary contraction of the diaphragm. When the diaphragm contracts, the vocal cords close quickly, causing the familiar noise associated with hiccups. Hiccups help the stomach get rid of a bit of gas, relieve the esophagus of an irritation, or resolve a temporary loss of coordination between the nerves controlling the movement of the diaphragm. A bout of hiccups can be brought on by eating or drinking too fast or fatigue or nervousness. Although hiccups generally go away on their own after a few minutes, several remedies have been recommended to cure one of hiccups: a loud distraction to scare the sufferer of hiccups, swallowing a spoonful of sugar, blowing into a paper bag, or holding one's breath.

NERVES AND SENSES

What are the **functions** of the **nervous system**?

The nervous system is one of the major regulatory systems of the body maintaining homeostasis. Its functions are to: 1) monitor the body's internal and external environments; 2) integrate sensory information; and 3) direct or coordinate the responses of other organ systems to the sensory input.

What are the **two subsystems** of the **nervous system**?

The nervous system is divided into the central nervous system and the peripheral nervous system. The central nervous system consists of the brain and spinal cord, while the peripheral nervous system consists of all the nerve tissue in the body, excluding the brain and spinal cord. Communication between the central nervous system and the rest of the body is via the peripheral nervous system. Specialized cells of the peripheral nervous system allow communication between the two systems.

What are the **two types** of **cells** found in the **peripheral nervous system**?

The peripheral nervous system consists of afferent (sensory) neurons and efferent (motor) neurons. The afferent nerve cells (from the Latin, *ad*, meaning "toward," and *ferre*, meaning "to bring") carry sensory information from the peripheral to the central nervous system. They have their cell bodies in ganglia and send a process into the central nervous system. The efferent nerve cells (from the Latin *ex*, meaning "away from," and *ferre*, meaning "to bring") carry information away from the central nervous system to the effectors (muscles and tissues). They have cell bodies in the central nervous system and send axons into the periphery.

How do **neurons transmit information** to other neurons?

Most neurons communicate with other neurons or muscle by releasing chemicals called neurotransmitters. These transmitters influence receptors on other neurons. In a few specialized places, neurons communicate directly with other neurons via pores called "gap junctions."

Which **skills** are controlled by the **left cerebral hemisphere** and which are controlled by the **right cerebral hemisphere**?

The left side of the brain controls the right side of the body, as well as spoken and written language, logic, reasoning, and scientific and mathematical abilities. In contrast, the right side of the brain controls the left side of the body and is associated with imagination, spatial perception, recognition of faces, and artistic and musical abilities.

In addition to left- or right-handedness, what **other left-right preferences** do people have?

Most people have a preferred eye, ear, and foot. In one study, for example, 46 percent were strongly right-footed, while 3.9 percent were strongly left-footed (similar to being right-handed); furthermore, 72 percent were strongly right-handed and 5.3 percent strongly left-handed. Estimates vary about the proportion of left-handers to right-handers, but it may be as high as one in ten. Some 90 percent of healthy adults use the

right hand for writing; two-thirds favor the right hand for most activities requiring
coordination and skill. There is no male-female difference in these proportions.

Where is the **spinal cord located**?

The spinal cord lies inside the vertebral column. It extends from the occipital bone of
the skull to the level of the first or second lumbar vertebra. In adults, the spinal cord
is 16 to 18 inches (42 to 45 centimeters) long and 0.5 inches (1.27 centimeters) in
diameter. The spinal cord is the connecting link between the brain and the rest of the
body. It is the site for integration of the spinal cord reflexes.

What is a **reflex**?

A reflex is a predictable, involuntary response to a stimulus. It was given this name in
the eighteenth century because it appeared that the stimulus was reflected off of the
spinal cord to generate the response, just as light is reflect by a mirror. Reflexes allow
the body to respond quickly to internal and external changes in the environment in
order to maintain homeostasis. Reflexes that involve the skeletal muscles are called
somatic reflexes. Reflexes that involve responses of smooth muscle, cardiac muscle, or
a gland are called visceral or autonomic reflexes.

How is the **autonomic nervous system organized**?

The autonomic nervous system regulates "involuntary" activity, which is not controlled on a conscious level. It consists of two divisions: the sympathetic nervous system and the parasympathetic nervous system. The sympathetic division is often called
the "fight or flight" system because it usually stimulates tissue metabolism, increases
alertness, and generally prepares the body to deal with emergencies. The parasympathetic division is considered the "rest and repose" division because it conserves energy
and promotes sedentary activities, such as digestion.

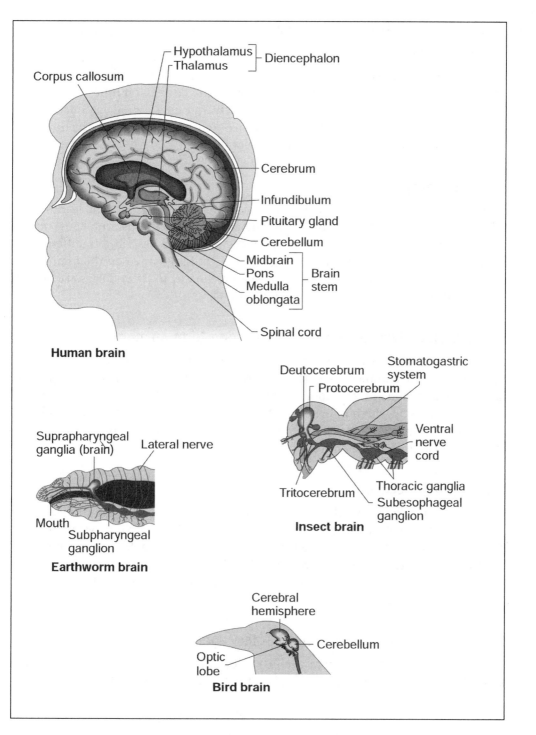

Human brain

- Corpus callosum
- Hypothalamus ⎱ Diencephalon
- Thalamus ⎰
- Cerebrum
- Infundibulum
- Pituitary gland
- Cerebellum
- Midbrain ⎱
- Pons ⎰ Brain stem
- Medulla oblongata
- Spinal cord

Earthworm brain

- Suprapharyngeal ganglia (brain)
- Lateral nerve
- Mouth
- Subpharyngeal ganglion

Insect brain

- Deutocerebrum
- Stomatogastric system
- Protocerebrum
- Ventral nerve cord
- Tritocerebrum
- Thoracic ganglia
- Subesophageal ganglion

Bird brain

- Cerebral hemisphere
- Cerebellum
- Optic lobe

A comparison between human, earthworm, insect, and bird brains.

Which **parts** of the **brain** are involved in **higher order functions**, such as learning and memory?

Higher order functions, such as learning and memory, involve complex interactions among areas of the cerebral cortex and between the cortex and other areas of the brain. Information is processed both consciously and unconsciously. Since higher order functions are not part of the programmed "wiring" of the brain, the functions are subject to modification and adjustment over time.

What is **memory**?

Memory is the ability to recall information and experiences. Memory and learning are related because in order to be able to remember something it must first be "learned." Memories may be facts or skills. Memory "traces" have been described traditionally as concrete things that are formed during learning and imprinted on the brain when neurons record and store information. However, the way that memories are formed and represented in the brain is not well understood.

How does **short-term memory** differ from **long-term memory**?

Short-term memory, also called primary memory, refers to small bits of information that can be recalled immediately. The recalled information has no permanent importance, such as a name or telephone number that is only used once. Long-term memory is the process by which information that for some reason is interpreted as being important is remembered for a much longer periods. Short-term memories may be converted to long-term memories.

Why do people **need sleep**?

Scientists do not know exactly why people need sleep, but studies show that sleep is necessary for survival. Sleep appears to be necessary for the nervous system to work properly. While too little sleep one night may leave us feeling drowsy and unable to concentrate the next day, a prolonged period of too little sleep leads to impaired memory and physical performance. Hallucinations and mood swings may develop if sleep deprivation continues.

Is it true that people **need less sleep** as they **age**?

As a person ages, the time spent each day sleeping decreases. The following table shows how much sleep is generally needed at night, depending on age:

Age	Sleep Time (hours)
1–15 days	16–22
6–23 months	13

Age	Sleep Time (hours)
3–9 years	11
10–13 years	10
14–18 years	9
19–30 years	8
31–45 years	7.5
45–50 years	6
50+ years	5.5

What is **REM sleep**?

REM sleep is rapid eye movement sleep. It is characterized by faster breathing and heart rates than NREM (non-rapid eye movement) sleep. The only people who do not have REM sleep are those who have been blind from birth. REM sleep usually occurs in four to five periods, varying from five minutes to about an hour, growing progressively longer as sleep continues.

When does **dreaming** occur during the **sleep cycle**?

Almost all dreams occur during REM sleep. Scientists do not understand why dreaming is important. One theory is that the brain is either cataloging the information it acquired during the day and discarding the data it does not want, or is creating scenarios to work through situations causing emotional distress. Regardless of its function, most people who are deprived of sleep or dreams become disoriented, unable to concentrate, and may even have hallucinations.

Why do people **snore** and **how loud** can snoring be?

Snoring is produced by vibrations of the soft palate, usually caused by any condition that hinders breathing through the nose. It is more common while sleeping on the back. Research has indicated that a snore can reach 69 decibels, as compared to 70–90 decibels for a pneumatic drill.

What are the **major senses**?

As early as 300 B.C.E., the five senses were recognized to include smell, taste, sight, hearing, and touch. More recently, scientists categorize the senses into two major groups. One group is the special senses, which are produced by highly localized sensory organs and include the senses of smell, taste, sight, hearing, and balance. The other group is the general senses, which are more widely distributed throughout the body and include such senses as touch, pressure, pain, temperature, and vibration.

How many **types** of **sensory receptors** have been identified?

Five types of sensory receptors, each responding to a different type of stimulus, have been identified.

Chemoreceptors—Respond to chemical compounds such as odor molecules

Photoreceptors—Respond to light

Thermoreceptors—Respond to changes in temperature

Mechanoreceptors—Respond to changes in pressure or movement

Pain receptors—Respond to stimuli that result in the sensation of pain

Which **organ** is **most sensitive** to touch, temperature, and pain?

The tongue is more sensitive to touch, temperature, and pain than any other part of the body.

How does the **sense of smell work**?

The sense of smell is associated with sensory receptor cells in the upper nasal cavity. The smell, or olfactory, receptors are chemoreceptors. Chemicals that stimulate olfactory receptors enter the nasal cavity as airborne molecules called gases. They must dissolve in the watery fluids that surround the cilia of the olfactory receptor cells before they can be detected. These specialized cells, the olfactory receptor neurons, are the only parts of the nervous system that are in direct contact with the outside environment. The odorous gases then waft up to the olfactory cells, where the chemicals bind to the cilia that line the nasal cavity. That action initiates a nerve impulse being sent through the olfactory cell, into the olfactory nerve fiber, to the olfactory bulb, and to the brain. The brain then knows what the chemical odors are.

Do **humans or bloodhounds** have a **keener** sense of **smell**?

Humans smell the world using about 12 million olfactory receptor cells, whereas bloodhounds have 4 billion such cells and, therefore, a much better sense of smell.

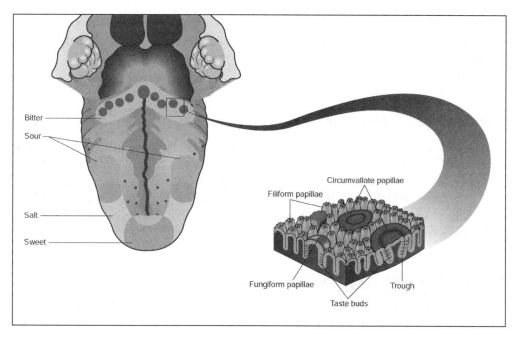

Each taste bud on the tongue is sensitive to a certain type of taste, such as sweet, bitter, salty, or sour.

For example, the trace of sweat that seeps through shoes and is left in footprints is a million times more powerful than the bloodhound needs to track down someone.

What are the **special organs** of **taste**?

The special organs of taste are the taste buds located primarily on the surface of the tongue, where they are associated with tiny elevations called papillae surrounded by deep folds. A taste bud is a cluster of approximately 100 taste cells representing all taste sensations and 100 supporting cells that separate the taste cells. Taste buds can also be found on the roof of the mouth and in the throat. An adult has approximately 10,000 taste buds.

What is the **average lifespan** of a human **taste bud**?

Each taste bud lives for seven to ten days.

Are certain **areas of the tongue** associated with a **particular taste** sensation?

All taste buds are able to detect each of the four basic taste sensations. However, each taste bud is usually most sensitive to one type of taste stimuli. The stimulus type to which each taste bud responds most strongly is related to its position on the tongue.

555

Sweet receptors are concentrated at the tip of the tongue, while sour receptors are more common at the sides of the tongue. Salt receptors occur most frequently at the tip and front edges of the tongue. Bitter receptors are most numerous at the back of the tongue.

Which **two special senses** are **closely related**?

The special senses of smell and taste are very closely related, both structurally and functionally. Experimental evidence shows that taste is partially dependent on the sense of smell. Most subjects are unable to distinguish between an onion and an apple on a blind taste test when their sense of smell is blocked. This also explains why food is "tasteless" when you have a cold because the olfactory receptor cells are covered with a thick mucus blocking the sense of smell.

What **two functions** are performed by the **ear**?

The ear has two functions: hearing and maintaining equilibrium or balance.

What are the **major parts** of the **ear**?

The major parts of the ear are the external ear, middle ear, and inner ear.

What are the **three bones** in the **middle ear**?

The three bones in the middle ear are the malleus (hammer), the incus (anvil), and the stapes (stirrup). The bones look somewhat like the objects for which they are named. These three tiny bones in the middle ear bridge the eardrum and the inner ear, transmitting sound vibrations.

What is **sound** and what **unit measures** it?

Sound is the vibration of air or other matter. Sound intensity refers to the energy of the sound waves, and loudness refers to the interpretation of sound as it reaches the ears. Both intensity and loudness are measured in logarithmic units called decibels (dB). A sound that is barely audible has an intensity of zero decibels. Every 10 decibels indicates a tenfold increase in sound intensity. A sound is ten times louder than threshold at 10 dB and 100 times louder at 20 dB.

What are some **common levels of sound** and how do they **affect hearing**?

Some common levels of sound and their effect on hearing are listed in the following table:

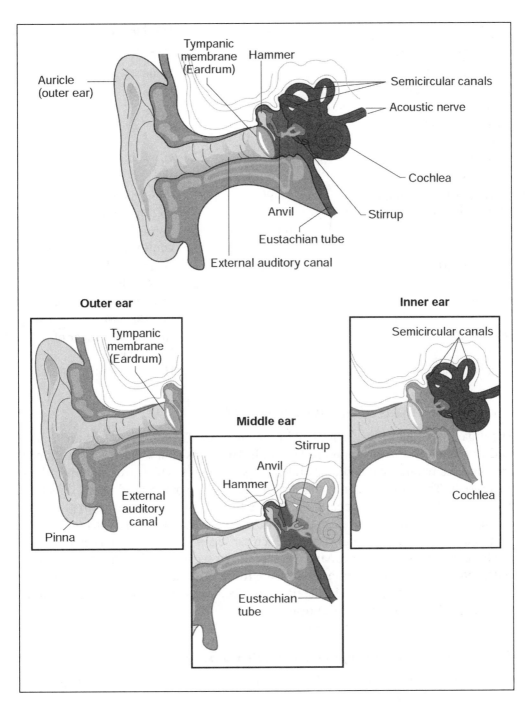

The parts of the outer, middle, and inner human ear.

Sound	Decibel Level	Effect on Hearing
Lowest audible sound	Just above 0	None
Rustling leaves	20	None
Quiet library or quiet office	30–40	None
Normal conversation; refrigerator; light traffic at a distance	50–60	None
Busy traffic, vacuum cleaner; noisy restaurant	70	None
Heavy city traffic; subway; shop tools; power lawn mower	80–90	Some damage if continuous for 8 hours
Chain saw	100	Some damage if continuous for 2 hours
Rock concert	110–120	Definite risk of permanent hearing loss
Gunshot	140	Immediate danger of hearing loss
Jet engine	150	Immediate danger of hearing loss
Rocket launching pad	160	Hearing loss inevitable

Where are the **organs of equilibrium** located?

The organs of equilibrium are located in the inner ear. The otolith organs are located in the vestibule of the membranous labyrinth. They consist of sheets of hair cells covered by a membrane that contains otoliths ("ear stones"), which are calcium carbonate crystals. The otolith organs sense linear acceleration of the head in any direction, such as acceleration due to changing the position of your head relative to gravity or acceleration in a car or amusement ride. The inner ear also contains horizontal, posterior, and anterior semicircular canals, which sense angular motions (acceleration) of the head. Each semicircular canal has a specialized sensory region that contains hair cells, and each canal is important for sensing rotation of the head in a different primary direction. For example, the horizontal semicircular canal receptors are sensitive to rotating the head leftward and rightward.

What are the **parts** of the **eye** and their **functions**?

The major parts of the eye and their functions are summarized in the following chart:

Structure	Function
Sclera	Maintains shape of eye; protects eyeball; site of eye muscle attachment

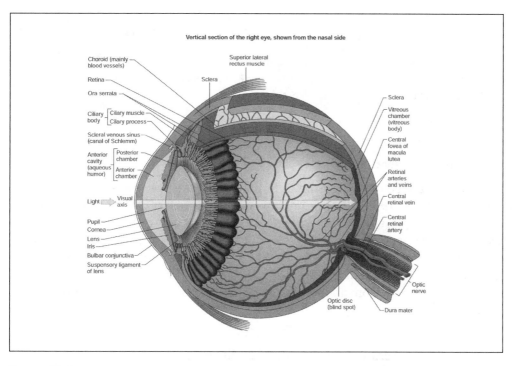

The parts of the human eye.

Structure	Function
Cornea	Refracts incoming light; focuses light on the retina
Pupil	Admits light
Iris	Regulates amount of incoming light
Lens	Refracts and focuses light rays
Aqueous humor	Helps maintain shape of eye; maintains introcular pressure; nourishes and cushions cornea and lens
Ciliary body	Holds lens in place; changes shape of lens
Vitreous humor	Maintains intraocular pressure; transmits light to retina; keeps retina firmly pressed against choroids
Retina	Absorbs light; stores vitamin A; forms impulses that are transmitted to brain
Optic nerve	Transmits impulses to the brain
Choroid	Absorbs stray light; nourishes retina

The accessory structures of the eye include the eyebrows, eyelids, eyelashes, conjunctiva, and lacrimal apparatus. These structures have several functions including protecting the anterior portion of the eye, preventing the entry of foreign particles, and keeping the eyeball moist.

What determines **eye color**?

Variations in eye color range from light blue to dark brown and are inherited. Eye color is chiefly determined by the amount and distribution of melanin within the irises. If melanin is present only in the epithelial cells that cover the posterior surface of the iris, the iris appears blue. When this condition exists together with denser than usual tissue within the body of the iris, it looks gray. When melanin is present within the body of the iris, as well as the epithelial covering, the iris appears brown. Green and hazel eyes result from an increase in the amount of a combination of yellow and black melanin.

What are the **floaters** that move around on the eye?

Floaters are semi-transparent specks perceived to be floating in the field of vision. Some originate with red blood cells that have leaked out of the retina. The blood cells swell into spheres, some forming strings, and float around the areas of the retina. Others are shadows cast by the microscopic structures in the vitreous humor, a jellylike substructure located behind the retina. A sudden appearance of a cloud of dark floaters, if accompanied by bright light flashes, could indicate retinal detachment.

What is the difference in the functions of the **rods and cones** found in the eyes?

Rods and cones contain photoreceptors that convert light first to chemical energy and then into electrical energy for transmission to the vision centers of the brain via the optic nerve. Rods are specialized for vision in dim light; they cannot detect color, but they are the first receptors to detect movement and register shapes. There are about 125 million rods in each eye. Cones provide acute vision, functioning best in bright daylight. They allow us to see colors and fine detail. Cones are divided into three different types, which absorb wavelengths in the short (blue), middle (green), and long (red) ranges. There are about 7 million cones in each eye.

How long does it take for a person to **adapt to dim light**?

Rods give us vision in dim light but not in color and not with sharp detail. They are
hundreds of times more sensitive to light than cones, letting us detect shape and

> ## Who invented bifocal lenses?
>
> The original bifocal lens was invented in 1784 by Benjamin Franklin (1706–1790). At that time, the two lenses were joined in a metallic frame. In 1899, J.L. Borsch welded the two lenses together. One-part bifocal lenses were developed in 1910 by researchers at the Carl Zeiss Company.

movement in dim light. This type of photoreceptor takes about 15 minutes to fully adapt to very dim light.

How **often** does a person **blink**?

The rate of blinking varies, but on average the eye blinks once every five seconds (12 blinks per minute). Assuming the average person sleeps eight hours a day and is awake for 16 hours, he or she would blink 11,520 times a day, or 4,204,800 times a year. The average blink of the human eye lasts about 0.05 seconds.

What is **nearsightedness**?

Nearsightedness, or myopia, is the ability to see close objects but not distant ones. It is a defect of the eye in which the focal point is too near the lens and the image is focused in front of the retina when looking at distant objects. This condition is corrected by concave lenses (eyeglasses or contact lenses) that diffuse the light rays coming to the eyes so that when the light is focused by the eyes it reaches the proper spot on the retinas. Approximately 25 percent of Americans are nearsighted.

What is **farsightedness**?

Farsightedness, or hyperopia, is the ability to see distant objects but not close ones. It is a disorder in which the focal point is too far from the lens, and the image is focused "behind" the retina when looking at a close object. In this condition, the lens must thicken to bring somewhat distant objects into focus. Farsightedness is corrected by a convex lens that causes light rays to converge as they approach the eye to focus on the retina. Farsightedness is less common than nearsightedness, affecting only five to ten percent of Americans.

What does it mean to have **20/20 vision**?

Many people think that 20/20 vision equals perfect eyesight, but it actually means that the eye can see clearly at 20 feet (6 meters) what a normal eye can see clearly at that distance. Some people can see even better—20/15, for example. With their eyes, they can view objects from 20 feet (6 meters) away with the same sharpness that a normal-sighted person would have to move in to 15 feet (4.5 meters) to achieve.

What are the **similarities** and **differences** between the **nervous system** and the **endocrine system**?

The endocrine and nervous are both regulatory systems that permit communication between cells, tissues, and organs. Both systems are devoted to maintaining homeostasis by coordinating and regulating the activities of other cells, tissues, organs, and systems. Both systems are regulated by negative feedback mechanisms. Chemical messengers are important in both systems, although their method of transmission and release differs in the two systems.

A major difference between the endocrine system and nervous system is the rate of response to a stimulus. In general, the nervous system responds to a stimulus very rapidly, often within a few milliseconds, while it may take the endocrine system seconds and sometimes hours or even days to offer a response. Furthermore, the chemical signals released by the nervous system typically act over very short distances (a synapse), while hormones in the endocrine system are generally carried by the blood to target organs. Finally, the effects of the nervous system generally last only a brief amount of time, while those of the endocrine system are longer lasting.

What are the **organs** of the **endocrine system**?

The endocrine system consists of glands and other hormone-producing tissues. Glands are specialized cells that secrete hormones into the interstitial fluid. Hormones are then transported to the capillaries and circulated via the blood. The major endocrine glands are the pituitary, thyroid, parathyroid, pineal, and adrenal glands. Other hormone-secreting organs are the central nervous system (hypothalamus), kidneys, heart, pancreas, thymus, ovaries, and testes. Some organs, such as the pancreas, secrete hormones as an endocrine function but have other functions also.

What are **hormones**?

Hormones are chemical messengers that are secreted by the endocrine glands into the blood. They produce a specific effect on the activity of cells that are remotely located from their point of origin. Hormones are transported via the bloodstream to reach specific cells, called target cells, in other tissues. Target cells have special receptors on their outer membranes that allow the individual hormones to bind to the cell. The hormones and receptors fit together much like a lock and key.

Which **endocrine glands** produce which **hormones**?

Each endocrine gland produces specific hormones, as explained in the below table.

Who discovered the first hormone?

The British physiologists William Bayliss (1860–1924) and Ernest Starling (1866–1927) discovered secretin in 1902. They used the term "hormone" (from the Greek word *horman*, meaning "to set in motion") to describe the chemical substance they had discovered that stimulated an organ at a distance from the chemical's site of origin. Their famous experiment using anesthetized dogs demonstrated that dilute hydrochloric acid, mixed with partially digested food, activated a chemical substance in the duodenum (the upper part of the small intestine). This activated substance (secretin) was released into the bloodstream and came into contact with cells of the pancreas. In the pancreas it stimulated secretion of digestive juice into the intestine through the pancreatic duct.

Endocrine Glands and Their Hormones

Gland	Hormone(s) Produced
Pituitary	
Anterior	Thyroid-stimulating hormone (TSH)
	Adrenocorticotropic hormone (ACTH)
	Follicle-stimulating hormone (FSH)
	Luteinizing hormone (LH)
	Prolactin (PRL)
	Growth hormone (GH)
	Melanocyte-stimulating hormone (MSH)
Posterior	Antidiuretic hormone (ADH)
	Oxytocin
Thyroid	Thyroxine (T_4)
	Triiodothyronine (T_3)
	Calcitonin (CT)
Parathyroid	Parathyroid hormone (PTH)
Pineal	Melatonin
Adrenal	
Cortex	Mineralocorticoids, primarily aldosterone
	Glucocorticoids, mainly cortisol (hydrocortisone), corticosterone, cortisone
Medulla	Epinephrine (E)
	Norepinephrine (NE)
Pancreas	Insulin
	Glucagon

Gland	Hormone(s) Produced
Thymus	Thymosins
Ovaries	Estrogens and progesterone
Testes	Androgens, mainly testosterone

What are **endorphins**?

Endorphins and closely related chemicals called enkephalins are part of a larger group called opiods, which have properties simliar to drugs such as heroin or morphine. They can act not only as pain killers but also can induce a sense of well-being or euphoria. Clinical applications of endorphin research include possible treatments for some forms of mental illness; treatment or control of pain for chronic pain sufferers; development of new anesthetics; and the development of non-addictive, safe, and effective pain relievers.

DIGESTION

What are the **steps** in the **digestive process**?

There are five major steps in the process of digestion.

1. Ingestion: the eating of any food
2. Peristalsis: the involuntary muscle contractions that move the ingested food through the digestive tract
3. Digestion: the conversion of the food molecules into nutrients that can then be used by the body
4. Absorption: the passage of the nutrients into the bloodstream and/or lymphatic system to be used by the body's cells
5. Defecation: the elimination of the undigested and unabsorbed ingested materials

What are the **major organs** of the **digestive system**?

The digestive system consists of the upper gastrointestinal tract, the lower gastrointestinal tract, and the accessory organs. The organs of the upper gastrointestinal tract are the oral cavity, esophagus, and stomach. The organs of the lower gastrointestinal tract are the small intestine and large intestine (also called the colon). The accessory organs are the salivary glands, the liver, gallbladder, and pancreas.

What was the **likely purpose** of the human **appendix**?

Experts can only theorize on its use. It may have had the same purpose it does in present-day herbivores, where it harbors colonies of bacteria that help in the digestion of

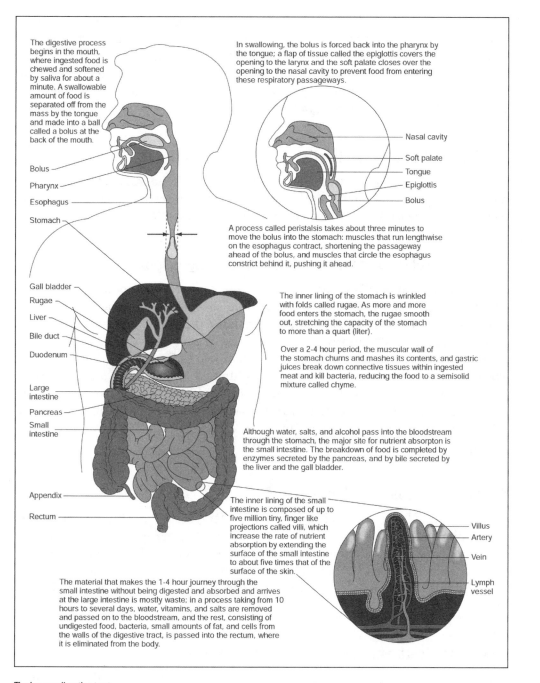

The digestive process begins in the mouth, where ingested food is chewed and softened by saliva for about a minute. A swallowable amount of food is separated off from the mass by the tongue and made into a ball called a bolus at the back of the mouth.

Bolus

Pharynx

Esophagus

Stomach

Gall bladder

Rugae

Liver

Bile duct

Duodenum

Large intestine

Pancreas

Small intestine

Appendix

Rectum

In swallowing, the bolus is forced back into the pharynx by the tongue; a flap of tissue called the epiglottis covers the opening to the larynx and the soft palate closes over the opening to the nasal cavity to prevent food from entering these respiratory passageways.

Nasal cavity

Soft palate

Tongue

Epiglottis

Bolus

A process called peristalsis takes about three minutes to move the bolus into the stomach: muscles that run lengthwise on the esophagus contract, shortening the passageway ahead of the bolus, and muscles that circle the esophagus constrict behind it, pushing it ahead.

The inner lining of the stomach is wrinkled with folds called rugae. As more and more food enters the stomach, the rugae smooth out, stretching the capacity of the stomach to more than a quart (liter).

Over a 2-4 hour period, the muscular wall of the stomach churns and mashes its contents, and gastric juices break down connective tissues within ingested meat and kill bacteria, reducing the food to a semisolid mixture called chyme.

Although water, salts, and alcohol pass into the bloodstream through the stomach, the major site for nutrient absorpton is the small intestine. The breakdown of food is completed by enzymes secreted by the pancreas, and by bile secreted by the liver and the gall bladder.

The inner lining of the small intestine is composed of up to five million tiny, finger like projections called villi, which increase the rate of nutrient absorption by extending the surface of the small intestine to about five times that of the surface of the skin.

Villus

Artery

Vein

Lymph vessel

The material that makes the 1-4 hour journey through the small intestine without being digested and absorbed and arrives at the large intestine is mostly waste; in a process taking from 10 hours to several days, water, vitamins, and salts are removed and passed on to the bloodstream, and the rest, consisting of undigested food, bacteria, small amounts of fat, and cells from the walls of the digestive tract, is passed into the rectum, where it is eliminated from the body.

The human digestive tract.

cellulose in plant material. Another theory suggests that tonsils and the appendix might manufacture the antibody-producing white blood cells called B lymphocytes; however, B lymphocytes could also be produced by the bone marrow. The third theory is that the appendix may "attract" body infections to localize the infection in one spot that is not critical to body functioning. The earliest surgical removal of the appendix was by Claudries Amyand (1680–1740) in England in 1736.

How long is the digestive tract?

The digestive tract, also called the alimentary canal, is approximately 30 feet (9 meters) long from the mouth to the anus. The small intestine is about 22 feet (7 meters) long. The large intestine is about 5 feet (1.5 meters) long.

Who performed some of the earliest studies on digestion?

William Beaumont (1785–1853), an army surgeon, performed some of the earliest studies on digestion. In 1822, Alexis St. Martin (1794–1880), was accidentally wounded by a shotgun blast. Beaumont began treatment of the wound immediately. St. Martin's recuperation lasted nearly three years, and the enormous wound healed except for a small opening leading into his stomach. A fold of flesh covered this opening; when this was pushed aside the interior of the stomach was exposed to view. Through the opening, Beaumont was able to extract and analyze gastric juice and stomach contents at various stages of digestion, observe changes in secretions, and note the stomach's muscular movements. The results of his experiments and observations formed the basis of our modern knowledge of digestion.

What is the purpose of primary teeth?

Primary teeth, also known as baby, deciduous, temporary, or milk (for their milk-white color) teeth serve many of the same purposes as permanent teeth. They are needed for chewing and they are necessary for speech development. They also prepare the mouth for the permanent teeth by maintaining space for the permanent teeth to emerge in proper alignment. Each individual has 20 primary teeth followed by 32 permanent teeth.

How are the teeth and tongue involved in chewing?

The first stage of mechanical digestion is mastication, or chewing. Initially, the teeth tear and shred large pieces of food into smaller units. The muscles of the tongue, cheeks, and lips help keep the food on the surfaces of the teeth. The tongue then compacts the food into a small round mass of material called the bolus. The salivary glands help lubricate the food with secretions.

How much **force** does a **human bite** generate?

All the jaw muscles working together can close the teeth with a force as great as 55 pounds (25 kilograms) on the incisors or 200 pounds (90.7 kilograms) on the molars. A force as great as 268 pounds (122 kilograms) for molars has been reported.

How much saliva does a person produce in a **day**?

Saliva is a mixture of mucus, water, salts, and the enzymes that break down carbohydrates. Awake individuals secrete saliva at a rate of approximately 0.5 milliliters per minute, or an average of 480 milliliters of saliva in a 16-hour waking day. Various activities such as exercise, eating, drinking, and speaking increase salivary volume.

When a **person swallows** solid or liquid food, what **prevents** it from going **down the windpipe**?

Once food is chewed, voluntary muscles move it to the throat. In the pharynx (throat), automatic, involuntary reflexes take over. The epiglottis closes over the larynx (voice box), which leads to the windpipe. A sphincter at the top of the esophagus relaxes, allowing the food to enter the digestive tract.

How does the **volume** of the **stomach change** from when it is **empty** to when it is **full**?

The inner mucous membrane of the stomach contains branching wrinkles called rugae (from the Latin, meaning "folds"). As the stomach fills, the rugae flatten until they almost disappear when the stomach is full. An empty stomach has a volume of only 0.05 quarts (50 milliliters). A full stomach expands to contain 1 to 1.5 quarts (a little less than 1 to 1.5 liters) of food in the process of being digested.

How **long** does it take **food to digest**?

The stomach holds a little under 2 quarts (1.9 liters) of semi-digested food that stays in the stomach three to five hours. The stomach slowly releases food to the rest of the digestive tract. Fifteen hours or more after the first bite started down the alimentary canal, the final residue of the food is passed along to the rectum and is excreted through the anus as feces.

What is the role of the **small intestine** in **nutrient processing**?

The small intestine is the site of most nutrient processing in the body. The first step is to breakdown the large complex structures of all nutrients, including carbohydrates, lipids, proteins, and nucleic acids, into smaller units. Most absorption of these nutrients also takes place in the small intestine.

How many distinct **regions** are in the **large intestine**?

The large intestine is mostly a storage site for undigested materials until they are eliminated from the body via defecation. It consists of three distinct regions: 1) the cecum, 2) the colon, and 3) the rectum. The cecum is the first section of the large intestine below the ileocecal valve. The appendix is attached to the cecum. Since the colon (ascending, transverse, descending, and sigmoid colon) is the largest region of the large intestine, the term "colon" is often applied to the entire large intestine. The rectum (rectum, anal canal, and anus) is the final region of the large intestine and the end of the digestive tract. Although most absorption has occurred in the small intestine, water, and electrolytes are still absorbed through the large intestine.

Is the **large intestine essential** for life?

Since the role of the large intestine is mainly as a storage site for fecal material and the elimination of it from the body, it is not essential for life. Individuals who suffer from colon cancer or other diseases will often have their colon removed. The end of the ileum is brought to the abdominal wall. Food residues are eliminated directly from the ileum into a sac attached to the abdominal wall on the outside of the body. Alternatively, the ileum may be connected directly to the anal canal.

What is **feces**?

Feces (from the Latin *faex*, meaning "dregs") is the remaining portion of undigested food. Approximately 5.3 ounces (150 grams) of feces are produced daily. Feces normally consists of 3.5 ounces (100 grams) of water and 1.7 ounces (50 grams) of solid material. The solid material is composed of fat, nitrogen, bile pigments, undigested food such as cellulose, and other waste products from the blood of intestinal wall. The normal brown color of feces is caused by bilirubin. Blood and foods containing a large amount of iron will darken the feces. Excessive fat from the diet causes feces to be a more pale color.

What are the **functions** of the **urinary system**?

The functions of the urinary system include regulation of body fluids, removal of metabolic waste products, regulation of volume and chemical make-up of blood plasma, and excretion of toxins.

What are the **major parts** of the **urinary system**?

The major parts of the urinary system are the kidneys, the urinary bladder, two ureters, and the urethra. Each component of the urinary system has a unique function. Urine is manufactured in the kidneys. The urinary bladder serves as a temporary storage reservoir for urine. The ureters transport urine from the kidney to the bladder, while the urethra transports urine from the bladder to the outside of the body.

What is **urea** and where is it produced?

During the process of metabolizing proteins, the body produces ammonia. Ammonia combines with carbon dioxide to form urea. The human body can tolerate 100,000 times more urea than ammonia. It is the most abundant organic waste produced in the body and it is eliminated by the kidneys. Humans generate about 0.75 ounces (21 grams) of urea each day.

How much **fluid** do the **kidneys remove** from the blood?

The kidneys filter about 48 gallons (182 liters) of blood daily and produce about 4 ounces of filtrate per minute. About 1.5 to 2 quarts (1.4 to 1.9 liters) of urine is eventually excreted per day. The entire blood supply is filtered through the kidneys 60 times per day. The kidneys in a person living 73 years filter almost 1.3 million gallons of blood.

REPRODUCTION AND HUMAN DEVELOPMENT

What is the **function** of the **reproductive system**?

The function of the reproductive system is to produce new offspring. The reproductive system is essential for the survival of the species and guarantees the continuity of the human species.

What are the **male reproductive organs** and structures?

The male reproductive organs and structures are the testes, a duct system that includes the epididymis and the vas deferens; the accessory glands, including the seminal vesicles and prostate gland; and the penis. The testes are the male gonads. They produce the male reproductive cells called sperm.

At what **age** does **sperm production** begin?

Sperm production begins with the onset of puberty, usually between ages 11 to 14 in boys. It continues throughout the life of an adult male. It is estimated that during his lifetime a normal male will produce as many as 300 million sperm per day.

How **large** is the **prostate gland**?

The prostate gland in healthy adult males is about the size of a walnut. It is located in front of the rectum and under the bladder. The prostate surrounds the urethra. When it becomes enlarged, it squeezes the urethra, restricting the flow of urine.

What are the **female reproductive organs**?

The organs of the female reproductive system include the ovaries, the uterine tubes, the uterus, the vagina, the external organs called the vulva, and the mammary glands. The paired ovaries are the female gonads. They produce the female gametes, called ova, and secrete the female sex hormones.

How **large** is the **uterus**?

The uterus (from the Latin, meaning "womb") is an inverted, pear-shaped structure located between the urinary bladder and the rectum. The uterus is normally 3 inches (7.5 centimeters) long and 2 inches (5 centimeters) wide. It weighs 1 to 1.4 ounces (30 to 40 grams). During pregnancy, implantation of a fertilized ovum occurs in the uterus. The uterus then houses, nourishes, and protects the developing fetus. It can increase three to six times in size during the course of pregnancy. At the end of gestation, the uterus is usually 12 inches (30 centimeters) long and weighs 2.4 pounds (1,100 grams). The total weight of the uterus and its contents (fetus and fluid) at the end of pregnancy is approximately 22 pounds (10 kilograms).

What is the **female reproductive cycle**?

The female reproductive cycle is a general term to describe both the ovarian cycle and the uterine cycle, as well as the hormonal cycles that regulate them. The ovarian cycle is the monthly series of events that occur in the ovaries related to the maturation of an oocyte. The menstrual cycle is the monthly series of changes that occur in the uterus as it awaits a fertilized ovum.

How **long** the does female **reproductive cycle** last?

The reproductive cycle averages 28 days, but it may last from 24 to 35 days. A cycle of 20 to 45 days is still considered within the range of normalcy. The menstrual phase lasts five to seven days. The preovulatory phase lasts 8 to 11 days, and the postovulatory phase lasts an average of 14 days.

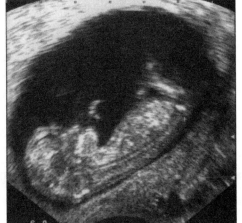

When does **menarche typically begin**?

Menarche (from the Greek, meaning "beginning the monthly") is the first

An ultrasound image of a fetus at about 11 weeks of development.

menses, or menstrual cycle, of a girl's life. It usually begins between the ages of 11 and 12.

What is **menopause**?

Menopause is the cessation of ovulation and menstrual periods. The supply of follicles in the ovaries is depleted, increasing the amount of follicle-stimulating hormone (FSH), while decreasing the amount of estrogen and progesterone. The process may take one to two years. The years preceding the final menstrual period are known as perimenopause. Menopause usually occurs between ages 45 and 55; the average age in the United States is 51 to 52 years.

How is **pregnancy confirmed**?

Early external signs of pregnancy include a missed menstrual period, bleeding or spotting, fatigue, and tender, swollen breasts. A pregnancy test is needed to confirm pregnancy. Pregnancy tests look for the hormone human chorionic gonadotropin (hCG), also called the pregnancy hormone, in either urine or blood. This hormone is produced when the fertilized egg implants in the uterus. It is only present in pregnant women.

When do **significant changes** occur during the **early stages** of **prenatal development**?

Several significant changes occur during the first two weeks of prenatal development, as shown in the accompanying table.

571

Significant Changes in the First Two Weeks of Prenatal Development

Time Period	Developmental Stage
12–24 hours following ovulation	Fertilized ovum
30 hours to third day	Cleavage
Third to fourth day	Morula (solid ball of cells is formed)
Fifth day through second week	Blastocyst
End of second week	Gastrula (germ layers form)

When does the **heart begin to beat** in a developing **embryo**?

The heart begins to beat by the twenty-fifth day of development.

What are some major **developmental events** during the **embryonic period**?

At the end of the embryonic period (eighth week of development), all of the major external features (ears, eyes, mouth, upper and lower limbs, fingers, and toes) are formed and the major organ systems are nearing completion.

Major Developmental Events during the Embryonic Period

Time Period	Major Developments
Week three	Neural tube, primitive body cavities and cardiovascular system form
Week four	Heart is beating; upper limb buds and primitive ears visible; lower limb buds and primitive eye lenses appear shortly after ears
Week five	Brain develops rapidly; head grows disproportionately; hand plates develop
Week six	Limb buds differentiate noticeably; retinal pigment accentuates eyes
Week seven	Limbs differentiate rapidly
Week eight	Embryo appears human; external ears are visible; fingers, toes lengthen; external genitalia are visible, but are not distinctly male or female

How large is the **fetus** at the end of the **first trimester**?

At the end of the first three months of pregnancy, the fetus is nearly 3 inches (7.6 centimeters) long and weighs about 0.8 ounces (23 grams).

What are some major **developmental events** during the **second trimester** of pregnancy?

The second trimester of a pregnancy lasts from weeks 13 through 27. Each week brings changes and new developments in the fetus.

Major Developmental Events during the Second Trimester

Time Period	Major Developments
Week 13	Baby begins to move, although the movements are too weak to be felt by the mother; ossification of bones begins
Week 14	Prostate gland develops in boys; ovaries move from the abdomen to the pelvis in girls
Week 15	Skin and hair (including eyebrows and eyelashes) begins to form; bone and marrow continues to develop; eyes and ears are nearly in their final location
Week 16	Facial muscles are developing allowing for facial expressions; hands can form a fist; eggs are forming in the ovaries in girls
Week 17	Brown fat tissue begins to develop under the skin
Week 18	Fetus is able to hear such things as the mother's heartbeat
Week 19	Lanugo and vernix cover the skin; fetal movement is usually felt by the mother
Week 20	Skin is thickening and developing layers; fetus has eyebrows, hair on the scalp and well-developed limbs; fetus often assumes the fetal position of head bent and curved spine
Week 21	Bone marrow begins making blood cells
Week 22	Taste buds begin to form; brain and nerve endings can process the sensation of touch; testes begin to descend from the abdomen in boys; uterus and ovaries (with the lifetime supply of eggs) are in place in girls
Week 23	Skin becomes less transparent; fat production increases; lungs begin to produce surfactant, which will allow air sacs to inflate; may begin to practice breathing
Week 24	Footprints and fingerprints begin to form; inner ear is developed, controlling balance
Week 25	Hands are developed, although the nerve connections are not yet fully developed
Week 26	Eyes are developed; eyebrows and eyelashes are well-formed; hair on head becomes fuller and longer
Week 27	Lungs, liver, and immune system are developing

What are some major **developmental events** during the **third trimester** of pregnancy?

During the third trimester of a pregnancy the fetus continues to grow, while the organ systems continue to develop to the point of being fully functional. Fetal movements become stronger and more frequent.

Major Developmental Events during the Third Trimester

Time Period	Major Developments
Week 28	Eyes begin to open and close; fetus has wake and sleep cycles
Week 29	Bones are fully developed, but still pliable; fetus begins to store iron, calcium, and phosphorus
Week 30	Rate of weight gain increases to 0.5 pounds (227 grams) per week; fetus practices breathing; hiccups are not uncommon
Week 31	Testes begin to descend into scrotum in boys; lungs continue to mature
Week 32	Lanugo begins to fall off
Week 33	Pupils in the eyes constrict, dilate, and detect light; lungs are nearly completely developed
Week 34	Vernix becomes thicker; lanugo has almost disappeared
Week 35	Fetus stores fat all over the body; weight gain continues
Week 36	Sucking muscles are developed
Week 37	Fat continues to accumulate
Week 38	Brain and nervous system are ready for birth
Week 39	Placenta continues to supply nutrients and antibodies to fight infection
Week 40	Fetus is fully developed and ready for birth

How much does the **fetus grow** during **each month** of **pregnancy**?

During the early weeks of development there is great change from embryo to fetus but the overall size of the embryo is very small. As the pregnancy continues, weight gain and overall size become much more significant. Until the twentieth week of pregnancy, length measurements are from the crown (or top) of the head to the rump. After the twentieth week, the fetus is less curled up, and measurements are from the head to toe.

Average Size of the Fetus during Pregnancy

Gestational Age	Size	Weight
8 weeks	0.63 inches (1.6 cm)	0.04 ounce (1 gram)
12 weeks	2.13 inches (5.4 cm)	0.49 ounce (14 grams)

Gestational Age	Size	Weight
16 weeks	4.57 inches (11.6 cm)	3.53 ounces (100 grams)
20 weeks	6.46 inches (16.4 cm)	10.58 ounces (300 grams)
24 weeks	11.81 inches (30 cm)	1.32 pounds (600 grams)
28 weeks	14.80 inches (37.6 cm)	2.22 pounds (1,005 grams)
32 weeks	16.69 inches (42.4 cm)	3.75 pounds (1,702 grams)
36 weeks	18.66 inches (47.5 cm)	5.78 pounds (2,622 grams)
40 weeks	20.16 inches (51.2 cm)	7.63 pounds (3,462 grams)

What are the **stages of labor**?

The goal of labor is the birth of a new baby. Labor is divided into three stages: 1) dilation, 2) expulsion, and 3) placental. Delivery of the fetus occurs during expulsion.

How **common** are **multiple births**?

In 2007, 138,961 twins were born in the United States. This represented 3.2 percent of all live births (a birth ratio of 32.2 per 1,000 births). There were 5,967 triplet births, 369 quadruplet births, and 91quintuplets and other higher order births in 2007. The birth rate for triplets or higher-order multiple births was 148.9 per 100,000 live births.

What are the major **developmental milestones** during **infancy**?

A normal infant will double his or her birth weight by five or six months of age and triple his or her birth weight during the first year of life. Major developmental milestones during infancy are summarized in the table below. There is considerable variation between individuals, but these are within the normal range.

Major Developmental Milestones during Infancy

Age	Major Milestones in Average Infant
End of first month	Bring hands to face; move head from side to side while lying on stomach; can hear very well and often recognize parents' voices
End of third month	Raise head and chest while lying on stomach; open and shut hands; bring hands to mouth; smile; recognize familiar objects and people
End of seventh month	Roll over stomach to back and back to stomach; sit up; reach for objects with hand; support whole weight on legs when supported and held up; enjoy playing peek-a-boo; begin to babble

Age	Major Milestones in Average Infant
End of first year	Sit up without assistance; get into the hands and knees position; crawl; walk while holding on; some babies are able to take a few steps without support; use the pincer grasp; use simple gestures, e.g., nodding head, waving bye-bye
End of second year	Walk alone; begin to run; walk up and down stairs; pull a toy behind them; say single words (15–18 months); use simple phrases and two-word sentences (18–24 months); scribble with a crayon; build a tower with blocks

What are the **stages** of **postnatal development**?

The five life stages of postnatal development are: 1) neonatal, 2) infancy, 3) childhood, 4) adolescence, and 5) maturity. The neonatal period extends from birth to one month. Infancy begins at one month and continues to two years of age. Childhood begins at two years of age and lasts until adolescence. Adolescence begins at around 12 or 13 years of age and ends with the beginning of adulthood. Adulthood, or maturity, includes the years between ages 18 to 25 and old age. The process of aging is called senescence.

What is the average age when **puberty begins**?

The average age when puberty begins in the United States today is around 12 years in boys and 11 years in girls. The normal range is 10 to 15 years in boys and 9 to 14 years in girls.

In the **United States**, what is the **average height** and **weight** for a man and a woman?

The average female is 5 feet, 3.75 inches (1.62 meters) tall and weighs 152 pounds (69.09 kilograms). The average male is 5 feet, 9 inches (1.75 meters) tall and weighs 180 pounds (81.82 kilograms). Between 1960 and 2000 the average American male became 2 inches (5 centimeters) taller and 45 pounds (20.45 kilograms) heavier, while the average American woman also grew 2 inches (5 centimeters) taller and gained 18 pounds (8.18 kilogram).

How is **senescence** defined?

Senescence (from the Latin *senex*, meaning "old") is the process of aging. Physiological changes continue to occur even after complete physical growth is attained at maturity. As people age, the body is less able to and less efficient in adapting to environmental changes. Maintaining homeostasis becomes harder and harder, especially when the body is under stress. Ultimately, death occurs when the combination of stresses cannot be overcome by the body's existing homeostatic mechanisms.

What are some general **effects of aging** on the human body?

The aging process affects every organ system. Some changes begin as early as ages 30 to 40. The aging process becomes more rapid between ages 55 and 60.

Effects of Aging

Organ System	Effect of Aging
Integumentary	Loss of elasticity in the skin tissue, producing wrinkles and sagging skin; oil glands and sweat glands decrease their activity, causing dry skin; hair thins
Skeletal	Decline in the rate of bone deposition, causing weak and brittle bones; decrease in height
Muscular	Muscles begin to weaken; muscle reflexes become slower
Nervous	Brain size and weight decreases; fewer cortical neurons; rate of neurotransmitter production declines; short-term memory may be impaired; intellectual capabilities remain constant unless disturbed by a stroke; reaction times are slower
Sensory	Eyesight is impaired with most people becoming far-sighted; hearing, smell, and taste are reduced
Endocrine	Reduction in the production of circulating hormones; thyroid becomes smaller; production of insulin is reduced
Cardiovascular	Pumping efficiency of the heart is reduced; blood pressure is usually higher; reduction in peripheral blood flow; arteries tend to become more narrow

Organ System	Effect of Aging
Lymphatic	Reduced sensitivity and responsiveness of the immune system; increased chances of infection and/or cancer
Respiratory	Breathing capacity and lung capacity are reduced due to less elasticity of the lungs; air sacs in lungs are replaced by fibrous tissue
Digestive	Decreased peristalsis and muscle tone; stomach produces less hydrochloric acid; intestines produce fewer digestive enzymes; intestinal walls are less able to absorb nutrients
Excretory	Glomerular filtration rate is reduced; decreased peristalsis and muscle tone; weakened muscle tone often leads to incontinence
Reproductive	Ovaries decrease in weight and begin to atrophy in women; reproductive capabilities cease with menopause in women; sperm count decreases in men

How **many people** living in the **United States** are **centenarians**?

The number of centenarians in the United States has increased steadily over the decades. The 2000 census recorded more than 50,000 centenarians, up from 37,000 in the 1990 census. The Census Bureau estimated there were 96,548 centenarians in the United States on November 1, 2008. The United Nations projects there will be 2.2 million centenarians, or one of every 5,000 people, world-wide by 2050.

HEALTH AND MEDICINE

HEALTH HAZARDS AND RISKS

Which **risk factors affect** one's **health**?

Such characteristics as age, gender, work, family history, behavior, and body chemistry are some of the factors to consider when deciding whether one is at risk for various conditions. Some risk factors are statistical, describing trends among large groups of people but not giving information about what will happen to individuals. Other risk factors might be described as causative—exposure to them has a direct effect on whether or not the person will become sick.

What are the leading **causes** of **stress**?

In 1967, when they conducted a study of the correlation between significant life events and the onset of illness, Dr. Thomas H. Holmes (1918–1989) and Dr. Richard H. Rahe (1936–) from the University of Washington compiled a chart of the major causes of stress with assigned point values. They published their findings on stress effects as "The Social Readjustment Scale," printed in *The Journal of Psychosomatic Research*. The researchers calculated that a score of 150 points indicated a 50/50 chance of the respondent developing an illness or a "health change." A score of 300 would increase the risk to 90 percent.

This type of rating scale continues to be used to help individuals determine their composite stress level within the last year. Since 1967 other researchers have adapted and modified the checklist, but the basic checklist has remained constant. Of course, many factors enter into an individual's response to a particular event, so this scale, partially represented below, can only be used as a guide.

Event Point	Value
Death of spouse	100
Divorce	73
Marital separation	65
Jail term or death of close family member	63
Personal injury or illness	53
Marriage	50
Fired at work	47
Marital reconciliation or retirement	45
Pregnancy	40
Change in financial state	38
Death of close friend	37
Change in employment	36
Foreclosure of mortgage or loan	30
Outstanding personal achievement	28
Trouble with boss	23
Change in work hours or conditions or change in residence or schools	20
Vacation	13
Christmas	12
Minor violations of the law	11

What are the **leading causes of death** in the United States?

According to the most recent figures available (2007), there were 2,423,712 deaths in the United States. The leading cause of death was heart disease followed by cancer, stroke, and chronic respiratory diseases.

Rank	Cause of Death	Number	Percentage of Total
1	Heart disease	616,067	25.4
2	Cancer	562,875	23.2
3	Stroke	135,952	5.6
4	Chronic lower respiratory disease	127,924	5.3

What are the **odds** against being **struck by lightning**?

The National Oceanic and Atmospheric Administration estimates the odds of being struck by lightning at 1 in 750,000 in a given year. However, the odds drop to 1 in 500,000 based on the number of unreported lightning strikes. The odds of being struck by lightning in one's lifetime are 1 in 6,250. During the past 30 years, there has been an average of 58 reported lightning fatalities per year.

> ## Which direction of impact results in the greatest number of fatalities in automobile crashes?
>
> Frontal crashes are responsible for the largest percentage of fatalities in passenger cars.

Are men or women more accident-prone?

Women drive and even cross the street more safely than men. Since 1980, men have accounted for 70 percent of pedestrian fatalities. Between the ages of 18 and 45, males outnumber females as fatal crash victims by almost three to one, according to the National Highway Traffic Safety Administration. Accidental deaths of all types—from falls, firearms, drownings, fires, even food and other poisonings—also are more common among men than women.

What health risks are associated with smoking?

Smoking harms nearly every organ in the body. Tobacco smoking, especially cigarettes, is linked to lung cancer, which is the most common cause of cancer death in most countries. Eighty-seven percent of lung cancer cases are caused by cigarette smoking. More deaths are caused each year by tobacco use than by all deaths from human immunodeficiency virus (HIV), illegal drug use, alcohol use, motor vehicle injuries, suicides, and murders combined. For each pack of cigarettes smoked per day, life expectancy is decreased by seven years. A two-pack-a-day smoker will likely die at 60, compared to 74 for a nonsmoker in the United States. There is an increased incidence of other lung diseases, heart and blood vessel diseases, stroke, and cataracts in smokers. Women who smoke face increased problems with infertility, preterm delivery, stillbirths, babies born with low birth weight, or having a baby die from sudden infant death syndrome (SIDS).

What is the composition of cigarette smoke?

Cigarette smoke contains about 4,000 chemicals. Carbon dioxide, carbon monoxide, methane, and nicotine are some of the major components, with lesser amounts of acetone, acetylene, formaldehyde, propane, hydrogen cyanide, toluene, and many others.

Why does the risk of cancer diminish rapidly after one quits the cigarette smoking habit?

Exposure of a premalignant cell to a promoter converts the cell to an irreversibly malignant state. Promotion is a slow process, and exposure to the promoter must be

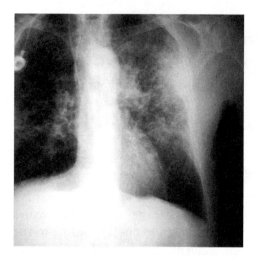

An X ray showing a smoker suffering from lung cancer.

sustained for a certain period of time. This requirement explains why the risk of cancer diminishes rapidly after one quits smoking; both cancer initiators and promoters appear to be contained in tobacco smoke.

Which **sport** has the **highest rate of injuries** and what kind of injury is most common?

Football players suffer more injuries than other athletes, collectively. They have 12 times as many injuries as do basketball players, who have the next highest rate of injury. Knee problems are the most common type of injury, with two-thirds of basketball players' injuries and one-third of football players' injuries being knee-related.

What are the **most frequent sports-related injuries** requiring visits to the **emergency room** for **children** and **young adults**?

Injuries associated with basketball and cycling are the most frequent sports-related injuries requiring visits to emergency rooms among children ages 5 to 14.

Approximate Number of Children's Sports Injuries per Year Requiring Visits to Emergency Rooms

Sport	Number of Emergency Room Visits
Basketball and cycling	475,000
Playground injuries	206,900
Football	194,000
Baseball and softball	117,000
Ice or roller skating, in-line skating, and skateboarding	109,600
Soccer	75,000
Trampolines	80,000
Sledding and skiing/snowboarding	44,000
Gymnastics	23,500

The most common types of sport-related injuries in children are sprains (mostly ankle), muscle strains, bone or growth plate injuries, repetitive motion injuries, and heat-related illness. Baseball has the highest fatality rate among sports for children ages 5–14 with three or four children dying from baseball injuries each year.

How **frequently** is traumatic **brain injury** associated with **sports** injuries?

Traumatic brain injury (TBI) is defined as a blow or jolt to the head or a penetrating head injury that disrupts normal brain function. Nearly one-quarter of all traumatic brain injuries in children and adolescents are the result of a sports-related injury. TBI is the leading cause of death from sports-related injuries in children and adolescents. According to the U.S. Consumer Product Safety Commission, there were 351,922 sports-related head injuries treated at hospital emergency rooms in 2008. The following charts the top sports/recreational activities with the greatest incidence of TBI:

Sport/Recreational Activity	Estimated Number of Head Injuries Treated at Hospital Emergency Rooms (Adults)
Cycling	70,802
Football	40,825
Basketball	27,583
Baseball/softball	26,964
Powered recreational vehicles (ATVs, off-road vehicles, etc.)	25,970

Sport/Recreational Activity	Estimated Number of Head Injuries Treated at Hospital Emergency Rooms (children under 14)
Cycling	34,366
Football	16,902
Skateboards/powered scooters	11,727
Baseball/softball	11,672
Basketball	11,359

How is human **exposure** to **radiation measured**?

The radiation absorbed dose (rad) and the roentgen equivalent man (rem) were used for many years to measure the amount and effect of ionizing radiation absorbed by humans. While officially replaced by the gray and the sievert, both are still used in many reference sources. The rad equals the energy absorption of 100 ergs per gram of irradiated material (an erg is a unit of work or energy). The rem is the absorbed dose of ionizing radiation that produces the same biological effect as one rad of X rays or gamma rays (which are equal). The rem of X rays and gamma rays is therefore equal to the rad; for each type of radiation, the number of rads is multiplied by a specific factor to find the number of rems. The millirem, 0.001 rems, is also frequently used; the average radiation dose received by a person in the United States is about 360 millirems per year. Natural radiation accounts for about 82 percent of a person's yearly exposure, and manufactured sources for 18 percent. Indoor radon has only recently been recognized as a significant source of natural radiation, with 55 percent of the natural radiation coming from this source.

In the SI system (*Système International d'Unités*, or International System of Units), the gray and the sievert are used to measure radiation absorbed; these units have largely superseded the older rad and rem. The gray (Gy), equal to 100 rads, is now the base unit. It is also expressed as the energy absorption of one joule per kilogram of irradiated material. The sievert (Sv) is the absorbed dose of radiation that produces the same biological effect as one gray of X rays or gamma rays. The sievert is equal to 100 rems, and has superseded the rem. The becquerel (Bq) measures the radioactive strength of a source, but does not consider effects on tissue. One becquerel is defined as one disintegration (or other nuclear transformation) per second.

What is the **effect** of **radiation** on humans?

When ionizing radiation penetrates living tissue, random collisions with atoms and molecules in its path cause the formation of ions and reactive radicals. These ions and reactive radicals break chemical bonds and cause other molecular changes that produce biological injury. At the cellular level, radiation exposure inhibits cell division, and produces chromosomal damage and gene mutations as well as various other changes. Large enough doses of ionizing radiation will kill any kind of living cell.

How much **radiation** does the average **dental X ray** emit?

Dental examinations are estimated to contribute 0.15 millirems per year to the average genetically significant dose, a small amount when compared to other medical X rays.

How does the U.S. Environmental Protection Agency (**EPA**) **classify carcinogens**?

A carcinogen is an agent that can produce cancer (a malignant growth or tumor that spreads throughout the body, destroying tissue). The EPA classifies chemical and physical substances according to their toxicity to humans.

EPA Classification System for Carcinogens

Group A. Human carcinogen

This classification indicates that there is sufficient evidence from epidemiological studies to support a cause-effect relationship between the substance and cancer.

Group B. Probable human carcinogen

B_1: Substances are classified as B_1 carcinogens on the basis of sufficient evidence from animal studies, and limited evidence from epidemiological studies.

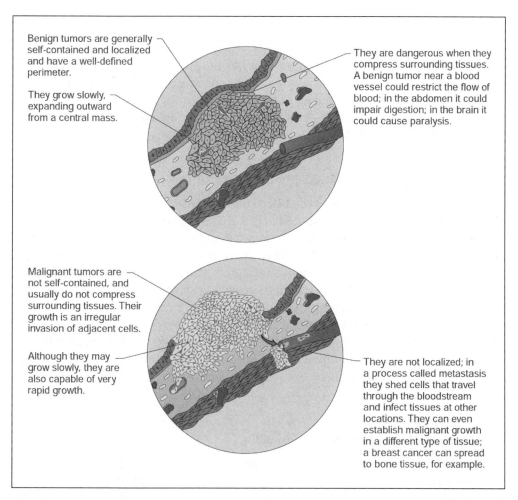

A comparison of benign and malignant tumor characteristics.

B$_2$: Substances are classified as B2 carcinogens on the basis of sufficient evidence from animal studies, with inadequate or nonexistent epidemiological data.

Group C. Possible human carcinogen

For this classification, there is limited evidence of carcinogenicity from animal studies and no epidemiological data.

Group D. Not classifiable as to human carcinogenicity

The data from human epidemiological and animal studies are inadequate or completely lacking, so no assessment as to the substance's cancer-causing hazard is possible.

Group E. Evidence of noncarcinogenicity for humans

Substances in this category have tested negative in at least two adequate (as defined by the EPA) animal cancer tests in different species and in adequate epidemiological and animal studies. Classification in group E is based on available evidence; substances may prove to be carcinogenic under certain conditions.

The National Toxicology Program also has published a report on carcinogens (RoC). It identifies substances that are either "known to be human carcinogen" or "reasonably anticipated to be human carcinogen". The most recent RoC, the eleventh, released in 2005, includes 246 entries; 58 of which are "known to be human carcinogen" and 188 which are "reasonably anticipated to be human carcinogen".

Does the use of **plastic products** present a **health risk?**

The chemical Bisphenol A, also called BPA, has been used to make lightweight, hard plastics and epoxy resins since the 1960s. BPA may be found in many food and beverage packaging including baby bottles, infant feeding cups, water bottles, plastic dinnerware, toys, and other products. The epoxy resins are used in the inside lining of metal food cans, bottle tops, floorings, paints, and water supply pipes. In general, plastic products made with BPA will have a #7 recycling symbol on them or contain the letters "PC" near the recycling symbol. Although many studies have shown BPA to be safe, some recent studies have reported subtle effects of low doses of BPA in laboratory animals. It has not been proven that BPA is harmful to adults or children, but certain precautions have been recommended while research continues. These precautions include not using scratched baby bottles and infant feeding cups and not using BPA containers to heat food.

Are **cell phones dangerous** to your health?

Scientific research has failed to show an association between exposure to radiofrequency energy from a cell phone and health problems. The results of one of the largest studies, released in 2010, of the health effects of cell phone usage found there was little or no health risk, including the development of brain tumors, from cell phone use.

How is body mass index **(BMI) calculated?**

The National Heart, Lung, and Blood Institute (NHLBI), in cooperation with the National Institute of Diabetes and Digestive and Kidney Diseases, released guidelines for weight for adults in 1998. These guidelines define degrees of overweight and obesity in terms of body mass index (BMI). Body mass index is based on weight and height and is strongly correlated with total body fat content. It is used to assess an individual's weight-related level of risk for heart disease, diabetes, and high blood pressure. Very muscular individuals, such as athletes, may have a high BMI without health risks.

The formula to calculate BMI is: BMI = (weight in pounds $\times$ 703)/(height in inches)^2. For example, an individual who is 5 feet, 10 inches and weighs 185 pounds, the calculation would be as follows:

185 pounds $\times$ 703 = 130,055
70^2 = 4,900
130,055/4,900 = 26.5

Evaluating BMI in Adults

BMI	Weight Status
Below 18.5	Underweight
18.5–24.9	Normal
25.0–29.9	Overweight
30.0 and above	Obese

The BMI for children and teens (ages 2 to 20) is age- and gender-specific. Healthcare professionals use established percentile cutoff points to identify underweight and overweight children. The percentile indicates the relative position of the child's BMI relative to children of the same sex and age.

Evaluating BMI in Children and Teenagers Ages 2 to 20

BMI	Weight Status
BMI-for-age < 5th percentile	Underweight
BMI-for-age 5th percentile to < 85th percentile	Normal
BMI-for-age 85th percentile to < 95th percentile	At risk of being overweight
BMI-for-age > 95th percentile	Overweight

BMI	19	20	21	22	23	24	25	26	27	28	29	30
Height (inches)					Body Weight (pounds)							
58	91	96	100	105	110	115	119	124	129	134	138	143
59	94	99	104	109	114	119	124	128	133	138	143	148
60	97	102	107	112	118	123	128	133	138	143	148	153
61	100	106	111	116	122	127	132	137	143	148	153	158
62	104	109	115	120	126	131	136	142	147	153	158	164
63	107	113	118	124	130	135	141	146	152	158	163	169
64	110	116	122	128	134	140	145	151	157	163	169	174
65	114	120	126	132	138	144	150	156	162	168	174	180
66	118	124	130	136	142	148	155	161	167	173	179	186
67	121	127	134	140	146	153	159	166	172	178	186	191
68	125	131	138	144	151	158	164	171	177	184	190	197
69	128	135	142	149	155	162	169	176	182	189	196	203
70	132	139	146	153	160	167	174	181	188	195	202	209
71	136	143	150	157	165	172	179	186	193	200	208	215

Height											
72	140	147	154	162	169	177	184	191	199	206	213	221
73	144	151	159	166	174	182	189	197	204	212	219	227
74	148	155	163	171	179	186	194	202	210	218	225	233
75	152	160	168	176	184	192	200	208	216	224	232	240
76	156	164	172	180	189	197	205	213	221	230	238	246

BMI	31	32	33	34	35	36	37	38	39	40	41	42
Height (inches)					**Body Weight (pounds)**							
58	148	153	158	162	167	172	177	181	186	191	196	201
59	153	158	163	168	173	178	183	188	193	198	203	208
60	158	163	168	174	179	184	189	194	199	204	209	215
61	164	169	174	180	185	190	195	201	206	211	217	222
62	169	175	180	186	191	196	202	207	213	218	224	229
63	175	180	186	191	197	203	208	214	220	225	231	237
64	180	186	192	197	204	209	215	221	227	232	238	244
65	186	192	198	204	210	216	222	228	234	240	246	252
66	192	198	204	210	216	223	229	235	241	247	253	260
67	198	204	211	217	223	230	236	242	249	255	261	268

BMI	31	32	33	34	35	36	37	38	39	40	41	42
Height (inches)					**Body Weight (pounds)**							
68	203	210	216	223	230	236	243	249	256	262	269	276
69	209	216	223	230	236	243	250	257	263	270	277	284
70	216	222	229	236	243	250	257	264	271	278	285	292
71	222	229	236	243	250	257	265	272	279	286	293	301
72	228	235	242	250	258	265	272	279	287	294	302	309
73	235	242	250	257	265	272	280	288	295	302	310	318
74	241	249	256	264	272	280	287	295	303	311	319	326
75	248	256	264	272	279	287	295	303	311	319	327	335
76	254	263	271	279	287	295	304	312	320	328	336	344

BMI	43	44	45	46	47	48	49	50	51	52	53	54
Height (inches)					**Body Weight (pounds)**							
58	205	210	215	220	224	229	234	239	244	248	253	258
59	212	217	222	227	232	237	242	247	252	257	262	267
60	220	225	230	235	240	245	250	255	261	266	271	276
61	227	232	238	243	248	254	259	264	269	275	280	285
62	235	240	246	251	256	262	267	273	278	284	289	295
63	242	248	254	259	265	270	278	282	287	293	299	304
64	250	256	262	267	273	279	285	291	296	302	308	314

Height (inches)	Body Weight (pounds)											
65	258	264	270	276	282	288	294	300	306	312	318	324
66	266	272	278	284	291	297	303	309	315	322	328	334
67	274	280	287	293	299	306	312	319	325	331	338	344
68	282	289	295	302	308	315	322	328	335	341	348	354
69	291	297	304	311	318	324	331	338	345	351	358	365
70	299	306	313	320	327	334	341	348	355	362	369	376
71	308	315	322	329	338	343	351	358	365	372	379	386
72	316	324	331	338	346	353	361	368	375	383	390	397
73	325	333	340	348	355	363	371	378	386	393	401	408
74	334	342	350	358	365	373	381	389	396	404	412	420
75	343	351	359	367	375	383	391	399	407	415	423	431
76	353	361	369	377	385	394	402	410	418	426	435	443

What health **risks** are associated with **obesity**?

People who are obese are more likely to develop a variety of health problems, including:

- Hypertension
- Dyslipidemia (for example, high total cholesterol or high levels of triglycerides)
- Type 2 diabetes
- Coronary heart disease
- Stroke
- Gall bladder disease
- Osteoarthritis
- Sleep apnea and respiratory problems
- Some cancers (endometrial, breast, and colon)

What is **"good"** and **"bad"** cholesterol?

Chemically a lipid, cholesterol is an important constituent of body cells. This fatty substance, produced mostly in the liver, is involved in bile salt and hormone formation, and in the transport of fats in the bloodstream to the tissues throughout the body. Both cholesterol and fats are transported as lipoproteins (units having a core of cholesterol and fats in varying proportions with an outer wrapping of carrier protein [phospholoids and apoproteins]). An overabundance of cholesterol in the bloodstream can be an inherited trait, can be triggered by dietary intake, or can be the result of a metabolic disease, such as diabetes mellitus. Fats (from meat, oil, and dairy products) strongly affect the cholesterol level. High cholesterol levels in the blood may lead to a narrowing of the inner lining of the coronary arteries from the buildup of a fatty tissue called atheroma. This increases the risk of coronary heart disease or stroke. However, if most cholesterol in the blood is in the form of high density lipoproteins (HDL), then it seems to protect against arterial disease. HDL picks up cholesterol in the arteries and brings it back to the liver for excretion or reprocessing. HDL is referred to as "good cholesterol." Conversely, if most cholesterol is in the form of low-density lipoproteins (LDL), or very-low-density lipoproteins (VLDL), then arteries can become clogged. "Bad cholesterol" is the term used to refer to LDL and VLDL.

How does **blood alcohol level affect** the body and behavior?

The effects of drinking alcoholic beverages depend on body weight and the amount of actual ethyl alcohol consumed. The level of alcohol in the blood is calculated in milligrams (one milligram equals 0.035 ounce) of pure (ethyl) alcohol per deciliter (3.5 fluid ounces), commonly expressed in percentages.

Number of Drinks	Blood Alcohol Level	Effect of Drinks
1	0.02–.03%	Changes in behavior, coordination, and ability to think clearly
2	0.05%	Sedation or tranquilized feeling
3	0.08–0.10%	Legal intoxication in many states
5	0.15–0.20%	Person is obviously intoxicated and may show signs of delirium
12	0.30–0.40%	Loss of consciousness
24	0.50%	Heart and respiration become so depressed that they cease to function and death follows

FIRST AID AND POISONS

Who **discovered cardiopulmonary resuscitation** (CPR) as a method to resuscitate an individual whose heart had stopped?

Cardiopulmonary resuscitation (CPR) is a first-aid technique that combines mouth-to-mouth resuscitation and rhythmic compression of the chest to a person whose heart has stopped. The Scottish surgeon William Tossach (c. 1700–after 1771) first performed mouth-to-mouth resuscitation in 1732. The technique was not further developed (or widely used) for many centuries until Dr. Edward Schafer (1850–1935) developed a method of chest pressure to stimulate respiration. In 1910 the American Red Cross adopted and began to teach Schafer's method. A team of specialists at Johns Hopkins Medical School, Orthello R. Langworthy (1897–1996), R.D. Hooker, and William B. Kouwenhoven (1886–1975), attempted to improve on the technique. Kouwenhoven realized that chest compression could maintain blood flow in a person whose heart had stopped. In 1958 Kouwenhoven's method of chest compression was used on a two-year-old child whose heart had stopped. The American Red Cross endorsed the technique in 1963.

What is the "ABCD" survey first responders use to evaluate an emergency?

- "A" stands for airway. It is important to be certain the airway from the mouth or nose to the lungs is clear. The airway can be opened by tilting the head back and lifting the chin.
- "B" stands for breathing. Be certain the person is breathing or perform rescue breathing (CPR) to ensure a supply of oxygen.
- "C" stands for circulation. If a pulse cannot be found, then there is no blood circulating. Emergency personnel can attempt to get the heart to resume breathing by performing rhythmic chest thrusts (CPR). Adults require 15 chest compressions for every two rescue breaths. It also means to check for profuse bleeding which must be controlled.
- "D" stands for disability. It involves checking for consciousness and the likelihood of spinal cord or neck injury.

What is the Heimlich maneuver?

This effective first-aid technique to resuscitate choking and drowning victims was introduced by Dr. Henry J. Heimlich (1920–) of Xavier University in Cincinnati, Ohio. It is a technique for removing a foreign body from the trachea or pharynx where it is preventing flow of air to the lungs. When the victim is in the vertical position, the maneuver consists of applying subdiaphragmatic pressure by wrapping one's arms around the victim's waist from behind, making a fist with one hand and placing it against the victim's abdomen between the navel and the rib cage, clasping one's fist with the other hand, and pressing in with a quick, forceful thrust. Repeat several times if necessary. When the victim is in the horizontal position (which some experts recommend), the rescuer straddles the victim's thighs.

What safety rules should be observed during a thunderstorm?

These safety rules should be observed when lightning threatens:

1. Stay indoors. Seek shelter in buildings. If no buildings are available, the best protection is a cave, ditch, canyon, or under head-high clumps of trees in open

forest glades. If there is no shelter, avoid the highest object in the area. Keep away from isolated trees.

2. Get out of the water and off small boats.
3. Do not use the telephone.
4. Do not use metal objects like fishing rods and golf clubs.
5. Stay in your automobile if you are traveling.
6. Do not use plug-in electrical equipment like hair dryers, electric razors, or electric toothbrushes during the storm.

How is **activated charcoal** used **medically**?

Activated charcoal is an organic substance, such as burned wood or coal, that has been heated to approximately 1,000°F (537°C) in a controlled atmosphere. The result is a fine powder containing thousands of pores that have great absorbent qualities to rapidly absorb toxins and poisons. Activated charcoal is used medically in the treatment of drug overdoses and poisonings.

What is the **deadliest natural toxin**?

Botulinal toxin, produced by the bacterium *Clostridium botulinum*, is the most potent poison of humans. It has an estimated lethal dose in the bloodstream of 10^{-9} milligrams per kilogram. It causes botulism, a severe neuroparalytic disease that travels to the junctions of skeletal muscles and nerves, where it blocks the release of the neurotransmitter acetylcholine, causing muscle weakness and paralysis, and impairing vision, speech, and swallowing. Death occurs when the respiratory muscles are paralyzed; this usually occurs during the first week of illness. Mortality from botulism is about 25 percent.

Because the bacterium can form the toxin only in the absence of oxygen, canned goods and meat products wrapped in airtight casings are potential sources of botulism. The toxin is more likely to grow in low-acid foods, such as mushrooms, peas, corn, or beans, rather than high-acid foods like tomatoes. However, some new tomato hybrids are not acidic enough to prevent the bacteria from forming the toxin. Foods being canned must be heated to a temperature high enough and for a long enough time to kill the bacteria present. Suspect food includes any canned or jarred food product with a swollen lid or can. Ironically, this dreaded toxin in tiny doses is being used to treat disorders that bring on involuntary muscle contractions, twisting, etc. The United States Food and Drug Administration (FDA) has approved the toxin for the treatment of strabismus (misalignment of the eyes), blepharospasm (forcible closure of eyelids), and hemifacial spasm (muscular contraction on one side of the face).

What are some **common causes** of **poisoning**?

Poisoning is defined as the exposure to any substance in sufficient quantity to cause adverse health effects. Poisonings can be grouped into several different categories,

It's not too smart to play golf during a lightning storm. For the sake of safety, stay indoors and away from plug-in electronics.

including intentional, accidental, occupational and environmental, social and iatro-genic. Accidental poisonings are the most common, with more than 90 percent occurring in children at home. Intentional poisonings are usually suicide-related, with carbon monoxide being one of the most frequently used agents. Toxic chemical releases in industrial accidents are an occupational and environmental hazard.

How was **Mr. Yuk**, the symbol for dangerous or poisonous products, developed?

The Pittsburgh Poison Center developed the Mr. Yuk symbol in 1971. The depiction of a green, gagging face with an extended tongue is used by Children's Hospital to promote prevention education through affiliated hospitals and poison centers. In the testing program, children at day care centers selected Mr. Yuk as the symbol that represented the most unappealing product. Among the other symbols used were a red stop sign and a skull and crossbones. Interestingly, the children found the skull and crossbones the most appealing symbol.

How deadly is **strychnine**?

The fatal dose of strychnine or deadly nightshade (the plant from which it is obtained) is 0.0005 to 0.001 ounces (15 to 30 milligrams). It causes severe convulsions and respiratory failure. If the patient lives for 24 hours, recovery is probable.

593

The most common opinion today is that Napoleon Bonaparte (1769–1821), Emperor of France from 1804 to 1815, died of a cancerous, perforated stomach. A significant minority of doctors and historians have made other claims ranging from various diseases to benign neglect to outright homicide. A Swedish toxicologist, Sten Forshufvud (1903–1985), advanced the theory that Napoleon died from arsenic administered by an agent of the French Royalists who was planted in Napoleon's household during his final exile on the island of St. Helena.

Which part of **mistletoe** is **poisonous**?

The white berries contain toxic amines, which cause acute stomach and intestinal irritation with diarrhea and a slow pulse. Mistletoe should be considered a potentially dangerous Christmas decoration, especially if children are around.

What is the **poison on arrows** used by **South American Indians** to kill prey and enemies?

The botanical poison used by the Aucas and similar tribes in the South American jungles is curare. It is a sticky, black mixture with the appearance of licorice and is processed from either of two different vines. One is a liana (*Chondodendron tomentosum*); the other is a massive, tree-like vine (*Strychonos quianensis*).

How deadly is *Amanita phalloides*?

The poisonous mushroom *Amanita phalloides* has a fatality rate of about 50 percent. Ingestion of part of one mushroom may be sufficient to cause death. Over 100 fatalities occur each year from eating poisonous mushrooms, with more than 90 percent caused by the *Amanita phalloides* group.

What **first-aid** remedies may be used for **bee stings**?

If a person is allergic to bee stings, he or she should seek professional medical care immediately. For persons not allergic to bee stings, the following steps may be taken: The stinger should be removed by scraping it with a knife, a long fingernail, or a credit card, rather than by trying to pull it out. A wet aspirin may be rubbed on the area of the sting to help neutralize some of the inflammatory agents in the venom (unless the person is allergic or sensitive to aspirin taken by mouth).

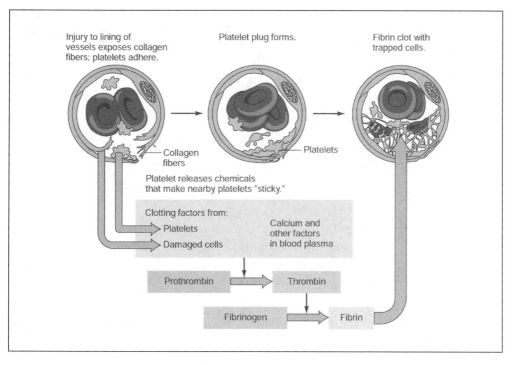

Injury to lining of vessels exposes collagen fibers; platelets adhere.

Platelet plug forms.

Fibrin clot with trapped cells.

Collagen fibers

Platelets

Platelet releases chemicals that make nearby platelets "sticky."

Clotting factors from:

Platelets

Damaged cells

Calcium and other factors in blood plasma

Prothrombin → Thrombin

Fibrinogen → Fibrin

Blood platelets are key to the body's ability to heal wounds.

A paste made of meat tenderizer (or another product that contains papain) mixed with water will relieve the pain. Adults may take an antihistamine along with a mild pain reliever such as aspirin, ibuprofen, or acetaminophen.

Which **first-aid** measures can be used for a bite by a **black widow spider**?

The black widow spider (*Latrodectus mactans*) is common throughout the United States. Its bite is severely poisonous, but no first-aid measures are of value. Age, body size, and degree of sensitivity determine the severity of symptoms, which include an initial pinprick with a dull numbing pain, followed by swelling. An ice cube may be placed over the bite to relieve pain. Between 10 and 40 minutes after the bite, severe abdominal pain and rigidity of stomach muscles develop. Muscle spasms in the extremities, ascending paralysis, and difficulty in swallowing and breathing follow. The mortality rate is less than one percent, but anyone who has been bitten should see a doctor; the elderly, infants, and those with allergies are most at risk, and should be hospitalized.

How do **wounds heal**?

Damage to tissue, such as a cut in the skin, begins to heal with the formation of a sticky lump known as a blood clot. Blood clots prevent blood and other fluids from leaking out.

595

Microscopic sticky threads of the clotting protein fibrin make a tangled mesh that traps blood cells. Within a short time the clot begins to take shape, harden, and become more solid. The clot turns into a scab as it dries and hardens. Skin cells beneath the scab multiply to repair the damage. When the scab falls off the wound will be healed.

How can the amount of **lead** in tap water be **reduced** in an older house having **lead-containing pipes**?

The easiest way is to let the tap run until the water becomes very cold before using it for human consumption. By letting the tap run, water that has been in the lead-containing pipes for awhile is flushed out. Also, cold water, being less corrosive than warm, contains less lead from the pipes. Lead (Pb) accumulates in the blood, bones, and soft tissues of the body as well as the kidneys, nervous system, and blood-forming organs. Excessive exposure to lead can cause seizures, mental retardation, and behavior disorders. Infants and children are particularly susceptible to low doses of lead and suffer from nervous system damage.

Another source of lead poisoning is old, flaking lead paint. Lead oxide and other lead compounds were added to paints before 1950 to make the paint shinier and more durable. Improperly glazed pottery can be a source of poisoning, too. Acidic liquids such as tea, coffee, wine, and juice can break down the glazes so that the lead can leak out of the pottery. The lead is ingested little by little over a period of time. People can also be exposed to lead in the air. Lead gasoline additives, nonferrous smelters, and battery plants are the most significant contributors of atmospheric lead emissions.

How did **lead contribute** to the **fall** of the **Roman Empire**?

Some believe Romans from the period around 150 B.C.E. may have been victims of lead poisoning. Symptoms of lead poisoning include sterility, general weakness, apathy, mental retardation, and early death. The lead could have been ingested in water taken from lead-lined water pipes or from food cooked in their lead-lined cooking pots or from wine served in lead-lined goblets. Unaware of its dangers, some ancient Romans unwittingly used lead as a sweetening agent or medical treatment for diarrhea. Lead poisoning could have caused infertility in women, leading to a subsequent long-term decline in the birth rate of the Roman upper classes. The effect of this inadvertent toxic food additive on Roman history, however, is only speculative.

DISEASES, DISORDERS, AND OTHER HEALTH PROBLEMS

Which **disease** is the **most common**?

The most common noncontagious disease is periodontal disease, such as gingivitis or inflammation of the gums. Few people in their lifetime can avoid the effects of tooth decay. The most common contagious disease in the world is coryza or the common cold. There are nearly 62 million cases of the common cold in the United States annually.

Which **disease** is the **deadliest**?

The most deadly infectious disease was the pneumonic form of the plague, the so-called Black Death of 1347–1351, with a mortality rate of 100 percent. Today, the disease with the highest mortality (almost 100 percent) is rabies in humans when it prevents the victim from swallowing water. This disease is not to be confused with being bitten by a rabid animal. With immediate attention, the rabies virus can be prevented from invading the nervous system and the survival rate in this circumstance is 95 percent. AIDS (acquired immunodeficiency syndrome), first reported in 1981, is caused by HIV (the human immunodeficiency virus). In 1993, HIV infection was the most common cause of death among persons aged 25 to 44 years. In 1999 alone, 14,802 U.S. residents died from the AIDS/HIV infection, according to the National Center for Health Statistics. Although still a significant cause of death among persons aged 25 to 44, it is no longer the most common cause of death.

What is the **difference** between human immunodeficiency virus (**HIV**) and **AIDS**?

The term AIDS applies to the most advanced stages of HIV infection. The Center for Disease Control (CDC) definition of AIDS includes all HIV-infected people who have fewer than 200 CD41 T cells per cubic millimeter of blood. (Healthy adults usually have CD41 T cell counts of 1,000 or more.) The definition also includes 26 clinical conditions (mostly opportunistic infections) that affect people with advanced HIV disease.

597

How many **individuals** are **infected** with **HIV/AIDS**?

At the end of 2009, it was estimated that 33.3 million people worldwide were living with HIV/AIDS. Adults accounted for 30.8 million cases of HIV/AIDS (51 percent or 15.9 million are women) and children under 15 accounted for 2.5 million cases. Since 1981, more than 25 million people have died from AIDS. The states/territories in the United States with the greatest incidence of AIDS cases are reported in the following table:

State	Number of Cumulative AIDS Cases*
New York	192,753
California	160,293
Florida	117,612
Texas	77,070
New Jersey	54,557
Georgia	38,300
Pennsylvania	38,217
Illinois	37,880
Maryland	35,725
Puerto Rico	32,463

*Through 2008 (most recent figures available).

What are the **symptoms** and signs of **AIDS**?

The early symptoms (AIDS-related complex, or ARC, symptoms) include night sweats, prolonged fevers, severe weight loss, persistent diarrhea, skin rash, persistent cough, and shortness of breath. The diagnosis changes to AIDS (acquired immunodeficiency syndrome) when the immune system is affected and the patient becomes susceptible to opportunistic infections and unusual cancers, such as herpes viruses (herpes simplex, herpes zoster, cytomegalovirus infection), *Candida albicans* (fungus) infection, *Cryptosporidium enterocolitis* (protozoan intestinal infection), *Pneumocystis carinii* pneumonia (PCP, a common AIDS lung infection), toxoplasmosis (protozoan brain infection), progressive multifocal leukoencephalopathy (PML, a central nervous system disease causing gradual brain degeneration), *Mycobacterium avium intracellulare* infection (MAI, a common generalized bacterial infection), and Kaposi's sarcoma (a malignant skin cancer characterized by blue-red nodules on limbs and body, and internally in the gastrointestinal and respiratory tracts, where the tumors cause severe internal bleeding).

The signs of AIDS are generalized swollen glands, emaciation, blue or purple-brown spots on the body, especially on the legs and arms, prolonged pneumonia, and oral thrush.

> ## What was the contribution of
> ## Dr. Gorgas to the building of the Panama Canal?
>
> **D**r. William C. Gorgas (1854–1920) brought the endemic diseases of Panama under control by destroying mosquito breeding grounds, virtually eliminating yellow fever and malaria. His work was probably more essential to the completion of the canal than any engineering technique.

How is the **term zoonosis** defined?

A zoonosis is any infectious disease or parasitic disease of animals that can be transmitted to humans. Lyme disease and Rocky Mountain spotted fever, for example, are indirectly spread to humans from an animal through the bite of a tick. Common household pets also can directly transmit diseases to humans unless preventive measures are taken. Cat-scratch fever and toxoplasmosis may be contracted from cats. Wild animals and dogs can transmit rabies. However, most zoonosis diseases are relatively rare and can be treated once detected. Such sensible actions as regularly vaccinating pets and wearing long-sleeved shirts and pants when hiking can prevent the spread of most zoonoses.

What is meant by **vectors** in medicine?

A vector is an animal that transmits a particular infectious disease. A vector picks up disease organisms from a source of infection, carries them within or on its body, and later deposits them where they infect a new host. Mosquitoes, fleas, lice, ticks, and flies are the most important vectors of disease to humans.

Which **species of mosquito** causes **malaria** and **yellow fever** in humans?

The bite of the female mosquito of the genus *Anopheles* can contain the parasite of the genus *Plasmodium*, which causes malaria, a serious tropical infectious disease affecting 200 to 300 million people worldwide. More than one million African babies and children die from the disease annually. The *Aedes aegypti* mosquito transmits yellow fever, a serious infectious disease characterized by jaundice, giving the patient yellowish skin; ten percent of the patients die.

What is **"mad cow disease"** and how does it affect humans?

Mad cow disease, bovine spongiform encephalopathy (BSE), is a cattle disease of the central nervous system. First identified in Britain in 1986, BSE is a transmissible spongiform encephalopathy (TSE), a disease characterized by the damage caused to

599

the brain tissue. The tissue is pierced with small holes like a sponge. The disease is incurable, untreatable, and fatal. Researchers believe BSE is linked to Creutzfeldt-Jakob disease (CJD) in humans through the consumption of contaminated bovine products. CJD is a fatal illness marked by brain tissue deterioration and progressive degeneration of the central nervous system.

How is **Lyme disease carried**?

The cause of Lyme disease is the spirochete *Borrelia burgdorferi* that is transmitted to humans by the small tick *Ixodes dammini* or other ticks in the Ixodidae family. The tick injects spirochete-laden saliva into the bloodstream or deposits fecal matter on the skin. This multisystemic disease usually begins in the summer with a skin lesion called erythema chronicum migrans (ECM), followed by more lesions, a malar rash, conjunctivitis, and urticaria. The lesions are eventually replaced by small red blotches. Other common symptoms in the first stage include fatigue, intermittent headache, fever, chills, and muscle aches.

In stage two, which can be weeks or months later, cardiac or neurologic abnormalities sometimes develop. In the last stage (weeks or years later) arthritis develops with marked swelling, especially in the large joints. If tetracycline, penicillin, or erythromycin is given in the early stages, the later complications can be minimized. High dosage of intravenously given penicillin can also be effective on the late stages.

When were the **first cases** of **West Nile virus reported** in the United States?

The first cases of West Nile virus were identified in 1999 in the New York City area. West Nile virus is primarily a disease of birds found in Africa, West Asia, and the Middle East. It is transmitted to humans mainly via mosquito bites (mainly from the species *Culex pipiens*). The female mosquito catches the virus when it bites an infected bird and then passes it on when it later bites a human. In humans it causes encephalitis, an infection of the brain that can be lethal.

Why is **Legionnaire's disease** known by that name?

Legionnaire's disease was first identified in 1976 when a sudden, virulent outbreak of pneumonia took place at a hotel in Philadelphia, Pennsylvania, where delegates to an American Legion convention were staying. The cause was eventually identified as a previously unknown bacterium that was given the name *Legionnella pneumophilia*. The bacterium probably was transmitted by an airborne route. It can spread through cooling tower or evaporation condensers in air-conditioning systems, and has been known to flourish in soil and excavation sites. Usually, the disease occurs in late summer or early fall, and its severity ranges from mild to life-threatening, with a mortality rate as high as 15 percent. Symptoms include diarrhea, anorexia, malaise, headache,

Who was Typhoid Mary?

Mary Mallon (1855–1938), a cook who lived in New York City at the turn of the century, was identified as a chronic carrier of the typhoid bacilli. Immune to the disease herself, she was the cause of at least three deaths and 51 cases of typhoid fever. She was confined to an isolation center on North Brother Island, near the Bronx, from 1907 to 1910 and from 1914 to 1938. The New York City Health Department released her after the first confinement on the condition that she never accept employment that involved handling food. But when a later epidemic occurred at two places where she had worked as a cook, the authorities returned her to North Brother Island, where she remained until her death from a stroke in 1938.

generalized weakness, recurrent chills, and fever accompanied by cough, nausea, and chest pain. Antibiotics such as Erythroycin™ are administered along with other therapies (fluid replacement, oxygen, etc.) that treat the symptoms.

Which name is now used as a **synonym for leprosy**?

Hansen's disease is the name of this chronic, systemic infection characterized by progressive lesions. Caused by a bacterium, *Mycobacterium leprae*, which is transmitted through airborne respiratory droplets, the disease is not highly contagious. Continuous close contact is needed for transmittal. Antimicrobial agents, such as sulfones (dapsone in particular), are used to treat the disease.

How many **types of herpes** virus are there?

There are five human herpes viruses:

Herpes simplex type 1—causes recurrent cold sores and infections of the lips, mouth, and face. The virus is contagious and spreads by direct contact with the lesions or fluid from the lesions. Cold sores are usually recurrent at the same sites and occur where there is an elevated temperature at the affected site, such as with a fever or prolonged sun exposure. Occasionally this virus may occur on the fingers with a rash of blisters. If the virus gets into the eye, it could cause conjunctivitis, or even a corneal ulcer. On rare occasions, it can spread to the brain to cause encephalitis.

Herpes simplex type 2—causes genital herpes and infections acquired by babies at birth. The virus is contagious and can be transmitted by sexual intercourse. The virus produces small blisters in the genital area that burst to leave small painful ulcers, which heal within ten days to three weeks. Headache, fever, enlarged

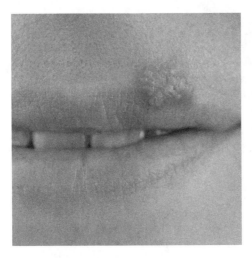

Cold sores are caused by herpes simplex type 1, which is quite common and usually not dangerous, except for rare instances when it can spread to the brain and cause encephalitis.

lymph nodes, and painful urination are the other symptoms.

Varicella-zoster (Herpes zoster)—causes chicken pox and shingles. Shingles can be caused by the dormant virus in certain sensory nerves that re-emerge with the decline of the immune system (because of age, certain diseases, and the use of immunosuppressants), excessive stress, or use of corticosteroid drugs. The painful rash of small blisters dry and crust over, eventually leaving small pitted scars. The rash tends to occur over the rib area or a strip on one side of the neck or lower body. Sometimes it involves the lower half of the face and can affect the eyes. Pain that can be severe and long-lasting affects about half of the sufferers and is caused by nerve damage.

Epstein-Barr—causes infectious mononucleosis (acute infection having high fever, sore throat, and swollen lymph glands, especially in the neck, which occurs mainly during adolescence) and is associated with Burkitt's lymphoma (malignant tumors of the jaw or abdomen that occur mainly in African children and in tropical areas).

Cytomegalovirus—usually results in no symptoms but enlarges the cells it infects; it can cause birth defects when a pregnant mother infects her unborn child.

Three other human herpes viruses are also known: Human herpes virus 6 (HHV-6), commonly associated with roseola, and human herpes viruses 7 and 8 (HHV 7/8), whose disease association is not yet understood. Herpes gestationis is a rare skin-blister disorder occurring only in pregnancy and is not related to the herpes simplex virus.

How are **warts caused**?

A wart is a lump on the skin produced when one of the 30 types of papillomavirus invades skin cells and causes them to multiply rapidly. There are several different types of warts: common warts, usually on injury sites; flat warts on hands, accompanied by itching; digitate warts having fingerlike projections; filiform warts on eyelids, armpits, and necks; plantar warts on the soles of the feet; and genital warts, pink cauliflower-like areas that, if occurring in a woman's cervix, could predispose her to cervical cancer. Each is produced by a specific virus, and most are usually symptomless. Wart viruses are spread by touch or by contact with the skin shed from a wart.

What is the difference between **Type I** and **Type II diabetes**?

Type I is insulin-dependent diabetes mellitus (IDDM) and Type II is non-insulin-dependent diabetes mellitus (NIDDM). In Type I diabetes there is an absolute deficiency of insulin. It accounts for approximately ten percent of all cases of diabetes and has a greater prevalence in children. In Type II diabetes, insulin secretion may be normal, but the target cells for insulin are less responsive than normal. The incidence of Type II diabetes increases greatly after age 40 and is normally associated with obesity and lack of exercise as well as genetic predisposition. The symptoms of Type II diabetes are usually less severe than Type I, but long-term complications are similar in both types.

What causes a **stomach ulcer**?

For decades, doctors thought that genetics or anxiety or even spicy foods caused stomach ulcers. Scientists now believe that stress and spicy foods only worsen the pain of an ulcer. The gastric ulcer itself is caused by a bacterium called *Helicobacter pylori*. Researcher Barry Marshall (1951–) of Australia observed that many ulcer patients had these bacteria present in their systems. In 1984, he designed an experiment to determine whether there was a link between *Helicobacter pylori* and stomach ulcers. He consumed a large amount of the bacteria. He developed ulcers ten days later. Ulcers are now treated with antibiotics. Marshall shared the 2005 Nobel Prize in Physiology or Medicine with J. Robin Warren (1937–) for their discovery of the bacterium *Helicobacter pylori* and its role in gastritis and peptic ulcer disease.

How **serious** is **osteoporosis**?

Osteoporosis (from the Greek *osteo*, meaning "bone," *por*, meaning "passageway," and *osis*, meaning "condition") is a condition that reduces bone mass because the rate of bone reabsorbtion is quicker than the rate of bone deposition. The bones become very thin and porous and are easily broken. Osteoporosis is most common in the elderly, who may experience a greater number of broken bones as a result of the mechanical stresses of daily living and not from accidents or other trauma. Generally, osteoporosis is more severe in women, since their bones are thinner and less massive than men's bones. In addition, estrogen helps to maintain bone mass, so the loss of estrogen in women after menopause contributes to more severe osteoporosis.

How **prevalent** is **back pain** in the United States?

Studies have found that low back pain is the most common source of pain. It is also the leading cause of disability in Americans under 45 years old. Back pain may be the result of an injury or associated with certain diseases and conditions. Some conditions and diseases which may cause back pain are scoliosis, arthritis, spinal stenosis, pregnancy, kidney stones, endometriosis, fibromyalgia, infections, tumors, or stress. Tense muscles,

603

spasms, ruptured disks, and other disk problems are also causes of back pain. Exercise, a diet to maintain a healthy weight, and learning to lift properly will prevent back pain.

What is **carpal tunnel syndrome**?

Carpal tunnel syndrome occurs when a branch of the median nerve in the forearm is compressed at the wrist as it passes through the tunnel formed by the wrist bones (or carpals), and a ligament that lies just under the skin. The syndrome occurs most often in middle age and more so in women than men. The symptoms are intermittent at first, then become constant. Numbness and tingling begin in the thumb and first two fingers; then the hand and sometimes the whole arm becomes painful. Treatment involves wrist splinting, weight loss, and control of edema; treatments for arthritis may help also. If not, a surgical procedure in which the ligament at the wrist is cut can relieve pressure on the nerve. Those who work continuously with computer keyboards are particularly vulnerable to carpal tunnel syndrome. To minimize the risk of developing this problem, operators should keep their wrists straight as they type, rather than tilting the hands up. It is also best to place the keyboard at a lower position than a standard desktop.

Which is the **most common** type of **arthritis**?

Arthritis (from the Greek *arthro*, meaning "joint," and *itis*, meaning "inflammation") is a group of diseases that affect synovial joints. Arthritis may originate from an infection, an injury, metabolic problems, or autoimmune disorders. The most common type of arthritis is osteoarthritis. Osteoarthritis is a chronic, degenerative disease most often beginning as part of the aging process. Often referred to as "wear and tear" arthritis because it is the result of life's everyday activities, it is a degradation of the articular cartilage that protects the bones as they move at a joint site. Osteoarthritis usually affects the larger, weight-bearing joints first, such as the hips, knees, and lumbar region of the vertebral column.

How does **tennis elbow** differ from **golfer's elbow**?

Tennis elbow and golfer's elbow are both injuries caused by overuse of the muscles of the forearm. Tennis elbow, also called lateral epicondylitis, affects the muscles of the forearm, which attach to the bony prominence on the outside of the elbow. Golfer's elbow, also called medial epicondylitis, affects the muscles of the forearm that attach to the inside of the elbow.

What are the **symptoms** of **Guillain-Barré syndrome**?

Guillain-Barré syndrome is an autoimmune disorder where the body's immune system attacks part of the peripheral nervous system. The immune system starts to destroy the myelin sheath that surrounds the axons of many peripheral nerves, or even the axons themselves. The loss of the myelin sheath surrounding the axons slows down the transmission of nerve signals and muscles begin to lose their ability to respond to the brain's commands. The first symptoms of this disorder include varying degrees of weakness or tingling sensations in the legs. In many instances the weakness and abnormal sensations spread to the arms and upper body. In severe cases the patient may be almost totally paralyzed since the muscles cannot be used at all. In these cases the disorder is life threatening because it potentially interferes with breathing and, at times, with blood pressure or heart rate. Such a patient is often put on a respirator to assist with breathing and is watched closely for problems such as an abnormal heart beat, infections, blood clots, and high or low blood pressure. Most patients, however, recover from even the most severe cases of Guillain-Barré syndrome, although some continue to have a certain degree of weakness.

How serious is **Bell's palsy**?

Bell's palsy, a form of temporary facial paralysis, is the result of damage or trauma to one of the cranial nerves. The nerve may be swollen, inflamed, or compressed, resulting in an interruption of messages from the brain to the facial muscles. Individuals with Bell's palsy may exhibit twitching, weakness, or paralysis on one or both sides of the face; drooping of the eyelid and corner of the mouth; drooling; dryness of the eye or mouth; impairment of taste; and excessive tearing in one eye. Although the symptoms appear suddenly, individuals begin to recover within two weeks and return to normal function within three to six months.

What is **Lou Gehrig's disease**?

Amyotrophic lateral sclerosis (ALS), sometimes called Lou Gehrig's (1903–1941) disease after the New York Yankees baseball player who retired from baseball in 1939 after being diagnosed with ALS, is a motor neuron disease of middle or late life. It results from a progressive degeneration of nerve cells controlling voluntary motor functions that ends in death three to ten years after onset. There is no cure for it. At the beginning of the disease, the patient notices weakness in the hands and arms, with involuntary muscle quivering and possible muscle cramping or stiffness. Eventually all four extremities become involved. As nerve degeneration progresses, disability occurs and physical independence declines until the patient, while mentally and intellectually aware, can no longer move, swallow, or breathe.

What are the **causes** of **epilepsy**?

Epilepsy is a brain disorder in which clusters of neurons in the brain sometimes signal abnormally. Epilepsy may develop because of an abnormality in brain wiring, an imbalance of neurotransmitters, or some combination of these factors. During an epileptic seizure, neurons may fire as many as 500 times a second, much faster than the normal rate of about 80 times a second. When the normal pattern of neuronal activity becomes disturbed, strange sensations, emotions and behavior, convulsions, muscle spasms, and loss of consciousness may be experienced.

What is a **concussion**?

A concussion is an injury to the brain caused by a blow or jolt to the head that disrupts the normal functioning of the brain. Concussions are usually not life-threatening. Since the brain is very complex, there is great variation in the signs and symptoms of a concussion. Some people lose consciousness; others never lose consciousness. Some symptoms may appear immediately, while others do not appear for several days or even weeks. Symptoms include:

- Headaches or neck pain that will not go away
- Difficulty with mental tasks such as remembering, concentrating, or making decisions
- Slowness in thinking, speaking, acting, or reading
- Getting lost or easily confused
- Feeling tired all of the time, having no energy or motivation
- Mood changes (feeling sad or angry for no reason)
- Changes in sleep patterns (sleeping a lot more or having a hard time sleeping)
- Light-headedness, dizziness, or loss of balance
- Urge to vomit (nausea)
- Increased sensitivity to lights, sounds, or distractions
- Blurred vision or eyes that tire easily
- Loss of sense of smell or taste
- Ringing in the ears

What are the **two forms** of **stroke**?

There are two forms of stroke: ischemic and hemorrhagic. Ischemic stroke is the blockage of a blood vessel that supplies blood to the brain. Ischemic strokes account for 80 percent of all strokes. Hemorrhagic stroke is bleeding into or around the brain.

Hemorrhagic strokes account for 20 percent of all strokes. The symptoms of stroke appear suddenly and include numbness or weakness, especially on one side of the body; confusion or trouble speaking or understanding speech; trouble seeing in one or both eyes; trouble walking, dizziness, or loss of balance or coordination; or severe headache with no known cause. Often more than one of these symptoms will be present, but they all appear suddenly.

What are **two** of the most **common forms** of **dementia**?

The term "dementia" describes a group of symptoms that are caused by changes in brain function. The two most common forms of dementia in older people are Alzheimer's disease and multi-infarct dementia (sometimes called vascular dementia). There is no cure for these types of dementia. In Alzheimer's disease nerve-cell changes in certain parts of the brain result in the death of a large number of cells. Some researchers believe there is a genetic origin to Alzheimer's disease. An estimated 5.3 million Americans of all ages have Alzheimer's. It is the sixth leading cause of death in the United States and was reported as the cause of death for 74,632 people in 2007. The symptoms of Alzheimer's disease range from mild forgetfulness to serious impairments in thinking, judgment, and the ability to perform daily activities.

In multi-infarct dementia a series of small strokes or changes in the brain's blood supply may result in the death of brain tissue. The location in the brain where the small strokes occur determines the seriousness of the problem and the symptoms that arise. Symptoms that begin suddenly may be a sign of this kind of dementia. People with multi-infarct dementia are likely to show signs of improvement or remain stable for long periods of time, then quickly develop new symptoms if more strokes occur. In many people with multi-infarct dementia, high blood pressure is to blame.

What are the **seven warning signs** of **Alzheimer's disease**?

The seven warning signs of Alzheimer's disease are:

1. Asking the same question over and over again
2. Repeating the same story, word for word, again and again
3. Forgetting how to cook, how to make repairs, how to play cards, or any other activities that were previously done with ease and regularity
4. Losing one's ability to pay bills or balance one's checkbook
5. Getting lost in familiar surroundings, or misplacing household objects
6. Neglecting to bathe, or wearing the same clothes over and over again, while insisting that one has taken a bath or that clothes are still clean
7. Relying on someone else, such as a spouse, to make decisions

It is important to understand that even if someone has several or even most of these symptoms, it does not mean they definitely have Alzheimer's disease. It does

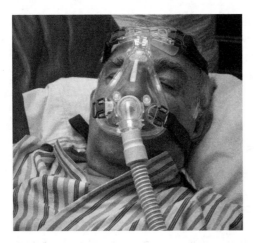

A man suffering from sleep apnea wears an oxygen mask to help alleviate his symptoms.

mean they should be thoroughly examined by a medical specialist trained in evaluating memory disorders, such as a neurologist or a psychiatrist, or by a comprehensive memory disorder clinic with an entire team of experts knowledgeable about memory problems.

What are some **sleep disorders**?

The most common sleep disorder is insomnia. Insomnia is ongoing difficulty in falling asleep, staying asleep, or restless sleep. Technically, insomnia is a symptom of other sleep disorders. Consequently, treatment for insomnia depends on the primary cause of insomnia, which may be stress, depression, or too much caffeine or alcohol.

Hypersomnia is extreme sleepiness during the day even with adequate sleep the night before. Hypersomnia has been mistakenly blamed on depression, laziness, boredom, or other negative personality traits.

Narcolepsy is characterized by falling asleep at inappropriate times. The sleep may last only a few minutes and is often preceded by a period of muscular weakness. Emotional events may trigger an episode of narcolepsy. Some individuals with narcolepsy experience a state called sleep paralysis. They wake up to find their body is paralyzed except for breathing and eye movement. In other words, the brain is awake but the body is still asleep.

Sleep apnea is a breathing disorder in which an individual briefly wakes up because breathing has been interrupted and may even stop for a brief period of time. Obstructive sleep apnea (OSA) is the most common form of sleep apnea. It occurs when air cannot flow into or out of the person's nose or mouth as they breathe.

What is **meningitis**?

Meningitis is an infection or inflammation of the meninges. Meningitis is most often caused by a bacterial or viral infection, although certain fungal infections and tumors may also cause meningitis. The usual symptoms and signs of meningitis are sudden fever, severe headache, and a stiff neck. In more severe cases, neurological symptoms may include nausea and vomiting, confusion and disorientation, drowsiness, sensitivity to bright light, and poor appetite. Early treatment of bacterial meningitis with antibiotics is important to reduce the risk of dying from the disease.

Can **meningitis** be **prevented**?

The introduction and widespread use of *Hemophilus influenzae* type b and *Strepto-coccus pneumoniae* conjugated vaccines has dramatically reduced the incidence of meningitis caused by these bacteria. In 2005, the Centers for Disease Control recommended routine vaccination of adolescents and college freshmen with the new meningococcal vaccine, which prevents four types of meningococcal disease caused by the bacteria *Neisseria meningitides*.

What is **presbycusis**?

Presbycusis is the scientific name for age-related sensorineural hearing loss. The first symptom is an inability to hear sounds at the highest frequencies and can occur as early as age 20. Around age 60, there is considerable variation in how well people hear. Some have had significant loss of hearing since age 50, while others have no hearing problems into their 90s. In general, men seem to experience hearing loss more often and more severely than women. One explanation for this difference may be that men's occupations are usually associated with prolonged exposure to louder noises.

What is the scientific name for **"ringing in the ears"**?

Tinnitus is the perception of sound in the ears or head where no external source is present. It is often referred to as "ringing in the ears." In almost all cases, tinnitus is a subjective noise, meaning that only the person who has tinnitus can hear it. Persistent tinnitus usually indicates the presence of hearing loss. The exact cause of tinnitus is not known, but there are several likely sources, all of which are known to trigger or worsen the condition. They include noise-induced hearing loss, wax buildup in the ear canal, medicines that are toxic to the ear, ear or sinus infections, jaw misalignment, and head and neck trauma.

How does **jet lag affect** one's body?

The physiological and mental stress encountered by airplane travelers when crossing four or more time zones is commonly called jet lag. Patterns of hunger, sleep, and elimination, along with alertness, memory, and normal judgment, may all be affected. More than 100 biological functions that fluctuate during the 24-hour cycle (circadian rhythm) can become desynchronized. Most people's bodies adjust at a rate of about one hour per day. Thus after four time zone changes, the body will require about four days to return to its usual rhythms. Flying eastward is often more difficult than flying westward, which adds hours to the day.

What is the **"Christmas factor"**?

In the clotting of blood, factor IX, or the Christmas factor, is a coagulation factor present in normal plasma, but deficient in the blood of persons with hemophilia B or

Christmas disease. It was named after a man named Christmas who, in 1952, was the first patient in whom this genetic disease was shown to be distinct from hemophilia (another genetic blood-clotting disease in which the blood does not have factor VIII).

What is the **medical term** for a **heart attack**?

Myocardial infarction is the term used for a heart attack in which part of the heart muscle's cells die as a result of reduced blood flow through one of the main arteries (many times due to arteriosclerosis). The outlook for the patient is dependent on the size and location of the blockage and extent of damage, but 33 percent of patients die within 20 days after the attack; it is a leading cause of death in the United States. Also, almost half of sudden deaths due to myocardial infarction occur before hospitalization. However, the possibility of recovery improves if vigorous treatment begins immediately.

How are the forms of **cancer classified**?

The over 150 different types of cancer are classified into four major groups:

1. Carcinomas—Nine in ten cancers are carcinomas, which involve the skin and skin-like membranes of the internal organs.
2. Sarcomas—Involve the bones, muscles, cartilage, fat, and linings of the lungs, abdomen, heart, central nervous system, and blood vessels.
3. Leukemias—Develop in blood, bone marrow, and the spleen.
4. Lymphomas—Involve the lymphatic system.

What are **dust mites**?

Dust mites are microscopic arachnids (members of the spider family) commonly found in house dust. Dust mite allergen is probably one of the most important causes of asthma (breathlessness and wheezing caused by the narrowing of small airways of the lungs) in North America, as well as the major cause of common allergies (exaggerated reactions of the immune system to exposure of offending agents).

Thorough, regular cleaning of the home, including the following measures, will help control dust mites:

1. Clean all major appliances such as furnaces and air conditioners, and change filters as recommended by the manufacturer.
2. Launder bedding every seven to ten days in hot water. Use synthetic or foam rubber mattress pads and pillows. Cover mattresses with dust-proof covers. Clean or replace pillows regularly.
3. Keep moist surfaces in kitchen and bathroom clean and free of mold.
4. Vacuum and dust often. A high-efficiency particulate air filter (HEPA) vacuum is especially effective.

If the sap of the **poison ivy plant** touches the skin, will a rash develop?

Studies show that 85 percent of the population will develop an allergic reaction if exposed to poison ivy, but this sensitivity varies with each individual according to circumstance, age, genetics, and previous exposure. The poison comes mainly from the leaves whose allergens touch the skin. A red rash with itching and burning will develop, and skin blisters will usually develop within six hours to several days after exposure. Washing the affected area thoroughly with mild soap within five minutes of exposure can be effective; sponging with alcohol and applying a soothing and drying lotion, such as calamine lotion, is the prescribed treatment for light cases. If the affected area is large, fever, headache, and generalized body weakness may develop. For severe reactions, a physician should be consulted to prescribe a corticosteroid drug. Clothing that touched the plants should also be washed.

What is **dyslexia** and what causes it?

Dyslexia covers a wide range of language difficulties. In general, a person with dyslexia cannot grasp the meaning of sequences of letters, words, or symbols or the concept of direction. The condition can affect people of otherwise normal intelligence. Dyslexic children may reverse letter and word order, make bizarre spelling errors, and may not be able to name colors or write from dictation. It may be caused by minor visual defects, emotional disturbance, or failure to train the brain. New evidence shows that a neurological disorder may be the underlying cause. Approximately 90 percent of dyslexics are male.

The term *dyslexia* (of Greek origin) was first suggested by Professor Rudolph Berlin of Stuttgart, Germany, in 1887. The earliest references to the condition date as far back as 30 C.E. when Valerius Maximus and Pliny described a man who lost his ability to read after being struck on the head by a stone.

What are the **essential nutrients**?

There are six essential nutrients: carbohydrates, fats, proteins, water, vitamins, and minerals. Energy nutrients are those that provide the body with the majority of the energy needed for daily metabolic reactions. Carbohydrates, fats, and proteins are energy nutrients.

What are **vitamins** and **minerals**?

A vitamin is an organic, nonprotein substance that is required by an organism for normal metabolic function but cannot be synthesized by that organism. In other words, vitamins are crucial molecules that must be acquired from outside sources. While most vitamins are present in food, vitamin D, for example, is produced as a precursor in our skin and converted to the active form by sunlight. Minerals, such as calcium and iron, are inorganic substances that also enhance cell metabolism. Vitamins may be fat- or water-soluble.

Functions and Sources of Vitamins

Vitamin	Fat- Or Water-Soluble	Major Sources	Major Functions
A	Fat-Soluble	Animal products; plants contain only development; particularly vitamin A building blocks	Aids normal cell division and maintenance of visual health
B-complex	Water-soluble	Fruits and vegetables, meat (thiamine, niacin, vitamin B6, and B12); milk (riboflavin, B12)	Energy metabolism; promotes harvesting energy from food

Vitamin	Fat- Or Water-Soluble	Major Sources	Major Functions
C	Water-soluble	Fruits and vegetables, particularly citrus, strawberries, spinach, and broccoli	Collagen synthesis; antioxidant benefits; promotes resistance to infection
D	Fat-soluble	Egg yolks; liver; fatty fish; sunlight	Supports bone growth; maintenance of muscular structure and digestive function
E	Fat-soluble	Vegetable oils; spinach; avocado; shrimp; cashews	Antioxidant
K	Fat-soluble	Leafy, green vegetables; cabbage	Blood clotting

Are **vitamin supplements** necessary?

Vitamin supplements may be a useful addition to the diet of individuals who do not receive all of the nutrients they need from their diet. These individuals cannot or do not eat enough of a variety of healthy foods.

What is the **Food Guidance System**?

The Food Guidance System (MyPyramid) is an educational tool to help individuals implement the *Dietary Guidelines for Americans* published jointly by the Department of Health and Human Services (HHS) and the U.S. Department of Agriculture (USDA). The current system maintains the shape of the familiar food guide pyramid, but it is personalized for age, sex, and individual physical activity levels. The new pyramid symbol in MyPyramid features six vertical color bands representing the five food groups and oils. Each food group narrows toward the top to indicate moderation. The stylized symbol of a person climbing the steps is to encourage physical activity.

The revised Food Pyramid created by the U.S. Department of Agriculture in 2005 divides the food groups into proportions shown from left to right: grains, vegetables, fruits, fat and oils, milk, and meat and beans.

How many **calories** does a person **burn while sleeping**?

A 150-pound (68-kilogram) person burns one calorie per minute during bed rest. The approximate caloric expenditure of other activities for a person weighing 150 pounds are given below. Actual numbers may vary, depending on the vigor of the exercise, air temperature, clothing, etc.

Activity	Calories Used per Hour
Aerobic dance	684
Basketball	500
Bicycling (5.5 mph)	210
Bicycling (13 mph)	660
Bowling	220–270
Calisthenics	300
Circuit weight training	756
Digging	360–420
Gardening	200
Golfing (using power cart)	150–220
Golfing (pulling cart)	240–300
Golfing (carrying clubs)	300–360
Football	500
Handball (social)	600–660
Handball (competitive)	> 660
Hoeing	300–360
Housework	180
In-line skating	600
Jogging (5–10 mph)	500–800
Lawn mowing (power)	250
Lawn mowing (hand)	420–480
Racquetball	456
Raking leaves	300–360
Rowing machine	415
Sitting	100
Skiing (cross-country)	600–660
Skiing (downhill)	570
Snow shoveling	420–480
Square dancing	350
Standing	140
Swimming moderately	500–700
Tennis (doubles)	300–360

Activity	Calories Used per Hour
Tennis (singles)	420–480
Vacuuming	240–300
Volleyball	350
Walking (2 mph)	150–240
Walking (3.5 mph)	240–300
Walking (4 mph)	300–400
Walking (5 mph)	420–480

Which **foods** contain **trans fats**?

Trans fatty acids, or trans fats, are made when manufacturers add hydrogen to liquid vegetable oil—a process called hydrogenation—creating solid fats like shortening and hard margarine. Hydrogenation increases the shelf life and flavor stability of foods containing these fats. Diets high in trans fat raise the LDL (low density lipoprotein) or "bad" cholesterol, increasing the risk for coronary heart disease.

Cakes, crackers, cookies, snack foods, and other foods made with or fried in partially hydrogenated oils are the largest source (40 percent) of trans fats in the American diet. Animal products and margarine are also major sources of trans fats. Since January 2006, the U.S. government has directed that the amount of trans fat in a product must be included in the Nutrition Facts panel on food labels. In 2008, California became the first state to enact legislation to gradually phase trans fat out of foods served in food facilities and in baked goods. Similarly, New York City is phasing in a ban on trans fat in all city restaurants. Other states have proposed legislation to ban the use of trans fats in restaurants or schools or baked goods.

What is **lactose intolerance**?

Lactose, the principal sugar in cow's milk and found only in dairy products, requires the enzyme lactase for human digestion. Lactose intolerance occurs when the lining of the walls of a person's small intestine does not produce normal amounts of this enzyme. Lactose intolerance causes abdominal cramps, bloating, diarrhea, and excessive gas when more than a certain amount of milk is ingested. Most people are less able to tolerate lactose as they grow older.

A person having lactose intolerance need not eliminate dairy products totally from the diet. Decreasing the consumption of milk products, drinking milk only during meals, and getting calcium from cheese, yogurt, and other dairy products having lower lactose values are options. Another alternative is to buy a commercial lactose preparation that can be mixed into milk. These preparations convert lactose into simple sugars that can be easily digested.

What is **anorexia**?

Anorexia simply means a loss of appetite. Anorexia nervosa is a psychological disturbance that is characterized by an intense fear of being fat. It usually affects teenage or young adult women. This persistent "fat image," however untrue in reality, leads the patient to self-imposed starvation and emaciation (extreme thinness) to the point where one-third of the body weight is lost. There are many theories on the causes of this disease, which is difficult to treat and can be fatal. Between five and ten percent of patients hospitalized for anorexia nervosa later die from starvation or suicide. Symptoms include a 25 percent or greater weight loss (for no organic reason) coupled with a morbid dread of being fat, an obsession with food, an avoidance of eating, compulsive exercising and restlessness, binge eating followed by induced vomiting, and/or use of laxatives or diuretics.

What type of **diet** is recommended for individuals with **celiac disease**?

A gluten-free diet is the only treatment for individuals with celiac disease. Celiac disease is an autoimmune digestive disease that damages the small intestine and interferes with absorption of nutrients. The villi in the small intestine are damaged or destroyed whenever sufferers of celiac disease eat products that contain gluten. Gluten is found in wheat, rye, and barley. Once the villi are damaged, they are not able to allow nutrients to be absorbed by the bloodstream, leading to malnutrition.

What is **progeria**?

Progeria is premature old age. There are two distinct forms of the condition, both of which are extremely rare. In Hutchinson-Gilford syndrome, aging starts around the age 4, and by 10 or 12, the affected child has all the external features of old age, including gray hair, baldness, and loss of fat, resulting in thin limbs and sagging skin on the trunk and face. There are also internal degenerative changes, such as atherosclerosis (fatty deposits lining the artery walls). Death usually occurs at puberty. Werner's syndrome, or adult progeria, starts in early adult life and follows the same rapid progression as the juvenile form with individuals developing disorders usually associated with aging as early as in their twenties and thirties. Most individuals with Werner's syndrome live until their late forties or early fifties. The cause of progeria is unknown.

What is **pelvic inflammatory disease**?

Pelvic inflammatory disease (PID) is a term used for a group of infections in the female organs, including inflammations of the Fallopian tubes, cervix, uterus, and ovaries. It is the most common cause of female infertility today. PID is most often found in sexually active women under the age of 25 and almost always results from gonorrhea or chlamydia, but women who use intrauterine devices (IUDs) are also at risk.

Why do deep-sea divers get the bends?

Bends is a painful condition in the limbs and abdomen. It is caused by the formation and enlargement of bubbles of nitrogen in blood and tissues as a result of rapid reduction of pressure. This condition can develop when a diver ascends too rapidly after being exposed to increased pressure. Severe pain will develop in the muscles and joints of the arms and legs. More severe symptoms include vertigo, nausea, vomiting, choking, shock, and sometimes death. Bends is also known as decompression sickness, caisson disease, tunnel disease, and diver's paralysis.

A variety of organisms have been shown to cause PID, including *Neisseria gonorrhoeae* and such common bacteria as staphylococci, chlamydiae, and coliforms (*Pseudomonas* and *Escherichia coli*). Signs and symptoms of PID vary with the site of the infection, but usually include profuse, purulent vaginal discharge, low-grade fever and malaise (especially with *N. gonorrhoeae* infections), and lower abdominal pain. PID is treated with antibiotics, and early diagnosis and treatment will prevent damage to the reproductive system.

Severe, untreated PID can result in the development of a pelvic abscess that requires drainage. A ruptured pelvic abscess is a potentially fatal complication, and a patient who develops this complication may require a total hysterectomy.

Which medical condition is caused by the **abnormal location** of **endometrial tissue**?

Endometriosis is the abnormal location of uterine endometrial tissue. Under normal conditions, endometrial tissue only lines the uterus, but in some instances the tissue migrates to remote locations of the body such as the ovaries, pelvic peritoneum, the vagina, the bladder, or even the small intestine and lining of the chest cavity. Endometriosis is associated with dysmenorrhea (painful menstruation), pelvic pain, and infertility.

How are **burns classified**?

Type	Causes and Effects
First-degree	Sunburn; steam. Reddening and peeling. Affects epidermis (top layer of skin). Heals within a week.
Second-degree	Scalding; holding hot metal. Deeper burns causing blisters. Affects dermis (deep skin layer). Heals in two to three weeks.

Type	Causes and Effects
Third-degree	Fire. A full layer of skin is destroyed. Requires a doctor's care and grafting.
Circumferential	Any burns (often electrical) that completely encircle a limb or body region (such as the chest), which can impair circulation or respiration; requires a doctor's care; fasciotomy (repair of connective tissues) is sometimes required.
Chemical	Acid, alkali. Can be neutralized with water (for up to half an hour). Doctor's evaluation recommended.
Electrical	Destruction of muscles, nerves, circulatory system, etc., below the skin. Doctor's evaluation and ECG monitoring required.

If more than ten percent of body surface is affected in second-and third-degree burns, shock can develop when large quantities of fluid (and its protein) are lost. When skin is burned, it cannot protect the body from airborne bacteria.

What is the **phobia** of the **number 13** called?

Fear of the number 13 is known as tridecaphobia, tredecaphobia, or triskaidekaphobia. Persons may fear any situation involving this number, including a house number, the floor of a building, or the thirteenth day of the month. Many buildings omit labeling the thirteenth floor as such for this reason. A phobia can develop for a wide range of objects, situations, or organisms. The list below demonstrates the variety:

Phobia Subject	Phobia Term
Animals	Zoophobia
Beards	Pogonophobia
Books	Bibliophobia
Churches	Ecclesiaphobia
Dreams	Oneirophobia
Fish	Ichthyophobia
Flowers	Anthophobia
Food	Sitophobia
Graves	Taphophobia
Infection	Nosemaphobia
Lakes	Limnophobia
Leaves	Phyllophobia
Lightning	Astraphobia
Men	Androphobia
Money	Chrometophobia
Music	Musicophobia

Phobia Subject	Phobia Term
Birds	Ornithophobia
Sex	Genophobia
Shadows	Sciophobia
Spiders	Arachnophobia
Sun	Heliophobia
Touch	Haptophobia
Trees	Dendrophobia
Walking	Basiphobia
Water	Hydrophobia
Women	Gynophobia
Work	Ergophobia
Writing	Graphophobia

A fear of animals is called zoophobia; a fear of birds, specifically, is called ornithophobia.

HEALTH CARE

Which **symbol** is used to **represent medicine**?

The staff of Aesculapius has represented medicine since 800 B.C.E. It is a single serpent wound around a staff. The caduceus, the twin-serpent magic wand of the god Hermes or Mercury, came into use after 1800 and is commonly used today. The serpent has traditionally been a symbol of healing, and it is an old belief that eating part of a serpent would bring the power of healing to the ingester. Early Greeks saw in the serpent regenerative powers expressed by the serpent's periodic sloughing of its skin, and venerated the serpent. Later, the Greek god of medicine, Asklepius, called Aesculapius by the Romans, performed his functions in the form of a serpent. Sometimes this god is represented in art as an old man with a staff, around which is coiled a serpent.

Who is generally regarded as the **father of medicine**?

Hippocrates (c. 460–c. 377 B.C.E.), a Greek physician, holds this honor. Greek medicine, previous to Hippocrates, was a mixture of religion, mysticism, and necromancy. Hippocrates established the rational system of medicine as a science, separating it from religion and philosophy. Diseases had natural causes and natural laws: they were not the "wrath of the gods." Hippocrates believed that the four elements (earth, air, fire, and water) were represented in the body by four body fluids (blood, phlegm, black bile, and yellow bile) or "humors." When they existed in harmony within the body, the body was in good health. The duty of the physician was to help nature to restore the body's harmony. Diet, exercise, and moderation in all things kept the body well, and psychological healing (good attitude toward recovery), bed rest, and quiet were part of

619

his therapy. Hippocrates was the first to recognize that different diseases had different symptoms; and he described them in such detail that the descriptions generally would hold today. His descriptions not only included diagnosis but prognosis.

What is the **Hippocratic Oath**?

The Hippocratic Oath is the traditional oath physicians take when they graduate from medical school. It can be traced back to the Greek physician and teacher Hippocrates (c. 460 B.C.E.–c. 377 B.C.E.). The oath reads as follows:

> "I swear by Apollo the physician, by Aesculapius, Hygeia, and Panacea, and I take to witness all the gods, all the goddesses, to keep according to my ability and my judgment the following Oath:

> "To consider dear to me as my parents him who taught me this art; to live in common with him and if necessary to share my goods with him; to look upon his children as my own brothers, to teach them this art if they so desire without fee or written promise; to impart to my sons and the sons of the master who taught me and the disciples who have enrolled themselves and have agreed to rules of the profession, but to these alone, the precepts and the instruction. I will prescribe regimes for the good of my patients according to my ability and my judgment and never to harm anyone. To please no one will I prescribe a deadly drug, nor give advice which may cause his death. Nor will I give a woman a pessary to procure abortion. But I will preserve the purity of my life and my art. I will not cut for stone, even for patients in whom the disease is manifest; I will leave this operation to be performed by practitioners (specialists in this art). In every house where I come I will enter only for the good of my patients, keeping myself far from all intentional ill-doing and all seduction, and especially from the pleasures of love with women or with men, be they free or slaves. All that may come to my knowledge in the exercise of my profession or outside of my profession or in daily commerce with men, which ought not to be spread abroad, I will keep secret and will never reveal. If I keep this oath faithfully, may I enjoy life and practice my art, respected by all men and in all times; but if I swerve from it or violate it, may the reverse be my lot."

The oath varies slightly in wording among different sources, and many medical schools have adopted modern versions of the oath.

Who was the **founder of modern medicine**?

Thomas Sydenham (1624–1689), who was also called the English Hippocrates, reintroduced the Hippocratic method of accurate observation at the bedside, and recording of observations, to build up a general clinical description of individual diseases. He is also considered one of the founders of epidemiology and was among the first to describe scarlet fever and Sydenham's chorea.

What was the **first medical college** in the United States?

The College of Philadelphia Department of Medicine, now the University of Pennsylvania School of Medicine, was established on May 3, 1765. The first commencement was held June 21, 1768, when medical diplomas were presented to the ten members of the graduating class.

Who was the **first woman physician** in the United States?

Elizabeth Blackwell (1821–1910) received her degree in 1849 from Geneva Medical College in New York. After overcoming many obstacles, she set up a small practice that expanded into the New York Infirmary for Women and Children, which featured an all-female staff.

When and where did the **first blood bank** open?

Several sites claim the distinction. Some sources list the first blood bank as opening in 1940 in New York City under the supervision of Dr. Richard C. Drew (1904–1950). Others list an earlier date of 1938 in Moscow at the Sklifosovsky Institute (Moscow's central emergency service hospital) founded by Professor Sergei Yudin. The term blood bank was coined by Bernard Fantus (1874–1940), who set up a centralized storage depot for blood in 1937 at the Cook County Hospital in Chicago, Illinois.

How does **conventional medicine** differ from **complementary and alternative medicine** (CAM)?

Conventional medicine, also called allopathic medicine, is based on scientific knowledge of the body and uses treatments that are based on scientific research. It is practiced by healthcare practitioners who hold an M.D. (doctor of medicine) or D.O. (doctor of osteopathy). Conventional medicine is also referred to as Western medicine. Complementary and alternative medicine are healthcare systems, practices and products that are generally not considered to be in the mainstream of conventional medicine. There may be some scientific evidence for some CAM therapies, but for most CAM therapies there are still questions that have not been answered through well-designed scientific studies. CAM therapies are often based on the belief that a healthcare provider must treat the whole person—body, mind, and spirit—as a unit. Most CAM therapies are less invasive than conventional medicine.

How does **complementary medicine** differ from **alternative medicine**?

Complementary medicine is used together with conventional medicine. An example of complementary medicine is using aromatherapy following conventional surgery. Alternative medicine is used in place of conventional medicine. One example of alternative medicine is treating cancer with a special diet instead of surgery, radiation, and/or chemotherapy. Integrated or integrative medicine combines treatments from conventional medicine with CAM therapies and techniques for which there is evidence of safety and effectiveness.

What is the difference between osteopathic medicine, homeopathy, naturopathy, and chiropractic medicine?

Osteopathic medicine was developed in the United States by Dr. Andrew Taylor Still (1828–1917). It recognizes the role of the musculoskeletal system in healthy function of the human body. The physician is fully licensed and uses manipulation techniques as well as traditional diagnostic and therapeutic procedures. Osteopathy is practiced as part of conventional Western medicine.

Developed by a German physician, Christian F. S. Hahnemann (1755–1843), the therapy of homeopathy treats patients with small doses of 2,000 substances. Based on the principle that "like cures like," the medicine used is one that produces the same symptoms in a healthy person that the disease is producing in the sick person.

Chiropractic medicine is based on the belief that disease results from the lack of normal nerve function. Relying on physical manipulation and adjustment of the spine for therapy, rather than on drugs or surgery, this therapy, used by the ancient Egyptians, Chinese, and Hindus, was rediscovered in 1895 by American osteopath Daniel David Palmer (1845–1913).

Naturopathy is based on the principle that disease is due to the accumulation of waste products and toxins in the body. Practitioners believe that health is maintained by avoiding anything artificial or unnatural in the diet or in the environment.

What are the different domains of complementary and alternative medicine?

Complementary and alternative medicine practices may be divided into four domains: 1) mind-body medicine, 2) biologically based practices, 3) manipulative, and body-based practices, and 4) energy medicine.

Mind-body medicine uses a variety of techniques designed to enhance the mind's capacity to affect bodily function and symptoms. Mind-body techniques include meditation, prayer, mental healing, and therapies that use creative outlets such as art, dance, and music. Patient-support groups and cognitive-behavioral therapy were once considered CAM techniques, but are currently considered mainstream.

Biologically based practices use substances found in nature, such as herbs, food, and vitamins, including dietary supplements and herbal products. One example of a natural product to treat a medical disorder is using shark cartilage to treat cancer.

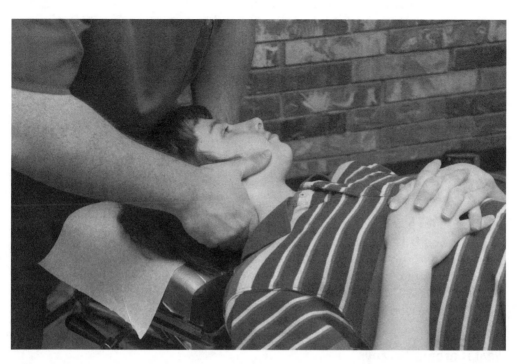

The philosophy behind chiropractic medicine is that problems with the spine can affect the central nervous system, which, in turn, can adversely affect one's health. Correcting a misaligned spine will alleviate problems ranging from headaches and back pain to immune system issues.

Manipulative and body-based practices are based on manipulation and/or movement of one or more body parts. Manipulation may be part of whole medical systems, such as chiropractic medicine or osteopathic medicine. It often includes massage therapy to increase the flow of blood and oxygen to the targeted areas.

The two types of energy fields used in energy therapies are biofield therapies and bioelectromagnetic-based therapies. Biofield therapies are intended to affect energy fields that purportedly surround and penetrate the human body. The existence of such fields has not yet been scientifically proven. Some forms of energy therapy manipulate biofields by applying pressure and/or manipulating the body by placing the hands in, or through, these fields. Bioelectromagnetic-based therapies involve the unconventional use of electromagnetic fields, such as pulsed fields, magnetic fields, or alternating-current or direct-current fields.

What are some **examples** of **biofield therapies**?

Qi gong, Reiki, and therapeutic touch are examples of biofield therapies. Qi gong, a part of traditional Chinese medicine, combines movement, meditation, and controlled breathing. The goal is to improve blood flow and the flow of qi.

623

Reiki, based on the spiritual teachings of Mikao Usui (1865–1926) of Japan, is used to promote overall health and well-being. Practitioners of Reiki seek to transmit a universal energy to a person, either from a distance or by placing their hands on or near the person. The goal is to heal the spirit and thus the body.

Therapeutic touch is a therapy in which practitioners pass their hands over another person's body with the intent to use their own perceived healing energy to identify energy imbalances and promote health.

What is **reflexology**?

Reflexology is the application of specific pressures to reflex points in the hands and feet. The reflex points relate to every organ and every part of the body. Massaging of the reflex points is done to prevent or cure diseases. Believed to have been used in Asian cultures as long as 2,000 years ago, reflexology was introduced to the United States at the turn of the century by Dr. William Fitzgerald (1872–1942) and Eunice D. Ingham (1889–1974). Today nearly 25,000 certified practitioners can be found throughout the world.

What is **aromatherapy**?

Holistic medicine looks at the health of the whole individual, and treatments emphasize the connection of mind, body, and spirit. Aromatherapy involves using particular scents derived from essential oils to influence emotions and to treat and cure minor ailments. It is based on the fact that the olfactory and emotional centers of the body are connected. By inhaling different aromas, emotional concerns as well as physical complaints are said to be eased. The term "aromatherapy" was first used by René-Maurice Gattefossé (1881–1950), a French perfume chemist. He discovered the healing powers of lavender oil following a laboratory accident during which he burned his hand. Gattefossé began to investigate the properties of lavender oil and other essential oils and published a book on plant extracts. During aromatherapy treatments, essential oils are absorbed through breath or the pores of the skin; this process triggers certain physiological responses.

What is the difference between the degrees **doctor of dental surgery** (DDS) and **doctor of medical dentistry** (DMD)?

The title depends entirely on the school's preference in terminology. The degrees are equivalent.

How many people **visit** a **dentist regularly**?

Only about half the population visits a dentist as often as once a year. Children, ages 2 to 17, visit the dentist more frequently than adults.

Why do some dentists treat the molars and premolars of children with sealants?

Sealants, a soft plastic coating applied to the tooth surface, can protect a child's first and second permanent molars from decay by filling in the pits and fissures where food and bacteria might otherwise accumulate. The plastic is hardened with a special light or chemical.

Age	Percent Who Visited a Dentist in 2007
2–17	77
18–64	63
65 and over	58

What is the difference between an **ophthalmologist, optometrist, and optician**?

An ophthalmologist is a physician who specializes in care of the eyes. Ophthalmologists conduct examinations to determine the quality of vision and the need for corrective glasses or contact lenses. They also check for the presence of any disorders, such as glaucoma or cataracts. Ophthalmologists may perform surgery or prescribe glasses, contact lenses, or medication, as necessary.

An optometrist is a specialist trained to examine the eyes and to prescribe, supply, and adjust glasses or contact lenses. Because they are not physicians, optometrists may not prescribe drugs or perform surgery. An optometrist refers patients requiring these types of treatment to an ophthalmologist.

An optician is a person who fits, supplies, and adjusts glasses or contact lenses. Because their training is limited, opticians may not examine or test eyes or prescribe glasses or drugs.

DIAGNOSTIC EQUIPMENT, TESTS, AND TECHNIQUES

What is the **meaning** of the medical abbreviation **NYD**?

Not yet diagnosed.

How is **blood pressure measured**?

A sphygmomanometer is the device used to measure blood pressure. It was invented in 1881 by an Austrian named Von Bash. It consists of a cuff with an inflatable bladder that is wrapped around the upper arm, a rubber bulb to inflate the bladder, and a device that indicates the pressure of blood. Measuring arterial tension (blood pressure) of a person's circulation is achieved when the cuff is applied to the arm over the artery and pumped to a pressure that occludes or blocks it. This gives the systolic measure, or the maximum pressure of the blood, which occurs during contraction of the ventricles of the heart. Air is then released from the cuff until the blood is first heard passing through the opening artery (called Korotkoff sounds). This gives diastolic pressure, or the minimum value of blood pressure that occurs during the relaxation of the arterial-filling phase of the heart muscle.

What is the **meaning** of the **numbers** in a **blood pressure** reading?

When blood is forced into the aorta, it exerts a pressure against the walls; this is referred to as blood pressure. The upper number, the systolic, measures the pressure during the period of ventricular contraction. The lower number, the diastolic, measures the pressure when blood is entering the relaxed chambers of the heart. While these numbers can vary due to age, sex, weight, and other factors, the normal blood pressure is around 110/60 to 140/90 millimeters of mercury.

Is there a name for the **heart-monitoring machine** that people sometimes wear for a day or two while carrying on their normal activities?

A portable version of the electrocardiograph (ECG) designed by J.J. Holter is called a Holter monitor. Electrodes attached to the chest are linked to a small box containing a recording device. The device records the activity of the heart.

What are the **normal** test ranges for **total cholesterol**, low-density lipoproteins (**LDL**), and high-density lipoproteins (**HDL**)?

The National Cholesterol Education Program has drawn up these guidelines:

	Desirable	Borderline	High Risk
Total Blood Cholesterol	Less than 200 mg/dl	200–239 mg/dl	240 mg/dl or more
LDL	Less than 130 mg/dl	130–159 mg/dl	160 mg/dl or more
HDL	45–65 mg/dl	35–45 mg/dl	Below 35 mg/dl

mg/dl = milligrams per deciliter

Who **invented** the **pacemaker**?

Paul Zoll (1911–1999) invented an electric stimulator device to deliver electrical impulses to the heart externally. In 1958, biomedical engineer Wilson Greatbatch (1919–), in cooperation with doctors William M. Chardack (1915–2006) and Andrew A. Gage (1922–), invented the first internal pacemaker. It was a small, flat, plastic disk powered by a battery. It was implanted into the body and connected by wires sewn directly onto the heart. The wires emitted rhythmic electric impulses to trigger the heart's action. Pacemaker batteries now last from six to ten years.

What are **X rays**?

X rays are electromagnetic radiation with short wavelengths (10^{-3} nanometers) and a great amount of energy. They were discovered in 1898 by William Conrad Roentgen (1845–1923). X rays are frequently used in medicine because they are able to pass through opaque, dense structures such as bone and form an image on a photographic plate. They are especially helpful in assessing damage to bones, identifying certain tumors, and examining the chest—heart and lungs—and abdomen. A major disadvantage of X rays as a diagnostic tool is that they provide little information about the soft tissues. Since they only show a flat, two-dimensional picture, they cannot distinguish between the various layers of an organ, some of which may be healthy while others may be diseased.

What is **nuclear magnetic resonance imaging**?

Magnetic resonance imaging (MRI), sometimes called nuclear magnetic resonance imaging (NMR), is a non-invasive, non-ionizing diagnostic technique. It is useful in detecting small tumors, blocked blood vessels, or damaged vertebral disks. Because it does not involve the use of radiation, it can often be used where X rays are dangerous. Large magnets beam energy through the body causing hydrogen atoms in the body to resonate. This produces energy in the form of tiny electrical signals. A computer detects these signals, which vary in different parts of the body and according to whether an organ is healthy or not. The variation enables a picture to be produced on a screen and interpreted by a medical specialist.

What distinguishes MRI from computerized X-ray scanners is that most X-ray studies cannot distinguish between a living body and a cadaver, while MRI "sees" the difference between life and death in great detail. More specifically, it can discriminate between healthy and diseased tissues with more sensitivity than conventional radiographic instruments like X rays or CAT scans. CAT (computerized axial tomography) scanners have been around since 1973 and are actually glorified X-ray machines. They offer three-dimensional viewing but are limited because the object imaged must remain still.

The concept of using MRI to detect tumors in patients was proposed by Raymond Damadian (1936–) in a 1972 patent application. The fundamental MRI imaging con-

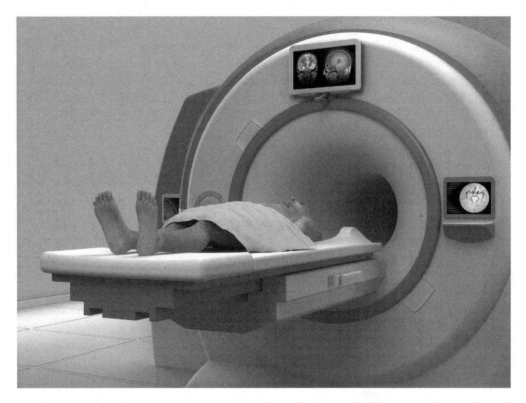

MRI scanners can create images of the inside of a patient's body without the use of dangerous X rays.

cept used in all present-day MRI instruments was proposed by Paul Lauterbar (1929–2007) in an article published in *Nature* in 1973. Lauterbur and Peter Mansfield (1933–) were awarded the Nobel Prize in Physiology or Medicine in 2003 for their discoveries concerning magnetic resonance imaging. The main advantages of MRI are that it not only gives superior images of soft tissues (like organs), but can also measure dynamic physiological changes in a non-invasive manner (without penetrating the body in any way). A disadvantage of MRI is that it cannot be used for every patient. For example, patients with implants, pacemakers, or cerebral aneurysm clips made of metal cannot be examined using MRI because the machine's magnet could potentially move these objects within the body, causing damage.

What is **ultrasound**?

Ultrasound, also called sonography, is another type of 3-D computerized imaging. Using brief pulses of ultrahigh frequency acoustic waves (lasting 0.01 second), it can produce a sonar map of the imaged object. The technique is similar to the echolocation used by bats, whales, and dolphins. By measuring the echo waves, it is possible to determine the size, shape, location, and consistency (whether it is solid, fluid-filled, or

> ## What is the instrument a doctor uses to check reflexes?
>
> **A** plessor or plexor or percussor is a small hammer, usually with a soft rubber head, used to tap the part directly. Also called a reflex hammer or a percussion hammer, it is used by a doctor to elicit reflexes by tapping on tendons. In the most common test, the patient sits on a surface high enough to allow his legs to dangle freely, and the physician lightly taps the patellar tendon, just below the kneecap. This stimulus briefly stretches the quadriceps muscle on top of the thigh. The stretch causes the muscle to contract, which makes the leg kick forward. The time interval between the tendon tap and the start of the leg extension is about 50 microseconds. That interval is too short for the involvement of the brain and is totally reflexive. This test indicates the status of an individual's reflex control of movement.

both) of an object. It is frequently used in obstetrics because unlike X rays, it does not use ionizing radiation to produce an image. It causes no health problems for the mother or unborn fetus and may be repeated as often as necessary.

How are **CAT** or **CT scans used** to study the human body?

CAT or CT scans (computer-assisted tomography or simply computerized tomography), are specialized X rays that produce cross-sectional images of the body. They are used to study many parts of the body, including the chest, abdomen and pelvis, extremities (arms and legs), and internal organs, such as pancreas, liver, gall bladder, and kidneys. CT scans of the head and brain may detect an abnormal mass or growth, stroke damage, area of bleeding, or blood vessel abnormality. Patients complaining of pain may have a CT scan to determine the source of the pain. Sometimes a CT scan will be used to further investigate an abnormality found on a regular X ray.

Dr. Allan M. Cormack (1924–1998) and Godfrey N. Hounsfield (1919–2004) independently discovered and developed computer assisted tomography in the early 1970s. They shared the 1979 Nobel Prize in Physiology or Medicine for their research.

When were **hearing aids invented**?

Specially designed hearing aids were described as early as 1588 by Giovanni Battista Porta (1535–1615) in his book *Natural Magick*. These hearing aids were made out of wood in the shape of the ears of animals with a sharp sense of hearing. During the 1700s, speaking tubes and ear trumpets were developed. Bone conduction devices to transmit sound vibrations from outside to the bones of the ear, first suggested in 1550 by Gerolamo Cardano (1501–1576), were further developed during the 1800s. The first battery-powered hearing aid in the United States was made by the Dictagraph Company in 1898. Miller Reese Hutchison (1876–1944) filed the patent for the first electric hearing aid in 1901. Refinements were made to the hearing aids throughout the twentieth and twenty-first centuries. Miniaturization and microchips made it possible for hearing aids to become so small they now fit invisibly inside the ear.

DRUGS AND MEDICINES

What is **pharmacognosy**?

It is the science of natural drugs and their physical, botanical, and chemical properties. Natural products derived from plant, vegetable, animal, and mineral sources have been a part of medical practice for thousands of years. Today about 25 percent of all prescriptions dispensed in pharmacies contain active ingredients that are extracted from higher plants, and many more are found in over-the-counter products.

Why do physicians use the **symbol Rx** when they write their prescriptions?

There are several explanations for the symbol Rx. One common explanation is that it comes from the Latin word *recipi* or *recipere*, which means "take" and is abbreviated as Rx. The symbol can also be traced to the sign of Jupiter, which was found on ancient prescriptions to appeal to the Roman god Jupiter. In ancient medical books the crossed R has been found wherever the letter R occurred.

Others believe that the origin of the Rx symbol can be found in the Egyptian myth about two brothers, Seth and Horus, who ruled over Upper and Lower Egypt as gods. Horus's eye was injured in a battle with Seth and healed by another god, Thoth. The eye of Horus consisted of the sun and the moon, and it was the moon eye that was damaged. This explained the phases of the moon—the waning of the moon was the eye being damaged and the waxing, the healing. The eye of Horus became a powerful symbol of healing in the eyes of the Egyptians. In Egyptian art, the eye of Horus strongly resembles the modern Rx of the physician.

What is the **meaning** of the **abbreviations** often used by a doctor when writing a **prescription**?

Latin Phrase	Shortened Form	Meaning
quaque hora	qh	every hour
quaque die	qd	every day

Latin Phrase	Shortened Form	Meaning
bis in die	bid	twice a day
ter in die	tid	three times a day
quarter in die	qid	four times a day
pro re nata	prn	as needed
ante cibum	a.c.	before meals
post cibum	p.c.	after meals
per os	p.o.	by mouth
nihil per os	n.p.o.	nothing by mouth
signetur	sig	let it be labeled
statim	stat	immediately
ad libitum	ad lib	at pleasure
hora somni	h.s.	at bedtime
cum	c	with
sine	s	without
guttae	gtt	drops
semis	ss	a half
et	et	and

What are the **common medication measures**?

Approximate equivalents of apothecary measures are given below.

Apothecary Volume

Volume	Equivalent
1 minim	0.06 milliliters or 0.02 fluid drams or 0.002 fluid ounces
1.5 minims	0.1 milliliters
15 minims	1 milliliter
480 minims	1 fluid ounce
1 dram	3.7 milliliters or 60 minims
1 t (teaspoon)	60 drops
3 t (teaspoons)	0.5 ounce
1 T (tablespoon)	0.5 ounce
2 T (tablespoons)	1 ounce
1 C (cup)	8 ounces or 30 milliliters

Apothecary Weights

Weight	Equivalent
1 grain	60 milligrams or 0.5 dram
60 grains	1 dram or 3.75 grams
8 drams	1 ounce or 30 grams

Many traditional herbal medicines have been found to be effective in treating a variety of ailments. Ginseng (center), for example, is used to boost energy and immunity and even increase libido.

What is meant by the term **orphan drugs**?

Orphan drugs are intended to treat diseases that affect fewer than 200,000 Americans. With little chance of making money, a drug company is not likely to undertake the necessary research and expense of finding drugs that might treat these diseases. Also, if the drug is a naturally occurring substance, it cannot be patented in the United States, and companies are reluctant to invest money in such a medication when it cannot be protected against exploitation by competing drug companies. Encouragingly, the Orphan Drug Act of 1983 offers a number of incentives to drug companies to encourage development of these drugs. The act has provided hope for millions of people with rare and otherwise untreatable conditions.

How many of the **medications** used today are **derived from plants**?

Of the more than 250,000 known plant species, less than one percent have been thoroughly tested for medical applications. Yet out of this tiny portion have come 25 percent of our prescription medicines. The U.S. National Cancer Institute has identified 3,000 plants from which anti-cancer drugs are or can be made. This includes ginseng (*Panax quinquefolius*), Asian mayapple (*Podophyllum hexandrum*), western yew (*Taxus brevifolia*), and rosy periwinkle. Seventy percent of these 3,000 come from rain forests, which also are a source of countless other drugs for diseases and infections. Rain forest plants are rich in so-called secondary metabolites, particularly alkaloids, which biochemists believe the plants produce to protect them from disease and insect attack. However, with the current rate of rain forest destruction, raw materials for future medicines are certainly being lost. Also, as tribal groups disappear, their knowledge of the properties and uses of these plants species will be lost.

What are some **medications** that have been obtained **from rain forest** plants, animals, and microorganisms?

Drug	Medicinal Use	Source
Allantoin	Wound healer	Blowfly larva
Atropine	High blood pressure	Bee venom

Drug	Medicinal Use	Source
Cocaine	Analgesic	Coca bush
Cortisone	Anti-inflammatory	Mexican yam
Cytarabine	Leukemia	Sponge
Diosgenin	Birth control	Mexican yam
Erythromycin	Antibiotic	Bacterium
Morphine	Analgesic	Opium poppy
Quinine	Malaria	Chincona tree bark
Reserpine	Hypertension	Rauwolfia plant
Tetracycline	Antibiotic	Bacterium
Vinblastine	Hodgkin's disease and leukemia	Rosy periwinkle plant

From what **plant** is **taxol** extracted?

Taxol is produced from the bark of the western or Pacific yew (*Taxus brevifolia*). It has been shown to inhibit the growth of HeLa cells (human cancer cells) and is a promising new treatment for several kinds of cancer. Originally it was a scarce drug, but in 1994 two groups of researchers announced its synthesis. The synthesis is a formidable challenge, and better procedures and modifications remain to be developed. Since taxol is developed now from needles instead of tree bark, the natural source is more available, but the synthetic version will be needed to devise modified or "designer" taxols whose cancer-fighting ability may prove more effective.

What are some of the most **common herbal remedies**?

Herbal medicine treats disease and promotes health with plant material. For centuries herbal medicines were the primary methods to administer medicinally active compounds.

Herb	Botanical Name	Common Use
Aloe	*Aloe vera*	Skin, gastritis
Black cohosh	*Cimicifuga racemosa*	Menstrual, menopause
Dong quai	*Angelica sinensis*	Menstrual, menopause
Echinacea	*Echinacea angustifolia*	Colds, immunity
Ephedra (ma huang)	*Ephedra sinica*	Asthma, energy, weight loss
Evening primrose oil	*Oenothera biennis*	Eczema, psoriasis, premenstrual syndrome, breast pain
Feverfew	*Tanacetum parthenium*	Migraine
Garlic	*Allium sativum*	Cholesterol, hypertension
Ginger	*Zingiber officinale*	Nausea, arthritis
Ginkgo biloba	*Ginkgo biloba*	Cerebrovascular insufficiency, memory

Herb	Botanical Name	Common Use
Ginseng	*Panax ginseng, Panax quinquifolius, Panax pseudoginseng, Eleuthero-coccus senticosus*	Energy, immunity, mentation, libido
Goldenseal	*Hydrastis candensis*	Immunity, colds
Hawthorne	*Crateaegus laeviagata*	Cardiac function
Kava kava	*Piper methysticum*	Anxiety
Milk thistle	*Silybum marianum*	Liver disease
Peppermint	*Mentha piperita*	Dyspepsia, irritable bowel syndrome
Saw palmetto	*Serona repens*	Prostate problems
St. John's Wort	*Hypericum perforatum*	Depression, anxiety, insomnia
Tea tree oil	*Melaleuca alternifolia*	Skin infections
Valerian	*Valeriana officinalis*	Anxiety, insomnia

What is a **dietary supplement**?

According to the Dietary Supplement Health and Education Act (DSHEA), a dietary supplement is a product taken by mouth that contains a "dietary ingredient" intended to supplement the diet. The dietary ingredient may include: vitamins, minerals, herbs or other botanicals, amino acids, a dietary substance to supplement the diet by increasing the total dietary intake, or a concentrate, metabolite, constituent, or extract. They may be in any form—tablets, capsules, liquids, gelcaps, or powders. The DSHEA places dietary supplements in a special category under the general umbrella of "foods," not drugs, and requires that every supplement be labeled a dietary supplement.

When was the term **"antibiotic" first used**?

Antibiotics are chemical products or derivatives of certain organisms that inhibit the growth of or destroy other organisms. The term "antibiotic" (from the Greek *anti*, meaning "against," and *biosis*, meaning "life") refers to its purpose in destroying a life form. In 1889 Paul Vuillemin (1861–1932) used the term "antibiosis" to describe bacterial antagonism. He had isolated pyocyanin, which inhibited the growth of bacteria in test tubes but was too lethal to be used in disease therapy. In was not until the mid–1940s that Selman Waksman (1888–1973) used the term "antibiotic" to describe a compound that had therapeutic effects against disease.

Who **discovered** the antibiotic **streptomycin**?

The Russian-born microbiologist Selman A. Waksman (1888–1973) discovered streptomycin in 1943. Streptomycin was the first antibiotic effective against tuberculosis.

> ## Who developed the poliomyelitis vaccine for polio?
>
> Immunologist Jonas E. Salk (1914–1995) developed the first vaccine (made from a killed virus) against poliomyelitis and is known as the man who defeated polio. In 1952 he prepared and tested the vaccine, and in 1954 massive field tests were successfully undertaken. Two years later immunologist Albert Sabin (1906–1993) developed an oral vaccine made from inactivated live viruses of three polio strains. Currently, an inactivated polio vaccine (IPV) is given as an injection at ages 2 months, 4 months, 6 to 18 months, and a booster at 4 to 6 years.

In 1944 Merck and Company agreed to produce it to be used against tuberculosis and tuberculosis meningitis. Streptomycin ultimately proved to have some human toxicity and was supplanted by other antibiotics, but its discovery changed the course of modern medicine. In addition to its use in treating tuberculosis, it was also used to treat bacterial meningitis, endocarditis, pulmonary and urinary tract infections, leprosy, typhoid fever, bacillary dysentery, cholera, and bubonic plague. Streptomycin saved countless lives, and its development led scientists to search the microbial world for other antibiotics and medicines. Waksman received the Nobel Prize in Physiology or Medicine in 1952 for his discovery of streptomycin.

Who was the **first physician** to use **chemotherapy** as a medical treatment?

Chemotherapy is the use of chemical substances to treat diseases, specifically malignant diseases. The drug must interfere with the growth of bacterial, parasitic, or tumor cells, without significantly affecting host cells. Especially effective in types of cancer such as leukemia and lymphoma, chemotherapy was introduced in medicine by the German physician Paul Ehrlich (1854–1915).

What are **monoclonal antibodies**?

Monoclonal antibodies are artificially produced antibodies designed to neutralize a specific foreign protein (antigen). Cloned cells (genetically identical) are stimulated to produce antibodies to the target antigen. Most monoclonal antibody work so far has used cloned cells from mice infected with cancer. In some cases they are used to destroy cancer cells directly; in others they carry other drugs to combat the cancer cells.

How are **anabolic steroids harmful** to those who use them?

Anabolic (protein-building) steroids are drugs that mimic the effects of testosterone and other male sex hormones. They can build muscle tissue, strengthen bone, and

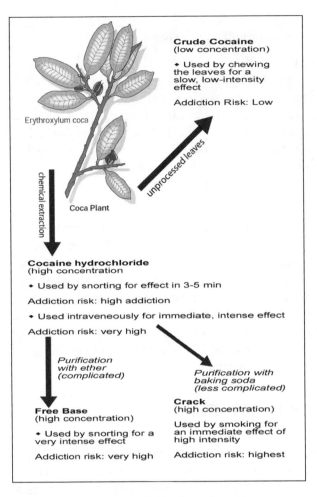

Crude Cocaine
(low concentration)

• Used by chewing the leaves for a slow, low-intensity effect

Addiction Risk: Low

Erythroxylum coca

unprocessed leaves

chemical extraction

Coca Plant

Cocaine hydrochloride
(high concentration

• Used by snorting for effect in 3-5 min

Addiction risk: high addiction

• Used intraveneously for immediate, intense effect

Addiction risk: very high

Purification with ether (complicated)

Purification with baking soda (less complicated)

Free Base
(high concentration)

Crack
(high concentration)

• Used by snorting for a very intense effect

Used by smoking for an immediate effect of high intensity

Addiction risk: very high

Addiction risk: highest

Various forms of cocaine and the addiction risks associated with them.

speed muscle recovery following exercise or injury. They are sometimes prescribed to treat osteoporosis in post-menopausal women and some types of anemia. Some athletes use anabolic steroids to build muscle strength and bulk and to allow a more rigorous training schedule. Weightlifters, field event athletes, and body-builders are most likely to use anabolic steroids. The drugs are banned from most organized competitions because of the dangers they pose to health and to prevent an unfair advantage.

Adverse effects include hypertension, acne, edema, and damage to the liver, the heart, and adrenal glands. Psychiatric symptoms can include hallucinations, paranoid delusions, and manic episodes. In men, anabolic steroids can cause infertility, impotence, and premature balding. Women can develop masculine characteristics, such as excessive hair growth, male-pattern balding, disruption of menstruation, and deepening of the voice. Children and adolescents can develop problems in growing bones, leading to short stature.

How is **patient-controlled analgesia** (PCA) administered?

This is a drug delivery system that dispenses a preset intravenous (IV) dose of a narcotic analgesic for reduction of pain whenever a patient pushes a switch on an electric cord. The device consists of a computerized pump with a chamber containing a syringe holding up to 60 milliliters of a drug. The patient administers a dose of narcotic when the need for pain relief arises. A lockout interval device automatically inactivates the system if the patient tries to increase the amount of narcotic within a preset time period.

What were the **birth defects** caused by the drug **Thalidomide**?

In the early 1960s, Thalidomide was marketed as a sedative and anti-nausea drug. It was found to cause birth defects in babies whose mothers had taken the drug for morning sickness. Some babies were born without arms or legs. Others were born blind or deaf or with heart defects or intestinal abnormalities. Although some were mentally retarded, most were of normal intelligence. This tragedy led to much stricter laws regulating the sale and testing of new drugs.

How does **RU-486** cause an abortion?

A pill containing RU-486 (mifepristone) deprives a fertilized embryo of a compatible uterine environment, terminating a pregnancy within 49 days of fertilization. It was approved for use in the United States on September 28, 2000. It is taken under the supervision of a qualified physician in a clinical setting or physician's office. Sometimes it is used as an emergency contraceptive ("morning-after pill") after unprotected sexual intercourse even when more than 49 days have passed since the woman's last menstrual period.

What are **designer drugs**, such as China White?

Designer drugs are synthesized chemicals that resemble such available narcotics as fentanyl and meperidine. China white (3-methyl-fentanyl) is one of these drugs and is an analogue of fentanyl. It is 3,000 times more potent than morphine. Even small amounts can be fatal, and it has been responsible for more than 100 overdose deaths in California.

What is a **controlled substance**?

The Comprehensive Drug Abuse Prevention and Control Act of 1971 was designed to control the distribution and use of all depressant and stimulant drugs and other drugs of abuse or potential abuse. Centrally acting drugs are divided into five classes called Schedule I through V.

Schedule	Examples
Schedule I	Experimental and illegal drugs (Heroin, LSD, Peyote) are not prescribable and do not have an acceptable medical use.
Schedule II	Like Schedule I drugs, Schedule II (Amphetamine, Cocaine) drugs can be abused. However, they (Codeine, Morphine) have acceptable medical uses. Prescriptions cannot be renewed.
Schedule III	Less likely to be abused than Aspirin with codeine, Schedule II drugs. Prescriptions can (Methylprylon, Phendimetrazine) be refilled up to five times in six months.
Schedule IV	Lower potential for abuse than (Chloral hydrate, Diazepam) Schedule I-III. Usually they fall under Phenobarbital; similar refill regulations as Schedule III drugs.

637

Schedule V   Low potential for abuse. May contain Lomotil, Parapectolin, small amounts of narcotics. Regulated (Cheracol, Robitussin) in the same manner as non-scheduled prescription drugs.

Which American **states** allow **medical marijuana usage?**

Since 1996, 15 states and the District of Columbia have passed laws allowing the use of marijuana for medical purposes. These include Alaska, Arizona, California, Colorado, District of Columbia, Hawaii, Maine, Michigan, Montana, Nevada, New Jersey, New Mexico, Oregon, Rhode Island, Vermont, and Washington. Depending on the state, patients may posess between 1 ounce (Alaska, Montana, Nevada) to 24 ounces (Oregon, Washington) of usable marijuana and grow their own plants. States also vary on which diseases are approved for treatment using marijuana.

SURGERY AND OTHER TREATMENTS

How does **minimally invasive surgery** differ from traditional, major, open surgery?

Traditional, major, open surgery requires a major incision in the body, often several inches long, allowing surgeons to physically place their hands inside the body to work. In minimally invasive surgery, the incision is very small and surgeons do not place their hands inside the body. Using a laparoscope, a narrow wand containing a video camera, surgeons are able to insert tools into tiny surgical openings to remove diseased tissue. Laparoscopy was introduced in the 1970s for gynecological treatment and gall bladder removal. At least half of all surgeries are now minimally invasive (laparoscopic or arthroscopic) with a wide range of applications—gall bladder removal, appendix removal, hernia repair, gynecological, colon removal, partial lung removal, spleen removal, and surgery for chronic heartburn or reflux disease.

> ## Does catgut really come from cats?
>
> Catgut, an absorbable sterile strand, is obtained from collagen derived from healthy mammals. It was originally prepared from the submucosal layer of the intestines of sheep. It is used as a surgical ligature.

The major advantage of minimally invasive surgery is that it is less traumatic to the patient. There is less scarring, and recovery time is much quicker. Hospital stays are shorter.

How many surgical procedures are performed each year?

According to the latest figures (2007), 45 million surgical procedures were performed on an inpatient basis. The most frequently performed procedures include:

Procedure	Procedures per Year
Arteriography and angiocardiography using contrast material (used to evaluate blockages in the arteries that supply blood to the heart)	1.9 million
Cesarean section	1.3 million
Cardiac catheterizations	1.1 million
Endoscopy of small intestine with or without biopsy	1.0 million

Who received the first heart transplant?

On December 3, 1967, in Capetown, South Africa, Dr. Christiaan Barnard (1922–2001) and a team of 30 associates performed the first heart transplant. In a five-hour operation the heart of Denise Anne Darvall (1943–1967), age 25, an auto accident victim, was transplanted into the body of Louis Washansky (1913–1967), a 55-year-old wholesale grocer. Washansky lived for 18 days before dying from pneumonia.

The first heart transplant performed in the United States was on a 2.5-week-old baby boy at Maimonides Hospital, Brooklyn, New York, on December 6, 1967, by Dr. Adrian Kantrowitz (1918–2008). The baby boy lived 6.5 hours. The first adult to receive a heart transplant in the United States was Mike Kasperak (1914–1968), age 54, at the Stanford Medical Center in Palo Alto, California, on January 6, 1968. Dr. Norman Shumway (1923–2006) performed the operation. Mr. Kasperak lived 14 days.

Almost no transplants were done in the 1970s because of the problem of rejection of the new heart by the recipient's immune system. In 1969, Jean-François Borel (1933–) discovered the anti-rejection drug cyclosporine, but it was not widely used until 1983 when the FDA granted approval. Today heart transplantation is an established medical procedure with 2,212 transplants performed in 2009. The percentage of patients surviving three years is more than 80 percent.

Is it possible to use **animal organs** as replacements for **human organ transplantation**?

In 1984, a 12-day-old infant, Baby Fae, received a baboon's heart. She lived for 20 days before her body rejected the transplanted heart. Since the supply of human organs is far less than the need for human organs, researchers continue to search for alternative sources of organs. In 1999, the Food and Drug Administration (FDA) prohibited the use of nonhuman primate organs for human transplantation unless researchers could assess the risk of disease. A major concern had been that viruses and other diseases that were harmful to the animals would be deadly to humans. Researchers continue to search for alternate sources of organs, including pigs.

When was the **first artificial heart** used?

On December 2, 1982, Dr. Barney B. Clark (1921–1983), a 61-year-old retired dentist, became the first human to receive a permanently implanted artificial heart. It was known as the Jarvik-7 after its inventor, Dr. Robert Jarvik (1946–). The 7.5-hour operation was performed by Dr. William DeVries (1943–), a surgeon at the University of Utah Medical Center. Dr. Clark died on March 23, 1983, 112 days later. In Louisville, Kentucky, William Schroeder (1923–1986) survived 620 days with an artificial heart (November 25, 1984, to August 7, 1986). On January 11, 1990, the U.S. Food and Drug Administration (FDA) recalled the Jarvik-7, which had been the only artificial heart approved by the FDA for use.

Two models of a total artificial heart (TAH) are now available that can replace the ventricles (the two lower chambers) of the heart. Eligible patients must have "end-stage" heart failure, meaning all treatments except a heart transplant have failed. A TAH may keep patients alive while they wait for heart transplants or may keep patients alive who are not eligible for a heart transplant. The two brands are the CardioWest and AbioCor. The CardioWest is connected to a power source outside of the body via tubes that run from inside the chest to the outside through holes in the abdomen. The AbioCor is a battery-powered TAH that is completely contained in the chest. The battery is charged through the skin with a special magnetic charger.

How does a **cardiac pacemaker** differ from an **implantable defibrillator**?

Both pacemakers and implantable defibrillators are used to treat arrhythmias in the heart rate. A pacemaker monitors the electrical impulses in the heart and delivers electrical pulses as necessary to make the heart beat in a more normal rhythm. An

> ## Why are eye transplants not available?
>
> The eye's retina is part of the brain, and the retina's cells are derived from the optic nerve in the brain. Retinal cells and the cells that connect them to the brain are the least amenable to being manipulated outside the body.

implantable cardioverter defibrillator monitors heart rhythms. When it senses irregular rhythms, it delivers shocks so the heart beats in a regular rhythm.

Who was the **first test-tube baby**?

Louise Brown, born on July 25, 1978, is the first baby produced from fertilization done in vitro—outside the mother's body. Patrick Steptoe (1913–1988) was the obstetrician and Robert Edwards (1939–) the biologist who designed the method for in-vitro fertilization and early embryo development.

In-vitro fertilization occurs in a glass dish, not a test tube, where eggs from the mother's ovary are combined with the father's sperm (in a salt solution). Fertilization should occur within 24 hours, and when cell division begins these fertilized eggs are placed in the mother's womb (or possibly another woman's womb).

What is **lithotripsy**?

Lithotripsy is the use of ultrasonic or shock waves to pulverize kidney stones (calculi), allowing the small particles to be excreted or removed from the body. There are two different methods: extracorporeal shock wave lithotripsy (ESWL) and percutaneous lithotripsy. The ESWL method, used on smaller stones, breaks up the stones with external shock waves from a machine called a lithotripter. This technique has eliminated the need for more invasive stone surgery in many cases. For larger stones, a type of endoscope, called a nephroscope, is inserted into the kidney through a small incision. The ultrasonic waves from the nephroscope shatter the stones, and the fragments are removed through the nephroscope.

Prior to the **condom**, what was the **main contraceptive** practice?

Contraceptive devices have been used throughout recorded history. The most traditional of such devices was a sponge soaked in vinegar. The condom was named for its English inventor, the personal physician to Charles II (1630–1685), who used a sheath of stretched, oiled sheep intestine to protect the king from syphilis. Previously penile sheaths were used, such as the linen one made by Italian anatomist Gabriel Fallopius (1523–1562), but they were too heavy to be successful.

What is **synthetic skin**?

The material consists of a very porous collagen fiber bonded to a sugar polymer (glycoaminoglycan) obtained from shark cartilage. It is covered by a sheet of silicon rubber. It was developed by Ioannis V. Yannas (1935–) and his colleagues at Massachusetts Institute of Technology around 1985. Soon after its introduction, synthetic skin was used to successfully treat more than 100 severely burned victims.

Who developed **psychodrama**?

Psychodrama was developed by Jacob L. Moreno (1892–1974), a psychiatrist born in Romania, who lived and practiced in Vienna until he came to the United States in 1927. He soon began to conduct psychodrama sessions and founded a psychodrama institute in Beacon, New York, in 1934. Largely because of Moreno's efforts it is practiced worldwide.

Psychodrama is defined as a method of group psychotherapy in which personality makeup, interpersonal relationships, conflicts, and emotional problems are explored by means of special dramatic methods.

Index

Note: (ill.) indicates photos and illustrations.

647

649